FUNDAMENTALS
of
COMPUTER NUMERICAL ANALYSIS

Menahem Friedman
Nuclear Research Center—Negev
Ben-Gurion University
Beer-Sheva, Israel

Abraham Kandel
Department of Computer Science
and Engineering
University of South Florida
Tampa, Florida

CRC Press
Boca Raton   Ann Arbor   London   Tokyo

Library of Congress Cataloging-in-Publication Data

Friedman, Menahem.
  Fundamentals of computer numerical analysis / authors, Menahem Friedman, Abraham Kandel.
    p.   cm.
  Includes bibliographical references and index.
  ISBN 0-8493-8637-3
  1. Electronic data processing. 2. Numerical analysis—Data processing. I. Kandel, Abraham. II. Title.
QA76.F757   1993
519.4–dc20   92-39853
                                                                                                               CIP

©1994 by CRC Press, Inc.
No claim to original U.S. Government works
International Standard Book Number 0-8493-8637-3
Library of Congress Card Number 92-39853
Printed in the United States of America 1 2 3 4 5 6 7 8 9 0
Printed on acid-free paper

Preface

This text is an introduction to numerical analysis, meant to be used by undergraduate students in mathematics and computer science as well as in related areas in science and engineering. The only prerequisite for using this book is a 1-year course in calculus of functions of one variable. Superficial knowledge of multivariable functions and linear algebra is advantageous. While most of this text is related to one-variable numerical analysis, some attention is also given to multivariable (mainly two) problems contained in topics such as Newton's method, numerical integration, and differential equations. It is felt that a new book on numerical analysis, totally restricted to one-variable problems, algorithms, and computations, would leave the student somewhat unprepared to play an active role in analyzing and solving real-world problems. We expect the students to be familiar with at least one scientific programming language such as Fortran or Pascal. This is eventually a necessity, because to solve any nontrivial problem in numerical analysis one should program a specific algorithm and run it on a digital computer.

A student who takes a course in numerical analysis is generally motivated by one of the following reasons:

1. The need to fulfill the requirements of a degree program

2. A desire to expand one's horizons of knowledge

3. The possibility of successfully applying the studied numerical methods to various fields

Although the book is primarily intended for use by students who qualify for classes (2) and (3), it is also written to be at least partially favored by students who belong to class (1).

The purpose of this book is to teach the student the classical topics in numerical analysis, demonstrate the theory by solving real problems with emphasis on obtaining intuitive understanding of both the problem and its associated algorithm, and prepare the student for relatively independent work in the field. A vast accumulated teaching experience in numerical analysis shows that intuitive understanding of algorithms by the student is almost always a guarantee for the student to successfully repeat and apply these algorithms later.

In numerical analysis, far more than in any other scientific discipline, extensive practice is the ultimate key to success. Throughout the book, each topic is followed by a few examples solved in detail. Then a large variety of problems is given at the end of each section.

Twenty-seven Fortran 77 programs are also included with the book, on a separate $5\frac{1}{4}$-in. floppy disk. Their goal is mainly to demonstrate and teach proper programming habits and occasionally let the student concentrate on the algorithm rather than on coding. However, if a problem requires an algorithm that does not fall within the range of the existing package, it is up to the student to program, debug, and run that particular algorithm.

In order for this text to suit and serve students of various disciplines, emphasis has not been placed on providing complete proofs of all theorems and algorithms. However, whenever we believe that the student will benefit from such a detailed proof, the proof is given in the most rigorous manner. In most cases, a theorem or algorithm is first formulated, followed by a discussion of its main aspects in a way that could inspire the motivation, imagination, and understanding of the student. Finally, the applicability of the algorithm is demonstrated through detailed examples illustrating "real–life" applications and practical situations in which numerical methods should be applied.

The first chapter is introductory, containing Taylor series, complex numbers, and partial derivatives. In the second chapter we investigate the concept of error, its sources and propagation. These two chapters are the basics that are essential and should be taught with special care. Any student who has mastered both should later have no problem with the rest of the material.

Even though the book is aimed at undergraduate students, topics such as partial differential equations and finite elements are also covered, although not in extreme depth.

It should be noted that the content of this text was successfully tested on many classes of students in mathematics, computer science, and electrical engineering since 1970. We thank many of them for their constructive suggestions, endless reviews, and enormous support. We would like to give special thanks to Dean Michael G. Kovac, Mrs. Cay Pelc, Mrs. Lisa Croy, and Mrs. Lorie Miros from the College of Engineering at the University of South Florida, and to Mrs. Saskia Beeser from the Ben-Gurion University of the Negev, Israel, for their encouragement and help during the final stages of preparing the manuscript.

Finally, our gratitude to Mr. Russ Hall, formerly of CRC Press, and Miss Janice M. Morey of CRC Press for their continued support throughout the completion of this project.

July 1993   **M. Friedman**

Tampa, Florida   **A. Kandel**

The Authors

Menahem Friedman, born in Jerusalem, received his M.Sc. in Mathematics from the Hebrew University in Jerusalem and his Ph.D. in Applied Mathematics from the Weizmann Institute of Science in Rehovot. He is a Senior Research Scientist at the Nuclear Research Center—Negev and an Associate Professor in Mathematics and Computer Science at the Ben-Gurion University of the Negev. Dr. Friedman has published numerous papers and has contributed to books in numerical analysis and artificial intelligence.

Abraham Kandel received his B.Sc. degree in Electrical Engineering from the Technion, Israel Institute of Technology, his M.Sc. from the University of California, and his Ph.D. in Electrical Engineering and Computer Science from the University of New Mexico. Dr. Kandel is a Professor and Endowed Eminent Scholar in Computer Science and Engineering and is the Chairman of the Department of Computer Science and Engineering at the University of South Florida. Previously, he was Professor and Chairman of the Computer Science Department at Florida State University, as well as the Director of the Institute of Expert Systems and Robotics at FSU and the Director of the State University System Center for Artificial Intelligence. He is a Fellow of the Institute of Electrical and Electronics Engineers, a member of the Association for Computing Machines, and an Advisory Editor to the international journals *Fuzzy Sets and Systems, Information Sciences, Expert Systems,* and *Engineering Applications of Artificial Intelligence.* Dr. Kandel has published over 250 technical research papers and 19 books on computer science and engineering.

Contents

4   LINEAR DIFFERENCE EQUATIONS

5   INTERPOLATION AND APPROXIMATION

6 NUMERICAL INTEGRATION AND DIFFERENTIATION 249

7 LINEAR EQUATIONS 297

8 NUMERICAL SOLUTIONS OF DIFFERENTIAL EQUATIONS ... 381

9 NUMERICAL SOLUTIONS OF PARTIAL DIFFERENTIAL EQUATIONS ... 447

And he made a molten sea, ten cubits from the one brim to the other:
it was round all about, and its height was five cubits:
and a line of thirty cubits did compass it round about.

I Kings 7:23

To Lynn, Danny, and Rinat
Nurit, Sharon, Gill, and Adi
with love

FUNDAMENTALS
of
COMPUTER
NUMERICAL
ANALYSIS

Introduction

1.1 What Is Numerical Analysis?

The mathematical discipline called *numerical analysis* deals with the development of techniques for approximating solutions to mathematical problems as well as to real-world problems that can be represented by mathematical *models*.

Unlike other mathematical disciplines such as linear algebra and differential equations, which concentrate on showing existence and uniqueness of solutions to given problems, numerical analysis provides *constructive* methods for calculating these solutions. A "constructive method" is a procedure that generates the solution of a mathematical problem with an arbitrary *precision* in a *finite* number of *steps*, performed in a predetermined order. Each step is a finite sequence of arithmetical or logical operations, performed sequentially.

A set of instructions to perform mathematical operations, designed to obtain the solution to a given problem, is called an *algorithm*. An algorithm is said to be finite, if it terminates after a finite number of steps. Algorithms designed for solving problems in algebra are usually finite. For example, the student can easily obtain an algorithm that determines whether a given integer is a prime number. However, many problems in analysis cannot be solved in a finite number of steps. Algorithms designed to solve such problems are therefore infinite, that is, they consist of an infinite number of mathematical operations. Because only a finite number of these operations can be carried, one needs to specify a desired accuracy of the calculated solution, which will cause the computation to terminate after some finite number of steps. Generally, if the accuracy increases, a larger number of operations will be performed before the computation is successfully completed.

An example of an infinite algorithm is the scheme based on Newton's method (Section 3.5) to calculate $\sqrt{a}$ for any given $a > 0$. Given an initial guess x_0, the scheme is

$$x_{n+1} = \frac{1}{2}\left(x_n + \frac{a}{x_n}\right), \quad n \geq 0$$

and satisfies $x_n \to \sqrt{a}$ as $n \to \infty$. Because usually $x_n \neq \sqrt{a}$ for all n, the algorithm is infinite. If, however, we wish to terminate the computation when

$$|x_n - \sqrt{a}| < 10^{-6}$$

for the first time, only a finite number of *iterations* must be performed.

Once an algorithm is formulated, we wish to know the conditions under which the algorithm indeed yields the solution to the given problem. Theorems that guarantee *convergence* of algorithms, under certain requirements, often exist. For example, sufficient conditions for the standard iteration method for solving $x = f(x), a \leq x \leq b$ (Section 3.1) to converge to the unique solution are

1. $f(x)$ is continuous for $a \leq x \leq b$.

2. $a \leq f(x) \leq b, \quad a \leq x \leq b$.

3. $|f'(x)| \leq L < 1, \quad a \leq x \leq b$.

Once the requirements for convergence are known, one is interested in the performance of the algorithm, that is, how *fast* the algorithm is. Another question in this category concerns how big the error is if the algorithm is terminated after some given number of steps. There is of course a significant difference between searching for an *error bound* or trying to evaluate the error *approximately* using an *asymptotic formula*. The second approach is more practical, whenever feasible.

For an algorithm already known to converge to be useful, it must also be *numerically stable*. Because all digital computers use only a finite number of digits, *round-off* errors are constantly introduced and may accumulate. Each individual rounding error is negligent, but a sequence of billions of operations may divert us significantly from the exact solution. Thus, for an algorithm to be useful, it must be immune to the accumulation of round-off errors. If this is the case, the algorithm is said to be *stable*. An *error propagation* study may provide the conditions under which a given algorithm is numerically stable.

1.2 Algebraic Representation and Arithmetic with Complex Numbers

The development of the *number* concept did not take place in one day, but rather evolved over a few centuries. It all started with the set $N = \{1, 2, 3, \ldots\}$ of the natural numbers. It was simply there, as L. Kronecker once stated: "God created the natural numbers. All the rest is man-made."

Addition and multiplication within the set N can be carried without restriction. For all $m, n \in N$, the numbers $m + n$, mn are defined and belong to N. This is not the case with respect to subtraction and division. Simple equations such as

$$x + 5 = 3 \tag{1.2.1}$$

and

$$2y = 7 \tag{1.2.2}$$

can be easily interpreted as "mathematical models" for real problems, but do not have solutions within the set N. This situation is not desirable. It calls for the first *extension* of the *set of numbers* and leads to the legalization of the negative integers and of zero.

One should bear in mind that a *proper* extension B of a given set A must be accompanied by a simultaneous extension of the set of operators defined over A. Furthermore, if P_A, an operator defined over A, is extended to an operator P_B defined over B, then

$$xP_By = xP_Ay \tag{1.2.3}$$

must hold for all $x, y \in A$. Also, because we are dealing with numbers, we would like the extended operators over the extended set to obey the commutative laws

$$x + y = y + x, \quad xy = yx$$

the associative laws

$$(x + y) + z = x + (y + z)$$

$$(xy)z = x(yz)$$

and the distributive law

$$(x + y)z = xz + yz$$

for all $x, y, z \in B$.

Within the new set I of all the integers $0, \pm1, \pm2, \ldots$ addition, multiplication, and subtraction are now well defined for any pair of numbers. Division, however, is still limited and Eq. (1.2.2), for instance, does not possess a solution either in N or in I.

The next extension of the set of numbers takes care of this problem and introduces all the *rational* numbers m/n, where $m, n \in I$ and $n \neq 0$. Consequently, one can now cope with a far larger class of problems. Actually, because computers take and provide only rational numbers, one could assume that further extensions of the set of numbers would be artificial and unnecessary. A strong indication against such an attitude is given simply by showing that some *legitimate numbers* are still missing.

For example, although it is quite natural to accept the existence of a number x whose square is 2, one can easily show that a rational number x satisfying

$$x^2 = 2 \tag{1.2.4}$$

does not exist. It is clear, then, that the set of all rational numbers is incomplete and needs to be further extended to include all the *irrational numbers* such as $\sqrt{2}$, π, and e, which *cannot* be expressed as m/n. One is therefore led to define the set of all real numbers R, formally written as

$$R \equiv \{x| -\infty < x < \infty\} \tag{1.2.5}$$

and containing all the rational and irrational numbers.

And yet although the four basic operators of arithmetic are now well defined without any restriction (other than division by 0) over R, another operator, quite commonly used, is still significantly limited. The *square root* of a negative x, for instance $\sqrt{-1}$, does not exist, that is, the equation

$$x^2 + 1 = 0 \tag{1.2.6}$$

does not have a solution within R.

It is obvious that we may ignore the idea of introducing complex numbers, and concentrate on solving problems that provide only real solutions. However, there are some good arguments against this approach:

1. Without using complex numbers one faces major difficulties in trying to cope with various fundamental problems in most sciences.

2. Complex numbers are often needed even to establish the real solutions of various problems.

We therefore choose to introduce one last and most important extension of the set of numbers and investigate a new notation—the *complex number*.

Formally, only one new symbol, $i = \sqrt{-1}$, is needed to materialize the new extension. The complex number is then defined as

$$z \equiv a + ib \tag{1.2.7}$$

where a and b are real numbers. The *real* and *imaginary* parts of z are defined as

$$\begin{aligned} \Re z &\equiv a \\ \Im z &\equiv b \end{aligned} \tag{1.2.8}$$

respectively. A real number is identified with a complex number for which $b = 0$.

Two basic "rules" must be kept while performing arithmetic with complex numbers:

Rule 1: i^2 can be replaced by -1.

Rule 2: For all computational purposes, i is treated *as if it were a real number*.

For example,

$$i^5 = i^4 i = i^2 i^2 i = (-1)(-1)i = i$$
$$2i + 3 + i = 3 + 3i$$

These rules play a leading role in establishing *sensible* definitions of the four basic arithmetical operations between complex numbers. Once the arithmetic is defined, one must show that the present extension is *proper* and that the commutative, associative, and distributive laws are valid within the set of complex numbers.

Let $z = a + ib, w = c + id$. We start by defining

$$z \equiv w \quad \text{if and only if} \quad a = c, b = d \tag{1.2.9}$$

Thus two complex numbers given in the form of Eq. (1.2.7) are equal if and only if they are *identical*.

We shall now start and perform addition and subtraction between z, w while *obeying* Rule 2.

$$z + w = (a + ib) + (c + id) = a + ib + c + id = a + c + ib + id = (a + c) + i(b + d)$$
$$z - w = (a + ib) - (c + id) = a + ib - c - id = a - c + ib - id = (a - c) + i(b - d)$$

Thus we *must* define

$$\begin{aligned} (a + ib) + (c + id) &\equiv (a + c) + i(b + d) \\ (a + ib) - (c + id) &\equiv (a - c) + i(b - d) \end{aligned} \tag{1.2.10}$$

Next, we multiply z by w, using Rule 1 and Rule 2, to obtain

$$zw = (a + ib)(c + id) = ac + ibc + aid + ibid = ac + ibc + iad + i^2bd$$
$$= (ac - bd) + i(bc + ad)$$

This leads us to define

$$(a + ib)(c + id) \equiv (ac - bd) + i(bc + ad) \tag{1.2.11}$$

We finally turn to define z/w. If one assumes that z/w is a complex number and that Rule 1 and Rule 2 hold, then

$$\frac{z}{w} = \frac{a + ib}{c + id} = \frac{(a + ib)(c - id)}{(c + id)(c - id)} = \frac{(ac + bd) + i(bc - ad)}{c^2 + d^2}$$
$$= \frac{ac + bd}{c^2 + d^2} + i\frac{bc - ad}{c^2 + d^2}$$

Therefore we should define

$$\frac{a + ib}{c + id} \equiv \frac{ac + bd}{c^2 + d^2} + i\frac{bc - ad}{c^2 + d^2} \tag{1.2.12}$$

This definition is proper, provided that $c + id \neq 0 + i0$. We naturally refer to $0 + i0$ as the complex number zero because for all z

$$z + (0 + i0) = z, \quad z(0 + i0) = 0 + i0 \tag{1.2.13}$$

The complex number $0 + i0$ is actually identified with the real number 0. Thus z/w is well defined provided that $w \neq 0$.

Example 1.2.1.

The following three equations involve complex numbers:

1. $(5 - i)(3 + 2i) = (15 + 2) + i(-3 + 10) = 17 + 7i$

2. $\dfrac{2 + i}{1 - 5i} = \dfrac{(2 + i)(1 + 5i)}{(1 - 5i)(1 + 5i)} = \dfrac{(2 - 5) + i(1 + 10)}{26} = -\dfrac{3}{26} + \dfrac{11}{26}i$

3. $\dfrac{1}{1 + [1/(i - 1)]} = \dfrac{i - 1}{i} = 1 - \dfrac{1}{i} = 1 - \dfrac{i}{i^2} = 1 + i$ □

Example 1.2.2.

Let us now solve a quadratic equation with real coefficients, given as

$$ax^2 + bx + c = 0; \quad a, b, c \in R; \quad a \neq 0 \tag{1.2.14}$$

The formal solutions of Eq. (1.2.14) are given by

$$x_1 = \frac{-b + \sqrt{b^2 - 4ac}}{2a}, \quad x_2 = \frac{-b - \sqrt{b^2 - 4ac}}{2a}$$

and can be real or complex, depending on the quantity $\Delta = b^2 - 4ac$:

$$\Delta > 0: \text{two different real solutions}$$
$$\Delta = 0: \text{a multiple real root}$$
$$\Delta < 0: \text{two different complex roots}$$

For example, the equation $x^2 + x + 1 = 0$ possesses two complex solutions:

$$x_1 = -\frac{1}{2} + i\frac{\sqrt{3}}{2}, \quad x_2 = -\frac{1}{2} - i\frac{\sqrt{3}}{2} \qquad \square$$

At this point one may wonder whether we are through with extensions of the set of numbers, or if there is still room for more. We shall give a partial answer by showing that a quadratic equation, regardless of the type of its coefficients, always has two complex roots, suggesting that a further extension of the set of numbers is unnecessary.

If the coefficients a, b, c in Eq. (1.2.14) are complex numbers, then $b^2 - 4ac$ would generally be a complex number, that is, $b^2 - 4ac = A + iB$. Thus one must show that for any given complex number $A + iB$, real numbers x, y that satisfy

$$A + iB = (x + iy)^2 \tag{1.2.15}$$

can be found. By comparing the real and imaginary parts of both sides one obtains two equations with two variables x, y:

$$x^2 - y^2 = A, \quad 2xy = B \tag{1.2.16}$$

Algebra leads to

$$x = \pm \left(\frac{A + \sqrt{A^2 + B^2}}{2} \right)^{1/2}$$
$$y = \frac{B}{2x}$$

provided that $x \neq 0$.

A vanishing x is possible only if $B = 0$, $A \leq 0$. In this particular case $y = \pm\sqrt{-A}$. Thus there are always exactly two complex solutions z_1, z_2 to Eq. (1.2.15) and they satisfy $z_2 = -z_1$. This is an expected extension of the real case, where every positive number has exactly two roots with the same property.

Example 1.2.3.

1. $\sqrt{3+4i} = x + iy$: $x = \pm[(3 + \sqrt{9+16})/2]^{1/2} = \pm 2$, $y = \pm 4/2x = \pm 1$;
   therefore $z_1 = 2 + i$; $z_2 = -2 - i$.

2. $\sqrt{-9} = x + iy$: $x = 0$, $y = \pm 3$;
   therefore $z_1 = 3i$, $z_2 = -3i$. $\qquad\qquad\qquad\qquad\qquad\qquad\square$

PROBLEMS

1. Express the following expressions as $a + ib$:

   (a) $(2 - i) + (3 + 4i)$

   (b) $\dfrac{4+i}{1+i+(1/i)}$

   (c) $\sqrt{1-i}$

   (d) $\left(\dfrac{\cos\phi + i\sin\phi}{\cos\phi - i\sin\phi}\right)^3$

2. Find all z for which $(1 + i + z)/(1 - i + z)$ is real.

3. Find all the solutions of $z^4 + z^2 + i = 0$.

4. Let $z_n = 1 + i/\{1 + [i/(1 + \cdots + i)]\}(n\ i\ \text{'s})$ and assume that $\lim_{n\to\infty} z_n = z$. Compute z.

1.3 Geometric Interpretation

"Geometric interpretation" requires the establishment of a one-to-one correspondence between the set of numbers and some "set of points," and then the determination of the geometric meaning of each of the arithmetic operations.

In the case of the set of all real numbers one can establish a one-to-one correspondence between this set and all the points of an infinite straight line, for example, the x axis spreading from $-\infty$ to $+\infty$.

The "object" that can be used for geometrically interpreting the set of all complex numbers is the whole xy plane, which consists of all the points (a, b), where a, b are both real and satisfy $-\infty < a, b < \infty$. This "object" is referred to as the *complex plane*.

To each complex number $a + ib$ we assign the point (a, b). This is obviously a one-to-one correspondence between the set of all complex numbers and all the points of the xy plane (Fig. 1.3.1). The x, y axes are the real and the imaginary axes, respectively.

The complex number $a + ib$ can also be associated with the two-dimensional *vector* whose components in the x and y directions are a and b, respectively. However, because two parallel vectors of identical lengths are defined as equal, each complex number is identified with an infinite set of vectors, all parallel, having the same length and varying only by their endpoints. If each such set is represented by a unique vector, for example the one that starts at the origin, a one-to-one correspondence between all the complex numbers and a *representative subset* of all the two-dimensional vectors is maintained.

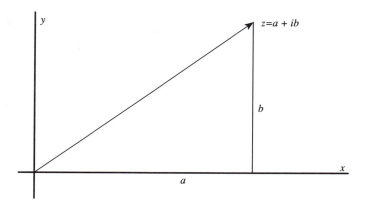

Figure 1.3.1.   The complex plane.

Now, let $z = a + ib, w = c + id$ be any two given complex numbers and $\mathbf{z} = (a, b), \mathbf{w} = (c, d)$ their corresponding vectors. The sum definitions between complex numbers and vectors are

$$z + w = (a + c) + i(b + d)$$

and

$$\mathbf{z} + \mathbf{w} = (a + c, b + d)$$

It is therefore clear that $\mathbf{z} + \mathbf{w}$ is the vector corresponding to the complex number $z + w$ as expected and desired.

If $\mathbf{z} = \mathbf{OA}, \mathbf{w} = \mathbf{OB}, z + w = OC$ (Fig. 1.3.2), simple geometry guarantees $BC \parallel OA$ and $AC \parallel OB$. The polygon $OACB$ is thus a parallelogram and therefore $\overrightarrow{BC} = \overrightarrow{OA}$, $\overrightarrow{AC} = \overrightarrow{OB}$.

This suggests a way to add n consecutive complex numbers $z_1, z_2, \ldots, z_n$. Suppose that $s_k = z_1 + z_2 + \cdots + z_k$ has been already computed and represented by a corresponding vector $\mathbf{s}_k$. To calculate $\mathbf{s}_{k+1}$, one should draw the vector $\mathbf{z}_{k+1}$, starting at the endpoint of $\mathbf{s}_k$, then connect the new endpoint to the origin. The procedure for $n = 3$ is shown in Fig. 1.3.3.

To compute $z_1 - z_2$ one can draw the vector $-\mathbf{z}_2$, which is opposite to $\mathbf{z}_2$, then add it to $\mathbf{z}_1$ using the parallelogram rule (Fig. 1.3.4).

The geometric interpretation of multiplication and division between complex numbers is more complicated. To maintain a suitable interpretation one should first introduce *polar coordinates*.

Let $z = a + ib$ be a complex number. The polar coordinates of z are defined as (r, ϕ) (Fig. 1.3.5) where

$$a = r \cos \phi, \quad b = r \sin \phi \tag{1.3.1}$$

The *absolute value* of z is defined as

$$|z| = \sqrt{a^2 + b^2}$$

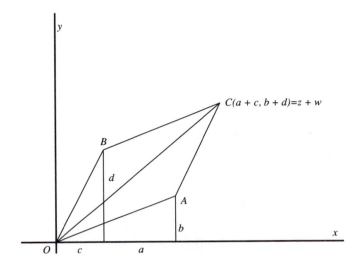

Figure 1.3.2. Complex numbers: addition.

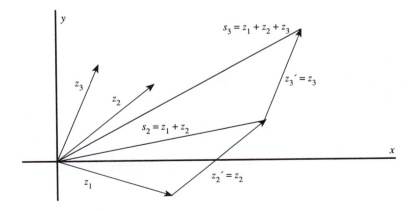

Figure 1.3.3. Calculating $\sum_{i=1}^{n} z_i$.

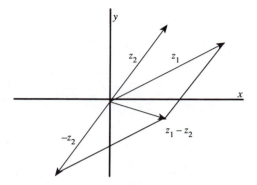

Figure 1.3.4. Complex numbers: subtraction.

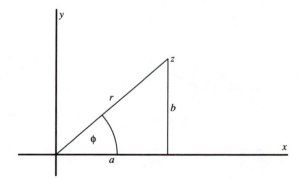

Figure 1.3.5. Polar representation of z.

The *argument* of z, denoted by $\arg(z)$, is defined as the angle between the positive x axis and the vector **z**. The angle is positive if the positive real axis is rotated counterclockwise to coincide with **z**. Otherwise it is negative. Thus $|z| = r$, $\arg(z) = \phi$. One exception is the case $z = 0$, for which $|z| = 0$ but $\arg(z)$ is not defined.

One should note that the argument of a complex number is not uniquely defined. If $\arg(z) = \phi$, so is $\phi + 2k\pi$ for any integer k.

One way to obtain $\phi = \arg(z)$ for $z = a + ib \neq 0$ is first to compute its absolute value and then use

$$\cos \phi = \frac{a}{r}, \quad |\phi| \leq \pi$$

The sign of ϕ is determined by

$$\text{sign}(\phi) = \text{sign}(b)$$

This is not valid for $b = 0$. For this particular case, $\phi = 0$ if $a = r$ and $\phi = \pm\pi$ if $a = -r$. But if one restricts ϕ (as we usually do) by $-\pi < \phi \leq \pi$, then $\phi = -\pi$ is not an admissible value.

Example 1.3.1.

Let $z = 1 - 2i$. The absolute value of z is

$$|z| = \sqrt{1^2 + (-2)^2} = \sqrt{5}$$

The argument of z, ϕ satisfies

$$\cos \phi = \frac{1}{\sqrt{5}}$$

leading to $|\phi| = 63.4°$. Because $b = -2$ is negative, one finally obtains $\phi = -63.4°$. $\square$

Example 1.3.2.

Let $z = 2 + 3i$, $w = 4 - 5i$. The sum of these complex numbers is $z + w = 6 - 2i$. Thus $|z + w| = \sqrt{40}$ and $|\arg(z + w)| = |\cos^{-1}(6/\sqrt{40})| = 18.4°$. Because $I_m(z + w) < 0$ we have $\arg(z + w) = -18.4°$. $\square$

By using Eq. (1.3.1), each complex number can be expressed in terms of r, ϕ as

$$z = r\cos\phi + ir\sin\phi = r(\cos\phi + i\sin\phi) \tag{1.3.2}$$

This polar representation allows the geometric interpretation of multiplication and division between complex numbers. Let $z = r(\cos\phi + i\sin\phi)$, $w = R(\cos\theta + i\sin\theta)$, be two complex numbers. By using the algebraic rule for multiplying, one obtains

$$zw = rR[(\cos\phi\cos\theta - \sin\phi\sin\theta) + i(\cos\phi\sin\theta + \sin\phi\cos\theta)]$$
$$= rR[\cos(\phi + \theta) + i\sin(\phi + \theta)]$$

The right-hand side is the polar representation of a complex number with absolute value rR and argument $\phi + \theta$. We have thus proved

$$|zw| = |z||w|, \quad \arg(zw) = \arg(z) + \arg(w) \tag{1.3.3}$$

If $\arg(z) + \arg(w)$ exceeds the range $(-\pi, \pi]$ it is simply replaced by $\arg(z) + \arg(w) + 2k\phi$, where k is either 1 or -1, so that $\arg(z) + \arg(w) + 2k\pi$ is an admissible value. We now assume that $w \neq 0$ and turn to evaluate z/w.

$$\begin{aligned}
\frac{z}{w} &= \frac{r(\cos\phi + i\sin\phi)}{R(\cos\theta + i\sin\theta)} = \frac{r(\cos\phi + i\sin\phi)(\cos\theta - i\sin\theta)}{R(\cos\theta + i\sin\theta)(\cos\theta - i\sin\theta)} \\
&= \frac{r}{R}[(\cos\phi\cos\theta + \sin\phi\sin\theta) + i(\sin\phi\cos\theta - \cos\phi\sin\theta)] \\
&= \frac{r}{R}[\cos(\phi - \theta) + i\sin(\phi - \theta)]
\end{aligned}$$

The right-hand side is a polar representation of a complex number whose absolute value and argument are r/R and $\phi - \theta$, respectively. We have thus proved

$$\left|\frac{z}{w}\right| = \frac{|z|}{|w|}, \quad \arg\left(\frac{z}{w}\right) = \arg(z) - \arg(w) \tag{1.3.4}$$

provided that $w \neq 0$.

Example 1.3.3.
Let $z = 2(\cos 30° + i\sin 30°)$, $w = 1.5(\cos 45° + i\sin 45°)$. Then

$$\begin{aligned}
zw &= (2 \cdot 1.5)[\cos(30° + 45°) + i\sin(30° + 45°)] \\
&= 3(\cos 75° + i\sin 75°)
\end{aligned}$$
$\square$

Example 1.3.4.
Let $z = 2(\cos 30° + i\sin 30°)$, $w = 2.5(\cos 60° + \sin 60°)$. Then

$$\begin{aligned}
\frac{z}{w} &= \frac{2}{2.5}[\cos(30° - 60°) + i\sin(30° - 60°)] \\
&= 0.8\left[\cos(-30°) + i\sin(-30°)\right]
\end{aligned}$$
$\square$

Let $z = a + ib$ be a complex number. Its reflection (Fig. 1.3.6) with respect to the real axis is called the *complex conjugate* of z and is denoted by $\bar{z}$. Thus

$$\bar{z} = a - ib, \quad |\bar{z}| = |z|, \quad \arg(\bar{z}) = -\arg(z) \tag{1.3.5}$$

The real and the imaginary parts of $z = a + ib$ satisfy

$$a = \frac{z + \bar{z}}{2}, \quad b = \frac{z - \bar{z}}{2i} \tag{1.3.6}$$

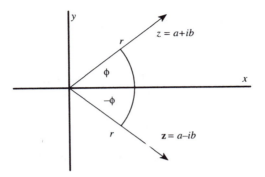

Figure 1.3.6.   Complex conjugate.

Other properties of the conjugate complex are

$$\overline{z_1 \pm z_2} = \bar{z}_1 \pm \bar{z}_2 \qquad (1.3.7)$$

$$\overline{z_1 z_2} = \bar{z}_1 \bar{z}_2 \qquad (1.3.8)$$

$$\overline{z_1 / z_2} = \frac{\bar{z}_1}{\bar{z}_2} \qquad (1.3.9)$$

$$z\bar{z} = |z|^2 \qquad (1.3.10)$$

all of which are results of straightforward algebra.

An extremely important rule applying to complex numbers is the *triangle inequality*. It basically states that for any given triangle, each side could never exceed the sum of the remaining two. Algebraically, it is formulated as follows: for any given complex numbers z_1, z_2

$$|(|z_1| - |z_2|)| \leq |z_1 + z_2| \leq |z_1| + |z_2| \qquad (1.3.11)$$

This follows from the properties of the complex conjugate. We first write

$$
\begin{aligned}
|z_1 + z_2|^2 &= (z_1 + z_2)(\overline{z_1 + z_2}) = (z_1 + z_2)(\bar{z}_1 + \bar{z}_2) \\
&= z_1\bar{z}_1 + z_2\bar{z}_1 + z_1\bar{z}_2 + z_2\bar{z}_2 = |z_1|^2 + |z_2|^2 + z_2\bar{z}_1 + \overline{z_2\bar{z}_1} \\
&= |z_1|^2 + |z_2|^2 + 2R_e(\bar{z}_1 z_2)
\end{aligned}
$$

and because for any complex number $z = a + ib, |R_e z| \leq |z|$ ($|a| \leq \sqrt{a^2 + b^2}$) one obtains

$$|z_1 + z_2|^2 \geq |z_1|^2 + |z_2|^2 - 2|\bar{z}_1 z_2| = (|z_1| - |z_2|)^2$$

Thus, for any given z_1, z_2,

$$|(|z_1| - |z_2|)| \leq |z_1 + z_2|$$

In particular,

$$|z_1 + z_2| - |-z_2| \leq |z_1 + z_2 + (-z_2)| = |z_1|$$

or

$$|z_1 + z_2| \leq |z_1| + |-z_2| = |z_1| + |z_2|$$

which concludes the proof.

Example 1.3.5.
Let $z = 3 + 4i$. Then $\bar{z} = 3 - 4i$, $z\bar{z} = 3^2 + 4^2 = 25$. If now $w = 2 - 3i$ then

$$|w + z| = |3 + 4i + 2 - 3i| = \sqrt{26} \leq |z| + |w| = 5 + \sqrt{13} \qquad \square$$

By applying Eq. (1.3.3) for the case $z = w$, we obtain

$$[r(\cos \phi + i \sin \phi)]^2 = r^2(\cos 2\phi + i \sin 2\phi)$$

and further, by using induction,

$$[r(\cos \phi + i \sin \phi)]^n = r^n [\cos(n\phi) + i \sin(n\phi)] \qquad (1.3.12)$$

The special case $r = 1$ yields the well-known De Moivre formula

$$(\cos \phi + i \sin \phi)^n = \cos(n\phi) + i \sin(n\phi) \qquad (1.3.13)$$

which is valid for all ϕ. By using Newton's binomial formula

$$(A + B)^n = A^n + \binom{n}{1} A^{n-1}B + \binom{n}{2} A^{n-2}B^2 + \cdots$$

$$+ \binom{n}{n-2} A^2 B^{n-2} + \binom{n}{n-1} AB^{n-1} + B^n \qquad (1.3.14)$$

where $\binom{n}{k}$ is the binomial coefficient defined as

$$\binom{n}{k} = \frac{n!}{k!(n-k)!} \qquad (1.3.15)$$

we can expand the left-hand side of Eq. (1.3.13), equate the real and imaginary parts to $\cos(n\phi)$ and $\sin(n\phi)$, respectively, and obtain the following identities:

$$\cos(n\phi) = \cos^n \phi - \binom{n}{2} \cos^{n-2} \phi \sin^2 \phi + \binom{n}{4} \cos^{n-4} \phi \sin^4 \phi - \cdots$$

$$\sin(n\phi) = \binom{n}{1} \cos^{n-1}\phi \sin\phi - \binom{n}{3} \cos^{n-3}\phi \sin^3\phi + \cdots$$

This is a beautiful example of nontrivial trigonometric identities that are easily derived using complex numbers.

Example 1.3.6.

For $n = 2$ we obtain

$$\cos(2\phi) = \cos^2\phi - \sin^2\phi$$
$$\sin(2\phi) = 2\cos\phi\sin\phi$$

□

Example 1.3.7.

For $n = 3$ we obtain

$$\cos(3\phi) = \cos^3\phi - 3\cos\phi\sin^2\phi$$
$$\sin(3\phi) = 3\cos^2\phi\sin\phi - \sin^3\phi$$

□

De Moivre's formula can be used to find the nth roots of any given complex number $w = r(\cos\phi + i\sin\phi)$. A complex number z will be called an nth root of w if the relation $z^n = w$ holds. Let $z = R(\cos\theta + i\sin\theta)$. Then

$$R^n\left[\cos(n\theta) + i\sin(n\theta)\right] = r(\cos\phi + i\sin\phi) \qquad (1.3.16)$$

implying

$$R^n = r; \quad n\theta = \phi + 2k\pi, \quad k = 0, \pm 1, \pm 2, \ldots$$

Thus

$$R = r^{(1/n)}; \quad \theta = \frac{\phi}{n} + \frac{2k\pi}{n}, \quad k = 0, \pm 1, \pm 2, \ldots$$

To avoid repetitions one can restrict k to $0, 1, 2, \ldots, n-1$ and obtain the n distinct nth roots with absolute value $r^{(1/n)}$ and arguments $\theta = \phi/n, (\phi + 2\pi)/n, \ldots, [\phi + 2(n-1)\pi]/n$.

Example 1.3.8.

Let $w = 1 + i$, $z^3 = w$. Then $r = \sqrt{2}$, $\phi = \pi/4$. The three distinct third roots of w share the same absolute value, $2^{(1/6)}$, and their arguments are $\pi/12$, $[(\pi/4) + 2\pi]/3$, $[(\pi/4) + 4\pi]/3$. Thus,

$$z_1 = 2^{(1/6)}\left(\cos\frac{\pi}{12} + i\sin\frac{\pi}{12}\right)$$
$$z_2 = 2^{(1/6)}\left(\cos\frac{3\pi}{4} + i\sin\frac{3\pi}{4}\right)$$

$$z_3 = 2^{(1/6)} \left(\cos \frac{17\pi}{12} + i \sin \frac{17\pi}{12} \right)$$

If we agree that $-\pi < \arg(z) \leq \pi$, then z_3 should be written as

$$z_3 = 2^{(1/6)} \left(\cos \frac{-7\pi}{12} + i \sin \frac{-7\pi}{12} \right) \qquad \square$$

PROBLEMS

1. Express the following expressions in polar coordinates
   - (a) $1 + i$
   - (b) $\dfrac{i + 1}{i}$
   - (c) $\dfrac{3 - 4i}{1 + 2i}$

2. Evaluate the complex conjugate of
   - (a) $(2 + i)(1 - i)$
   - (b) $\dfrac{3 + 5i + i}{2 - 1/i}$
   - (c) $\sin \phi - i \cos(2\phi)$

3. Using induction, prove that for arbitrary complex numbers $z_1, \ldots, z_n$

$$|z_1 + z_2 + \cdots + z_n| \leq |z_1| + |z_2| + \cdots + |z_n|$$

4. For arbitrary integers m, n calculate $i^m + i^n$.

5. Calculate the nth roots of 1.

6. Evaluate $z = (1 + i)^{(1/3)} + (1 - i)^{(1/2)}$ with the smallest absolute value.

1.4 Polynomials

Within the vast family of functions, special attention is given to one small subset in particular—the *polynomials*. The main reasons for this are as follow:

1. Polynomials are easy to manipulate and compute.

2. Polynomials are continuous and possess continuous derivatives of any order.

3. Polynomials can be used to approximate any given piecewise continuous function.

Definition 1.4.1. *(Polynomial of order n). A function of the form*

$$p_n(x) = a_n x^n + a_{n-1} x^{n-1} + \cdots + a_1 x + a_0, \quad a_n \neq 0 \tag{1.4.1}$$

is called a polynomial of order (degree) n with coefficients $a_0, a_1, \cdots, a_n$. Clearly all polynomials are well defined for $-\infty < x < \infty$, but the argument may also be complex.

If all $a_i, 0 \leq i \leq n$ and x are real numbers, the polynomial is a *real polynomial*. If, however, the coefficients and/or the argument are complex, we have a *complex polynomial*.

For example, $p_2(x) = x^2 + 1$, $0 \leq x \leq 1$ is a second order real polynomial defined over the interval $[0, 1]$. On the other hand the expression $p_3(z) = iz^3 - (i+1)z^2 + 1$, $|z| < 2$ is a third order complex polynomial defined inside the circle $|z| = 2$. To compute these particular polynomials is straightforward. For example, direct substitution gives

$$p_2(1) = 1 \cdot 1 + 1 = 2$$
$$p_3(i) = i \cdot i^3 - (i+1)i^2 + 1 = 3 + i$$

However, if it is necessary to design a general algorithm to evaluate $p_n(x)$, a direct computation is not recommended. Instead, one should rewrite the polynomial as

$$p_n(x) = \{\ldots [(a_n x + a_{n-1})x + a_{n-2}]x + \cdots + a_1\}x + a_0 \tag{1.4.2}$$

and compute it from left to right, performing the operations sequentially. It may be seen that by doing that, one needs exactly n multiplications and n additions for one complete evaluation. If we follow the primitive routine, compute $x, x^2, \ldots, x^n$, multiply them by the corresponding coefficients, and add, then, $n(n+1)/2$ multiplications and n additions are needed. Thus, by using Eq. (1.4.2), known as Horner's algorithm, one reduces the computing time for evaluating polynomials by *one order of magnitude*, from $[n(n+1)]/2 + n$ to $2n$.

The representation of a polynomial originally defined by Eq. (1.4.1) can be simplified, once its *roots* (i.e., the argument values at which the polynomial vanishes) are known. A basic theorem from linear algebra states the following.

Theorem 1.4.1.
(Roots of polynomial). *Any given polynomial $p_n(z)$ of order $n \geq 1$ must have a root z within the complex plane.*

The domain of definition of every polynomial may be extended to include the whole complex plane. Theorem 1.4.1 guarantees the existence of at least one complex number z_1, for which $p_n(z_1) = 0$.

Once a single root z_1 of a polynomial is known to exist, it may be seen that $p_n(z)$ must possess n complex roots $z_1, \ldots, z_n$ and that

$$p_n(z) = a_n(z - z_1)(z - z_2) \cdots (z - z_n) \tag{1.4.3}$$

Theorem 1.4.2.
Let $p_n(z)$ be an nth order polynomial, $n \geq 1$ and $p_n(z_1) = 0$. Then, there exists an $(n-1)$th order

polynomial $q_{n-1}(z)$ such that

$$p_n(z) = (z - z_1)q_{n-1}(z) \tag{1.4.4}$$

Proof.

Because

$$p_n(z) = a_n z^n + a_{n-1} z^{n-1} + \cdots + a_1 z + a_0$$
$$p_n(z_1) = a_n z_1^n + a_{n-1} z_1^{n-1} + \cdots + a_1 z_1 + a_0 = 0$$

we may subtract the second equation from the first to obtain

$$p_n(z) = a_n(z^n - z_1^n) + a_{n-1}(z^{n-1} - z_1^{n-1}) + \cdots + a_1(z - z_1)$$

Because each term $z^k - z_1^k$ can be rewritten as

$$z^k - z_1^k = (z - z_1)(z^{k-1} + z^{k-2}z_1 + \cdots + zz_1^{k-2} + z_1^{k-1}) = (z - z_1)r_{k-1}(z)$$

where $r_{k-1}(z)$ is a $(k-1)$th order polynomial, $p_n(z)$ may be expressed as $(z - z_1)q_{n-1}(z)$, where

$$q_{n-1}(z) = a_n r_{n-1}(z) + a_{n-1}r_{n-2}(z) + \cdots + a_1$$

is an $(n-1)$th order polynomial. This completes the proof.   □

By implementing Theorem 1.4.1 on $q_{n-1}(z)$ and by using mathematical induction, one finally obtains Eq. (1.4.3).

It should be noted that Theorem 1.4.1 does not provide us with an algorithm for computing the roots of a polynomial. It only guarantees the *existence* of such roots. The computation itself is discussed later in Chapters 3 and 4.

A root z_0 of an nth order polynomial $p(z)$ is called *multiple* and of multiplicity k if

$$p(z) = (z - z_0)^k q(z), \quad q(z_0) \neq 0$$

where $q(z)$ is an $(n-k)$th order polynomial. It can be shown that z_0 is a multiple root of order k if and only if

$$p(z_0) = p'(z_0) = \cdots = p^{(k-1)}(z_0) = 0, \quad p^{(k)}(z_0) \neq 0 \tag{1.4.5}$$

If $k = 1, z_0$ is a *simple* root.

Example 1.4.1.

The general fourth order polynomial whose roots are $i, i + 1, 0, -i$ is

$$p(z) = a(z - i)(z - 1 - i)(z - 0)(z + i) = az(z^2 + 1)(z - 1 - i)$$ □

Example 1.4.2.

Let $p(z) = z^5 - 3iz^4 - 3z^3 + iz^2$. The first three derivatives of $p(z)$ are

$$p'(z) = 5z^4 - 12iz^3 - 9z^2 + 2iz$$
$$p''(z) = 20z^3 - 36iz^2 - 18z + 2i$$
$$p'''(z) = 60z^2 - 72iz - 18$$

By substituting $z = i$ one obtains

$$p(i) = p'(i) = p''(i) = 0$$
$$p'''(i) = -6 \neq 0$$

Thus i is a multiple root of $p(z)$ with multiplicity 3. □

If one should find the real roots of a given polynomial, the following result may be applicable.

Theorem 1.4.3.

Any odd order polynomial $p_n(z)$, with real coefficients, must have at least one real root.

The proof is simple and is based on the following:

· $p_n(x)$ changes signs when x goes from $-\infty$ to $+\infty$.
· A continuous function whose values at a and b have opposite signs must vanish at some interim point c:

$$-\infty < a < c < b < \infty$$

PROBLEMS

1. Find the roots and the corresponding multiplicities of the following polynomials.
   (a) $z^4 - 1$
   (b) $2(z-1)^2(z+i)^3$
   (c) $z^5 - z^2$
2. Let $p(z)$ be a polynomial with a root z_0, that is, $p(z_0) = 0$. Find a sufficient condition for $p(\bar{z}_0) = 0$ to hold as well. Show that this condition is not necessary.
3. Explain why the polynomial $p(x) = x^4 + x^3 + 10x^2 - 2x$ must have at least two real roots.

So far we have concentrated on some of the basic properties of polynomials. This next section is a detailed discussion of what is probably the most popular application of polynomials—Taylor's approximations and series.

1.5 Taylor Series

The concept of approximating functions by polynomials of any order originates from the recognition that polynomials are far easier to manipulate than other functions. The most commonly used forms of polynomial approximation, the *Taylor's approximations*, are based on *Taylor's theorem*.

We start by introducing the set of functions $C^n[a, b]$.

Definition 1.5.1.   *A real function* $f(x)$ *defined over the interval*

$$[a, b] \equiv \{x | a \leq x \leq b\}$$

is said to belong to $C^n[a, b]$ *if* $f(x), f'(x), f''(x), \ldots, f^{(n)}(x)$ *exist and are continuous over* $[a, b]$.
If $f^{(n)}(x)$ exists and is continuous over $[a, b]$ for all $n \geq 0$, we write

$$f(x) \in C^\infty[a, b]$$

The definition of C^n over open and semiopen intervals is similar.

Example 1.5.1.

The functions $\sin x, \cos x$ possess continuous derivatives of any order. Therefore, $\sin x, \cos x \in C^\infty(-\infty, \infty)$.   □

Example 1.5.2.

The function defined as

$$f(x) = \begin{cases} x^3 \sin(1/x), & x \neq 0 \\ 0, & x = 0 \end{cases}$$

and its first derivative are continuous over $(-\infty, \infty)$. Clearly, this is true for all $x \neq 0$. At $x = 0$, $f(x)$ is continuous too, because

$$\lim_{h \to 0} h^3 \sin \frac{1}{h} = 0$$

The existence and continuity of $f'(x)$ for $x \neq 0$, that is, of $3x^2 \sin 1/x - x \cos 1/x$ are also trivial. At $x = 0$,

$$f'(0) = \lim_{h \to 0} \frac{f(h) - f(0)}{h} = \lim_{h \to 0} h^2 \sin \frac{1}{h} = 0$$

and

$$\lim_{h \to 0} f'(h) = \lim_{h \to 0} \left(3h^2 \sin \frac{1}{h} - h \cos \frac{1}{h} \right) = 0$$

Thus $f'(x)$ exists and is continuous at $x = 0$ as well.

Hence, $f(x) \in C^1(-\infty, \infty)$. However, it can be shown that $f''(0)$ does not exist, that is, $f(x)$ is not in $C^2(-\infty, \infty)$.   □

A major theorem from basic calculus is Theorem 1.5.1:

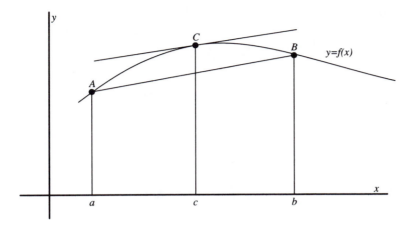

Figure 1.5.1. The mean-value theorem.

Theorem 1.5.1.

(The first mean-value theorem). *Let $f(x) \in C[a, b]$ and $f'(x)$ exist. Then*

$$f(b) - f(a) = (b - a)f'(c) \tag{1.5.1}$$

for some c between a and b. The geometric interpretation of this statement is clear (Figure 1.5.1): For any "smooth" curve $y = f(x)$, connecting two points A and B, there must be a third point C "somewhere" between them, at which the tangent to the curve is parallel to the secant AB.

A significant result from the mean-value theorem is the following.

Theorem 1.5.2.

(Taylor's theorem). *Let a real-valued function $f(x)$ defined over the interval $[a, b]$ satisfy*

1. $f(x) \in C^n[a, b]$
2. $f^{(n+1)}(x)$ *exists for* $x : a < x < b$

Then, for any x_0, x within $[a, b]$,

$$f(x) = f(x_0) + \frac{f'(x_0)}{1!}(x - x_0) + \frac{f''(x_0)}{2!}(x - x_0)^2 + \cdots + \frac{f^{(n)}(x_0)}{n!}(x - x_0)^n$$
$$+ \frac{f^{(n+1)}(c_x)}{(n + 1)!}(x - x_0)^{n+1} \tag{1.5.2}$$

where c_x is some unknown point between x_0 and x. If x_0 is fixed, we relate to Eq. (1.5.2) as the nth order "Taylor's expansion" of $f(x)$ about x_0. The expression

$$f(x_0) + \frac{f'(x_0)}{1!}(x - x_0) + \frac{f''(x_0)}{2!}(x - x_0)^2 + \cdots + \frac{f^{(n)}(x_0)}{n!}(x - x_0)^n$$

is called a "Taylor polynomial" of order n, and the last term

$$R_n = \frac{f^{(n+1)}(c_x)}{(n+1)!}(x - x_0)^{n+1}$$

is referred to as the "remainder". The unknown point is generally x dependent and therefore specified as c_x.

Example 1.5.3.

Let $f(x) = e^x, x_0 = 0$. Because $e^x \in C^\infty(-\infty, \infty)$, and for all $n \geq 0 : f^{(n)}(0) = 1$, we obtain

$$e^x = 1 + x + \frac{x^2}{2!} + \cdots + \frac{x^n}{n!} + \frac{e^{c_x}}{(n+1)!}x^{n+1} \qquad \square$$

Example 1.5.4.

Let $f(x) = \cos x, -\pi < x < \pi, x_0 = 0$. Then, $f(x_0) = 1, f'(x_0) = 0$, and $f''(x_0) = -1$. The three-term Taylor expansion of $\cos x$ about $x_0 = 0$ is therefore

$$\cos x = 1 - \frac{x^2}{2!} + \frac{\sin(c_x)}{3!}x^3$$

where c_x is an interim point between 0 and x.

If we choose $x_0 = \pi/4$, then $f(x_0) = \sqrt{2}/2, f'(x_0) = -\sqrt{2}/2, f''(x_0) = -\sqrt{2}/2$ and the three-term expansion is

$$\cos x = \frac{\sqrt{2}}{2} - \frac{\sqrt{2}}{2}(x - \frac{\pi}{4}) - \frac{1}{2!}\left[\frac{\sqrt{2}}{2}\right](x - \frac{\pi}{4})^2 + \frac{\sin(c_x)}{3!}(x - \frac{\pi}{4})^3$$

where c_x (different from the previous c_x) is some interim point between x and $\pi/4$. $\qquad \square$

Example 1.5.5.

Define

$$f(x) = \begin{cases} x^4 \sin(1/x), & x > 0 \\ \\ 0 & , x = 0 \end{cases}$$

Using arguments similar to those in Example 1.5.2, one can show that $f'(0) = f''(0) = 0$, whereas $f'''(x)$ exists everywhere except at $x = 0$. Thus

$$f(x) = \frac{f'''(c_x)}{3!}x^3$$

where

$$f'''(x) = 24x \sin\frac{1}{x} - 18\cos\frac{1}{x} - \frac{6}{x}\sin\frac{1}{x} + \frac{1}{x^2}\cos\frac{1}{x}, \quad 0 < c_x < x \qquad \square$$

One is naturally tempted to relate to a Taylor polynomial as an "approximation" to $f(x)$ "near" x_0. If this can be justified, the remainder R_n can be related to as the "error term."

Let us search for an nth order polynomial

$$p_n(x) = a_0 + a_1(x - x_0) + \cdots + a_n(x - x_0)^n$$

that approximates $f(x)$ near $x = x_0$. There are $(n+1)$ free parameters to be determined, namely, the coefficients a_i, $0 \le i \le n$. One way of doing it is to require

$$p_n^{(j)}(x_0) = f^{(j)}(x_0), \quad 0 \le j \le n \tag{1.5.3}$$

[defining $f^{(0)}(x) = f(x)$].

Clearly, a polynomial for which Eq. (1.5.3) holds can be considered an "approximation" to $f(x)$, near x_0, because $f(x)$ and $p_n(x)$ and all their derivatives up to order n coincide at x_0. Thus

$$f(x) \approx p_n(x) \quad \text{near } x_0$$

If we enforce Eq. (1.5.3) we obtain

$$a_j = \frac{f^{(j)}(x_0)}{j!}, \quad 0 \le j \le n \tag{1.5.4}$$

which means that the Taylor polynomial previously defined by Taylor's theorem is indeed an approximation to $f(x)$. Table 1.5.1 provides a comparison between $p_1(x)$, $p_2(x)$, $p_3(x)$, and e^x for $-2 < x < 2, x_0 = 0$. It demonstrates two facts:

1. For a fixed x the accuracy increases with the degree of the polynomial.

2. For a fixed polynomial, the accuracy decreases when $|x - x_0|$ increases.

Table 1.5.1. Taylor Approximations to e^x

x	$p_1(x)$	$p_2(x)$	$p_3(x)$	e^x
-2.0	-1.0	1.000	-0.33333	0.13534
-1.0	0.0	0.500	0.33333	0.36788
-0.50	0.5	0.625	0.60417	0.60653
-0.1	0.9	0.905	0.90483	0.90484
0.0	1.0	1.000	1.00000	1.00000
0.1	1.1	1.105	1.10517	1.10517
0.5	1.5	1.625	1.64583	1.64872
1.0	2.0	2.500	2.66667	2.71828
2.0	3.0	5.000	6.33333	7.38906

The main problem related to Taylor's theorem is whether Taylor's polynomials converge to $f(x)$, as $n \to \infty$. From Eq. (1.5.2) it is clear that convergence is guaranteed, provided that

$$R_n = \frac{f^{(n+1)}(c_x)}{(n+1)!}(x - x_0)^{n+1} \to 0 \quad \text{as } n \to \infty \tag{1.5.5}$$

If Eq. (1.5.5) holds, $f(x)$ is not only approximated by its Taylor's polynomials but can also be replaced by its infinite *Taylor's series*, that is,

$$f(x) = \sum_{n=0}^{\infty} \frac{f^{(n)}(x_0)}{n!}(x - x_0)^n \tag{1.5.6}$$

The *only* test that one needs to perform before replacing $f(x)$ by an approximating Taylor's polynomial is whether or not R_n is small enough. To replace $f(x)$ by an equivalent Taylor series, one *must* show that $R_n \to 0$ as $n \to \infty$.

Because the behavior of $(x - x_0)^{n+1}/(n+1)!$ is well known, we need only find the behavior of $f^{(n+1)}(x)$. This is often quite difficult, if not impossible.

Example 1.5.6.

Let $f(x) = e^x$, $x_0 = 0$. Here $f^{(n)}(x) = f(x) = e^x$ for all n. The remainder R_n satisfies

$$|R_n| = \left| \frac{e^c}{(n+1)!} x^{n+1} \right| \le e^x \frac{x^{n+1}}{(n+1)!} \to 0 \quad \text{as } n \to \infty$$

for all x, and therefore

$$e^x = 1 + x + \frac{x^2}{2!} + \cdots + \frac{x^n}{n!} + \cdots = \sum_{n=0}^{\infty} \frac{x^n}{n!}, \quad -\infty < x < \infty$$

Moreover, for any finite interval the convergence is *uniform*, that is, given $a \le x \le b$ and $\epsilon > 0$, one can find an x-independent $N = N(\epsilon)$ that depends only on a, b, ϵ such that

$$\left| e^x - \left(1 + x + \frac{x^2}{2!} + \cdots + \frac{x^n}{n!} \right) \right| \le \epsilon \quad \text{for } n > N$$

for all $x \in [a, b]$. For $0 \le x \le 1$, the error is bounded by

$$|R_n| \le \frac{e^1 \cdot 1^{n+1}}{(n+1)!} < \frac{3}{(n+1)!}$$

To obtain, for example, a Taylor's approximation to e^x over $[0, 1]$ with an error $R_n \le 10^{-8}$, one should find the smallest n for which the inequality

$$\frac{3}{(n+1)!} \le 10^{-8}$$

holds. The quantity $n!$ increases extremely fast with n. Therefore the smallest n to satisfy the last inequality cannot be too large.

To evaluate it, one simply computes $1 \cdot 2 \cdot 3 \ldots n$ until one exceeds $3 \cdot 10^8$. If, however, it is necessary to approximate a complicated expression that includes factorials, one should use Stirling's asymptotic formula (see Chapter 5). □

Example 1.5.7.

Let $f(x) = \sin x, \pi < x < \pi$, $x_0 = 0$. Clearly $|f^{(n)}(x)| \leq 1$ for all x and n because each derivative is either $\cos x$ or $\sin x$. The error R_n is therefore bounded by $x^n/n!$, which converges to 0 for all x as $n \to \infty$. Because

$$\sin^{(2n)}(0) = 0, \quad \sin^{(2n+1)}(0) = (-1)^n \quad \text{for } n \geq 0$$

we obtain

$$\sin x = x - \frac{x^3}{3!} + \frac{x^5}{5!} - \frac{x^7}{7!} + \cdots, \quad -\infty < x < \infty$$

To approximate $\sin x, -\pi \leq x \leq \pi$, for example by a third order polynomial, one should calculate an upper bound to $|R_4|$. This is easily obtained because

$$|R_4| \leq \left| \frac{x^5}{5!} \right| \leq \frac{\pi^5}{120} \approx 2.6$$

This "discouraging" upper bound for the error is expected. Because as previously stated, Taylor's polynomial best approximates $f(x)$ near x_0, we cannot establish a reasonable approximation over the whole interval unless we use a higher order polynomial. For example, by choosing

$$\sin x \approx x - \frac{x^3}{3!} + \frac{x^5}{5!} - \frac{x^7}{7!} + \frac{x^9}{9!} = p_9(x)$$

we obtain

$$|\sin x - p_9(x)| \leq \frac{\pi^{11}}{11!} \approx 0.007, \quad |x| \leq \pi$$

But if one is merely interested in computing $\sin x$ between, say, $x = 0$ and $x = 0.2$, the previous third order approximation is already quite accurate, because

$$|R_4| \leq \left| \frac{x^5}{5!} \right| \leq \frac{0.2^5}{120} < 3 \cdot 10^{-6}$$ □

Example 1.5.8.

Consider the function

$$f(x) = \begin{cases} e^{-1/x^2}, & x \neq 0 \\ 0, & x = 0 \end{cases}$$

which satisfies $f^{(n)}(0) = 0$ for all $n \geq 0$. To show that, we first establish

$$\lim_{x \to 0} \frac{e^{-1/x^2}}{x^n} = 0, \quad n \geq 0 \tag{1.5.7}$$

Next, by using mathematical induction we express $f^{(n)}(x)$ as a finite sum of terms similar to that of Eq. (1.5.7). Thus $f(x) \in C^\infty(-\infty, \infty)$, and satisfies

$$f^{(n)}(0) = 0, \quad n = 0, 1, 2, \ldots$$

The Taylor series of the function about $x_0 = 0$ is therefore identically zero! Yet $f(x)$ is nonzero for $x \neq 0$. The ultimate conclusion is that R_n *does not converge* to 0 as $n \to \infty$. Thus the existence and convergence of Taylor's series,

$$T_f(x) \equiv \sum_{n=0}^{\infty} \frac{f^{(n)}(0)}{n!} x^n$$

do not guarantee $f(x) = T_f(x)$, unless $R_n \to 0$ as $n \to \infty$.   □

Example 1.5.9.

Consider $f(x) = 1/(1-x)$, $x \neq 1$, $x_0 = 0$. Clearly $f(0) = 1$, and by using induction we obtain

$$f^{(n)}(x) = \frac{n!}{(1-x)^{n+1}}, \quad x \neq 1, \quad n \geq 1$$

so that

$$\frac{1}{1-x} = 1 + x + x^2 + \cdots + x^n + \frac{x^{n+1}}{(1-c)^{n+1}}$$

We can also write

$$\frac{1}{1-x} = 1 + x + x^2 + \cdots + x^n + \frac{x^{n+1}}{1-x} \tag{1.5.8}$$

Thus, $x^{n+1}/(1-c)^{n+1} = x^{n+1}/(1-x)$. This is one rare case where c can be easily computed!   □

Although Eq. (1.5.2) is the standard way to define Taylor's polynomials, there are often other convenient ways to establish these polynomials and the remainder. For example, consider the function $\ln(1-x)$ about $x_0 = 0$. Rather than using Eq. (1.5.2), integrate Eq. (1.5.8) to obtain

$$-\ln(1-x) = \int_0^x \frac{1}{1-t} \, dt = \int_0^x \left(1 + t + t^2 + \cdots + t^n + \frac{t^{n+1}}{1-t}\right) dt$$

$$= x + \frac{x^2}{2} + \frac{x^3}{3} + \cdots + \frac{x^{n+1}}{n+1} + \int_0^x \frac{t^{n+1}}{1-t} \, dt$$

By applying an integral mean-value theorem (Theorem 1.5.3), we obtain

$$\int_0^x \frac{t^{n+1}}{1-t}\, dt = \frac{1}{1-c}\int_0^x t^{n+1}\, dt = \frac{x^{n+2}}{(1-c)(n+2)}, \qquad 0 < c < x$$

and finally

$$\ln(1-x) = -\left(x + \frac{x^2}{2} + \frac{x^3}{3} + \cdots + \frac{x^{n+1}}{n+1}\right) - \frac{x^{n+2}}{(1-c)(n+2)}$$

Because $[\ln(1-x)]' = -1/(1-x)$ and Eq. (1.5.8) is the Taylor's expansion of $1/(1-x)$, the term $\{-x^{n+2}/[(1-c)(n+2)]\}$ must be the $(n+1)$th order remainder of $\ln(1-x)$.

In calculating R_{n+1} of $-\ln(1-x)$ we applied an integral mean-value theorem, which like Theorem 1.5.1 can be useful and is given here without a proof.

Theorem 1.5.3.

(An integral mean-value theorem). *Let* $f(x), g(x) \in C[a,b]$ *and let* $g(x)$ *maintain a constant sign throughout the interval. Then*

$$\int_a^b f(x)g(x)dx = f(c)\int_a^b g(x)dx$$

for some c between a and b.

Example 1.5.10.

We conclude this section by calculating a polynomial approximation to

$$I(x) = \int_0^x \frac{e^t - 1}{t}\, dt, \qquad 0 \le x \le 1 \tag{1.5.9}$$

We first replace e^t by its Taylor series to obtain

$$I(x) = \int_0^x \left(1 + \frac{t}{2!} + \frac{t^2}{3!} + \cdots + \frac{t^{n-1}}{n!} + \frac{e^{c_t}}{(n+1)!}t^n\right) dt, \qquad 0 < c_t < t$$

$$= x + \frac{x^2}{(2!)2} + \frac{x^3}{(3!)3} + \cdots + \frac{x^n}{(n!)n} + R_n$$

where

$$|R_n| = \left|\int_o^x e^{c_t}\frac{t^n dt}{(n+1)!}\right| \le \left|e^x \frac{x^{n+1}}{(n+1)!(n+1)}\right| \le \frac{e}{(n+1)!(n+1)}$$

If we denote the desired *tolerance* by ϵ, we need to find the smallest n such that

$$\frac{e}{(n+1)!(n+1)} \le \epsilon$$

or

$$(n+1)!(n+1) \geq \frac{e}{\epsilon}$$

For $\epsilon = 10^{-10}$, we obtain $n = 12$. $\square$

PROBLEMS

1. Define

$$f(x) = \begin{cases} x^3 \ln x, & x > 0 \\ 0, & x = 0 \end{cases}$$

and discuss the function and its derivatives. You may use the fact that $\lim_{x \to 0} x^\alpha \ln x = 0$ for all $\alpha > 0$.

2. Generate a table of comparison between $p_1(x), p_3(x), p_5(x)$, and $\sin x$ over the interval $-2\pi \leq x \leq 2\pi$. Choose $x_0 = 0$ and evaluate the functions at $x_i = -2\pi + (i-1)(\pi/4), 0 \leq i \leq 17$.

3. Find $p_1(x), p_2(x), p_3(x)$ in the following cases:
   (a) $f(x) = e^{-x}, \quad x_0 = 0$
   (b) $f(x) = e^x \cos x, \quad x_0 = 0$
   (c) $f(x) = \sqrt{1+x}, \quad x_0 = 1$
   (d) $f(x) = (1+x)/(1-x), \quad x_0 = -1$

4. Let $f(x) = x^{7/2}, x > 0$. Evaluate $p_2(x)$ for $x_0 = 0$.
   (a) Is it an approximation to $f(x)$?
   (b) What can be said about the remainder R_2?

5. Produce the Taylor series of $\ln [(1+x)/(1-x)]$ for $|x| < 1$ and $x_0 = 0$.

6. Calculate $p_3(x)$ for $f(x) = e^{x^2}, 0 \leq x \leq 1.5, x_0 = 0$ and an upper bound to the remainder.

7. (a) Find a Taylor polynomial and remainder of order n for $f(x) = 1/(1+x^2)$ about $x_0 = 0$. (*Hint:* substitute $u = x^2$.)
   (b) Use part (a) and the identity

$$\tan^{-1}(x) = \int_0^x \frac{dt}{1+t^2}$$

to obtain a Taylor polynomial and a remainder for $f(x) = \tan^{-1}(x)$ about $x_0 = 0$.

8. Use the results of problem 7 and the identity

$$\pi = 4 \tan^{-1}(1)$$

to obtain π as an infinite series. How many terms are needed to evaluate π, given a tolerance of 10^{-3}?

9. Let $f(x) \in C^\infty[-1, 1]$ and $|f''(x)| \le (n-1)!$. Let $p_n(x)$ be the Taylor polynomial of $f(x)$ about x_0. What should n be to guarantee $|R_n| \le 10^{-3}$?

1.6 Elementary Functions of a Complex Variable

A function of a complex variable is a single-valued transformation from one subset to another subset in the complex plane, that is:

$$w = f(z) \tag{1.6.1}$$

where z, w are complex numbers.

The most elementary functions of complex variable are the polynomials, for example, $w = z^2 - 1$, $w = iz^3 - (1+i)z$. Each polynomial $p_n(z)$ of order n, given as

$$p_n(z) = a_n z^n + a_{n-1} z^{n-1} + \cdots + a_1 z + a_0, \quad a_n \neq 0$$

has n roots and can also be represented as

$$p_n(z) = a_n(z - z_1)(z - z_2) \cdots (z - z_n)$$

where $z_i, 1 \le i \le n$ are the roots of the polynomial. We will now introduce exponential and trigonometric functions of complex variable. Because we already have

$$e^x = 1 + x + \frac{x^2}{2!} + \cdots + \frac{x^n}{n!} + \cdots, \quad -\infty < x < \infty$$

we start by defining

$$e^{iy} \equiv 1 + (iy) + \frac{(iy)^2}{2!} + \cdots + \frac{(iy)^n}{n!} + \cdots \tag{1.6.2}$$

for all real y. The right-hand side converges for any finite real y to some complex number, denoted by e^{iy}.

Example 1.6.1.

$$e^i \equiv 1 + i + \frac{i^2}{2!} + \frac{i^3}{3!} + \frac{i^4}{4!} + \cdots + \frac{i^n}{n!} + \cdots$$
$$= \left(1 - \frac{1}{2!} + \frac{1}{4!} - \cdots\right) + i\left(1 - \frac{1}{3!} + \frac{1}{5!} - \cdots\right) \qquad \square$$

We have in general

$$e^{iy} = 1 + iy + \frac{(iy)^2}{2!} + \cdots + \frac{(iy)^n}{n!} + \cdots$$

$$= \left(1 - \frac{y^2}{2!} + \frac{y^4}{4!} - \cdots\right) + i\left(y - \frac{y^3}{3!} + \frac{y^5}{5!} - \cdots\right)$$

that is,

$$e^{iy} = \cos y + i \sin y \qquad (1.6.3)$$

By choosing $y = 2\pi$ we obtain

$$e^{2\pi i} = \cos 2\pi + i \sin 2\pi = 1$$

Also, $e^{2\pi k i} = 1$ for all integers k, that is, e^{iy} is a periodic function of y with period $2\pi i$. This has no analog within the real numbers, because the equation $e^x = 1$ has a unique solution $x = 0$. Other properties of e^x hold for e^{iy} as well. For example, the relation

$$e^{x_1 + x_2} = e^{x_1} e^{x_2}, \qquad -\infty < x_1, x_2 < \infty \qquad (1.6.4)$$

has an analog for e^{iy},

$$e^{iy_1 + iy_2} = e^{iy_1} e^{iy_2}, \quad |y_1|, \ |y_2| \ < \infty$$

that holds because

$$\begin{aligned}
e^{iy_1} e^{iy_2} &= (\cos y_1 + i \sin y_1)(\cos y_2 + i \sin y_2) \\
&= \cos(y_1 + y_2) + i \sin(y_1 + y_2) = e^{i(y_1 + y_2)}
\end{aligned}$$

We next define $e^z = e^{x+iy}$ for any given complex number z. Because we would like Eq. (1.6.4) to hold for the complex case as well, it is necessary that we define

$$e^z = e^{x+iy} \equiv e^x e^{iy} = e^x(\cos y + i \sin y) \qquad (1.6.5)$$

Once defined by Eq. (1.6.5) the complex exponential function e^z satisfies

$$e^{z_1 + z_2} = e^{z_1} e^{z_2}$$

Moreover, one can show that

$$e^z = 1 + z + \frac{z^2}{2!} + \cdots + \frac{z^n}{n!} + \cdots, \quad |z| < \infty$$

This last equation is not a definition but a *theorem*, because both sides were already defined. Alternatively, we could define

$$e^z \equiv \sum_{n=0}^{\infty} \frac{z^n}{n!}$$

and then show that Eq. (1.6.5) holds.

Example 1.6.2.

$$|e^z| = |e^{x+iy}| = |e^x(\cos y + i\sin y)| = |e^x||\cos y + i\sin y| = e^x$$
$$\arg(e^z) = \arg\left[e^x(\cos y + i\sin y)\right] = y + 2k\pi$$

where k is an integer for which $-\pi < y + 2k\pi \le \pi$. For example,

$$|e^{1+i}| = e, \quad \arg(e^{1+i}) = 1 \quad \text{(radians)} \qquad \square$$

If we substitute $-y$ in Eq. (1.6.3), we obtain

$$e^{-iy} = \cos y - i\sin y$$

which together with Eq. (1.6.3) produces Euler's formulas

$$\cos y = \frac{e^{iy} + e^{-iy}}{2}, \quad \sin y = \frac{e^{iy} - e^{-iy}}{2i} \qquad (1.6.6)$$

These formulas are the inverse of De Moivre's equation [Eq. (1.3.13)]. Whereas De Moivre's equation enables us to express $\cos(ny)$ and $\sin(ny)$ in terms of powers of $\cos y$, $\sin y$, Euler's formulas provide expressions to $\cos^n y$ and $\sin^n y$ in terms of $\cos y$, $\sin y$, $\cos(2y)$, $\sin(2y), \ldots, \cos(ny)$, $\sin(ny)$.

Example 1.6.3.

For $n = 3$:

$$\sin^3 y = \left[\frac{e^{iy} - e^{-iy}}{2i}\right]^3 = \frac{i}{8}(e^{3iy} - 3e^{iy} + 3e^{-iy} - e^{-3iy})$$
$$= \frac{i}{8}\left[2i\sin(3y) - 6i\sin y\right] = \frac{1}{4}\left[3\sin y - \sin(3y)\right] \qquad \square$$

Example 1.6.4.

For $n = 4$:

$$\cos^4 y = \left[\frac{e^{iy} + e^{-iy}}{2}\right]^4 = \frac{1}{16}(e^{4iy} + 4e^{2iy} + 6 + 4e^{-2iy} + e^{-4iy})$$
$$= \frac{1}{16}\left[2\cos(4y) + 8\cos(2y) + 6\right] = \frac{1}{8}\left[\cos(4y) + 4\cos(2y) + 3\right] \qquad \square$$

We will now define the complex trigonometric functions $\cos z$ and $\sin z$ as

$$\cos z \equiv 1 - \frac{z^2}{2!} + \frac{z^4}{4!} - \cdots$$

$$\sin z \equiv z - \frac{z^3}{3!} + \frac{z^5}{5!} - \cdots \qquad (1.6.7)$$

for all finite complex numbers. Because these definitions satisfy

$$\cos(iz) - i\sin(iz) = e^z$$
$$\cos(iz) + i\sin(iz) = e^{-z}$$

we obtain

$$\cos(iz) = \frac{e^z + e^{-z}}{2}, \quad \sin(iz) = \frac{e^{-z} - e^z}{2i}$$

or

$$\cos z = \frac{e^{iz} + e^{-iz}}{2}, \quad \sin z = \frac{e^{iz} - e^{-iz}}{2i} \qquad (1.6.8)$$

which is an extension of Eq. (1.6.6). The complex trigonometric functions render properties similar to those of $\cos x, \sin x$, in particular

$$\cos(z_1 + z_2) = \cos z_1 \cos z_2 - \sin z_1 \sin z_2$$
$$\sin(z_1 + z_2) = \sin z_1 \cos z_2 + \cos z_1 \sin z_2$$

Example 1.6.5.

$$\cos(1 + i) = \frac{e^{i(1+i)} + e^{-i(1+i)}}{2} = \frac{e^{-1+i} + e^{1-i}}{2}$$
$$= \frac{1}{2}[e^{-1}(\cos 1 + i\sin 1) + e(\cos 1 - i\sin 1)] \approx 0.83 - 0.99i$$

Of course one can alternatively express $\cos(1 + i)$ as

$$1 - \frac{(1+i)^2}{2!} + \frac{(1+i)^4}{4!} - \cdots$$

and evaluate the infinite complex series. $\square$

PROBLEMS

1. Express $\sin^4 \phi \cos^4 \phi$ by $\{\cos(ny), \sin(ny)\}, 1 \le n \le 4$.

2. Calculate the following numbers:

   (a) e^{2i+3}

   (b) $\cos(2 + i) - \sin^2(1 - i)$

   (c) $\tan(1 + 2i)$ (Define $\tan z = \sin z / \cos z$)

3. Show that for all finite complex z

$$\sin^2 z + \cos^2 z = 1$$

4. Prove that $e^{\bar{z}} = \overline{e^z}$ for any given complex z.
5. Find all the solutions to the equation $\sin z = \cos z$.

1.7 Functions of Several Variables and Partial Derivatives

We start by defining the n-dimensional Euclidean vector space as the set

$$R^n \equiv \{x|x = (x_1,\ldots,x_n)^T, \quad -\infty < x_i < \infty, \quad i = 1,\ldots,n\} \tag{1.7.1}$$

where addition between vectors is defined by

$$\begin{pmatrix} x_1 \\ x_2 \\ \vdots \\ x_n \end{pmatrix} + \begin{pmatrix} y_1 \\ y_2 \\ \vdots \\ y_n \end{pmatrix} = \begin{pmatrix} x_1 + y_1 \\ x_2 + y_2 \\ \vdots \\ x_n + y_n \end{pmatrix}$$

and multiplication by an arbitrary real constant as

$$\alpha \begin{pmatrix} x_1 \\ x_2 \\ \vdots \\ x_n \end{pmatrix} = \begin{pmatrix} \alpha x_1 \\ \alpha x_2 \\ \vdots \\ \alpha x_n \end{pmatrix}$$

for all real α. The Euclidean *norm* of a vector $\mathbf{x}$ is defined as

$$||x|| = (x_1^2 + x_2^2 + \cdots + x_n^2)^{1/2} \tag{1.7.2}$$

and satisfies the three *axioms* that are expected to be satisfied by *any* norm, that is,

1. $||x|| \geq 0, ||x|| = 0$　　if and only if $x = 0$
2. $||\alpha x|| = |\alpha|||x||$　　for all x and arbitrary α
3. $||x + y|| \leq ||x|| + ||y||$　　(the inequality of the triangle)

A real function $f(x_1,\ldots,x_n)$ of n variables $x_1,\ldots,x_n$ is simply a single real-valued transformation from a subset in R^n into $R = R^1$, the set of all the real numbers.

Example 1.7.1.
The real function $(1 - x^2 - y^2)^{1/2}$ is a function of two variables defined over the region

$$D = \{(x,y)|x^2 + y^2 \leq 1\} \qquad \square$$

Example 1.7.2.
The real function $e^x \cos y + x/(1 - z)$ is a function of three variables defined everywhere except at $(x, y, 1)$. ☐

For the sake of simplicity, and without any loss of generality, we will now confine ourselves to functions of two variables.

Definition 1.7.1. (Neighborhood). *An ϵ neighborhood of a given point (x_0, y_0) is defined as the open circle centered at (x_0, y_0) with radius ϵ, that is, as the set*

$$\{(x, y)| \left[(x - x_0)^2 + (y - y_0)^2\right]^{1/2} < \epsilon\}$$

If $P \equiv (x_0, y_0)$, this neighborhood is denoted by P_ϵ.

Definition 1.7.2. (Continuity). *A function $f(x, y)$ is continuous at a given point (x_0, y_0), if and only if*

1. *$f(x, y)$ is defined at (x_0, y_0) and in some neighborhood of this point.*
2. *$\lim f(x, y) = f(x_0, y_0)$ as $x \to x_0$ and $y \to y_0$.*

One should realize that two-dimensional continuity is far more complex than the ordinary concept applied for functions of one variable. It is a direct result of not having a restriction on the path along which (x, y) approaches (x_0, y_0). Quite possibly

$$f(x, y) \to f(x_0, y_0)$$

when (x, y) approaches (x_0, y_0) from *several* directions. Yet $f(x, y)$ may not be continuous at (x_0, y_0)!

Example 1.7.3.
Define

$$f(x, y) = \begin{cases} 2xy/(x^2 + y^2), & (x, y) \neq (0, 0) \\ \\ 0 & , (x, y) = (0, 0) \end{cases}$$

$f(x, y)$ is well defined over the xy plane, and continuous at any given point $(x_0, y_0) \neq (0, 0)$. Let us study the behavior of the function at the origin by approaching it from three different directions:

1. $x = 0, y \to 0 \Rightarrow \lim f(x, y) = \lim 0 = 0$
2. $y = 0, x \to 0 \Rightarrow \lim f(x, y) = \lim 0 = 0$
3. $x = y \to 0 \Rightarrow \lim f(x, y) = \lim \left[2x^2/(x^2 + x^2)\right] = \lim 1 = 1$

Thus, $f(x, y)$ is not continuous at $(0, 0)$. ☐

A "partial" or "directional" derivative of functions of two variables is another concept to be defined and studied. When dealing with functions of one variable there is almost no need to use the term "directional derivative." However, even then the ordinary derivative is *directional:* the rate of change of the function is computed with respect to the positive x axis. If we want to calculate the derivative in the opposite direction (i.e., that of $-x$) we simply change signs. For example, let $f(x) = x^2$. Its derivatives in the directions of positive and negative x are $2x, -2x$, respectively.

We realize that, for a function of a single variable, there are only two possible directions for which a directional derivative can be calculated. For a function of two variables there is an infinite number of directions and the next definition is necessary.

Definition 1.7.3. (Directional derivative). *Let $f(x, y)$ satisfy the following requirements:*

1. $f(x, y)$ *is defined at $P(x_0, y_0)$ and at some P_ϵ.*

2. $\lim_{t \to 0}[f(x_0 + \alpha t, y_0 + \beta t) - f(x_0, y_0)]/t, t > 0$ *exists for some given $\alpha, \beta : \alpha^2 + \beta^2 = 1$.*

Then, $f(x, y)$ is said to have a directional derivative in the $s \equiv (\alpha, \beta)$ direction, and we write

$$\frac{\partial f}{\partial s} \equiv \lim_{t \to 0} \frac{f(x_0 + \alpha t, y_0 + \beta t) - f(x_0, y_0)}{t} \tag{1.7.3}$$

The directional derivatives in the positive x and y directions are

$$\frac{\partial f}{\partial x} = \lim_{h \to 0} \frac{f(x_0 + h, y_0) - f(x_0, y_0)}{h}, \quad (y = y_0 \text{ is held constant})$$

and

$$\frac{\partial f}{\partial y} = \lim_{h \to 0} \frac{f(x_0, y_0 + h) - f(x_0, y_0)}{h}, \quad (x = x_0 \text{ is held constant})$$

respectively.

Example 1.7.4.
For $f(x, y) = x^3 - 2xy$: $\partial f/\partial x = 3x^2 - 2y, \partial f/\partial y = -2x.$ □

Example 1.7.5.
For $f(x, y) = e^{xy}$: $\partial f/\partial x = ye^{xy}, \partial f/\partial y = xe^{xy}.$ □

The directional derivative defined by Eq. (1.7.3) is of the first order. Higher order directional derivatives are similarly defined. For example, to calculate $\partial^2 f/\partial x^2$ we hold y constant and calculate $\partial f/\partial x$, then again hold y constant and find $\partial^2 f/\partial x^2$.

Example 1.7.6.
Let $f(x, y) = \cos(xy) - x^2 \ln y + y^3$. Its derivatives up to the second order are

$$\frac{\partial f}{\partial x} = -y \sin(xy) - 2x \ln y, \quad \frac{\partial f}{\partial y} = -x \sin(xy) - \frac{x^2}{y} + 3y^2$$

$$\frac{\partial^2 f}{\partial x^2} = -y^2 \cos(xy) - 2 \ln y, \quad \frac{\partial^2 f}{\partial y^2} = -x^2 \cos(xy) + \frac{x^2}{y^2} + 6y$$

$$\frac{\partial^2 f}{\partial y \partial x} = \frac{\partial^2 f}{\partial x \partial y} = -\sin(xy) - xy \cos(xy) - \frac{2x}{y}$$

□

The mixed derivatives are not equal by coincidence, as the following theorem shows.

Theorem 1.7.1

(Mixed partial derivatives). *If $\partial^2 f/\partial x \partial y, \partial^2 f/\partial y \partial x$ exist and are continuous, then $\partial^2 f/\partial x \partial y = \partial^2 f/\partial y \partial x$.*

We will now extend the notation $C^n[a, b]$ and define the following.

Definition 1.7.4. *If $f(x, y)$ and all its partial derivatives of order $\leq n$ are defined and continuous over a given two-dimensional domain, $D, f(x, y)$ is said to belong to $C^n(D)$.*

Example 1.7.7.

Let $f(x, y) = e^x \cos y$. Because $f(x, y)$ and all its partial derivatives are defined and continuous at any given point (x_0, y_0), we conclude that $e^x \cos y \in C^\infty(R^2)$. ☐

Example 1.7.8.

The function $f(x, y) = x^{5/2} e^y$ is in $C^2(R^2)$ but not in $C^3(R^2)$. ☐

Example 1.7.9.

A derivative of any given complex function $f(z) = u + iv$ is defined as

$$f'(z) = \lim_{\Delta z \to 0} \frac{f(z + \Delta z) - f(z)}{\Delta z}$$

whenever the right-hand side exists. If indeed $f'(z)$ exists (that is, the right-hand side exists and is identical for *all possible directions* from which Δz approaches 0) for all $z \in D, f(z)$ is called *regular*. It can be shown that if $f(z)$ is regular over D, then $u(x, y), v(x, y) \in C^\infty(D)$. Furthermore, the *Cauchy–Riemann* equations

$$\frac{\partial u}{\partial x} = \frac{\partial v}{\partial y}, \quad \frac{\partial u}{\partial y} = -\frac{\partial v}{\partial x}$$

hold throughout D. It is then easily seen that both u and v satisfy Laplace's equation, that is,

$$\frac{\partial^2 u}{\partial x^2} + \frac{\partial^2 u}{\partial y^2} = 0, \quad \frac{\partial^2 v}{\partial x^2} + \frac{\partial^2 v}{\partial y^2} = 0$$ ☐

Frequently, a shorter notation is used for partial derivative—a subscript. For example,

$$\frac{\partial f}{\partial x} \equiv f_x, \quad \frac{\partial^2 f}{\partial y \partial x} \equiv f_{yx}$$

Thus, the real part $u(x, y)$ of a regular complex function yields $u_{xx} + u_{yy} = 0$.

As in the case of a single variable, functions of several variables can be approximated by polynomials. One way of doing it is by replacing the function by its equivalent multivariable Taylor expansion. In the case of two variables, if we denote $h = x - x_0, k = y - y_0$, we obtain the following expansion about (x_0, y_0):

$$f(x,y) = f(x_0, y_0) + hf_x(x_0, y_0) + kf_y(x_0, y_0)$$

$$+ \frac{1}{2!} \left[h^2 f_{xx}(x_0, y_0) + 2hk f_{xy}(x_0, y_0) + k^2 f_{yy}(x_0, y_0) \right] + \cdots$$

$$+ \frac{1}{n!} \left(h\frac{\partial}{\partial x} + k\frac{\partial}{\partial y} \right)^n f(x_0, y_0) + R_n$$

where

$$\left(h\frac{\partial}{\partial x} + k\frac{\partial}{\partial y} \right)^n f \equiv h^n \frac{\partial^n f}{\partial x^n} + \binom{n}{1} h^{n-1} k \frac{\partial^n f}{\partial x^{n-1} \partial y} + \binom{n}{2} h^{n-2} k^2 \frac{\partial^n f}{\partial x^{n-2} \partial y^2} + \cdots$$

$$+ \binom{n}{1} hk^{n-1} \frac{\partial^n f}{\partial x \partial y^{n-1}} + k^n \frac{\partial^n f}{\partial y^n}$$

and

$$R_n = \frac{1}{(n+1)!} \left(h\frac{\partial}{\partial x} + k\frac{\partial}{\partial y} \right)^{(n+1)} f(c_1, c_2)$$

where (c_1, c_2) is some point on the segment that connects between (x_0, y_0) and (x, y).

Example 1.7.10.

Let $f(x,y) = e^{x+y}$, $(x_0, y_0) = (0,0)$. Because

$$(e^{x+y})_x = (e^{x+y})_y = (e^{x+y})_{xx} = (e^{x+y})_{xy} = (e^{x+y})_{yy} = e^{x+y}$$

we obtain

$$e^{x+y} = 1 + x \cdot 1 + y \cdot 1 + \frac{1}{2!} (x^2 e^{cx+cy} + 2xy e^{cx+cy} + y^2 e^{cx+cy})$$

$$= 1 + x + y + \frac{e^{c(x+y)}}{2!} (x+y)^2, \quad 0 < c < 1$$

$[(cx, cy)$ is an unknown point between $(0,0)$ and (x, y) for some $c : 0 < c < 1$.] $\quad\square$

Example 1.7.11.

Let $f(x,y) = \ln(1 + xy)$, $(x_0, y_0) = (0,0)$. Then

$$f_x = \frac{y}{1+xy}, \quad f_y = \frac{x}{1+xy}, \quad f_{xx} = \frac{-y^2}{(1+xy)^2}, \quad f_{yy} = -\frac{x^2}{(1+xy)^2}$$

all vanish at (x_0, y_0). Because $f_{xy} = 1/(1+xy)^2$, the second order Taylor polynomial approximation to $f(x, y)$ is

$$f(x,y) \approx \ln(1+xy)\Big|_{(0,0)} + \frac{1}{2!} \frac{1}{(1+xy)^2}\Big|_{(0,0)} 2xy = xy$$

38 INTRODUCTION

This is expected because we already know that $\ln(1 + x) = x - x^2/2 + \cdots$. $\square$

PROBLEMS

1. Prove that

$$f(x, y) = \begin{cases} \sin(x + y)/(|x| + |y|), & (x, y) \neq (0, 0) \\ 0, & (x, y) = 0, 0 \end{cases}$$

is not continuous at $(0, 0)$. Is it possible to redefine $f(x, y)$ at the origin and make it continuous there?

2. Prove that

$$f(x, y) = \begin{cases} x^2y/(x^2 + y^2), & (x, y) \neq (0, 0) \\ 0, & (x, y) = (0, 0) \end{cases}$$

is continuous for any given (x, y).

3. Evaluate $f_x, f_y, f_{xx}, f_{yy}, f_{xy}$ for the following functions:
   (a) $f(x, y) = x^2y^2 - y^3$
   (b) $f(x, y) = \sin(xy) - x/y$
   (c) $f(x, y) = e^{x \cos y}$
   (d) $f(x, y) = 1/xy$

4. Find the directional derivatives of $f(x, y) = x \cos y$ at $(0, 0)$ in the directions that form the following angles with the positive x axis:
   (a) $45°$
   (b) $60°$
   (c) $135°$
   (d) $-45°$

5. Let the regular complex function $f(z) = z^4 - z^2$ be written as $f = u + iv$. Using direct computation show that

$$\nabla^2 u = \nabla^2 v = 0$$

where ∇^2 is the Laplace operator defined as

$$\nabla^2 g \equiv g_{xx} + g_{yy}, \quad g(x, y) \in C^2$$

6. Repeat and solve problem 5 in the case of $f(z) = \cos z$.

7. Calculate the second order Taylor polynomial and the remainder for $f(x, y) = e^x \cos y$. Is there a shortcut?

2

Error

2.1 Representation of Numbers and Conversions

For thousands of years, since humans started to count, we have been using solely the decimal system. Using this system, the symbol 2108 means

$$2 \cdot 10^3 + 1 \cdot 10^2 + 0 \cdot 10^1 + 8 \cdot 10^0$$

and −17.509 means

$$-(1 \cdot 10^1 + 7 \cdot 10^0 + 5 \cdot 10^{-1} + 0 \cdot 10^{-2} + 9 \cdot 0^{-3})$$

This is a *decimal representation* of numbers, interpreted as a sum of powers of 10. The number 10 is the *base* of the system and the digits that can be used for representing a number s are the digits 0, 1, ..., 9. Should, however, humans grow eight fingers, the chances are that base 8 would be the one commonly used, and the decimal number 217 would be replaced by

$$(331)_8 = 3 \cdot 8^2 + 3 \cdot 8^1 + 1 \cdot 8^0$$

In general, we may consider any base $N > 1$ and interpret $a_n a_{n-1} \cdots a_1 a_0$ as

$$a_n \cdot N^n + a_{n-1} \cdot N^{n-1} + \cdots + a_1 \cdot N + a_0 \tag{2.1.1}$$

provided that

$$0 \le a_i \le N - 1, \quad 0 \le i \le n \tag{2.1.2}$$

Computers, because of their basic structure, must work with base 2. This leads to a *binary system* and *binary representation* using two digits, 0 and 1. For example, the decimal number 39, represented in the binary system, is given by

$$(100111)_2 = 1 \cdot 2^5 + 0 \cdot 2^4 + 0 \cdot 2^3 + 1 \cdot 2^2 + 1 \cdot 2^1 + 1 \cdot 2^0$$

This may also be written as

$$(39)_{10} = (100111)_2$$

that is, the decimal 39 and the binary 100111 are identical. The digits of base 2, namely 0, 1 are called binary digits or, in short, *bits*.

The basic arithmetic in bases other than base 10 is *identical* to that performed using base 10. It seems awkward simply because we are not used to it.

Example 2.1.1.

Consider $(35)_{10} + (15)_{10} = (50)_{10}$. In base 8, one should write

$$(43)_8 + (17)_8 = (62)_8$$ □

Example 2.1.2.

Consider the decimal numbers 13 and 5, whose product is the decimal number 65. Using the binary system, we obtain

$$
\begin{array}{rccccccc}
 & & & 1 & 1 & 0 & 1 & \\
 & & \times & & 1 & 0 & 1 & \\
\hline
 & & & 1 & 1 & 0 & 1 & \\
 & & 0 & 0 & 0 & 0 & & \\
 & 1 & 1 & 0 & 1 & & & \\
\hline
(1 & 0 & 0 & 0 & 0 & 0 & 1)_2 & = & (65)_{10}
\end{array}
$$ □

Example 2.1.3.

Divide $(52)_{10}$ by $(4)_{10}$ using *octal* representation (base 8). We should first replace $(52)_{10}$ and $(4)_{10}$ by $(64)_8$ and $(4)_8$, respectively. Then we perform long division and obtain

$$
\begin{array}{r}
15 \\
4\,\overline{)64} \\
4 \\
\hline
24 \\
24 \\
\hline
0
\end{array}
$$

The result is $(15)_8 = (13)_{10}$. □

Example 2.1.4.

Table 2.1.1 is a multiplication table representing, in base 8, all the integers between 1 and 7.

Table 2.1.1. A Multiplication Table in Base 8

	1	2	3	4	5	6	7
1	1	2	3	4	5	6	7
2	2	4	6	10	12	14	16
3	3	6	11	14	17	22	25
4	4	10	14	20	24	30	34
5	5	12	17	24	31	36	43
6	6	14	22	30	36	44	52
7	7	16	25	34	43	52	61

□

2.1.1   Conversion

Consider an integer x represented in bases N, M as

$$x = a_n \cdot N^n + \cdots + a_1 \cdot N + a_0, \quad 0 \le a_i \le N - 1 \text{ for all } a_i$$
$$x = b_m \cdot M^m + \cdots + b_1 \cdot M + b_0, \quad 0 \le b_j \le M - 1 \text{ for all } b_j$$

respectively. To convert x from base N to base M, simply use $\{a_0, a_1, \ldots, a_n\}$ to find $\{b_0, b_1, \ldots, b_m\}$.

Case 1. Let $N > M$ and denote $x_0 = x$. Then use the following scheme: divide $(x_0)_N$ by $(M)_N$ (using base N arithmetic). Let the quotient and the remainder be x_1 and r_1, respectively. Define $b_0 = r_1$. If $b_0, b_1, \ldots, b_k$ are already defined, we divide $(x_{k+1})_N$ by $(M)_N$, to obtain a quotient x_{k+2} and a remainder r_{k+2}, that is,

$$(x_{k+1})_N = (x_{k+2})_N (M)_N + (r_{k+2})_N \tag{2.1.3}$$

and define $b_{k+1} = r_{k+2}$. Clearly, this procedure is finite and uniquely defines the sequence $\{b_0, b_1, \ldots, b_m\}$. It ends when $x_{k+1} = 0$.

Example 2.1.5.
To convert $(125)_{10}$ to binary we create Table 2.1.2, from which we conclude $(125)_{10} = (1111101)_2$.

Table 2.1.2. Integer Conversion from Decimal to Binary

k	x_k	r_k	b_k
0	125		1
1	62	1	0
2	31	0	1
3	15	1	1
4	7	1	1
5	3	1	1
6	1	1	1
7	0	1	

□

Example 2.1.6.

To convert $(87)_{11}$ to octal, we perform (as in the previous example) a long division of $(87)_{11}$ by 8, using base 11 arithmetic, and determine that $(87)_{11} = (137)_8$ (Table 2.1.3).

Table 2.1.3. Integer Conversion from Base 11 to Octal

k	x_k	r_k	b_k
0	87		7
1	10	7	3
2	1	3	1
3	0	1	

□

Case 2. If $M > N$, we represent x by a sum of powers of N and use base M arithmetic to obtain the b_i values.

Example 2.1.7.

Let $(x)_3 = 21022$. Then

$$(x)_{10} = 2 \cdot 3^4 + 1 \cdot 3^3 + 0 \cdot 3^2 + 2 \cdot 3^1 + 2 \cdot 3^0$$
$$= 162 + 27 + 6 + 2 = 197$$

□

Suppose now that x is a fraction, represented by $0.a_1 a_2 \ldots$ in base N, and by $0.b_1 b_2 \ldots$ in base M, and that a_i, $1 \leq i$ are known. Both sequences may be infinite. The b_i value can be calculated as follows:

Case 1. Let $N > M$. Denote $x_0 = x$ and define

$$t_1 = [(M)_N \cdot (x_0)_N]_N, \quad b_1 = \text{int}(t_1), \quad x_1 = \text{frac}(t_1)$$

where frac and int are abbreviations for "fractional part" and "integer part" of the specific numbers. If $b_1, \ldots, b_k$ are already defined, we construct

$$t_{k+1} = [(M)_N \cdot (x_k)_N]_N, \quad b_{k+1} = \text{int}(t_{k+1}), \quad x_{k+1} = \text{frac}(t_{k+1}) \tag{2.1.4}$$

This scheme produces a sequence $\{b_1, b_2, \ldots\}$ (which may be finite or infinite) that satisfies

$$(0.b_1 b_2 \ldots)_M = (0.a_1 a_2 \ldots)_N$$

If x is irrational, $b_1, b_2, \ldots$ are unique. If x is rational there are always two representations. For example

$$(0.5)_{10} = (0.499 \ldots)_{10}$$

Case 2. If $M > N$ we express x as

$$x = \frac{a_1}{N} + \frac{a_2}{N^2} + \cdots$$

and use base M arithmetic to compute the right-hand side.

Example 2.1.8.

Let $(x)_{10} = 0.625$. The scheme for determining the binary equivalent is given in Table 2.1.4:

Table 2.1.4. Fraction Conversion from Decimal to Binary

k	x_k	$(2 \cdot x_k)_{10}$	b_{k+1}
0	0.625	1.250	1
1	0.250	0.500	0
2	0.500	1.000	1
3	0		

Thus $(0.625)_{10} = (0.101)_2$.

Example 2.1.9.

Consider the number $x = (0.25)_{10}$. Its base 3 equivalent is constructed as follows (Table 2.1.5).

Table 2.1.5. Fraction Conversion from Decimal to Base 3

k	x_k	$(3 \cdot x_k)_{10}$	b_{k+1}
0	0.25	0.75	0
1	0.75	2.25	2
2	0.25	0.75	0
3	0.75	2.25	2
$\vdots$	$\vdots$	$\vdots$	$\vdots$

Thus $(0.25)_{10} = (0.020202\ldots)_3$.

Example 2.1.10.

Let $x = (0.42)_5$ (i.e., $x = 4/5 + 2/5^2$). Its octal equivalent is obtained by first expressing x as an octal fraction:

$$x = \left(\frac{4}{5} + \frac{2}{5^2}\right)_8 = \left(\frac{4 \cdot 5 + 2}{5^2}\right)_8 = \frac{(26)_8}{(31)_8}$$

and then dividing the two numbers, using octal arithmetic, to obtain:

$$\frac{(26)_8}{(31)_8} = 0.7024\ldots$$

2.1.2 Special Bases

The most commonly used bases for number representation are decimal, binary, octal, and hexadecimal. The base for the hexadecimal system is 16. The letters A, B, C, D, E, F are chosen to represent 10 to 15, that is,

$$A \equiv 10, \quad B \equiv 11, \quad \ldots, \quad F \equiv 15$$

For example,

$$(A8.BF)_{16} = (10 \cdot 16^1 + 8 \cdot 16^0 + 11 \cdot 16^{-1} + 15 \cdot 16^{-2})_{10} = 168.74609375$$

Because $2^4 = 16$ the conversion between binary and hexadecimal bases is straightforward. To convert from base 16 to base 2 we replace each hexadecimal digit by a 4-bit binary number. The opposite is accomplished by dividing a binary number to 4-bit packets, starting from the right, then replacing each packet by its equivalent hexadecimal digit.

Example 2.1.11.
The binary equivalent of $(2C8)_{16}$ is

$$0010|1100|1000 \equiv 1011001000 \qquad \square$$

Example 2.1.12.
To determine the hexadecimal representation of the number $x = 101011010110101$, we divide the binary string to 4-bit packets

$$x = 101|0110|1011|0101 \qquad \square$$

and find that $x = (56B5)_{16}$.

PROBLEMS

1. Give the binary representation of the following decimal numbers:
   (a) 105
   (b) 32.53
   (c) 14.001
   (d) 99.999

2. Convert the following binary numbers to decimal:
   (a) 11111
   (b) 101.10101
   (c) 111.00001

3. Convert the following numbers to base 7:
   (a) $(65)_{10}$
   (b) $(373)_8$
   (c) $(AAA)_{16}$

4. Give the binary representation of π and π^2 up to the seventh bit.

5. Convert the following numbers to binary:

(a) $(35)_7$

(b) $(-2.2)_3$

(c) $(1.5)_9$

6. Show that the 4-bit packet method applies to hexadecimal fractions as well as to integers.

7. Convert the following hexadecimal numbers to binary:

(a) ABCDEF

(b) F8.F8

(c) –B78.384

8. Find a simple algorithm to convert from octal to binary and find the binary equivalents of:

(a) $(7.55)_8$

(b) $(-30.707)_8$

(c) $(1.2345)_8$

9. Convert $(0.333\ldots)_{10}$ to binary and to base 3.

10. Use the result of problem 9 to show

$$\frac{5}{16} + \frac{5}{16^2} + \frac{5}{16^3} + \cdots = \frac{1}{3}$$

without computing the left-hand side.

2.2 Floating Point Numbers

Consider the decimal number $x = 17.5$. It can also be written as $(0.175) \cdot 10^2$. Similarly, we can write 0.0035 as 0.35×10^{-2} and -74.3 as $-0.743 \cdot 10^2$. In general, we define a floating point number as follows.

Definition 2.2.1.   (Floating point numbers). *A number x represented as*

$$x = s \cdot m \cdot M^e \tag{2.2.1}$$

where s is +1 or –1, m is a fraction in base M, satisfying $(1/M) \le m \le 1$, and e is an integer, is called a floating point *number and Eq. (2.2.1) is its floating point representation in base M, with sign s, mantissa m, and* exponent e.

Example 2.2.1.
The decimal number $x = 8.75$, written as $0.875 \cdot 10^1$, has $s = 1$, $m = 0.875$, $e = 1$. Its binary equivalent is

$$x = (1000.11)_2 = +(0.100011)_2 \cdot (2)^{(100)_2}$$

with $s = 1$, $m = (0.100011)_2$, and $e = (100)_2$.   □

s	e	m
1	2–12	13–60

Figure 2.2.1. A 60-bit CDC floating point number.

s	k	$a_1 \ldots a_k$	$(n-k)$ zeros

Figure 2.2.2.   Exact representation of $I < 2^n$.

Example 2.2.2.

The floating point representation of $(72.55)_8$ is $(0.7255)_8 \cdot 8^2$. The mantissa is $(0.7255)_8$ and the exponent is 2. □

Example 2.2.3.

The Control Data Corporation (CDC) 7600 "ex-computer" used a 60-bit word to represent floating point numbers. The sign is given 1 bit, the exponent 11 bits, and the mantissa 48 bits. This is shown in Fig. 2.2.1. One bit of the e field must be assigned for the sign of the exponent. Thus only 10 bits are left to represent the absolute value of the exponent. The largest number possible is

$$(0.11 \ldots 1)_2 \cdot 2^{(1023)_{10}} = \left(1 - \frac{1}{2^{48}}\right)_{10} \cdot 2^{(1023)_{10}}$$

The largest negative exponent is $-(1024)_{10}$ (why?). Thus the smallest number in absolute value that has a floating point representation is

$$(0.1)_2 \cdot 2^{(-1024)_{10}} = 2^{(-1025)_{10}}$$

Let n be the number of digits assigned to the mantissa. Then the floating point representation of any integer $I < 2^n$ is exact. Indeed, let

$$I = a_1 \ldots a_k, \quad k \leq n \tag{2.2.2}$$

be the binary representation of I. Then we may write $I = 0.a_1 \ldots a_k \cdot 2^k$ and I has an exact representation, as shown in Fig. 2.2.2. □

Example 2.2.4.

For $n = 48$ we have $2^n = 2^{48} \approx 2.815 \cdot 10^{14}$. Thus all the integers up to 14 decimal digits have an *exact* floating point representation. □

2.2.1 Rounding

If the mantissa m of x is represented by more than n bits, it is shortened, and x has an *approximate* floating point representation in the computer. The computer either *truncates* or *chops* the mantissa to n bits by ignoring all the bits from $(n+1)$ on, or (considering the *size* of m beyond n bits) *rounds* it to n bits: if the $(n+1)$th bit is zero, m is chopped to n bits; if it is one, m is chopped and 1 is added to the nth bit.

Example 2.2.5.

Consider a 14-bit mantissa, 0.10110111010011, and a computer for which $n = 12$. The possible representations of m are

$$\text{Chopping: } m \approx 0.101101110100$$
$$\text{Rounding: } m \approx 0.101101110101$$

□

Denote the computer representation of x by float(x). We define the "error" ϵ of this representation by the relation

$$\text{float}(x) = x(1 + \epsilon) \tag{2.2.3}$$

provided that $x \neq 0$.

Let $x = 0.a_1 \ldots a_n a_{n+1} \ldots a_N \cdot 2^e$. Then the chopping method implies float$(x)=0.a_1 \ldots a_n \cdot 2^e$. Thus

$$\epsilon = -\frac{0.0 \ldots 0 a_{n+1} \ldots a_N}{0.a_1 \ldots a_N}$$

leading to

$$-\frac{1}{2^{n-1}} < \epsilon \leq 0 \text{ (why?)} \tag{2.2.4}$$

Although ϵ is always ≤ 0, the "absolute" error defined as x-float(x) changes signs with x. In rounding we have two cases:

1. $x = 0.a_1 \ldots a_n 0 a_{n+2} \ldots a_N \cdot 2^e$. The rounding procedure yields
   float$(x) = 0.a_1 \ldots a_n \cdot 2^e$ and, by using the fact that the $(n+1)$th bit is zero, we find $-1/2^n < \epsilon \leq 0$.

2. $x = 0.a_1 \ldots a_n 1 a_{n+2} \ldots a_N \cdot 2^e$. Here we have

$$\text{float}(x) = \left(0.a_1 \ldots a_n + \frac{1}{2^n}\right) \cdot 2^e$$

and therefore

$$\left(0.a_1 \ldots a_n + \frac{1}{2^n}\right) \cdot 2^e = 0.a_1 \ldots a_n 1 a_{n+2} \ldots a_n \cdot 2^e (1 + \epsilon)$$

leading to

$$0 < \epsilon = \frac{1/2^n - 0.0 \ldots 01 a_{n+2} \ldots a_N}{0.a_1 \ldots a_n 1 a_{n+2} \ldots a_N} < \frac{1/2^n - 1/2^{n+1}}{1/2} = \frac{1}{2^n}$$

Adding this inequality to the previous one, we see that

$$-\frac{1}{2^n} < \epsilon < \frac{1}{2^n} \tag{2.2.5}$$

The absolute error x-float(x) is positive or negative, depending on sign(x) and a_{n+1}. Equations (2.2.4) and (2.2.5) indicate that, on average, the chopping error is twice as much as the rounding error. This, together with the even distribution of sign$[x-\text{float}(x)]$, are important features of the rounding scheme, in particular with respect to *error propagation*.

2.2.2   Floating Point Arithmetic

Suppose that $x = 0.a_1 \ldots a_n \cdot 2^{e_1}$, $y = 0.b_1 \ldots b_n \cdot 2^{e_2}$ are two floating point numbers that are processed by an n-bit mantissa binary computer. The arithmetic scheme between x and y is called floating point arithmetic and can be described as follows:

Case 1. Addition and subtraction: Let $e_1 = e_2 + k$ for some integer k. Then the mantissa of y is shifted k bits rightward and the exponent is adjusted accordingly. We replace y by

$$y_{\text{new}} = 0.0 \overbrace{\ldots}^{k \ times} 0b_1 \ldots b_{n-k} \cdot 2^{e_1} \tag{2.2.6}$$

and define

$$x \pm y = (0.a_1 \ldots a_n \pm 0.0 \ldots 0b_1 \ldots b_{n-k}) \cdot 2^{e_1} \tag{2.2.7}$$

The exponent of the sum is either e_1 or $(e_1 + 1)$ depending on whether $(0.a_1 \ldots a_n \pm 0.0 \ldots 0b_1 \ldots b_{n-k})$ exceeds one or not. Clearly, this scheme is not restricted to binary representation and is valid for any given base.

Example 2.2.6.
Consider a computer with binary floating point representation, a 6-bit mantissa, and a rounding mode. Let

$$x = 0.110110 \cdot 2^3, \quad y = 0.110111 \cdot 2^1$$

Then

$$x + y = (0.110110 + 0.001101) \cdot 2^3 = 1.000011 \cdot 2^3 = 0.100010 \cdot 2^4 \text{(after rounding)}$$

The result for a chopping mode is $0.100001 \cdot 2^4$.   □

Example 2.2.7.
Let us assume a decimal floating point arithmetic, a four-digit mantissa, and a chopping mode. If $x = 0.9907 \cdot 10^4$, $y = 0.4937 \cdot 10^3$ then

$$x + y = (0.9907 + 0.0493) \cdot 10^4 = 1.0400 \cdot 10^4 = 0.1040 \cdot 10^5$$
$$x - y = (0.9907 - 0.0493) \cdot 10^4 = 0.9414 \cdot 10^4$$ □

If the difference $e_1 - e_2$ is larger than the length of the mantissa, we obtain

$$x + y = x, \quad x - y = x \tag{2.2.8}$$

Case 2. Multiplication and division: Let x, y be represented as

$$x = 0.a_1 \ldots a_n \cdot 2^{e_1}, \quad y = 0.b_1 \ldots b_n \cdot 2^{e_2}$$

and let $(0.a_1 \ldots a_n) \cdot (0.b_1 \ldots b_1 \ldots b_n) = 0.c_1 \ldots c_{2n}$. Let c_k be the first left-most nonzero bit. We define

$$xy = 0.c_k \ldots c_{n+k-1} \cdot 2^{e_1+e_2+1-k} \qquad (2.2.9)$$

with a proper adjustment, which depends on the mode of the computer (chopping or rounding). Let

$$\frac{0.a_1 \ldots a_n}{0.b_1 \ldots b_n} = d_1.d_2 d_3 \ldots d_{n-1} d_n \ldots$$

and let d_k be the first left-most nonzero bit. Then

$$\frac{x}{y} = 0.d_k \ldots d_{n+k-1} \cdot 2^{e_1-e_2+2-k} \qquad (2.2.10)$$

with a possible minor adjustment.

Example 2.2.8.
Consider a decimal four-digit mantissa with rounding mode and let $x = 0.4052 \cdot 10^3$, $y = 0.1069 \cdot 10^{-2}$. The exact product $xy = 0.04331588 \cdot 10^1$ is represented as $0.4332 \cdot 10^0$. ☐

Example 2.2.9.
Consider the computer in the previous example and $x = 0.9543 \cdot 10^3$, $y = 0.1756 \cdot 10^6$. Then x/y is $5.43451025 \ldots \cdot 10^{-3}$ but is represented as

$$\frac{x}{y} = 0.5435 \cdot 10^{-2} \qquad \qquad ☐$$

In many applications, the mantissa is not long enough to guarantee sufficient accuracy. In that case, most computers provide another representation for floating point numbers with a longer mantissa (generally $2n$ bits instead of n) and allow *double-precision* arithmetic.

Example 2.2.10.
A double-precision number within a CDC computer using a 60-bit word is represented by 120 bits. The mantissa consists of 108 bits, equivalent to 32 decimal digits, because

$$2^{108} \approx 3.25 \cdot 10^{32} \qquad \qquad ☐$$

Suppose that the computer arithmetic is in base M, rather than binary, and that $M = 2^k$ for some integer k. Let length n of the mantissa be an integer multiple of k, that is, $n = lk$. The mantissa thus consists of exactly l digits of base M. To complete the scheme, assign 1 bit to the sign of the mantissa and p bits to the exponent. Then a floating point representation is given by Fig. 2.2.3.

A general *single-precision* floating point number is represented as

$$x = s \cdot m \cdot M^e \qquad (2.2.11)$$

where $1/M \leq m < 1$, $-2^{p-1} \leq e \leq 2^{p-1} - 1$. All the integers $< M^l$ have exact internal representation. Let ϵ denote the error of the representation, defined by Eq. (2.2.3). Then

s	e	m
1	p bits	$n = l \cdot k$ bits

Figure 2.2.3. General floating-point representation.

0	10 ... 01	10011001 ... 1001

Figure 2.2.4. Floating point representation of $(0.3)_{10}$.

$$-\frac{1}{M^{\ell-1}} < \epsilon \leq 0 \qquad\qquad (2.2.12)$$

for a chopping mode, whereas in the case of rounding we have

$$-\frac{1}{2} \cdot \frac{1}{M^{\ell-1}} < \epsilon < \frac{1}{2} \cdot \frac{1}{M^{\ell-1}} \qquad\qquad (2.2.13)$$

Example 2.2.11.

Let $M = 16$, $n = 24$, $p = 7$, $l = 6$. All numbers less than 16^6 have exact representation. The error of the representation satisfies $-16^{-5} < \epsilon \leq 0$ in the case of chopping, and $-0.5 \cdot 16^{-5} < \epsilon < 0.5 \cdot 16^{-5}$ for rounding. □

Because of chopping or rounding, the representation of a number may not be exact. For example, consider the decimal number 0.3. It is of finite length in base 10, but its binary representation is infinite, given by 0.01001100110011 .... Let this number be processed by a CDC computer with 60 bits and subjected to chopping. Its internal representation is shown in Fig. 2.2.4, and satisfies $x\text{-float}(x) = 0.3 \cdot 2^{-48}$.

Consider now the following Fortran code:

```
        A = 0.0
10 A = A + 0.3
        IF (A.EQ. 3.0) STOP
        GO TO 10
```

The number $3.0 = (0.11)_2 \cdot 2^2$ has an exact internal representation. But 0.3 is chopped and after executing statement 10 ten times, we must have $A < 3.0$. However, after the next execution, we obtain $A > 3.0$ and the loop never ends. To stop it, when $A \approx 3.0$, we must replace the IF statement. For example, we could write

```
        A = 0.0
10 A = A + 0.3
        IF (A.GT.2.9) STOP
        GO TO 10
```

PROBLEMS

1. Write the following decimal numbers in floating point representation:
   (a) 354.3
   (b) $\sqrt{2}$
   (c) $1/103$
   (d) –0.00078

2. Give the floating point representation of the following decimal numbers by a CDC computer using a 60-bit word, assuming a chopping mode:
   (a) 0.75
   (b) 0.3
   (c) 0.7

3. Repeat and solve problem 2, assuming a rounding mode representation.

4. Repeat and solve problem 2 for M, n, p, l of Example 2.2.11, using a rounding mode representation.

5. Consider a computer with a rounding mode and a decimal, four-digit mantissa. Compute $x+y$, xy for the cases:
   (a) $x = 0.1002 \cdot 10^3$, $y = 0.2109 \cdot 10^2$
   (b) $x = 3.4735$, $y = -14.353$
   (c) $x = \pi$, $y = 10^4\pi$

6. Consider a binary computer having a rounding floating point representation with a 12-bit mantissa, a 1-bit sign, and a 3-bit exponent. Give the single- and double-precision representations of the following numbers:
   (a) $1/3$
   (b) 0.3
   (c) 10^6
   (d) 0.1666

7. Find the smallest positive number x for which $1 + x > 1$, assuming:
   (a) CDC computer with a 60-bit word, single-precision mode
   (b) CDC computer with a 60-bit word, double-precision mode
   (c) Hexadecimal 32-bit computer of problem 4, single-precision mode

8. Consider the Fortran codes executed by a binary computer with a chopping mode, and predict the outcomes:
   (a) $X = 0.$
            $H = 1.E - 15 + .75$
        10 $X = X + H$
            IF $(X.LT.6)$ GO TO 10
   (b) $X = 0.$
            $H = 1./3.$
        10 $X = X + H$
            IF $(X.GT.1.)$ GO TO 10

2.3 Definitions and Sources

Let x_t be the true value of a given quantity and x_c a computed approximation of x_t. We define the *absolute error* or simply the *error* of x_c as

$$\text{er}(x_c) = x_t - x_c \tag{2.3.1}$$

and the *relative* error as

$$\text{rel}(x_c) = \frac{x_t - x_c}{x_t} \tag{2.3.2}$$

provided that $x_t \neq 0$.

Example 2.3.1.
Let $x_t = 5$, $x_c = 5.0013$. The absolute and relative errors of x_c are $er(x_c) = -0.0013$, $\text{rel}(x_c) = -0.00026$. $\quad\square$

Example 2.3.2.
A well-known approximation to π is the fraction $22/7$. The two errors are

$$\text{er}\left(\frac{22}{7}\right) = \pi - \frac{22}{7} \approx -0.00126$$

$$\text{rel}\left(\frac{22}{7}\right) = \frac{\pi - 22/7}{\pi} \approx -0.000402$$

A better approximation is the fraction $355/113$. Here we have

$$\text{er}\left(\frac{355}{113}\right) \approx -0.000000267$$

$$\text{rel}\left(\frac{355}{113}\right) = \frac{\pi - 355/113}{\pi} \approx -0.0000000849 \qquad\square$$

We shall now discuss the concept of "significant digits."

Definition 2.3.1. *Let $|x_t - x_c|$ be less than or equal to five units in the nth digit of x_t, counting rightward and starting from the first nonzero digit. Then x_c is said to have at least n significant digits of accuracy. If n_0 is the largest n for which this holds, x_c has n_0 significant digits.*

Example 2.3.3.
Let $x_t = 28.536, x_c = 28.124$. Then $|x_t - x_c| = 0.412$ and $n_0 = 3$. If $x_c = 28.515$, then $|x_t - x_c| = 0.021$ and $n_0 = 4$. $\quad\square$

Definition 2.3.1 is intuitively acceptable. However, because generally the relative error is more meaningful than the absolute error, we shall adopt the following definition for significant digits.

Definition 2.3.2. (Significant digits). *Let x_t and x_c, the true and computed values of a given quantity, satisfy*

$$\left| \frac{x_t - x_c}{x_t} \right| \leq 5 \cdot 10^{-n} \tag{2.3.3}$$

for some $n \geq 0$. Then x_c has at least n significant digits of accuracy. If n_0 is the largest n for which Eq. (2.3.3) holds, x_c has exactly n_0 significant digits.

This definition given for decimals can be extended for any given base.

Example 2.3.4.
Let $x_t = e$, $x_c = 2.71835$. The absolute and relative errors are

$$\mathrm{er}(x_c) = e - 2.71835 \approx -0.000068$$

$$\mathrm{rel}(x_c) = \frac{e - 2.71835}{e} \approx -0.000025$$

and x_c has five digits of accuracy. □

Example 2.3.5.
Let $x_t = 1353.05$, $x_c = 1257$. The errors are

$$\mathrm{er}(x_c) = 1353.05 - 1257 = 96.05$$

$$\mathrm{rel}(x_c) = \frac{1353.05 - 1257}{1353.05} \approx 0.071$$

and x_c has only one accurate digit. □

Example 2.3.6.
Let $x_t = 1000$. To find a maximum range for x_c, such that x_c has three digits of accuracy, we solve the inequality

$$\left| \frac{1000 - x_c}{1000} \right| \leq 5 \cdot 10^{-3}$$

or

$$|1000 - x_c| \leq 5$$

Hence,

$$995 \leq x_c \leq 1005$$ □

Example 2.3.7.

For $x_t = 0.0045325$, the maximum interval within which x_c has four accurate digits is determined by

$$\left| \frac{0.0045325 - x_c}{0.0045325} \right| \leq 5 \cdot 10^{-4}$$

which leads to $0.00453023 \leq x_c \leq 0.00453477$.   □

2.3.1 Sources of Error

A numerical procedure for solving a given problem will, in general, involve an error of one or several types. Although different sources initiate the error, they all cause the same effect: diversion from the exact answer. Some errors are small and may be neglected. Others may be devastating if overlooked. In all cases, an *error analysis* must accompany the computational scheme, whenever possible. The major sources of error are as follow.

1. **Human error**: This category refers to any error introduced during the process of solving the problem by the human(s) in charge. It could be, for example, changing signs in a formula or a simple programming error. We may detect such errors by running the program on test cases for which the exact solutions are *a priori* known.

2. **Truncation error**: This type of error occurs when we are unable to evaluate explicitly a given quantity, and replace it by an *approximation* that can be computed. Thus, we calculate B instead of A, but have *control* over the truncated term

$$C = A - B \tag{2.3.4}$$

For example, the function $\sin x$ can be replaced by

$$\sin x = x - \frac{x^3}{3!} + \frac{x^5}{5!} + E(x) \tag{2.3.5}$$

where

$$|E(x)| = \left| \frac{x^7 \cos(c_x)}{7!} \right| \leq \frac{x^7}{7!} \tag{2.3.6}$$

and c_x is between x and 0.

We may calculate $x - x^3/3! + x^5/5!$ as an approximation to $\sin x$, provided that the error term $E(x)$ is *sufficiently small*. If we consider the interval $-0.2 \leq x \leq 0.2$, the absolute error of the above approximation cannot exceed (in absolute value)

$$\frac{0.2^7}{7!} \approx 2 \cdot 10^{-9}$$

The truncation error in this case is $x^7 \cos(c_x)/7!$.

Another example involves evaluating the area S of a circle with a radius $R = \sqrt{3}$. Because the exact values of π and $\sqrt{3}$ cannot be calculated, we may truncate and replace them, for example, by 3.141 and 1.732, respectively. Because

$$|\pi - 3.141| \le 10^{-3}, \quad |\sqrt{3} - 1.732| \le 10^{-3}$$

we obtain $|\pi R^2 - 3.141 \cdot 1.732^2| \le 14 \cdot 10^{-3}$. The truncation error is thus bounded by $14 \cdot 10^{-3}$.

3. **Machine error**: A floating point number processed by a computer may not have an exact representation. Also, floating point arithmetic in general is not exact. For example, if the length of a mantissa is 4 bits, then

$$[(0.1101)_2 \cdot 2^2] (0.1011 \cdot 2) = \begin{cases} (0.1000)_2 \cdot 2^3, & \text{chopping mode} \\ (0.1001)_2 \cdot 2^3, & \text{rounding mode} \end{cases}$$

whereas the exact value of the left-hand side is $(0.10001111)_2 \cdot 2^3$. This is a *machine error*. Machine errors are inevitable because a mantissa has a *finite* length, and the set of numbers with exact representation is obviously finite.

4. **Inaccurate observations**: Many numerical processes involve physical quantities such as the speed of light, density of iron, or the constant of gravity. These quantities are provided by experiments and naturally introduce some experimental error. For example, the speed of light in vacuum is

$$c = (2.997925 + \epsilon) \cdot 10^{10} \text{ cm/sec}, \quad |\epsilon| \le 10^{-6}$$

Experimental or observational errors cannot be removed or even reduced, without improving the observational technique. These errors carry some similarity to those discussed in class 2.

5. **Modeling errors**: Constructing a *mathematical model* is the first step in the process of solving a problem. However, an equation or a system of equations that is expected to describe some phenomena will generally only approximate the physical reality. Occasionally a mathematical model may be solved successfully, and yet the computed and the experimental results are far apart. This may occur if an important physical aspect is overlooked and the mathematical model is unjustly simplified. For example, consider the problem of population growth. An oversimplified model is given by

$$N(t) = N_0 e^{kt} \tag{2.3.7}$$

where N_0 is the population at time $t = 0, k$ is some positive constant, and $N(t)$ is the population at time t. This model does not consider important factors such as the standard of living and birth control education. It is obviously limited and generally overestimates the population for large values of t.

We will now discuss in detail several types of errors that are clearly machine errors and thus belong to class 3.

2.3.2 Losing Significant Digits

Consider evaluating the function

$$f(n) = \sqrt{n^2 + n} - n, \quad n = 1, 2, \ldots \tag{2.3.8}$$

using a decimal six-digit calculator with a rounding mode. We first calculate $n^2 + n$, then $\sqrt{n^2 + n}$, and finally subtract n from the result. The approximate and exact values of $f(n)$ are given in Table 2.3.1. The last column is S.D., the number of accurate significant digits. As n increases, the number of

Table 2.3.1. Evaluating $f(n) = \sqrt{n^2 + n} - n$ on a Six-Digit Calculator

n	Computed $f(n)$	Exact $f(n)$	S.D.
1	0.414210	0.414214	5
10	0.488100	0.488088	5
100	0.499000	0.498756	4
1,000	0.500000	0.499875	4
10,000	0.500000	0.499988	5
100,000	0.000000	0.499999	0

accurate significant digits decreases. Let us, for example, follow step by step the evaluation of $f(100)$. While computing $\sqrt{100^2 + 100}$ one does not lose any accuracy and the result is 100.499. Next we calculate 100.499–100 = 0.499000. The exact value, rounded to six-digits, is 0.498756. The computed value 0.499 has four rather than six significant digits of accuracy, due to (1) the subtraction of nearly equal numbers, namely 100.499 and 100, and (2) the restriction of six decimal digits imposed by the calculator. If we increase n above 10^5, the calculated difference is

$$\sqrt{n^2 + n} - n = 0.000000$$

and yields no digits of accuracy. To avoid the loss of accuracy, one could rewrite $f(n)$ as

$$f(n) = \frac{(\sqrt{n^2 + n} - n)(\sqrt{n^2 + n} + n)}{\sqrt{n^2 + n} + n} = \frac{n}{\sqrt{n^2 + n} + n} = \frac{1}{1 + \sqrt{1 + 1/n}} \tag{2.3.9}$$

Using the same calculator we obtain the results presented in Table 2.3.2.

Table 2.3.2. Computing $f(n) = n / \left(\sqrt{n^2 + n} + n \right)$ on a Decimal Six-Digit Calculator

x	Computed $f(x)$	Exact $f(x)$	S.D.
1	0.414214	0.414214	6
10	0.488088	0.488088	6
100	0.498756	0.498756	6
1,000	0.499875	0.499875	6
10,000	0.499988	0.499988	6
100,000	0.499998	0.499999	6

Because a computer or a calculator is restricted to a finite number of decimal digits we should in general avoid subtracting two almost identical numbers.

Let us now calculate the function

$$f(x) = \frac{e^{x^2} - 1}{x^2}, \quad x \to 0 \tag{2.3.10}$$

assuming a decimal eight-digit calculator with rounding mode. The results for $x = 10^{-n}$, $0 \leq n \leq 5$ are summarized in Table 2.3.3.

Table 2.3.3. Computing $\left(e^{x^2} - 1\right)/x^2$ on an Eight-Digit Calculator

x	Computed $f(x)$	Exact $f(x)$	S.D.
1.0	1.7182818	1.7182818	8
0.1	1.0050200	1.0050167	6
0.01	1.0000000	1.0000500	5
0.001	1.0000000	1.0000005	7
0.0001	0	1.0000000	0

We notice that the loss of significant digits occurs for $x \leq 0.1$. There is an improvement near $x = 0.001$, due to the particular choice of $f(x)$, but then, when $x \leq 0.0001$, there are no accurate digits and the computed value of $f(x)$ is zero. It happens because once again we perform a subtraction between two almost equal numbers, namely e^{x^2} $(0 \leq x << 1)$ and 1. Instead, let us write

$$e^{x^2} = 1 + x^2 + \frac{x^4}{2!} + \frac{x^6}{3!} + \frac{x^8}{4!} + \frac{x^{10}}{5!}e^c$$

where c is between 0 and x^2. Thus,

$$\frac{e^{x^2} - 1}{x^2} = 1 + \frac{x^2}{2!} + \frac{x^4}{3!} + \frac{x^6}{4!} + \frac{x^8}{5!}e^c$$

and we can replace $(e^{x^2} - 1)/x^2$ by $1 + x^2/2! + x^4/3! + x^6/4!$. The truncation error is

$$\left|\frac{x^8}{5!}e^c\right| \leq \frac{0.1^8}{120}e^{0.01} < 10^{-10}, \quad 0 \leq x \leq 0.1$$

Table 2.3.4 consists of the new computed values of $f(x)$ and demonstrates the improvement with respect to Table 2.3.3.

Table 2.3.4. Replacing $\dfrac{e^{x^2} - 1}{x^2}$ by $1 + \dfrac{x^2}{2!} + \dfrac{x^4}{3!} + \dfrac{x^6}{4!}$.

x	Computed $f(x)$	Exact $f(x)$	S.D.
0.1	1.0050167	1.0050167	8
0.01	1.0000500	1.0000500	8
0.001	1.0000005	1.0000005	8
0.0001	1.0000000	1.0000000	8

Addition, too, may cause difficulties. Let us consider a computer with floating point arithmetic, six-digit decimal mantissa, one-digit decimal exponent, and rounding mode. Consider the sum

$$S = 0.500000 + 0.1 \cdot 10^{-6} + 0.1 \cdot 10^{-6} + \cdots + 0.1 \cdot 10^{-6}$$

where the term $0.1 \cdot 10^{-6}$ occurs 10^7 times. If we add from left to right, we first find (due to the floating point arithmetic) $0.500000 + 0.1 \cdot 10^{-6} = 0.500000$ and finally, by induction, $S = 0.500000$, when the actual sum is 1.5. This can be corrected if we add the numbers from right to left in groups of 10^6 numbers each. There are 10 groups and we now obtain the correct result:

$$0.5 + 0.1 + \cdots (10 \text{ times}) + 0.1 = 1.5$$

If we simply add from right to left, the result is 0.6.

The loss of significant digits may be expected if we subtract two nearly equal numbers. However, detecting such a subtraction may be difficult. Problems for which we can *anticipate* the loss of significant digits are, for example, those problems in which we compute a sum known *a priori* to be "small." For example, consider the following Taylor's expansion for $\sin x$:

$$\sin x = x - \frac{x^3}{3!} + \frac{x^5}{5!} - \cdots - \frac{x^{15}}{15!} + \frac{x^{17}}{17!} \cos(c_x) \tag{2.3.11}$$

where c_x is between 0 and x. Because Eq. (2.3.11) is valid for all x, let us substitute $x = p = 3.14$:

$$\sin p = p - \frac{p^3}{3!} + \frac{p^5}{5!} - \cdots - \frac{p^{15}}{15!} + \frac{p^{17}}{17!} \cos(c_p) = 0.00159265\ldots$$

The error term satisfies

$$\left| \frac{p^{17}}{17!} \cos(c_p) \right| \le \frac{p^{17}}{17!} < 10^{-6} \tag{2.3.12}$$

Suppose that we want to compute $\sin p$, using this particular 15th order Taylor's polynomial. Let us consider the computational procedure on a computer with floating point arithmetic, and four-digit decimal mantissa. We use Eq. (2.3.11) and add from the leftmost term. Rounding is performed after computing each whole term. The results are summarized in Table 2.3.5. The final sum is 0.001783 and yields only one accurate digit, whereas the exact value of the polynomial, due to Eq. (2.3.12), has at least three (almost four) significant digits of accuracy.

Table 2.3.5. Computing a Sum $\ll 1$

No.	Term	Sum
1	$0.3140 \cdot 10^1$	$0.3140 \cdot 10^1$
2	$-0.5160 \cdot 10^1$	$-0.2020 \cdot 10^1$
3	$0.2544 \cdot 10^1$	$0.5240 \cdot 10^0$
4	$-0.5971 \cdot 10^0$	$-0.7310 \cdot 10^{-1}$
5	$0.8177 \cdot 10^{-1}$	$0.8670 \cdot 10^{-2}$
6	$-0.7329 \cdot 10^{-2}$	$0.1341 \cdot 10^{-2}$
7	$0.4632 \cdot 10^{-3}$	$0.1804 \cdot 10^{-2}$
8	$-0.2175 \cdot 10^{-4}$	$0.1783 \cdot 10^{-2}$

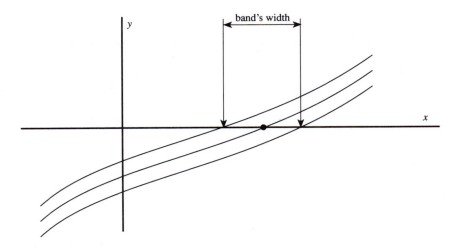

Figure 2.3.1. Noise band and false roots.

This means a net loss of two significant digits. A similar effect can be found in calculating e^x for large negative x, using Taylor's approximation. In this particular case one should calculate e^{-x} instead, using the polynomial approximation $(x < 0)$, and then use the relation

$$e^x = \frac{1}{e^{-x}}$$

2.3.3 Function Evaluation

Consider the problem of evaluating a continuous function. Using a computational device that possesses an infinite number of digits, we should naturally expect a continuous curve. Because this is not the case, floating point arithmetic and rounding (or chopping) cause inaccuracies, known as "noise." Rather than obtaining one continuous curve we have a "fuzzy" band of random values. In most cases the thickness of the band is quite small. But even then it may cause difficulties. Suppose, for example, that we are interested in finding a root of a given polynomial. The polynomial function may be quite flat within the neighborhood of the root, and due to the noise in the evaluation of the polynomial, we could detect false changes of sign and consequently false roots (Fig. 2.3.1).

2.3.4 Overflow and Underflow

The absolute value of any nonzero floating point number has upper and lower positive bounds. These depend on the field length assigned to the exponent. For example, a CDC computer with a 60-bit word and an 11-bit exponent yields an upper bound

$$M = 2^{1023} \approx 10^{308}$$

and a lower bound

$$N = 2^{-1024} = \frac{2}{M}$$

An attempt to create a number x smaller than N leads to *underflow* error. Naturally, most computers would set zero instead, and the calculation would not stop. The second case, when an *overflow* error occurs $(x > M)$, is clearly more serious, and the computation generally stops. If overflow happens, it could indicate division by zero somewhere along the computation. But this may not be the case. Suppose, for example, that we calculate

$$z = \sqrt{x^2 + y^2}$$

where x and y are known to satisfy

$$|x| \le \frac{M}{\sqrt{2}} , \quad |y| \le \frac{M}{\sqrt{2}}$$

for the previously defined upper limit M. Then $(x^2 + y^2)$ may exceed M and hence cause an overflow, although clearly $|z| \le M$. In this particular case, we can prevent the overflow by defining

$$z = \begin{cases} |x|\sqrt{1 + y^2/x^2}, & |y| < |x| \\ 0 & , \quad x = y = 0 \\ |y|\sqrt{1 + x^2/y^2}, & |x| \le |y| \end{cases}$$

and programming *accordingly*.

PROBLEMS

1. Let x_t and x_c be the true and the computed values. Find the absolute and the relative errors for the following cases:
   (a) $x_t = 1.5, \ x_c = 1.501$
   (b) $x_t = \pi^2, \ x_c = 10$
   (c) $x_t = 0, \ x_c = 0.01$
   (d) $x_t = 0.01, \ x_c = 0$

2. Find the number of significant digits of accuracy for parts (a) through (d) of problem 1.

3. If x_c has n digits of accuracy, it must satisfy $x_t - \delta \le x_c \le x_t + \delta$ for some $\delta > 0$. Find the maximal δ for the following cases:
   (a) $x_t = 2.05, \ n = 3$
   (b) $x_t = 1.2345, \ n = 5$
   (c) $x_t = \pi, \ n = 6$
   (d) $x_t = -0.000015, \ n = 3$

4. Solve $ax^2 + bx + c = 0$ and assume that $|ac| << b^2$. How would you compute the solutions x_1, x_2?

5. (a) Implement the conclusion of problem 4 and solve

$$x^2 - 50x + 1 = 0$$

   using a five-digit decimal computer with rounding.
   (b) Use the "classic" approach as well, and compare the results.

6. Rewrite the following expressions to avoid the loss of significant digits.

   (a) $\ln(x + a) - \ln x$, large x.

   (b) $\sin(x + a) - \sin a$, small x.

   (c) $1 + x + \cdots + x^n$, x near -1 and odd n.

7. Use Taylor polynomial approximations to calculate the following expressions for small x, without a loss of significant digits:

   (a) $\dfrac{e^x - 1}{x}$

   (b) $\dfrac{x - \tan x}{x^2}$

   (c) $\dfrac{\ln(1 + x) - x}{x^2}$

8. Use a Taylor approximation to calculate $\sin x$ for $0 \le x \le \pi$.

   (a) How many terms do you need to approximate $\sin x$ up to six significant digits of accuracy?

   (b) Calculate $\sin \pi$ using a five-digit rounding decimal calculator.

   (c) Explain the results.

9. Develop an algorithm to calculate $\sin x$, for all x, on a decimal six-digit rounding calculator with four accurate digits guaranteed. (*Hint:* Use the periodicity of $\sin x$.)

10. Write a program that causes both underflow and overflow errors.

11. Rearrange $(x - 1)^3 = x^3 - 3x^2 + 3x - 1$ and compute the polynomial by using Horner's algorithm. Assume a decimal six-digit rounding calculator and consider 20 evenly spaced points in the interval $[0.998, 1.002]$. Explain the results.

12. Use a decimal 12-digit rounding calculator and solve problem 11 for the polynomial

$$p(x) = (x - 9)(x - 10)(x - 11)(x - 12)$$
$$= x^4 - 42x^3 + 659x^2 - 4578x + 11880$$

at the interval $[10.9, 11.1]$.

2.4 Error Propagation

Let x_c, y_c be computed approximations to x_t, y_t, respectively. Denote by $*$ any of the four basic operations $+, -, \cdot, /$. We define the *propagated error* as

$$E = x_t * y_t - x_c * y_c \tag{2.4.1}$$

that is, the absolute error of $x_t * y_t$ approximated by $x_c * y_c$.

Example 2.4.1.

Let $x_t = 3.5$, $x_c = 3.41$, $y_t = 2.5$, $y_c = 2.2$. Then

$$E = 3.5 \cdot 2.5 - 3.41 \cdot 2.2 = 8.75 - 7.502 = 1.248 \qquad \square$$

Example 2.4.2.

Let $x_t = 3.5$, $x_c = 3$, $y_t = 1.5$, $y_c = 1.75$. The propagated error is

$$E = 3.5 \cdot 1.5 - 3 \cdot 1.75 = 5.25 - 5.25 = 0 \qquad \square$$

The second example demonstrates that "accidents" happen: both x_c, y_c are relatively poor approximations, yet the propagated error is by definition zero!

To obtain a bound for the propagated error, we can use the *interval arithmetic* technique: The bounds on $x_t - x_c$ and $y_t - y_c$ are used to determine an interval generally centered at $x_c * y_c$, which must contain $x_t * y_t$.

Example 2.4.3.

Let $|x_t - x_c| \leq \epsilon_1$, $|y_t - y_c| \leq \epsilon_2$. Then

$$|(x_t + y_t) - (x_c + y_c)| \leq |x_t - x_c| + |y_t - y_c| \leq \epsilon_1 + \epsilon_2$$

Hence, the propagated error of $x_c + y_c$ is bounded by $\epsilon_1 + \epsilon_2$. $\qquad \square$

Example 2.4.4.

For the same (x_t, x_c, ϵ_1) , (y_t, y_c, ϵ_2) we have

$$
\begin{aligned}
|x_t y_t - x_c y_c| &= |x_t y_t - x_t y_c + x_t y_c - x_c y_c| \\
&= |x_t(y_t - y_c) + y_c(x_t - x_c)| \leq |x_t|\epsilon_2 + |y_c|\epsilon_1 \\
&\leq (|x_c| + \epsilon_1)\epsilon_2 + |y_c|\epsilon_1 \approx |x_c|\epsilon_2 + |y_c|\epsilon_1
\end{aligned}
$$

provided that $\epsilon_1 \epsilon_2$ is indeed negligible. $\qquad \square$

Example 2.4.5.

Consider the previous example with

$$x_c = 1.538 , \; \epsilon_1 = 0.001 , \; y_c = 3.175 , \; \epsilon_2 = 0.005$$

Then an almost certain bound for the propagated error is $1.538 \cdot 0.005 + 3.175 \cdot 0.001 = 0.010865$, that is,

$$|x_t y_t - x_c y_c| \leq 0.010865$$

A *certain* bound is $0.010856 + \epsilon_1 \epsilon_2 = 0.010861$. $\qquad \square$

Example 2.4.6.

Let $x_t = e$, $x_c = 2.718$, $\epsilon_1 = 0.0003$; $y_t = \sqrt{2}$, $y_c = 1.414$, $\epsilon_2 = 0.0003$. The fraction x_t/y_t must satisfy

$$\frac{x_c - 0.0003}{y_c + 0.0003} \leq \frac{x_t}{y_t} \leq \frac{x_c + 0.0003}{y_c - 0.0003}$$

that is,

$$1.92158 \leq \frac{x_t}{y_t} \leq 1.92283$$

when actually $x_t/y_t = 1.9221\ldots$. ◻

Because any computation is bound to introduce rounding or chopping, the total error in approximating $x_t * y_t$, is

$$E' = x_t * y_t - float(x_c * y_c) \qquad (2.4.2)$$

Clearly

$$\begin{aligned} E' &= x_t * y_t - x_c * y_c + x_c * y_c - float(x_c * y_c) \\ &= E + x_c * y_c - float(x_c * y_c) \end{aligned}$$

Using Eq. (2.2.3) we have

$$float(x_c * y_c) = (x_c * y_c)(1 + \epsilon) \qquad (2.4.3)$$

where ϵ is machine dependent. Therefore,

$$E' = E - \epsilon(x_c * y_c) \qquad (2.4.4)$$

In most cases, however, $E' \approx E$.

2.4.1 Error Propagation in Function Evaluation

Consider evaluating a real-valued function $f(x) \in C^1[a, b]$ at a fixed point x_t. Two sources contribute to the total error: (1) the input is an approximation x_c rather than x_t; (2) noise due to rounding or chopping whose size is machine dependent. We will not consider the noise error, assuming it is negligible, and will concentrate only on $f(x_t) - f(x_c)$ which is the absolute error of the approximation $f(x_c)$ to $f(x_t)$. Because $f(x) \in C^1[a, b]$, we can use the mean-value theorem and obtain

$$f(x_t) - f(x_c) = (x_t - x_c)f'(\xi) \qquad (2.4.5)$$

where ξ is between x_t and x_c. Because $x_c \approx x_t$, we may express the absolute error as

$$f(x_t) - f(x_c) \approx (x_t - x_c)f'(x_t) \qquad (2.4.6)$$

and the relative error as

$$\text{rel}[f(x_c)] = \frac{f(x_t) - f(x_c)}{f(x_t)} \approx \frac{(x_t - x_c)f'(x_t)}{f(x_t)} \tag{2.4.7}$$

Hence

$$\text{rel}[f(x_c)] \approx \frac{f'(x_t)}{f(x_t)} x_t \, \text{rel}(x_c) \approx \frac{f'(x_c)}{f(x_c)} x_c \, \text{rel}(x_c) \tag{2.4.8}$$

Example 2.4.7.
Consider the function $y = e^x$. The relative error of e^{x_c} is

$$\text{rel}(e^{x_c}) \approx \frac{e^{x_t}}{e^{x_t}} x_t \, \text{rel}(x_c) = x_t \, \text{rel}(x_c)$$

For $x_t = 2.5$, $x_c = 2.49$ we have

$$\text{rel}(e^{2.49}) \approx 0.01 \qquad\qquad \square$$

Example 2.4.8.
Let $y = (\sin x)/x$, $x > 0$ and $x_t = \pi/4$. Consider a decimal six-digit calculator with a rounding mode. Then $x_t = 0.78539816\ldots$, $x_c = 0.785398$. The derivative of the function is

$$\left(\frac{\sin x}{x}\right)' = \frac{x \cos x - \sin x}{x^2}$$

and thus

$$\text{rel}\left(\frac{\sin x_c}{x_c}\right) \approx \frac{x_c \cos x_c - \sin x_c}{x_c^2} \cdot \frac{x_c}{\sin x_c} \cdot x_c \, \text{rel}(x_c)$$

$$= \left[x_c \cot(x_c) - 1\right] \text{rel}(x_c) \approx -4.4 \cdot 10^{-8}$$

Whereas x_c has seven significant digits of accuracy $[\text{rel}(x_c) \approx 2 \cdot 10^{-7}]$, $(\sin x_c)/x_c$ possesses eight accurate digits. $\qquad \square$

Equation (2.4.8) can also be written as

$$\text{rel}[f(x_c)] \approx K \, \text{rel}(x_c) \tag{2.4.9}$$

where $K \approx [f'(x_t)/f(x_t)]x_t$. This number is a *condition number* that connects between the "input" accuracy $\text{rel}(x_c)$ and the "output" accuracy $\text{rel}[f(x_c)]$. If, for example, $\text{rel}(x_c) = 10^{-6}, K = 1000$, then $\text{rel}[f(x_c)] = 10^{-3}$, which indicates a significant decrease of the accuracy from six to three significant digits.

PROBLEMS

1. Calculate the propagated error for $x_t \cdot y_t$, $(x_t + y_t)/x_t y_t$ in the following cases:
   (a) $x_t = 1.056$, $x_c = 1.057$; $y_t = \sqrt{2}$, $y_c = 1.414$
   (b) $x_t = \pi$, $x_c = 355/113$; $y_t = 9.99$, $y_c = 9.991$

2. Use given bounds for $|x_t - x_c|$ and $|y_t - y_c|$ to bound $(x_t/y_t) - (x_c/y_c)$.

3. Extend the result of Example 2.4.4: given $x_{1t}, \ldots, x_{nt}$, which are approximated by $x_{1c}, \ldots, x_{nc}$, respectively, use the bounds on $|x_{it} - x_{ic}|$, $i = 1, \ldots, n$ to bound the propagated error of $x_{1c} \cdots x_{nc}$.

4. Let x_c, y_c approximate x_t, y_t, respectively. Calculate $\text{rel}(x_c/y_c)$, $\text{rel}(x_c/y_c)$ and show

$$\text{rel}(x_c y_c) \approx \text{rel}(x_c) + \text{rel}(y_c)$$

$$\text{rel}\left(\frac{x_c}{y_c}\right) \approx \text{rel}(x_c) - \text{rel}(y_c)$$

5. Calculate $f(x_t) - f(x_c)$ and $\text{rel}[f(x_c)]$ in the following cases:
   (a) $f(x) = \ln x$, $x > 0$
   (b) $f(x) = \sqrt{x}$, $x > 0$
   (c) $f(x) = \tan x$
   (d) $f(x) = x^x$, $x > 0$

6. Find x_0 : $a \le x_0 \le b$, for which $|\text{rel}[f(x_{0c})]|$ is maximum:
   (a) $f(x) = \tan x$, $0 \le x \le \pi/4$
   (b) $f(x) = x^3 - x$, $0 \le x \le 1$
   (c) $f(x) = \sqrt{x}$, $1 \le x \le 2$

7. Assume that x_t is rounded to the number of digits shown, and bound the absolute error $f(x_t) - f(x_c)$ and $\text{rel}[f(x_c)]$:
   (a) $\ln(2)$
   (b) $\sin 1.345$
   (c) $1/6.444$
   (d) $e^{\sqrt{5.437}}$
   (e) $\sqrt{0.00012}$
   (f) 3.14159^{100}

2.5 Summation

We will now consider the problem of calculating the propagated error of a sum. In many applications we must add a large number of terms, and it is of interest to know the expected error and to try minimizing it, for example, by changing the order of summation. The problem is evaluating

$$S = x_1 + x_2 + \cdots + x_n \tag{2.5.1}$$

where x_i, $1 \le i \le n$ are given floating point numbers. To calculate S, we perform $(n-1)$ additions, each introducing either a rounding or a chopping error. We define

$$S_2 = \text{float}(x_1 + x_2)$$
$$S_3 = \text{float}(S_2 + x_3)$$
$$\vdots$$
$$S_n = \text{float}(S_{n-1} + x_n) \tag{2.5.2}$$

where S_n is the computed approximation to S. Recall Eq. (2.2.3), which yields $\text{float}(x) = x(1 + \epsilon)$, and write

$$S_2 = (x_1 + x_2)(1 + \epsilon_2)$$
$$S_3 = (S_2 + x_3)(1 + \epsilon_3)$$
$$\vdots$$
$$S_n = (S_{n-1} + x_n)(1 + \epsilon_n) \tag{2.5.3}$$

Suppose that the computation is done by a binary computer with m-bit mantissa. Then [Eqs. (2.2.4) and (2.2.5)]

$$|\epsilon_j| \le \begin{cases} \dfrac{1}{2^{m-1}} & \text{for chopping} \\[2mm] \dfrac{1}{2^m} & \text{for rounding} \end{cases} \tag{2.5.4}$$

The partial sum S_3 can be rewritten as

$$\begin{aligned} S_3 &= [(x_1 + x_2)(1 + \epsilon_2) + x_3](1 + \epsilon_3) \\ &= [(x_1 + x_2 + x_3) + \epsilon_2(x_1 + x_2)](1 + \epsilon_3) \\ &\approx (x_1 + x_2 + x_3) + x_1(\epsilon_2 + \epsilon_3) + x_2(\epsilon_2 + \epsilon_3) + x_3\epsilon_3 \end{aligned}$$

because $\epsilon_2\epsilon_3(x_1 + x_2)$ is negligible. Thus

$$(x_1 + x_2 + x_3) - S_3 \approx -x_1(\epsilon_2 + \epsilon_3) - x_2(\epsilon_2 + \epsilon_3) - x_3\epsilon_3$$

If we drop all the nonlinear terms in $\epsilon_2, \epsilon_3, \ldots, \epsilon_n$ we can show, using induction, that the absolute error of S_n is

$$S - S_n \approx -x_1(\epsilon_2 + \cdots + \epsilon_n)$$
$$-x_2(\epsilon_2 + \cdots + \epsilon_n)$$
$$-x_3(\epsilon_3 + \cdots + \epsilon_n)$$
$$-x_4(\epsilon_4 + \cdots + \epsilon_n)$$
$$\vdots$$
$$-x_n\epsilon_n \qquad (2.5.5)$$

This relation suggests a simple strategy for adding $x_1 + \cdots + x_n$: arrange the numbers as a monotonic increasing sequence in absolute value before adding them, that is,

$$|x_1| \le |x_2| \le \cdots \le |x_n| \qquad (2.5.6)$$

Thus the larger the number is, the lower the number of errors it will multiply.

Example 2.5.1.
Consider the sequence $x_j = \sqrt{10}j, \quad j = 1, \ldots, n$. Suppose that computing $\sum_{j=1}^{n} x_j$ is carried by a decimal, chopping, four-digit machine. The "best" and "worst" arrangements are

$$SB = \sqrt{10} + 2\sqrt{10} + 3\sqrt{10} + \cdots + n\sqrt{10}\text{(best)}$$
$$SW = n\sqrt{10} + (n-1)\sqrt{10} + \cdots + 2\sqrt{10} + \sqrt{10}\text{(worst)}$$

A comparison between the two sums for several values of n is given in Table 2.5.1.

Table 2.5.1. Error Propagation in Summation: Chopping Mode

N	SB	Exact	Error	SW	Exact	Error
5	47.42	47.434	0.014	47.41	47.434	0.024
10	173.8	173.925	0.125	173.5	173.925	0.425
15	379.1	379.473	0.373	378.7	379.473	0.773
20	663.4	664.078	0.678	663.1	664.078	0.978

If instead of chopping the arithmetic is carried via rounding, we still detect a difference between SB and SW, giving advantage to the strategy given by Eq. (2.5.6). In the particular case discussed here we obtain for $n = 10$: SB $= 173.9$, SW $= 174.0$, $S = 173.925$, error (SB) $= 0.025$, error (SW) $= -0.075$. $\qquad\square$

To understand the difference between the effects of rounding and chopping modes on error propagation, let us recall Eq. (2.5.5) and consider, for example, the first error term

$$E_1 = -x_1(\epsilon_2 + \cdots + \epsilon_n) \qquad (2.5.7)$$

For a decimal rounding four-digit machine we have

$$-0.0005 \leq \epsilon_i \leq 0.0005, \ 2 \leq i \leq n \tag{2.5.8}$$

Because ϵ_i can be treated as a random variable with standard deviation σ, the standard deviation of E_1 is

$$\sigma(E_1) = |x_1| \sigma \sqrt{n-1} \tag{2.5.9}$$

whereas its average is zero. Thus E_1 is likely not to grow faster than $\sqrt{n}$. If, however, the machine has a chopping mode, then

$$-0.001 \leq \epsilon_i \leq 0 \tag{2.5.10}$$

If ϵ_i is a random variable Eq. (2.5.9) holds, but for large n, E_1 can be approximated by its average, that is,

$$E_1 \approx -x_1(-0.0005)(n-1) \tag{2.5.11}$$

because $n >> \sqrt{n}$. Thus the error grows much faster. In the previous example we have for $n = 10$: 0.025 vs. 0.125 and -0.075 vs. 0.425.

An important case of error propagation in summation often occurs in DO loop sections of computer programs. Suppose that we want to calculate

$$x = x_0 + ih, \ \ 1 \leq i \leq n \tag{2.5.12}$$

and then substitute the computed value in some given function $f(x)$. This is generally done by a DO loop that can be coded either as

```
        DO 10 I = 1, N
        X = XO + I * H
        Y(I) = F(X)
     10 CONTINUE                          (code 1)
```

or

```
        X = XO
        DO 10 I = 1, N
        X = X + H
        Y(I) = F(X)
     10 CONTINUE                          (code 2)
```

The second code represents an alternative way for calculating x, by a repeated statement

$$x = x + h \tag{2.5.13}$$

We will show that code 1 is generally better and should lead to a more accurate x [and therefore to a more accurate $f(x)$]. Suppose that x is computed by Eq. (2.5.12). Then, by denoting $x_i = x$ and using Eq. (2.2.3), we obtain

$$\text{float}(x_i) = [x_0 + ih(1 + \epsilon_h)](1 + \epsilon_i) \qquad (2.5.14)$$

for some rounding or chopping errors ϵ_h, ϵ_i (provided that x_0 has an exact floating point representation). Thus

$$x_i - \text{float}(x_i) \approx -\epsilon_i x_i - ih\epsilon_h \qquad (2.5.15)$$

because nonlinear error terms are negligible. The other alternative is to compute the consecutive numbers $x_1 = x_0 + h$, $x_2 = x_1 + h$, $\ldots$, $x_i = x_{i-1} + h$. We have

$$\text{float}(x_1) \approx x_1(1 + \epsilon_1) + h\epsilon_h$$
$$\text{float}(x_2) \approx x_2(1 + \epsilon_2) + \epsilon_1 x_1 + 2h\epsilon_h$$
$$\text{float}(x_3) \approx x_3(1 + \epsilon_3) + \epsilon_1 x_1 + \epsilon_2 x_2 + 3h\epsilon_h$$
$$\vdots$$
$$\text{float}(x_i) \approx x_i(1 + \epsilon_i) + \epsilon_1 x_1 + \epsilon_2 x_2 + \cdots + \epsilon_{i-1} x_{i-1} + ih\epsilon_h \qquad (2.5.16)$$

where ϵ_i is defined by the relation

$$\text{float}(x_i) = [\text{float}(x_{i-1}) + h(1 + \epsilon_h)](1 + \epsilon_i) \qquad (2.5.17)$$

Therefore

$$x_i - \text{float}(x_i) \approx -(\epsilon_1 x_1 + \cdots + \epsilon_i x_i) - ih\epsilon_h \qquad (2.5.18)$$

Clearly, Eq. (2.5.18) produces in general a larger error than Eq. (2.5.15).

Sometimes, instead of performing a detailed analysis, we may overcome the accumulated error by using double-precision rather than single-precision computation.

Example 2.5.2.
Consider the *inner (scalar) product* of two vectors $(a_1, \ldots, a_n)^T$, $(b_1, \ldots, b_n)^T$ defined as

$$S = a_1 b_1 + \cdots + a_n b_n \qquad (2.5.19)$$

Between additions and multiplications we have a total of $(2n - 1)$ round-off or chopping errors. This can mount to a sizeable error for large n. However, by using double-precision arithmetic we round only once as we convert the double-word S to a single-precision value. □

PROBLEMS

1. Use a decimal three-digit machine to compute

$$S_n = \sum_{i=1}^{n} \frac{1}{i}$$

first from $i = 1$ to $i = n$, then backward. Assume both rounding and chopping modes and compare the results with the exact S_n for $n = 10, 50, 100, 500, 1000$.

2. Solve problem 1 for $S_n = \sum_{i=1}^{n} \sqrt{i}$.

3. Give an error analysis of code 1 and code 2, assuming that x_0 of Eq. (2.5.12) is also rounded (or chopped) by its floating point representation.

4. Analyze the propagated error of the inner product

$$S = a_1 b_1 + \cdots + a_n b_n$$

and design a strategy to compute S.

5. Use a decimal chopping four-digit machine, to calculate the sum

$$S_n = \sum_{i=1}^{n} \sin i$$

(a) From $i = 1$ to $i = n$

(b) After rearranging the terms in an increasing monotonic order

(c) After rearranging the terms in a decreasing monotonic order

Consider $n = 10, 20, 50$.

3

Iteration

Solving nonlinear equations is a major task of numerical analysis and the iteration method is a prototype of all procedures dealing with these equations.

In the first section of this chapter we provide a detailed discussion of a standard iteration method, used for solving a specific type of one-dimensional nonlinear equation. The basic theorems, related to the various algorithms, are rigorously stated and proved. This should motivate the reader to realize the importance of having a strong theoretical background as a basis for the algorithmic treatment. Also, many of the delicate features in these proofs can be successfully used in other, far more general and complicated problems.

3.1 Definition of an Iteration Problem and the Standard Iteration Method

We are concerned with solving an equation of the form

$$x = f(x) \tag{3.1.1}$$

where $f(x)$ is a real valued function defined over an interval $[a, b]$. The *standard iteration method* (SIM) for solving Eq. (3.1.1) is given by the following algorithm.

Algorithm 3.1.1.
(Standard iteration method).

Step 1. Choose a *first approximation* x_0, a tolerance ϵ, maximum number of iterations N, and set $n = 1$.

Step 2. Create the next *iteration* using the relation

$$x_n = f(x_{n-1}) \tag{3.1.2}$$

Step 3. If $|x_n - x_{n-1}| > \epsilon$ and $n < N$ set $n \leftarrow n + 1$ and go to step 2.

Step 4. If $|x_n - x_{n-1}| > \epsilon$ and $n \geq N$, output "maximum number of iterations exceeded" and stop.

Step 5. If the inequality

$$|x_n - x_{n-1}| < \epsilon \qquad\qquad (3.1.3)$$

holds, the iteration stops and x_n is considered an approximate solution to Eq. (3.1.1). Output "iteration completed successfully"; x_n and n.

If for some given $f(x)$ and x_0 there is even one $\epsilon = \epsilon_0$ for which the process diverges, then Algorithm 3.1.1 *does not converge* for this *specific* problem.

The subroutine SIM, which incorporates Algorithm 3.1.1, is found on the attached floppy disk.

Example 3.1.1.

The equation $x = e^{-x}$ has a unique solution $s = 0.567143\ldots$. Let $x_0 = 0.5$, $\epsilon = 10^{-3}$. Table 3.1.1 contains the values of x_n, $f(x_n)$, $|x_n - x_{n-1}|$ for $0 \le n \le 10$.

Table 3.1.1. Standard Iteration Method for $x = e^{-x}$, $x_0 = 0.5$, $\epsilon = 10^{-3}$

n	x_n	$f(x_n)$	$|x_n - x_{n-1}|$
0	0.500000	0.606531	
1	0.606531	0.545239	0.106531
2	0.545239	0.579703	0.061291
3	0.579703	0.560065	0.034464
4	0.560065	0.571172	0.019638
5	0.571172	0.564863	0.011108
6	0.564863	0.568438	0.006309
7	0.568438	0.566409	0.003575
8	0.566409	0.567560	0.002029
9	0.567560	0.566907	0.001150
10	0.566907		0.000652

Thus 10 iterations are needed to complete the computation, and

$$x_{10} = 0.566907, \quad |x_{10} - s| = 0.000236 \qquad\qquad \square$$

Example 3.1.2.

Let $f(x) = e^x - 1$. The equation $x = e^x - 1$ yields the unique solution $x = 0$. Table 3.1.2 contains the first five iterations for the choice of $x_0 = 0.5$, $\epsilon = 10^{-3}$.

Table 3.1.2. Standard Iteration Method Applied to
$x = e^x - 1$, $x_0 = 0.5$, $\epsilon = 10^{-3}$

n	x_n	$f(x_n)$	$|x_n - x_{n-1}|$
0	0.500000	0.648721	
1	0.648721	0.913093	0.148721
2	0.913093	1.492018	0.264372
3	1.492018	3.446060	0.578925
4	3.446060	30.376521	1.954042
5	30.376521		26.930461

By studying Table 3.1.2 one ultimately concludes that for the specific problem $x = e^x - 1$, the SIM algorithm, using a first approximation $x_0 = 0.5$, fails to converge. This could be due to either a bad choice of x_0 or, even worse, to the *behavior* of the function $f(x) = e^x - 1$.

Let us first try some different *initial* approximations. For $x_0 = 0.1$, the first few iterations, although close to x_0, are consistently increasing, and thus eventually

$$\lim_{n \to \infty} |x_n - x_{n-1}| = \infty$$

that is, the SIM diverges for $x_0 = 0.1$ as well. However, if we choose $x_0 = -0.1$ (or *any* other negative number!), Algorithm 3.1.1 converges, as demonstrated in Table 3.1.3. Because convergence to $s = 0$ is *slow*, only every fifth iteration is given.

Table 3.1.3. Performance of Standard Iteration Method for
$x = e^x - 1$, $x_0 = -0.1$, $\epsilon = 10^{-3}$

n	x_n	x_{n-1}	$|x_n - x_{n-1}|$
0	−0.100000		
5	−0.079758	−0.083118	0.003361
10	−0.066362	−0.068667	0.002305
15	−0.056835	−0.058514	0.001679
20	−0.049709	−0.050987	0.001278
25	−0.044176	−0.045181	0.001005
26	−0.043214	−0.044176	0.000962

Thus, for $x_0 = -0.1$, twenty-six iterations are performed to reach the tolerance of $\epsilon = 10^{-3}$.

Let x_{n_0} denote the final iteration. Then its *quality* is actually measured by $|x_{n_0} - s|$. In the first example (Table 3.1.1), one obtains $|x_{10} - s| = 0.000236$, which is superior to $|x_{10} - x_9| = 0.000652$. This is not always the case, as demonstrated by Table 3.1.3. Here the iteration stops when $|x_{26} - x_{25}| < 0.000962$, whereas $|x_{26} - s| = 0.043214$ is *relatively* quite large!

Because s is not known *a priori*, one would like to have some "guarantee" that if $|x_n - x_{n-1}|$ is small, so is $|x_n - s|$. $\square$

We will now introduce sufficient conditions for the *convergence* of Algorithm 3.1.1, that is,

$$\lim_{n \to \infty} x_n = s , \quad s = f(s) \tag{3.1.4}$$

Theorem 3.1.1.

[Existence theorem for $x = f(x)$]. *Let* $f(x)$, *a real-valued function defined over the closed interval* $[a, b]$, *satisfy:*

1. *For all* $x \in [a, b]$, $a \le f(x) \le b$

2. $f(x) \in C[a, b]$ *(continuity)*

Then, Eq. (3.1.1) possesses at least one solution within the interval $[a,b]$.

One should note that although condition (1) is needed to prove existence, it is also a *necessary* condition to ensure that the SIM algorithm given by Eq. (3.1.2) is *well defined*. If indeed for some m, the iteration x_m is not within $[a,b]$, $x_{m+1} = f(x_m)$ is not defined, and the procedure stops automatically.

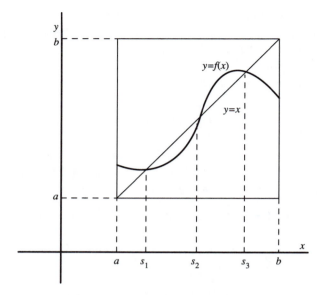

Figure 3.1.1.   Existence of a solution to $x = f(x)$.

Proof.

Define $g(x)$ as

$$g(x) = x - f(x) \tag{3.1.5}$$

Then

$$g(a) = a - f(a) \leq 0$$
$$g(b) = b - f(b) \geq 0$$

because $a \leq f(a)$, $f(b) \leq b$. If $g(a)$ or $g(b)$ equals zero, then a or b, respectively, is a solution to Eq. (3.1.1). If $g(a) < 0$ and $g(b) > 0$, then $g(x)$ must vanish for some s: $a < s < b$ because $g(x)$ as well as $f(x)$ is continuous over $[a, b]$ and changes signs at the end points. Thus $g(s) = 0$ or $s = f(s)$, which concludes the proof.   □

A typical case for which conditions (1) and (2) exist is shown in Fig. 3.1.1.
In this particular example, Eq. (3.1.1) is satisfied by three solutions s_1, s_2, s_3.

If $f(x)$ is not continuous, or if condition (1) is not satisfied, there *may not* be a solution to $x = f(x)$. This is shown in Fig. 3.1.2.

Yet a solution to Eq. (3.1.1) may exist even when both conditions of Theorem 3.1.1 are not satisfied. A typical example is shown in Fig. 3.1.3.

We therefore conclude that conditions (1) and (2) are *sufficient* but not *necessary*.

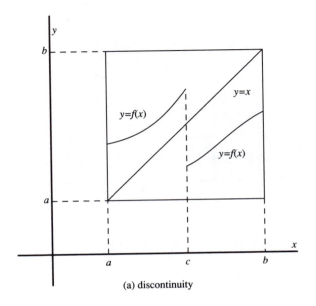

(a) discontinuity

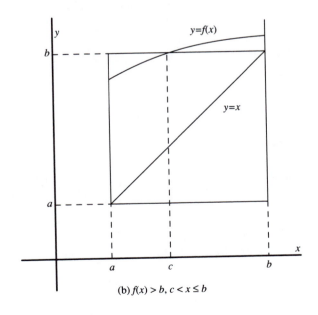

(b) $f(x) > b$, $c < x \leq b$

Figure 3.1.2. No solution to $x = f(x)$.

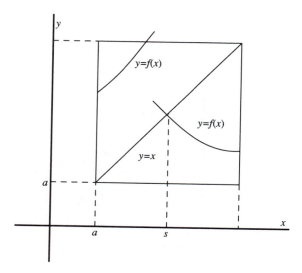

Figure 3.1.3. A jump discontinuity for $f(x)$, but $s = f(s)$.

Example 3.1.3.

The function $f(x) = e^{-x}$, defined over $[0, 1]$, is continuous and satisfies $0 \le e^{-x} \le 1$. Therefore there must exist a number s: $0 \le s \le 1$, for which $s = e^{-s}$. □

Example 3.1.4.

Let

$$f(x) = \frac{1 + x + x^2}{1 + 2x + (x^2/2)}, \quad 0 \le x \le 1$$

Clearly $f(x)$ is continuous over $[0, 1]$, and

$$0 \le f(x) = \frac{1 + x + x^2}{1 + 2x + (x^2/2)} = \frac{1 + 2x + (x^2/2) - [x - (x^2/2)]}{1 + 2x + (x^2/2)} \le 1, \quad 0 \le x \le 1$$

because $0 \le [x - (x^2/2)]$ for all $x \in [0, 1]$.

Thus there exists a number s for which

$$s = \frac{1 + s + s^2}{1 + 2s + (s^2/2)}, \quad 0 \le s \le 1$$ □

So far we have concentrated on conditions that would ensure the existence of a solution to Eq. (3.1.1). However, there could be several solutions, as shown in Fig. 3.1.1. To guarantee a *unique* solution, we must impose further restrictions on $f(x)$. We proceed with the following definition.

Definition 3.1.1.   (Lipschitz condition). *A function $f(x)$ defined over the interval $[a,b]$ is said to satisfy a* Lipschitz condition *if there exists a constant $L > 0$ for which*

$$|f(x_1) - f(x_2)| \leq L|x_1 - x_2|, \quad a \leq x_1, x_2 \leq b \tag{3.1.6}$$

The number L, small or large, is a constant and depends only on $f(x)$ and on the interval $[a,b]$. It *does not* depend on x_1 or x_2. Also, if $f(x)$ satisfies a Lipschitz condition [i.e., if Eq. (3.1.6) holds for some L], then $f(x)$ *must* be continuous. Indeed,

$$\lim_{x_1 \to x_2} |f(x_1) - f(x_2)| \leq \lim_{x_1 \to x_2} L|x_1 - x_2| = 0$$

that is, $f(x) \in C[a, b]$. The opposite is not necessarily true, that is, a continuous function $f(x)$ may not satisfy any Lipschitz condition. For example, $f(x) = \sqrt{x}$ is continuous for $0 \leq x \leq 1$. Yet a value of $L > 0$ for which

$$|\sqrt{x_1} - \sqrt{x_2}| \leq L|x_1 - x_2|, \quad 0 \leq x_1, x_2 \leq 1$$

does not exist, because $|\sqrt{x_1} - \sqrt{x_2}|/(x_1 - x_2) = 1/(\sqrt{x_1} + \sqrt{x_2})$ exceeds any given $L > 0$, if x_1 and x_2 are sufficiently small.

If $f(x) \in C^1[a, b]$, that is, if $f'(x)$ exists and is continuous, then $f(x)$ satisfies a Lipschitz condition as well. This is an immediate result of the continuity of $f'(x)$ and the mean-value theorem: $f'(x)$, being continuous over a closed interval, is also bounded, that is, there is an $L > 0$ such that

$$|f'(x)| \leq L, \quad a \leq x \leq b$$

Thus

$$|f(x_1) - f(x_2)| = |f'(c)(x_1 - x_2)| \leq L|x_1 - x_2|, \quad 0 \leq x_1, x_2 \leq b$$

where c is an interim point between x_1 and x_2, and $f(x)$ satisfies Eq. (3.1.6) with a Lipschitz constant L.

The opposite is generally not true, that is, a function $f(x)$ that satisfies Eq. (3.1.6) for some $L > 0$ need not possess a derivative. Still, the most practical way of checking whether a Lipschitz condition is satisfied by $f(x)$ would be by calculating a bound to its derivative.

Example 3.1.5.
The function $f(x) = xe^x$ defined over the interval $[0, 1]$ satisfies a Lipschitz condition with $L = 2e$, because its derivative yields

$$|f'(x)| = |e^x + xe^x| \leq |e^x| + |xe^x| \leq 2e, \quad 0 \leq x \leq 1 \qquad \square$$

Example 3.1.6.

Let

$$f(x) = \begin{cases} x^2 \ln x, & 0 < x \leq 1 \\ 0, & x = 0 \end{cases}$$

By using standard technique (Section 1.5, Example 1.5.2) it may be seen that $f(x) \in C^1[1,0]$. Therefore $f(x)$ satisfies Eq. (3.1.6) with

$$L = \max_{0 \leq x \leq 1} |f'| = \max_{0 \leq x \leq 1} |2x \ln x + x| = 1 \ (why?) \qquad \square$$

Example 3.1.7.

The function $f(x) = |x|$, $-2 \leq x \leq 2$ is not differentiable at $x = 0$. However, it does satisfy Eq. (3.1.6) with $L = 1$, because by the inequality of the triangle,

$$|\,|x_1| - |x_2|\,| \leq |x_1 - x_2| \qquad \square$$

Sometimes one is interested in finding the smallest $L = L_0$ for which Eq. (3.1.6) holds. Although in general this could be impossible to calculate, one could easily show that for any given $f(x) \in C^1[a,b]$:

$$L_0 = \max |f'(x)|, \quad a \leq x \leq b \qquad (3.1.7)$$

We will now discuss the issue of uniqueness for Eq. (3.1.1). A typical case for which there are several solutions, as shown in Fig. 3.1.1, indicates that a rapidly increasing $f(x)$ could result in producing more than a single solution to Eq. (3.1.1). The next theorem formulates just how "slowly" $f(x)$ is expected to increase in order to guarantee a unique solution s to $x = f(x)$.

Theorem 3.1.2.

(Uniqueness for $x = f(x)$). Let $f(x)$ satisfy conditions 1 and 2 of Theorem 3.1.1 (existence) and a Lipschitz condition with $L < 1$, that is,

$$|f(x_1) - f(x_2)| \leq L |x_1 - x_2|, \quad L < 1, \quad a \leq x_1, x_2 \leq b \qquad (3.1.8)$$

Then Eq. (3.1.1) has a unique solution within the interval [a,b].
The condition given by Eq. (3.1.8) is sufficient but not necessary, as shown in Fig. 3.1.4.

Proof.

Assume that there are two solutions s_1 and s_2 to the equation $x = f(x)$, namely, $s_1 = f(s_1)$, $s_2 = f(s_2)$. Then, due to the Lipschitz condition requirement,

$$|s_1 - s_2| = |f(s_1) - f(s_2)| \leq L |s_1 - s_2|$$

Dividing this last equation by $|s_1 - s_2|$ would lead to the contradiction $1 \leq L$, unless $s_1 = s_2$. $\quad\square$

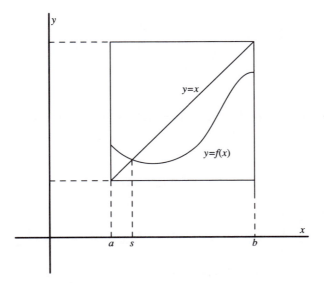

Figure 3.1.4. A unique solution for $x = f(x)$ with $L > 1$.

Example 3.1.8.

Let $f(x) = 1 - (x^2/3)$, $0 \le x \le 1$. Conditions 1 and 2 of Theorem 3.1.1 are satisfied, and because $\max_{0 \le x \le 1} |f'(x)| = 2/3$, Eq. (3.1.8) holds as well. Thus, there must exist a unique solution to $x = f(x)$ within $[0, 1]$:

$$s = 1 - \frac{s^2}{3} \quad \text{or} \quad s = \frac{-3 + \sqrt{21}}{2} \approx 0.8 \qquad \square$$

Example 3.1.9.

The function $f(x) = 0.5 + 0.9 \tan^{-1}x$, $0 \le x \le 2$ is continuous everywhere and satisfies

$$0.5 \le f(x) \le 0.5 + \frac{0.9\pi}{2} < 2, \quad 0 \le x \le 2$$

$$|f'(x)| = \frac{0.9}{1 + x^2} \le 0.9 < 1, \quad 0 \le x \le 2$$

Therefore, within the interval $[0, 2]$, a unique s for which $s = 0.5 + 0.9 \tan^{-1}s$ exists. $\qquad \square$

Turning back to the SIM algorithm, we are now able to formulate sufficient conditions under which this algorithm converges.

Theorem 3.1.3.

(Convergence of SIM). *Let $f(x)$ satisfy*

1. $a \le f(x) \le b$ *for all* x : $a \le x \le b$

2. $f(x) \in C[a,b]$

3. *There exists an L: $0 < L < 1$ for which*

$$|f(x_1) - f(x_2)| \leq L|x_1 - x_2|, \quad a \leq x_1, x_2 \leq b$$

Then the SIM algorithm stated by Eq. (3.1.2) converges to the unique solution s of Eq. (3.1.1), that is,

$$\lim_{n \to \infty} x_n = s \tag{3.1.9}$$

for any given x_0 within [a,b].

Proof.
First, due to Theorems 3.1.1 and 3.1.2, conditions 1–3 guarantee a unique solution s to Eq. (3.1.1).
   Next, for any given n,

$$|x_n - s| = |f(x_{n-1}) - f(s)| \leq L|x_{n-1} - s|$$

This can be repeated to estimate $|x_{n-1} - s|$, and after n repetitions one obtains

$$|x_n - s| \leq L^n|x_0 - s|$$

Because $0 \leq L < 1$, the right-hand side of the inequality approaches zero as $n \to \infty$. Thus,

$$\lim_{n \to \infty} |x_n - s| = 0$$

which is identical to Eq. (3.1.9).   □

Example 3.1.10.
Let $f(x) = e^{-x}, \; 0 \leq x \leq 1$. Conditions 1 and 2 of Theorem 3.1.3 hold. However, because

$$\max_{0 \leq x \leq 1} |f'(x)| = \max_{0 \leq x \leq 1} |-e^{-x}| = 1$$

condition 3 is violated. Although one may still show the uniqueness of the solution, there is no guarantee that the SIM algorithm converges for any given x_0. One way to overcome this obstacle is to replace the original interval $[0, 1]$, by a new subinterval $[a_1, b_1]$ (i.e., $0 \leq a_1 < b_1 \leq 1$) within which, hopefully, all of the conditions are satisfied.
   In this case s must be in $[a_1, b_1]$, and because

$$L = \max_{a_1 \le x \le b_1} |f'(x)| = e^{-a_1} < 1$$

the SIM algorithm must converge to s for any choice of x_0 such that $a_1 \le x_0 \le b_1$. Because e^{-x} is a decreasing monotonic function, a proper choice of $[a_1, b_1]$ should only fulfill

$$a_1 \le e^{-a_1} \le b_1, \quad a_1 \le e^{-b_1} \le b_1$$

for condition 1 to hold. One possible choice is $[a_1, b_1] = [1/2, 2/3]$ (verification is left for the reader). Now for a given x_0, the 10th iteration, for example, must satisfy

$$|x_{10} - s| \le (e^{-1/2})^{10} |x_0 - s| \le (e^{1/2})^{10} \left(\frac{2}{3} - \frac{1}{2}\right) < 0.0012$$

This can be validated for $x_0 = 0.5$, because $x_{10} = 0.566907$ (Table 3.1.1), $s = 0.567143$, $|x_{10} - s| = 0.000236$. $\square$

We define the *error* at the nth iteration as

$$\epsilon_n = x_n - s \tag{3.1.10}$$

Thus an immediate *error estimate* is given by the inequality

$$|\epsilon_n| \le L^n |x_0 - s| \tag{3.1.11}$$

One disadvantage in using this estimate is the presence of the unknown quantity s at the right-hand side. One can eliminate s from Eq. (3.1.11) by rewriting it as

$$|\epsilon_n| \le L^n |x_0 - s| \le L^n (b - a)$$

but this is rather artificial and insufficient because the length of the interval could be a relatively large number.

A different approach is to consider the inequality

$$|x_{n+1} - x_n| = |f(x_n) - f(x_{n-1})| \le L |x_n - x_{n-1}|$$

By using induction one can show that

$$|x_{n+1} - x_n| \le L^n |x_1 - x_0| \tag{3.1.12}$$

holds for any given n. Next, we find a bound for $|x_m - x_n|$, $m > n$. Implementing the triangle inequality on the relation

$$x_m - x_n = (x_m - x_{m-1}) + (x_{m-1} - x_{m-2}) + \cdots + (x_{n+1} - x_n)$$

leads to

$$|x_m - x_n| \le |x_m - x_{m-1}| + |x_{m-1} - x_{m-2}| + \cdots + |x_{n+1} - x_n|$$

and by applying Eq. (3.1.12), one obtains

$$\begin{aligned}|x_m - x_n| &\le (L^{m-1} + L^{m-2} + \cdots + L^n)|x_1 - x_0| \\ &\le L^n(1 + L + \cdots + L^{m-n-1})|x_1 - x_0|\end{aligned}$$

Because $0 < L < 1$, we can replace $(1 + L + \cdots + L^{m-n-1})$ by the converging infinite geometric series and obtain

$$|x_m - x_n| \le L^n(1 + L + L^2 + \cdots)|x_1 - x_0| = \frac{L^n}{1 - L}|x_1 - x_0|$$

Finally, by keeping n fixed and letting $m \to \infty$, we obtain a new error estimate, namely,

$$|\epsilon_n| = |x_n - s| \le \frac{L^n}{1 - L}|x_1 - x_0| \tag{3.1.13}$$

which is independent of s. It is particularly effective when L is not too close to 1, and $|x_1 - x_0|$ is small.

Example 3.1.11.

Let us refer to the previous example and reconsider the 10th iteration.

$$|\epsilon_{10}| \le \frac{(e^{-1/2})^{10}}{1 - e^{-1/2}}|0.606531 - 0.500000| = 0.00182$$

If one wishes to perform a minimal number of iterations, N, that for a given tolerance ϵ would guarantee $|x_n - s| < \epsilon$, one should solve the equation

$$\epsilon = \frac{L^N|x_1 - x_0|}{1 - L}$$

which yields $N = \ln\left[\epsilon(1 - L)/|x_1 - x_0|\right]/\ln L$. For the previous example, a tolerance of $\epsilon = 10^{-6}$ yields $N = 26$. $\square$

PROBLEMS

1. Which of the following functions satisfy condition 1 of Theorem 3.1.1:

   (a) $f(x) = \sin x, \quad 0 \le x \le \pi$

(b) $f(x) = \tan x$, $0 \le x \le \pi/4$

(c) $f(x) = 1/(1+x^2)$, $-1/2 \le x \le 1/2$

(d) $f(x) = |x+1| - |x-1|$, $-1/2 \le x \le 1$

2. Which of the following functions satisfy a Lipschitz condition? For those functions that do, determine a Lipschitz constant L (if possible, the smallest).

(a) $f(x) = x^3 + x^2 e^{-x}$, $0 \le x \le 2$

(b) $f(x) = \sqrt{1 - x^2}$, $0 \le x \le 1$

(c) $f(x) = x^2 \ln x$, $1/e \le x \le e$

(d) $f(x) = |1 - x| + |1 + x|$, $-1 \le x \le 2$

3. Prove that the equation

$$x = \frac{1}{2} e^{-x} \cos x$$

has a unique solution within the interval $I = [0, 1]$.

4. Let $f(x) = (1/2)e^{-x} \cos x$, $0 \le x \le 1$. Let $x_0 = 1/4$ be a first approximation to the unique solution s of $x = f(x)$. Given a tolerance $\epsilon = 10^{-5}$, calculate the minimal number of iterations N that would guarantee

$$|x_N - s| \le \epsilon$$

5. To solve the equation

$$x^3 - x^2 - x - 1 = 0$$

consider its equivalent form:

$$x = 1 + \frac{1}{x} + \frac{1}{x^2}$$

First, determine an interval $I = [a, b]$, within which this equation has a unique solution. Then choose $x_0 = (a + b)/2$ and iterate until an accuracy of $\epsilon = 10^{-5}$ is achieved.

6. Assume that $f(x)$ satisfies conditions 1–3 of Theorem 3.1.3 and that $f'(x) < 0$ within $I = [a, b]$. Prove that if $x_0 < s$ then

$$x_0 < x_2 < x_4 \cdots < s < \cdots < x_5 < x_3 < x_1$$

7. Let $f(x) = \alpha[x + (1/x)]$, $1 \le x \le 2$, $0 < \alpha$. Find all α for which there exists a unique solution to the equation $x = f(x)$.

8. Let $f(x) = \sum_{j=0}^{n} a_j x^j$, $0 \le x \le 1/2$ and assume $0 \le a_j < 1/4$, $j = 0, 1, 2, \ldots, n$.

(a) Show the existence of a unique solution to $x = f(x)$.

(b) How many iterations are needed to reach a tolerance of $\epsilon = 10^{-6}$, if $0 \leq a_j \leq 3/16$, $0 \leq j \leq n$, and $x_0 = 0$.

9. Use the SIM to solve $x = e^{-x}$, $1/2 \leq x \leq 2/3$, starting with $x_0 = 2/3$. Stop the iteration when $|x_{n+1} - x_n| \leq 10^{-6}$ first occurs. While computing, create the sequence

$$Q = \left\{ \frac{\epsilon_1}{\epsilon_0}, \frac{\epsilon_2}{\epsilon_1}, \ldots, \frac{\epsilon_N}{\epsilon_{N-1}} \right\}$$

where $\epsilon_n = x_{n+1} - x_n$ and N is the first integer n for which $|x_{n+1} - x_n| \leq 10^{-6}$. What can you say about the behavior of Q?

10. Let $g(x) \in C^1[0, 1]$ satisfy

(a) $g(0) < 0 < g(1)$

(b) $0 < a \leq g'(x) \leq b$, $0 \leq x \leq 1$

Determine a sufficient condition for M so that the equation $g(x) = 0$ can be solved by applying the SIM to $f(x) = x + Mg(x)$.

3.2 Rate of Convergence

Let us consider the equation $x = \cos x$, $0 \leq x \leq 1$. The three conditions previously discussed are satisfied and ensure a unique solution. Also, by virtue of Theorem 3.1.3, for any x_0 between 0 and 1, if $x_n = \cos(x_{n-1})$, then $\lim_{n\to\infty} x_n = s \approx 0.739085$.

Table 3.2.1 contains the numerical values of x_n, $|\epsilon_n|$, $|\epsilon_n/\epsilon_{n-1}|$ for $0 \leq n \leq 10$ generated from $x_0 = 0$. It indicates the convergence of $|\epsilon_n/\epsilon_{n-1}|$ to some number (≈ 0.67). This is no coincidence (compare with the results of problem 9 in Section 3.1), but rather presents a particular case of *linear convergence*.

Table 3.2.1. Convergence Pattern for $x = \cos x$

n	x_n	$|\epsilon_n|$	$|\epsilon_n/\epsilon_{n-1}|$
0	0.000000	0.739085	
1	1.000000	0.260915	0.3530
2	0.540302	0.198783	0.7619
3	0.857553	0.118468	0.5960
4	0.654290	0.084795	0.7158
5	0.793480	0.054395	0.6415
6	0.701369	0.037716	0.6934
7	0.763960	0.024875	0.6595
8	0.722102	0.016983	0.6827
9	0.750418	0.011333	0.6673
10	0.731404	0.007681	0.6778

We will now define, in general, the rate of convergence for a given sequence $\{x_n\}_{n=1}^{\infty}$.

Definition 3.2.1. (Rate of convergence). *Let* $\{x_n\}_{n=1}^{\infty}$ *be a sequence of iterations that converges to* s, *that is,*

$$\lim_{n \to \infty} x_n = s$$

If the error sequence $\{\epsilon_n = x_n - s, \ n = 1, 2, \ldots\}$ *satisfies*

$$\lim_{n \to \infty} \left| \frac{\epsilon_{n+1}}{|\epsilon_n|^{\alpha}} \right| = K, \quad K > 0 \tag{3.2.1}$$

for some fixed numbers α, K, *then* α *is the order of convergence by which* $\{x_n\}$ *converges, and* K *is the* convergence factor. *Both numbers yield the* rate of convergence *of* $\{x_n\}$.

It can be seen that α must always satisfy $\alpha \geq 1$, and if $\alpha = 1$ then $K < 1$. The particular cases $\alpha = 1$, $\alpha = 2$ correspond to *linear* and *quadratic* convergence, respectively.

The previously discussed SIM algorithm converges linearly, provided that some requirements are fulfilled.

Theorem 3.2.1.

(Rate of convergence of SIM). *Let* $f(x)$ *satisfy the following conditions:*

1. *For all* $a \leq x \leq b$: $a \leq f(x) \leq b$

2. $f(x) \in C^1[a, b]$

3. *For all* $a \leq x \leq b$: $0 < |f'(x)| < 1$

Then, for any given x_0, *the sequence* $\{x_n\}$ *created by the SIM converges linearly to the unique solution* s *of* $x = f(x)$.

Proof.

Condition 3 provides a Lipschitz constant $L = \max_{a \leq x \leq b} |f'(x)| < 1$ to ensure a unique solution. Furthermore, because for all x, $f'(x) \neq 0$, the iterative procedure never stops. Indeed, if a finite number of steps would be sufficient for the SIM to produce s, then for some $n = N$ one should have

$$x_N = f(x_N), \quad x_N = f(x_{N-1}), \quad x_N \neq x_{N-1}$$

By subtracting and applying the mean-value theorem we obtain

$$0 = x_N - x_N = f(x_N) - f(x_{N-1}) = (x_N - x_{N-1})f'(c)$$

for some c between x_N and x_{N-1}. Because $x_N \neq x_{N-1}$, $f'(c) = 0$, in contradiction with requirement 3. Now, by reusing the mean-value theorem one obtains

$$\epsilon_{n+1} = x_{n+1} - s$$
$$= f(x_n) - f(s) = (x_n - s)f'(c_n)$$
$$= \epsilon_n f'(c_n)$$

or

$$\frac{\epsilon_{n+1}}{\epsilon_n} = f'(c_n)$$

Because c_n is between s and x_n, it must converge to s (because x_n does) when $n \to \infty$. Because $f'(x)$ is continuous everywhere, in particular at $x = s$, we have

$$\lim_{n \to \infty} \frac{\epsilon_{n+1}}{\epsilon_n} = f'(s) \neq 0 \qquad (3.2.2)$$

                                    ⬚

We thus see that the error at the $(n+1)$th step is approximately equal to $f'(s)$ times the error at the nth step, provided that n is large enough and we are already close to s.

Example 3.2.1.

Let $f(x) = \cos x, \ \ 0 \leq x \leq 1$. The exact solution of $x = \cos x$ computed to six digits is $s = 0.739085$. Because $f'(x) = -\sin x$, we obtain

$$\lim_{n \to \infty} \frac{\epsilon_{n+1}}{\epsilon_n} = -\sin(s) = -0.673612$$

Thus, at each step (for sufficiently large n), the error in absolute value decreases by approximately one third.   □

Example 3.2.2.

The unique solution of $x = e^{-x}$, computed to six decimal digits, is $s = 0.567143$. Because $(e^{-x})' = -e^{-x}$, one finds that

$$\lim_{n \to \infty} \frac{\epsilon_{n+1}}{\epsilon_n} = -e^{-s} = -s = -0.567143 \qquad\qquad □$$

The conclusion from Eq. (3.2.2) is that a smaller $f'(s)$ is a necessary condition for a rapid convergence. If by chance $f'(s) = 0$, the speed of convergence changes dramatically. However, this particular case is not classified as a linear convergence-type case and will be discussed in detail in the next section.

Although it is true that the asymptotic behavior of the error depends solely on the numerical value of $f'(s)$, it is still important that we choose x_0 carefully. Once an algorithm converges, the scientist is primarily interested in the *number of iterations* needed for convergence. This number may strongly depend on x_0!

Assume now a general problem $x = f(x)$ that yields linear convergence with a random $f'(s)$, say, $f'(s) = 0.5$. If the iteration procedure is completely governed by Eq. (3.2.2), then approximately 20 iterations $(0.5^{20} \approx 10^{-6})$ are needed to obtain 6 accurate decimal digits. Computing $f(x)$ could be time consuming, and thus using the SIM many times could be costly. One should, therefore, try either to *accelerate* the existing algorithm, or use a different algorithm.

The Aitken algorithm is a scheme that significantly accelerates and improves the performance of the SIM. It is studied and discussed in detail in the next section. We will then introduce a class of Newton's methods whose *initial* type of convergence is quadratic. Other procedures whose rate of convergence, α, is not necessarily an integer will be presented as well.

PROBLEMS

1. Let $f(x) \in C^1[a, b]$ satisfy the conditions of Theorem 3.2.1. Define $\eta_n = x_n - x_{n-1}$, where $x_n = f(x_{n-1})$ and x_0 yields $a \leq x_0 \leq b$.

   (a) Prove that $\lim_{n \to \infty} (\eta_{n+1}/\eta_n) = f'(s)$.

   (b) What is the advantage of this result over that of Theorem 3.2.1?

   (c) For large n, express ϵ_n in terms of η_n.

2. Verify the results of problem 1 for the following:

   (a) $x = e^{-x}$, $1/2 \leq x \leq 2/3$, $x_0 = 1/2$ $(s = 0.567143\ldots)$

   (b) $x = \cos x$, $0 \leq x \leq 1$, $x_0 = 1$ $(s = 0.739085\ldots)$

3. Use the SIM to solve $x = \sin^2 x$, $-1/2 \leq x \leq 1/2$, $x_0 = 1/4$. Check the behavior of $(x_{n+1} - x_n)/(x_n - x_{n-1})$ as $n \to \infty$. Explain.

.3 Aitken's Method for Acceleration

We still consider the equation $x = f(x)$, where $f(x)$ fulfills the requirements of Theorem 3.2.1. Given x_0, let us construct the sequence $\{x_n\}$, $x_n = f(x_{n-1})$. Because Eq. (3.2.2) holds, we can write

$$\epsilon_{n+1} = (M + \theta_n)\epsilon_n \tag{3.3.1}$$

where $M = f'(s)$ satisfies $0 < |M| < 1$, and

$$\lim_{n \to \infty} \theta_n = 0 \tag{3.3.2}$$

Let us assume that for large n, θ_n is identically zero. Then, for such n,

$$x_{n+1} - s = M(x_n - s)$$
$$x_{n+2} - s = M(x_{n+1} - s)$$

By subtracting the first equation from the second we obtain

$$x_{n+2} - x_{n+1} = M(x_{n+1} - x_n)$$

or

$$M = \frac{x_{n+2} - x_{n+1}}{x_{n+1} - x_n}$$

Now, by substituting M in the first equation, we obtain

$$s = x_n - \frac{(x_{n+1} - x_n)^2}{x_{n+2} - 2x_{n+1} + x_n}$$

One should note that if for some large N, $\theta_n = 0$ for all $n > N$, then the last equation actually yields the exact solution to $x = f(x)$. However, because in general $\theta_n \neq 0$, one could only expect that because for large n, θ_n is significantly small, x_n', defined as

$$x_n' = x_n - \frac{(x_{n+1} - x_n)^2}{x_{n+2} - 2x_{n+1} + x_n} \tag{3.3.3}$$

is much closer to s than x_{n+2} is.

Let us now define the *forward difference operator* Δ, which operates on an infinite sequence of numbers $\{x_n\}_0^\infty$ by

$$\Delta x_n \equiv x_{n+1} - x_n, \quad n \geq 0 \tag{3.3.4}$$

$$\Delta \left(\sum_{j=1}^m \alpha_j x_{n_j} \right) \equiv \sum_{j=1}^m \alpha_j \Delta x_{n_j} \tag{3.3.5}$$

for all $m \geq 1$, real α_j and $n_j \geq 0$. Using induction we further define

$$\Delta^n E \equiv \Delta(\Delta^{n-1} E), \quad n \geq 2 \tag{3.3.6}$$

for any expression E for which the operator Δ^{n-1} is already defined. Because

$$\begin{aligned} \Delta^2 x_n &= \Delta(\Delta x_n) = \Delta(x_{n+1} - x_n) \\ &= \Delta x_{n+1} - \Delta x_n = x_{n+2} - 2x_{n+1} + x_n \end{aligned}$$

Eq. (3.3.3) can be written as

$$x_n' = x_n - \frac{(\Delta x_n)^2}{\Delta^2 x_n} \tag{3.3.7}$$

Thus, for any given sequence $\{x_n\}$ that converges to s and fulfills the requirements of Eqs. (3.3.1) and (3.3.2), one can apply the next algorithm, which is likely to speed up the convergence.

Algorithm 3.3.1.

[Aitken's Δ^2 method (ATKN)].

Step 1. Choose a first approximation x_0, a tolerance ϵ, and a maximum number of iterations N.

Step 2. Compute $x_1 = f(x_0)$, $x_2 = f(x_1)$; set $n = 0$ and compute x'_n [using Eq. (3.3.7)].

Step 3. Compute $x_{n+3} = f(x_{n+2})$ and x'_{n+1}.

Step 4. If $|x'_{n+1} - x'_n| \leq \epsilon$ set $s = x'_{n+1}$, output s, $(n+3)$ and stop.

Step 5. If $|x'_{n+1} - x'_n| > \epsilon$ and $n < N$ set $n \leftarrow n + 1$ and go to Step 3.

Step 6. Output "maximum number of iterations exceeded" and stop.

A subroutine ATKN that incorporates Algorithm 3.3.1 is found on the attached floppy disk.

Example 3.3.1.

Consider $x = e^{-x}$, $1/2 \leq x \leq 2/3$, $x_0 = 0.5$. Table 3.3.1 contains the first eight iterations, using the SIM, compared with eight iterations based on Aitken's scheme.

Table 3.3.1. $x = e^{-x}$ (Aitken's Method)

n	x_n	x'_n
0	0.500000	0.567624
1	0.606531	0.567299
2	0.545239	0.567193
3	0.579703	0.567159
4	0.560065	0.567148
5	0.571172	0.567145
6	0.564863	0.567144
7	0.568438	0.567143
8	0.566409	
$\vdots$		
22	0.567143	

We see that ATKN yields the first 6 digits of the exact solution after 8 iterations, compared with the 22 that are needed for the SIM (to provide the same accuracy). □

Example 3.3.2.

Let $x = \cos x$, $0 \leq x \leq 1$, $x_0 = 1$. Table 3.3.2 consists of the first 10 iterations performed by Algorithm 3.3.1.

Table 3.3.2. $x = \cos x$ (Aitken's Method)

n	x_n	x'_n
0	1.00000000	0.72801036
1	0.54030231	0.73366516
2	0.85755322	0.73690629
3	0.65428979	0.73805042
4	0.79348036	0.73863610
5	0.70136877	0.73887658
6	0.76395968	0.73899224
7	0.72210243	0.73904251
8	0.75041776	0.73906595
9	0.73140404	0.73907638
10	0.74423735	0.73908118
11	0.73560474	
$\vdots$		
26	0.73909441	

The exact solution computed to eight decimal digits is $s = 0.73908513$. Thus, to approximate s within an error tolerance $\epsilon = 10^{-5}$, only 11 iterations are needed by Aitken's method, a considerable reduction from the original 25 iterations performed by the SIM algorithm.   □

We will now rigorously prove that the conclusions derived from the previous examples demonstrate the general pattern and that the new sequence $\{x'_n\}$ *always* converges faster than the original sequence $\{x_n\}$.

Theorem 3.3.1.

(Aitken's acceleration procedure). *Let $\{x_n\}$ be a sequence of numbers that converges to s and satisfies Eqs. (3.3.1) and (3.3.2) for some M: $0 < |M| < 1$. Let $\{x'_n\}$ be the sequence generated by Eq. (3.3.7). Then*

$$\lim_{n \to \infty} \frac{x'_n - s}{x_n - s} = 0 \tag{3.3.8}$$

(that is, $x'_n \to s$ faster than $x_n \to s$ as $n \to \infty$).

Proof.

Define $\epsilon'_n = x'_n - s$. Then

$$\epsilon'_n = x'_n - s = x_n - s - \frac{(x_{n+1} - x_n)^2}{x_{n+2} - 2x_{n+1} + x_n}$$

$$= \epsilon_n - \frac{(\epsilon_{n+1} - \epsilon_n)^2}{\epsilon_{n+2} - 2\epsilon_{n+1} + \epsilon_n} = \frac{\epsilon_n \epsilon_{n+2} - \epsilon_{n+1}^2}{\epsilon_{n+2} - 2\epsilon_{n+1} + \epsilon_n}$$

Now, due to Eqs. (3.3.1) and (3.3.2) we have

$$\epsilon_{n+1} = (M + \theta_n)\epsilon_n$$

$$\epsilon_{n+2} = (M + \theta_{n+1})\epsilon_{n+1} = (M + \theta_{n+1})(M + \theta_n)\epsilon_n$$

where θ_n, $\theta_{n+1} \to 0$ as $n \to \infty$. Therefore,

$$
\begin{aligned}
\epsilon'_n &= \frac{(M + \theta_{n+1})(M + \theta_n)\epsilon_n^2 - (M + \theta_n)^2\epsilon_n^2}{(M + \theta_{n+1})(M + \theta_n)\epsilon_n - 2(M + \theta_n)\epsilon_n + \epsilon_n} \\
&= \frac{(M + \theta_n)}{(M - 1)^2 + M(\theta_n + \theta_{n+1}) + \theta_n(\theta_{n+1} - 2)}(\theta_{n+1} - \theta_n)\epsilon_n
\end{aligned}
$$

Hence

$$\frac{\epsilon'_n}{\epsilon_n} = \frac{M + \theta_n}{(M - 1)^2 + M(\theta_n + \theta_{n+1}) + \theta_n(\theta_{n+1} - 2)}(\theta_{n+1} - \theta_n)$$

and

$$\lim_{n\to\infty} \frac{\epsilon'_n}{\epsilon_n} = \frac{M}{(M-1)^2} \lim_{n\to\infty}(\theta_{n+1} - \theta_n) = 0 \tag{3.3.9}$$

$\square$

The advantages of ATKN are as follow:

1. There is no significant additional computing time beyond that consumed for the construction of $\{x_n\}$, because the time needed to produce x'_n, once x_n, x_{n+1}, x_{n+2} are already given, is negligible.

2. ATKN speeds up the convergence, for *any sequence* x_n, provided that Eqs. (3.3.1) and (3.3.2) hold.

Let us now figure the rate of convergence of this algorithm for the special case $x_n = f(x_{n-1})$. Because x_n converges linearly to s, and slower than x'_n, we may guess that if the Aitken generated sequence is governed by Eq. (3.2.1) then the relation

$$\lim_{n\to\infty}\left|\frac{\epsilon'_{n+1}}{|\epsilon'_n|^\alpha}\right| = K$$

holds for some $\alpha > 1$. This apparently is not the case. Aitken's method converges faster than the SIM, but still *linearly*! However, while using the SIM we have

$$\lim_{n\to\infty} \frac{\epsilon_{n+1}}{\epsilon_n} = M = f'(s)$$

the ATKN error behaves asymptotically, according to

$$\lim_{n\to\infty} \frac{\epsilon'_{n+1}}{\epsilon'_n} = M^2 \tag{3.3.10}$$

Because $|M| < 1$, M^2 *must* be smaller than $|M|$. This explains the superiority of ATKN over the SIM, and enables us to measure it.

Example 3.3.3.

Consider the equation $x = f(x) = \cos x$ and the results given in Table 3.3.2. Here $f'(x) = -\sin x$, $M = -\sin(0.739085) = -0.673612$, $M^2 \approx 0.454$. This is in agreement with

$$(\epsilon'_{n+1})/\epsilon'_n = 0.445, \ 0.459, \ 0.450, \ 0.456; \quad n = 5, \ 6, \ 7, \ 8 \qquad \square$$

Clearly, because $M^2 > 0$, x'_n, for sufficiently large n, must all fall on the same side of s. This is not the case for the sequence x_n, whenever $-1 < M < 0$.

The hypotheses needed to ensure Eq. (3.3.10) are given next.

Theorem 3.3.2.

(Aitken's method rate of convergence). *Let $f(x)$ satisfy the requirements of Theorem 3.3.1. If, further, $f(x) \in C^2[a, b]$ [i.e., $f(x)$ possesses a second continuous derivative] and $f''(x) \neq 0$, then x'_n converges linearly to s and*

$$\lim_{n \to \infty} \frac{\epsilon'_{n+1}}{\epsilon'_n} = M^2$$

where $M = f'(s)$.

Proof.

Following the proof of the previous theorem one may determine

$$\frac{\epsilon'_{n+1}}{\epsilon'_n} = \frac{(M + \theta_{n+1})[(M-1)^2 + M(\theta_n + \theta_{n+1}) + \theta_n(\theta_{n+1} - 2)]}{(M + \theta_n)[(M-1)^2 + M(\theta_{n+1} + \theta_{n+2}) + \theta_{n+1}(\theta_{n+2} - 2)]} \frac{\epsilon_{n+1}}{\epsilon_n} \frac{(\theta_{n+2} - \theta_{n+1})}{(\theta_{n+1} - \theta_n)}$$

Because $\lim_{n \to \infty} \theta_n = 0$, $\lim_{n \to \infty} (\epsilon_{n+1}/\epsilon_n) = M$, we have

$$\lim_{n \to \infty} \frac{\epsilon'_{n+1}}{\epsilon'_n} = M \lim_{n \to \infty} \frac{\theta_{n+2} - \theta_{n+1}}{\theta_{n+1} - \theta_n} \qquad (3.3.11)$$

provided the limit on the right-hand side exists. Because $f(x) \in C^2[a, b]$ we can use Taylor's theorem and write

$$\epsilon_{n+1} = x_{n+1} - s = f(x_n) - f(s) = (x_n - s)f'(s) + \frac{(x_n - s)^2}{2!} f''(c_n)$$

where c_n is between x_n and s. Thus, $\lim_{n \to \infty} c_n = s$, and due to the continuity of $f''(x)$

$$f''(c_n) = f''(s) + \theta'_n, \quad \lim_{n \to \infty} \theta'_n = 0$$

Hence

$$\epsilon_{n+1} = \epsilon_n M + \frac{\epsilon_n^2}{2}(M' + \theta'_n)$$

where $M' = f''(s)$. Also, because $\epsilon_{n+1} = (M + \theta_n)\epsilon_n$, we obtain

$$\theta_n = \frac{\epsilon_n}{2}(M' + \theta'_n) \tag{3.3.12}$$

and further, by using $f''(x) \neq 0$ (in particular, $M' \neq 0$), one obtains

$$\lim_{n \to \infty} \frac{\theta_{n+1}}{\theta_n} = \lim_{n \to \infty} \frac{\epsilon_{n+1}}{\epsilon_n} \frac{M' + \theta'_{n+1}}{M' + \theta'_n} = \lim_{n \to \infty} \frac{\epsilon_{n+1}}{\epsilon_n} = M$$

Therefore

$$\lim_{n \to \infty} \frac{\theta_{n+2} - \theta_{n+1}}{\theta_{n+1} - \theta_n} = \lim_{n \to \infty} \frac{[(\theta_{n+2}/\theta_{n+1}) - 1]}{[1 - (\theta_n/\theta_{n+1})]} = \frac{M - 1}{1 - (1/M)} = M$$

and Eq. (3.3.11) yields

$$\lim_{n \to \infty} \frac{\epsilon'_{n+1}}{\epsilon'_n} = M \cdot M = M^2 \qquad \square$$

The main corollary of Theorem 3.3.2 (see below) is that the ATKN algorithm reduces, roughly by half, the number of iterations needed by the SIM.

Corollary 3.3.1. *Given a tolerance $\epsilon > 0$, let N, N' denote the number of iterations required for convergence using the SIM and ATKN, respectively. Then,*

$$\lim_{\epsilon \to 0} \frac{N'}{N} = \frac{1}{2} \tag{3.3.13}$$

The proof, based on Eq. (3.3.10), is straightforward and is left as an exercise for the reader.

Example 3.3.4.
Let $f(x) = 1 - (x^3/8)$, $0 \leq x \leq 1$ (Fig. 3.3.1) and consider the equation $x = f(x)$, given a first approximation $x_0 = 0$ and a tolerance $\epsilon = 10^{-6}$. The exact solution computed to nine significant digits is $s = 0.906795303$. A comparison between the performances of the SIM and ATKN is given in Table 3.3.3.

Table 3.3.3. $x = 1 - (x^3/8)$, $0 \leq x \leq 1$, $x_0 = 0$, $\epsilon = 10^{-6}$

n	x_n	x'_n	$\lvert \epsilon_n/\epsilon_{n-1} \rvert$	$\epsilon'_n/\epsilon'_{n-1}$
0	0.000000000	0.888888889		
1	1.000000000	0.906020558	0.103	0.043
2	0.875000000	0.906717286	0.341	0.101
3	0.916259766	0.906788044	0.298	0.093
4	0.903846331	0.906794608	0.312	0.096
$\vdots$				
12	0.906796083			

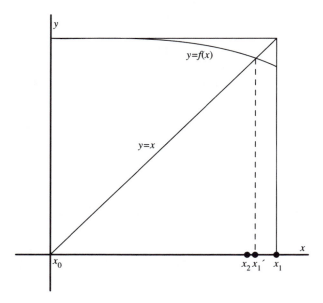

Figure 3.3.1.   Aitken's method for $x = f(x) = 1 - (x^3/8)$.

In this case $N = 11$, $N' = 5$, $|\epsilon_4/\epsilon_3| = 0.312 \approx |M| = (3/8)s^2 = 0.308\ldots$, $\epsilon_4'/\epsilon_3' = 0.096 \approx M^2 = 0.095\ldots$. $\quad\square$

It should be stressed that if the original sequence $\{x_n\}$ is not generated by the SIM, but merely satisfies the hypotheses of Theorem 3.3.1, nothing beyond Eq. (3.3.8) can be added with regard to the rate of convergence of the new sequence.

PROBLEMS

1. Let $\sum_{n=1}^{\infty} a_n = s$. Define $x_n = \sum_{j=1}^{n} a_j$ and give a formal expression to x_n'.

2. Apply the result of problem 1 to the converging series $\sum_{k=0}^{\infty} 2^{-k} = 2$ and show that Aitken's algorithm is applicable and finite. Explain.

3. Apply the result of problem 1 to $\sum_{n=1}^{\infty}(1/n^3)$, and make sure the requirements needed for implementing ATKN are fulfilled. Evaluate x_n, x_n' for $1 \leq n \leq 5$.

4. Apply the result of problem 1 to $\sum_{n=1}^{\infty}(1/n^2)$, and calculate x_n, x_n', $1 \leq n \leq 5$. Any conclusions?

5. Define $x_n = (1/2^n)+(1/2^{2^n})$, $n \geq 0$. Show that whereas x_n converges linearly to 0, the sequence x_n' generated by the ATKN algorithm converges quadratically.

6. Generate $\{x_n\}$, $\{x_n'\}$ for $x = e^{-x^2}, 0 \leq x \leq 1$, $x_0 = 1$ and a tolerance $\epsilon = 10^{-8}$.

   (a) How many iterations are needed?

   (b) What is the rate of convergence of either method?

3.4 Steffensen's Modification (STM)

So far we have discussed two algorithms for solving $x = f(x)$, that is, the standard iterative method and the Aitken procedure for acceleration, either of which converges linearly to the solution. It was assumed that $f'(x) \neq 0$ over the interval $I = [a, b]$, and in particular that $f'(s) \neq 0$.

Let us now investigate the behavior of the sequence $\{x_n\}$, $x_n = f(x_{n-1})$ in the case $f'(s) = 0$. We start by proving the following.

Theorem 3.4.1.

Let $f(x) \in C^1[a, b]$, $s = f(s)$ for some $a < s < b$ and $f'(s) = 0$. Then there exists a neighborhood $I_\epsilon = [s - \epsilon, s + \epsilon]$ inside the interval $I = [a, b]$ for which the equation $x = f(x)$ possesses a unique solution, and $\{x_n\}$ defined by $x_n = f(x_{n-1})$ converges to s for all $x_0 \in I_\epsilon$.

Proof.

We must show the existence of a neighborhood, I_ϵ, around s and within I, for which the hypotheses of Theorem 3.2.1 are fulfilled. Because $f'(x)$ is continuous and $f'(s) = 0$, we can find ϵ_0 smaller than $\min\{s - a, b - s\}$ for which

$$|f'(x)| < 1/2, \quad |x - s| < \epsilon_0 \tag{3.4.1}$$

Obviously, $I_{\epsilon_0} = [s - \epsilon_0, s + \epsilon_0] \subset I$, and for all $x \in I_{\epsilon_0}$

$$|f(x) - s| = |f(x) - f(s)| = |x - s||f'(c_x)| < \frac{1}{2}|x - s|$$

that is, $f(x) \in I_{\epsilon_0}$ (c_x is an x-dependent point between x and s). This, together with Eq. (3.4.1), concludes the proof. □

Example 3.4.1.

Consider $x = f(x) = (1/2)[x + (2/x)]$. Clearly $f(x) \in C^1[1/2, 2]$ and, at the solution point $s = \sqrt{2}$,

$$f' = \frac{1}{2}\left(1 - \frac{2}{x^2}\right)|_{\sqrt{2}} = 0$$

Thus there exists a neighborhood, for example $I_\epsilon = [\sqrt{2} - (1/4), \sqrt{2} + (1/4)]$ around $\sqrt{2}$, at which for any given $x_0 \in I_\epsilon$ the sequence

$$x_{n+1} = \frac{1}{2}\left(x_n + \frac{2}{x_n}\right), \quad n \geq 0$$

converges to $\sqrt{2}$. □

We now turn to discuss the rate of convergence of $\{x_n\}$, $x_n = f(x_{n-1})$, $x_0 \in I_{\epsilon_0}$.

Theorem 3.4.2.

[Rate of convergence for $f'(s) = 0$]. *Let $f(x)$ satisfy:*

1. *The requirements of Theorem 3.4.1*

2. $f''(x) \in C[a, b]$ *(i.e., $f(x) \in C^2[a, b]$)*

3. $f''(s) \neq 0$

Then the sequence $\{x_n\}$ converges quadratically to s.

Proof.

The error at the $(n + 1)$th step is

$$
\begin{aligned}
\epsilon_{n+1} = x_{n+1} - s &= f(x_n) - f(s) \\
&= (x_n - s)f'(s) + \frac{(x_n - s)^2}{2!}f''(c_n) \\
&= \frac{\epsilon_n^2}{2!}f''(c_{n+1})
\end{aligned}
$$

where c_n is between x_n and s. Thus

$$
\lim_{n \to \infty} \frac{\epsilon_{n+1}}{\epsilon_n^2} = \frac{1}{2}f''(s) \neq 0 \qquad\qquad \Box
$$

By applying an iterative method that converges quadratically, the number of correct digits is actually doubled after each iteration! This is demonstrated in Table 3.4.1, which contains the first three iterations for $f(x) = (1/2)[x + (2/x)]$ and $x_0 = 1.2$ $(s = 1.414213562\ldots)$.

Table 3.4.1. Operating with the Standard Iteration Method for $f'(s) = 0$

n	x_n	$\lvert \epsilon_n \rvert$
0	1.200000000	$0.2 \cdot 10^{0}$
1	1.433333333	$0.2 \cdot 10^{-1}$
2	1.414341085	$0.1 \cdot 10^{-3}$
3	1.414213568	$0.6 \cdot 10^{-8}$

In general, $f'(s) \neq 0$ and the convergence is linear. But even if $f'(s) = 0$ there is no way to know it *a priori*. However, one may always *replace* the original problem formulated by Eq. (3.1.1) and, having a unique solution s, by a new *equivalent* (i.e., having the *same* unique solution s) equation

$$
x = F(x) \tag{3.4.2}
$$

where $F(x)$ is *a priori* known to satisfy

$$F'(s) = 0 \qquad (3.4.3)$$

Thus the SIM algorithm implemented on Eq. (3.4.2) generates a sequence

$$\{x_n | x_n = F(x_{n-1}), \quad n \geq 1\}$$

that converges to s quadratically.

One way of constructing $F(x)$ is to modify Aitken's method in the following manner: we start with an initial approximation x_0 and define it as the initial value of both the Aitken's sequence and its modified version. Let x_n'' be the nth iteration already generated by the modified Aitken's scheme. We then use the SIM to construct $x_n^{(1)} = f(x_n'')$, $x_n^{(2)} = f[x_n^{(1)}]$ and define the $(n+1)$th iteration as

$$x_{n+1}'' = x_n'' - \frac{[x_n^{(1)} - x_n'']^2}{x_n^{(2)} - 2x_n^{(1)} + x_n''}$$

In other words, once x_n'' is computed, it is used as a first approximation for the SIM to generate two additional standard iterations. The three values are then substituted in Aitken's iterative scheme to define the $(n+1)$th iteration of the modified algorithm, which could possibly speed up the convergence to s.

Thus

$$x_{n+1}'' = x_n'' - \frac{[f(x_n'') - x_n'']^2}{f[f(x_n'')] - 2f(x_n'') + x_n''}, \quad n \geq 0$$

Let us define

$$G(x) = f[f(x)] - 2f(x) + x \qquad (3.4.4)$$

and

$$F(x) = \begin{cases} x - \{[f(x) - x]^2\}/G(x), & G(x) \neq 0 \\ x, & G(x) = 0 \end{cases} \qquad (3.4.5)$$

Then the new algorithm for solving Eq. (3.1.1) can be presented as follows.

Algorithm 3.4.1.
[Steffensen's method (STM)].

Step 1. Choose a first approximation x_0, a tolerance ϵ, a maximum number of iterations N, and set $n = 0$.

Step 2. Compute $x_{n+1} = F(x_n)$ [using Eqs. (3.4.4) and (3.4.5)]. If $G(x_n) = 0$, go to Step 5.

Step 3. If $|x_{n+1} - x_n| \leq \epsilon$ set $s = x_{n+1}$, output $(n+1)$, s and stop.

Step 4. If $|x_{n+1} - x_n| > \epsilon$ and $n < N$, set $n \leftarrow n+1$ and go to Step 2. If $n \geq N$ go to Step 6.

Step 5. Output "next iteration cannot be completed," n, x_n and stop.

Step 6. Output "maximum number of iterations exceeded," x_{n+1} and stop.

A subroutine STEF, which incorporates the STM algorithm, is found on the attached floppy disk.

From a heuristic point of view, Steffensen's method is expected to accelerate Aitken's algorithm. Still, we could have another linearly converging method with a smaller coefficient K [Eq. (3.2.1)]. The next theorem ensures the *quadratic nature* of STM, which is therefore superior to both the SIM and ATKN algorithms.

Theorem 3.4.3.

(Local convergence for STM). *Let $f(x)$ satisfy the following hypotheses:*

1. $f(x) \in C^3[a, b]$
2. $s = f(s), \quad a < s < b$
3. $f'(s) \neq 1$

and let $G(x)$, and $F(x)$ be defined by Eqs. (3.4.4) and (3.4.5), respectively. Then $F(x)$ is twice continuously differentiable near s, $s = F(s)$, and $F'(s) = 0$, that is, the sequence $\{x_n\}$ defined by $x_{n+1} = F(x_n)$ converges quadratically to s, provided that x_0 is sufficiently close to s.

Proof.

Because $f(x)$ is continuous, we can find and confine our discussion to a neighborhood $I_\epsilon = [s - \epsilon, s + \epsilon]$, where $x \in I_\epsilon \Rightarrow a \leq f(x) \leq b$. At this neighborhood, both $G(x)$ and $F(x)$ are well defined and $F(s) = s$. To check the smoothness of $F(x)$, we first define an auxiliary function $g(h)$ by

$$g(h) = \begin{cases} [f(s+h) - s]/h , & 0 < |h| \leq \epsilon \\ f'(s) , & h = 0 \end{cases} \tag{3.4.6}$$

Because $f(x) \in C^3(I_\epsilon)$, $g(h)$ is at least twice continuously differentiable at I_ϵ. Clearly, this is true for $h \neq 0$. At $h = 0$ the continuity of g follows immediately from

$$\lim_{h \to 0} \frac{f(s+h) - s}{h} = f'(s)$$

To show the existence of $g'(0)$ we apply Taylor's theorem to rewrite

$$g(h) = \frac{f(s+h) - f(s)}{h} = \frac{hf'(s) + (h^2/2)f''(s) + (h^3/6)f'''(c_1)}{h}, \quad h \neq 0$$

where c_1 is between $s + h$ and s. Thus,

$$\frac{g(h) - g(0)}{h} = \frac{g(h) - f'(s)}{h} = \frac{f''(s)}{2} + \frac{h}{6}f'''(c_1)$$

and hence

$$g'(0) = \lim_{h \to 0} \frac{g(h) - g(0)}{h} = \frac{f''(s)}{2}$$

To show the continuity of $g'(h)$ at $h = 0$, we express $g'(h)$ as

$$g'(h) = \frac{hf'(s+h) - [f(s+h) - s]}{h^2}$$

and use Taylor's theorem to obtain

$$\begin{aligned}
g'(h) &= \frac{h[f'(s) + hf''(s) + (h^2/2)f'''(c_2)] - [f(s+h) - f(s)]}{h^2} \\
&= \frac{h[f'(s) + hf''(s) + (h^2/2)f'''(c_2)] - [hf'(s) + (h^2/2)f''(s) + (h^3/6)f'''(c_1)]}{h^2} \\
&= \frac{f''(s)}{2} + h\left[\frac{f'''(c_2)}{2} - \frac{f'''(c_1)}{6}\right] \to g'(0) \text{ as } h \to 0
\end{aligned}$$

The existence of $g''(0)$ is now easily established because

$$g''(0) = \lim_{h \to 0} \frac{g'(h) - g'(0)}{h} = \lim_{h \to 0}\left[\frac{f'''(c_2)}{2} - \frac{f'''(c_1)}{6}\right] = \frac{f'''(s)}{3}$$

[because $f(x) \in C^3(I_\epsilon)$, $f'''(c_1)$, $f'''(c_2) \to f'''(s)$ as $h \to 0$].
The continuity of $g''(h)$ at $h = 0$ is left as an exercise for the reader.
We now return to $F(x)$. The definition of $g(h)$ implies (for $h = 0$ as well)

$$\begin{aligned}
f(s+h) &= s + hg(h) \\
f[f(s+h)] &= f[s + hg(h)] = s + hg(h)g[hg(h)]
\end{aligned}$$

Therefore,

$$\begin{aligned}
G(s+h) &= s + hg(h)g[hg(h)] - 2[s + hg(h)] + (s+h) \\
&= h\{1 - 2g(h) + g(h)g[hg(h)]\} = hH(h)
\end{aligned}$$

Because $g(h) \in C^2(I_\epsilon)$, so too is $H(h)$. Furthermore, by using $f'(s) \neq 1$, we have

$$H(0) = 1 - 2g(0) + [g(0)]^2 = [g(0) - 1]^2 = [f'(s) - 1]^2 \neq 0$$

Because $H(h)$ is continuous, $H(h) \neq 0$ and therefore $G(s+h) \neq 0$ provided that $h \neq 0$ is sufficiently small. Now

$$\begin{aligned}
F(s+h) &= s + h - \frac{[f(s+h) - (s+h)]^2}{hH(h)} \\
&= s + h - \frac{h[g(h) - 1]^2}{H(h)}
\end{aligned}$$

and, because $g(h)$, $H(h) \in C^2(I_\epsilon)$, so too is $F(s+h)$. Finally,

$$F'(s) = \lim_{h \to 0} \frac{F(s+h) - F(s)}{h} = \lim_{h \to 0} \left\{ 1 - \frac{[g(h) - 1]^2}{H(h)} \right\}$$

$$= 1 - \frac{[g(0) - 1]^2}{H(0)} = 1 - 1 = 0$$

Thus $F(s) = s$, $F'(s) = 0$, and $F(x) \in C^2(I_\epsilon)$ for some neighborhood I_ϵ of s. By Theorem 3.4.2 this establishes the quadratic convergence (at least) of Steffensen's method.   □

Example 3.4.2.

Let $x = \cos x$, $x_0 = 1$. Table 3.4.2 contains the first three iterations calculated using STM.

Table 3.4.2. Steffensen's Modification for $x = \cos x$, $x_0 = 1$.

n	x_n	$x_n^{(1)}$	$x_n^{(2)}$
0	1.000000	0.540302	0.857553
1	0.728010	0.746500	0.734070
2	0.739067	0.739097	0.739077
3	0.739085		

□

Example 3.4.3.

Let $x = e^{-x}$, $x_0 = 1/2$. To compute s within a tolerance of $\epsilon = 10^{-6}$ one needs 21 iterations using the SIM, 8 iterations using ATKN, but only 3 iterations using STM. The real comparison between the algorithms is clearly via the number of function evaluations: 21 for the SIM, 8 for ATKN, and 6 for STM.   □

The disadvantage of STM is that x_0 must be considerably close to s. Because there is generally no way of establishing that, the Steffensen's algorithm in its original form is not attractive. Theorem 3.4.4 enables us to choose *any* x_0, provided that certain other hypotheses hold.

Theorem 3.4.4.

(Nonlocal convergence for STM). *Let $f(x)$ be defined over $I = (a, \infty)$ and satisfy:*

1. $f(x) \in C^2(I)$

2. $f(x) > a$, $x \in I$

3. $f'(x) < 0$, $x \in I$

4. $f''(x) > 0$, $x \in I$

Then STM converges to the unique solution s of $x = f(x)$, for any initial x_0.

A typical problem for which Theorem 3.4.4 may be implemented is shown in Fig. 3.4.1. The last theorem imposes several restrictions on $f(x)$, but they are easier to cope with than the necessity of knowing *a priori* how close x_0 must be to s.

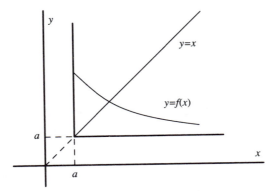

Figure 3.4.1.   A problem in which Steffensen's method converges.

Example 3.4.4.
Let $f(x) = e^{-x}$, $x > 0$. Then $f(x) > 0$, $f'(x) = -e^{-x} < 0$, $f''(x) = e^{-x} > 0$. Hence, STM converges to $s = e^{-s}$, independently of x_0. □

Example 3.4.5.
Let $f(x) = c/x$, $x > 0$, $c > 0$. Clearly, $f(x) > 0$, $f'(x) = -c/x^2 < 0$, $f''(x) = 2c/x^3 > 0$. Hence, the Steffensen's iteration method converges to the solution $s = \sqrt{c}$ for any choice of $x_0 > 0$.

Because $f(x_n) = c/x_n$, $f[f(x_n)] = x_n$ we have

$$x_{n+1} = x_n - \frac{[(c/x_n) - x_n]^2}{x_n - 2(c/x_n) + x_n} = \frac{1}{2}\left(x_n + \frac{c}{x_n}\right)$$

Table 3.4.3 contains the first four iterations for $c = 5$, $x_0 = 1$. The exact value is 2.2360679...

Table 3.4.3. Steffensen's Method for $x = c/x$

n	x_n	$|\epsilon_n|$
0	1.000000	$0.1 \cdot 10^{1}$
1	3.000000	$0.8 \cdot 10^{0}$
2	2.333333	$0.1 \cdot 10^{0}$
3	2.238095	$0.2 \cdot 10^{-2}$
4	2.236069	$0.1 \cdot 10^{-5}$

and the quadratic nature of the convergence is well demonstrated.   □

Example 3.4.6.
Let $f(x) = A + (B/x^n)$, $A > 0$, $B > 0$. Then, for $x > 0$, $f(x)$ is well defined and satisfies $f(x) > 0$. Furthermore,

$$f'(x) = -\frac{nB}{x^{n+1}} < 0, \quad x > 0$$

$$f'' = \frac{(n+1)nB}{x^{n+2}} > 0, \quad x > 0$$

Thus STM converges for any given $x_0 > 0$. If $A = B = 1$, we have

$$x = 1 + (1/x^2), \quad x > 0$$

that is, the polynomial equation $x^3 - x^2 - 1 = 0$ has a unique positive solution s, to which STM converges quadratically for any choice of $x_0 > 0$. ☐

PROBLEMS

1. Let $x = \cos x$. Start with $x_0 = 0.7$ and find the number of function evaluations needed to obtain the exact solution, within tolerance $\epsilon = 10^{-8}$, using the SIM, ATKN, and STM algorithms.

2. Find the exact rate of convergence of the STM algorithm [i.e., $(1/2)F''(s)$].

3. Follow Example 3.4.6 and prove that the polynomial equation

$$x^{n+1} - x^n - 1 = 0$$

has a unique solution. What would be a reasonable starting value x_0, for large n?

In the next section we introduce various iterative algorithms related to *Newton's method*, whose rate of convergence is generally quadratic.

3.5   Newton's Method: Advantages and Limitations

Let us consider the general problem of solving the equation

$$F(x) = 0 \qquad\qquad (3.5.1)$$

where $F(x)$ is a single real-valued function defined over an interval $[a, b]$ and twice continuously differentiable. Let s be a solution of Eq. (3.5.1), that is, $F(s) = 0$. If x_0 is within $[a, b]$, and $s = x_0 + h$, then by using Taylor's theorem we obtain

$$0 = F(x_0 + h) = F(x_0) + hF'(x_0) + \frac{h^2}{2}F''(c) \qquad\qquad (3.5.2)$$

where c is between x_0 and $x_0 + h$. If h is sufficiently small, we may neglect the last term of the right-hand side of Eq. (3.5.2) and obtain $F(x_0) + hF'(x_0) \approx 0$, or

$$h \approx -\frac{F(x_0)}{F'(x_0)} \qquad\qquad (3.5.3)$$

Thus, if x_0 is sufficiently close to s, one may expect the value

$$x_1 = x_0 - \frac{F(x_0)}{F'(x_0)}$$

to be a far better approximation to s than x_0. Based on this idea we can immediately formulate an iterative algorithm for solving Eq. (3.5.1) by writing

$$x_{n+1} = x_n - \frac{F(x_n)}{F'(x_n)} \tag{3.5.4}$$

The last equation is the Newton–Raphson method or simply the Newton's method for solving $F(x) = 0$.

Algorithm 3.5.1.

[Newton–Raphson method for solving $F(x) = 0$].

Step 1. Choose a first approximation x_0, maximum number of iterations N, a tolerance ϵ, and set $n = 0$.

Step 2. If $F'(x_n) = 0$, output "derivative vanishes," n, x_n and stop.

Step 3. Compute x_{n+1} from Eq. (3.5.4).

Step 4. If $|x_{n+1} - x_n| \leq \epsilon$, set $s = x_{n+1}$, output "iteration completed," n, s and stop.

Step 5. If $|x_{n+1} - x_n| > \epsilon$ and $n < N$, set $n \leftarrow n + 1$ and go to step 2. Otherwise, output "maximum number of iterations exceeded," x_{n+1} and stop.

The geometric interpretation of the Newton–Raphson method (NRM) is straightforward and is shown in Fig. 3.5.1. The $(n + 1)$th iteration is simply the intersection of the tangent at $P[x_n, F(x_n)]$ with the x axis, because

$$F'(x_n) = \tan \alpha = \frac{F(x_n)}{x_n - x_{n+1}}$$

A subroutine NEWTON, which incorporates the NRM, is found on the attached floppy disk.

Example 3.5.1.

Consider $F(x) = x - e^{-x} = 0$, $x_0 = 0$. The derivative of $F(x)$ is $1 + e^{-x}$, and the sequence constructed by NRM is defined by

$$x_{n+1} = x_n - \frac{x_n - e^{-x_n}}{1 + e^{-x_n}} = \frac{(1 + x_n)e^{-x_n}}{1 + e^{-x_n}}, \quad n \geq 0$$

Table 3.5.1 consists of the first three iterations.

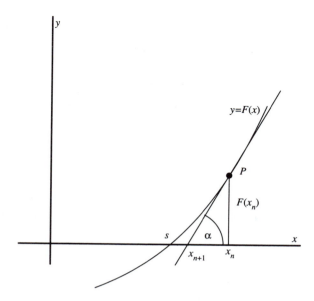

Figure 3.5.1. Newton–Raphson method.

Table 3.5.1. Newton–Raphson Method Applied to
$$x - e^{-x} = 0, \quad x_0 = 0$$

n	x_n	$|F(x_n)|$
0	0.0000000	$0.1 \cdot 10^1$
1	0.5000000	$0.1 \cdot 10^0$
2	0.5663110	$0.1 \cdot 10^{-2}$
3	0.5671432	$0.2 \cdot 10^{-6}$

The convergence is extremely fast and three iterations are sufficient to reach beyond a tolerance of $\epsilon = 10^{-6}$.   □

Example 3.5.2.

The problem of computing a square root of a positive number a can be formulated as finding a positive solution to

$$F(x) = x^2 - a = 0$$

Implementing the NRM algorithm in this case yields the relation

$$x_{n+1} = x_n - \frac{x_n^2 - a}{2x_n} = \frac{x_n^2 + a}{2x_n} = \frac{1}{2}\left(x_n + \frac{a}{x_n}\right)$$

Table 3.5.2 contains the first three iterations of the NRM applied for $a = 9, \quad x_0 = 2$.

Table 3.5.2. Newton–Raphson Method Applied to
$$x^2 - 9 = 0, \quad x_0 = 2$$

n	x_n	$|F(x_n)|$
0	2.0000000	$0.5 \cdot 10^{1}$
1	3.2500000	$0.2 \cdot 10^{1}$
2	3.0096154	$0.6 \cdot 10^{-1}$
3	3.0000154	$0.9 \cdot 10^{-4}$

□

Once more, the Newton–Raphson method provides fast convergence, far superior to the linear convergence of the standard iterative method and Aitken's procedure. However, the iterative scheme derived for the last example coincides with that of Steffensen's method (Example 3.4.5). Because Steffensen's scheme converges quadratically, so does Newton's method in the *particular case* $F(x) = x^2 - a = 0$. The next theorem confirms that in general, *whenever* the NRM converges, the convergence is quadratic.

Theorem 3.5.1.
(Convergence rate of Newton's method). *Let $F(x)$ be a real-valued function defined over the interval $[a, b]$ and satisfying the following requirements:*

1. $F(x) \in C^2[a, b]$
2. $F(s) = 0$ *for some s : $a < s < b$*
3. $F'(s) \neq 0$

Then there exists a neighborhood I_ϵ of s: $I_\epsilon = [s - \epsilon, \ s + \epsilon]$, such that for any given $x_0 \in I_\epsilon$, NRM converges at least quadratically to s.

Proof.
Let I_{ϵ_1} be a neighborhood of s at which $F'(x) \neq 0$. Such a neighborhood can be found because $F'(x)$ is continuous and $F'(s) \neq 0$ (hypotheses 1 and 3). Define

$$f(x) = x - \frac{F(x)}{F'(x)}, \quad s - \epsilon_1 \leq x \leq s + \epsilon_1$$

Then

$$f'(x) = 1 - \frac{F'^2(x) - F''(x)F(x)}{F'^2(x)} = \frac{F''(x)F(x)}{F'^2(x)}$$

and because $F(s) = 0$ we have $f'(s) = 0$. Thus, by implementing Theorem 3.4.1, we should be able to determine a neighborhood $I_{\epsilon_2} \subset I_{\epsilon_1}$ of s such that

$$x_{n+1} = f(x_n) \rightarrow s, \quad n \rightarrow \infty$$

for any given $x_0 \in I_{\epsilon_2}$. By choosing $\epsilon = \epsilon_2$ we have

$$x_{n+1} = x_n - \frac{F(x_n)}{F'(x_n)} \to s, \quad n \to \infty$$

for any given $x_0 \in I_\epsilon$.

Now, the quadratic nature of the convergence could be derived by showing the existence and continuity of $f''(x)$. Unfortunately, this requires $F(x) \in C^3[a, b]$ rather than $F(x) \in C^2[a, b]$. To bypass this obstacle, denote $\epsilon_n = x_n - s$ and replace x_0 by x_n in Eq. (3.5.2) to obtain

$$0 = F(x_n - \epsilon_n) = F(x_n) - \epsilon_n F'(x_n) + \frac{\epsilon_n^2}{2} F''(c_n)$$

where c_n is between x_n and s. Hence,

$$\frac{F(x_n)}{F'(x_n)} - \epsilon_n + \frac{\epsilon_n^2}{2} \frac{F''(c_n)}{F'(x_n)} = \frac{F(x_n)}{F'(x_n)} - x_n + s + \frac{\epsilon_n^2}{2} \frac{F''(c_n)}{F'(x_n)}$$

$$= -x_{n+1} + s + \frac{\epsilon_n^2}{2} \frac{F''(c_n)}{F'(x_n)} = -\epsilon_{n+1} + \frac{\epsilon_n^2}{2} \frac{F''(c_n)}{F'(x_n)} = 0$$

Because $\lim_{n\to\infty} x_n = s$, c_n must converge to s and thus

$$\lim_{n\to\infty} \frac{\epsilon_{n+1}}{\epsilon_n^2} = \frac{1}{2} \frac{F''(s)}{F'(s)} \tag{3.5.5}$$

Thus the convergence of the NRM is at least quadratic, depending on whether or not $F''(s) \neq 0$. $\square$

The main handicap of Theorem 3.5.1 is that it is a local convergence theorem: if x_0 is sufficiently close to s, then the NRM scheme converges at least quadratically to s. How close? There is no general, practical way to predict this unless one has *a priori* information regarding the location of s. Another general disadvantage of Newton's method is having $F'(x_n)$ in the denominator. If $F'(x_n) = 0$, the next iteration, that is, x_{n+1}, is not defined and the process terminates at least temporarily (Fig. 3.5.2). Moreover, if $F'(x_n)$ does not vanish but is small $[|F'(x_n)| << 1]$, the next iteration could occur *too far* from s, in which case the NRM may converge to the wrong solution (Fig. 3.5.3) s_1, rather than to s. In some cases the NRM may not converge at all. For example, let $F(x) = \cos x$, $s = \pi/2$. If we choose x_0 to be the solution s_0 of $2x + \cot x = \pi$, that is, $x_0 = 0.4052\ldots$, then

$$x_0 = x_2 = x_4 = \cdots, \quad x_1 = x_3 = x_5 = \cdots = \pi - x_0 \text{ (why?)}$$

and the NRM oscillates (Fig. 3.5.4). If $x_0 \neq s_0$ but $x_0 \approx s_0$, we still need an enormous number of iterations and may of course converge to the wrong solution.

The question of nonlocal convergence by the NRM is partially answered by the next result.

Theorem 3.5.2.

(Nonlocal convergence by Newton's method). *Let $F(x)$ be a real-valued function defined over the interval $[a, b]$ and satisfying the following requirements:*

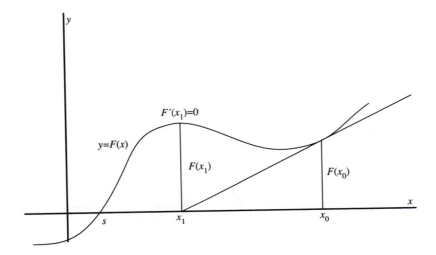

Figure 3.5.2. Newton's method, $F'(x) = 0$.

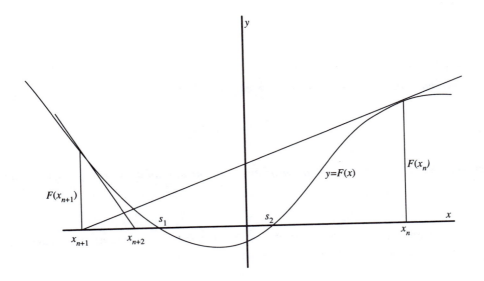

Figure 3.5.3. Newton–Raphson method: converging to the wrong solution.

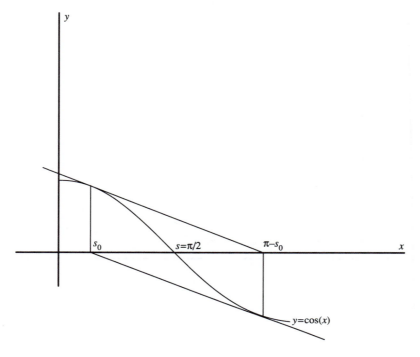

Figure 3.5.4. Newton–Raphson method: oscillation.

1. $F(x) \in C^2[a, b]$
2. $F(a)F(b) < 0$
3. $F'(x) \neq 0, \ x \in [a, b]$
4. $F''(x)$ *does not change sign within* $[a, b]$
5. *If c is the end point (either a or b) with the smaller value of* $|F'(x)|$, *then*

$$\left| \frac{F(c)}{F'(c)} \right| \leq b - a$$

Then, the NRM converges at least quadratically to the unique solution s of $F(x) = 0$ *for any first approximation* $x_0 \in [a, b]$.

Proof.
Hypothesis 2 ensures the existence of a solution s, because $F(x)$ is continuous and changes signs at the end points of the interval. If two different solutions s_1, s_2 were possible, then

$$F(s_1) - F(s_2) = (s_1 - s_2)F'(c)$$

where c is an intermediate point between s_1 and s_2. Thus $F'(c) = 0$ is in contradiction with hypothesis 3, and s must therefore be unique.

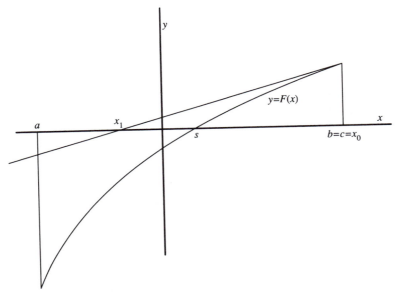

Figure 3.5.5. "Worst" first approximation.

Hypothesis 4 guarantees that the curve of the function is either concave ($F'' \geq 0$) or convex ($F'' \leq 0$). The last requirement ensures that the "worst" choice of x_0, that is, $x_0 = c$ would still provide an x_1 inside $[a, b]$ (Fig. 3.5.5).

We must consider the following four possibilities:

Case 1. $F(a) < 0, \quad F(b) > 0, \quad F''(x) \geq 0 \Rightarrow c = a$
Case 2. $F(a) < 0, \quad F(b) > 0, \quad F''(x) \leq 0 \Rightarrow c = b$
Case 3. $F(a) > 0, \quad F(b) < 0, \quad F''(x) \leq 0 \Rightarrow c = a$
Case 4. $F(a) > 0, \quad F(b) < 0, \quad F''(x) \geq 0 \Rightarrow c = b$

Cases 3 and 4 are reduced to cases 1 and 2, respectively, simply by replacing $F(x)$ by $-F(x)$. Let us consider case 2. If we define $G(x) \in C^2[-b, -a]$ as

$$G(x) = -F(-x), \quad -b \leq x \leq -a$$

we have $G(-a) = -F(a) > 0$, $G(-b) = -F(b) < 0$, $G(-s) = -F(s) = 0$, $G'(x) = F'(-x) \neq 0$, $G''(x) = -F''(-x) \geq 0$. Furthermore, hypothesis 5 is fulfilled with $c = -b$. Thus, case 2 for F is reduced to case 1 for G and $-x_n \to -s$ implies $x_n \to s$ with identical rate of convergence. It therefore suffices to consider and prove the theorem for case 1, shown in Fig. 3.5.6.

We start by assuming $s \leq x_0 \leq b$ and show that the NRM sequence, $\{x_n\}$, is monotonically decreasing to the right of s, that is,

$$s \leq x_n, \quad x_n \leq x_{n-1}; \quad n = 0, 1, 2, \dots$$

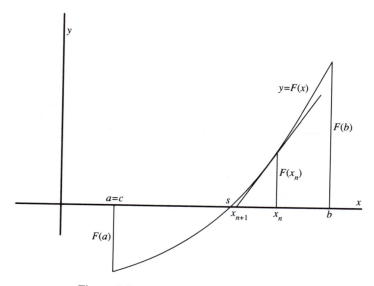

Figure 3.5.6.   Newton's method for Case 1.

This is done by induction:

1. $s \leq x_0$ by choice.

2. Let $s \leq x_n$ for some n. Then $F(x_n) \geq 0$, and because $F'(x_n) > 0$ $[F'(x) > 0$ for case 1] we have

$$x_{n+1} = x_n - \frac{F(x_n)}{F'(x_n)} \leq x_n$$

To show $s \leq x_{n+1}$, we use the mean-value theorem and write

$$F(x_n) = F(x_n) - F(s) = (x_n - s)F'(\xi), \quad s \leq \xi \leq x_n$$

The condition $F''(x) \geq 0$ implies that $F'(x)$ is a monotonic increasing function. In particular, $F'(\xi) \leq F'(x_n)$. Because $x_n - s \geq 0$ we obtain

$$F(x_n) = (x_n - s)F'(\xi) \leq (x_n - s)F'(x_n)$$

which implies

$$F(x_n)/F'(x_n) \leq x_n - s$$

or

$$s \leq x_n - \frac{F(x_n)}{F'(x_n)} = x_{n+1}$$

and the induction is completed.

If now $a \leq x_0 \leq s$, we again use $F(s) = 0$ and the mean-value theorem to find

$$x_1 - s = x_0 - \frac{F(x_0)}{F'(x_0)} - s = \frac{(x_0 - s)F'(x_0) - [F(x_0) - F(s)]}{F'(x_0)}$$

$$= \frac{(x_0 - s)[F'(x_0) - F'(\xi_1)]}{F'(x_0)}$$

where $x_0 \leq \xi_1 \leq s$. Because $x_0 \leq s$, $F'(x_0) \leq F'(\xi_1)$, $0 < F'(x_0)$ we have $s \leq x_1$. To show $x_1 \leq b$ we define

$$G(x) = x - \frac{F(x)}{F'(x)}$$

which implies

$$G'(x) = F(x)F''(x)/F'^2(x)$$

Thus $G'(x) \leq 0$ for $a \leq x \leq s$, indicating that $G(x)$ is monotonically decreasing there. In particular, $G(a) \geq G(x_0)$, that is,

$$a - \frac{F(a)}{F'(a)} \geq x_0 - \frac{F(x_0)}{F'(x_0)} = x_1$$

By using hypothesis 5, that is, $-F(a)/F'(a) \leq b - a$, we find that $x_1 \leq b$. Therefore, due to the previous discussion, $s \leq x_n \leq b$, $x_{n+1} \leq x_n$ for all $n \geq 1$.

We thus find that, for any starting value x_0, each x_n is within $[a, b]$ and the whole sequence (excluding x_0, perhaps) is monotonic, decreasing to the right of s. Using basic calculus we have

$$\lim_{n \to \infty} x_n = s_1$$

for some s_1: $s \leq s_1 \leq b$. Because $x_{n+1} = x_n - [F(x_n)/F'(x_n)]$, then by letting $n \to \infty$ we have $F(s_1) = 0$, implying $s_1 = s$. Finally, because $F(x) \in C^2[a, b]$ the convergence is at least quadratic, as was previously asserted and proved [Eq. (3.5.5)].   □

Example 3.5.3.

Let $F(x) = x^2 - k^2$, $k > 0$. Choose $I = [\epsilon, b]$ where $b > k$ and $0 < \epsilon < k$. Clearly, $F(\epsilon) < 0$, $F(b) > 0$, $F'(x) = 2x \neq 0$, $F''(x) = 2 \geq 0$, $F(x) \in C^2[\epsilon, b]$. Thus, hypotheses 1 through 4 of Theorem 3.5.2 are satisfied. Condition 5 holds as well, provided that

$$\left|\frac{F(\epsilon)}{F'(\epsilon)}\right| = \frac{k^2 - \epsilon^2}{2\epsilon} \le b - \epsilon$$

or $(k^2 + \epsilon^2)/2\epsilon \le b$. Thus, for such b, the NRM sequence converges for any given $x_0 \in [\epsilon, b]$. For example, let $x^2 - 5 = 0$, $\epsilon = 1/100$, $b = 300$. The NRM sequence defined by

$$x_{n+1} = \frac{1}{2}\left(x_n + \frac{5}{x_n}\right), \quad x_0 \in \left[\frac{1}{100}, 300\right]$$

converges quadratically to $\sqrt{5} \approx 2.236$.   □

Example 3.5.4.

Consider the function $F(x) = x^6 + x - 1$, $0 \le x \le 1$. Clearly, $F(x) \in C^2(-\infty, \infty)$, $F(0) = -1 < 0$, $F(1) = 1$, $F'(x) = 6x^5 + 1 > 0$, $F''(x) = 30x^4 \ge 0$. Also,

$$\left|\frac{F(0)}{F'(0)}\right| = 1 \le 1 - 0 = 1$$

Thus, for any given $x_0 : \ 0 \le x_0 \le 1$, the sequence

$$x_{n+1} = x_n - \frac{x_n^6 + x_n - 1}{6x_n^5 + 1} = \frac{5x_n^6 + 1}{6x_n^5 + 1}, \quad n \ge 0$$

converges to the unique solution s of $F(x) = 0$. Table 3.5.3 contains the first seven iterations of the NRM needed to reach a tolerance of $\epsilon = 10^{-8}$, given $x_0 = 0$.

Table 3.5.3. Newton–Raphson Method Applied to
$x^6 + x - 1 = 0$, $x_0 = 0$, $\epsilon = 10^{-8}$

n	x_n	$\|x_n - x_{n-1}\|$
0	0.000000000	
1	1.000000000	$0.1 \cdot 10^1$
2	0.857142857	$0.1 \cdot 10^0$
3	0.789951850	$0.7 \cdot 10^{-1}$
4	0.778372711	$0.1 \cdot 10^{-1}$
5	0.778089761	$0.3 \cdot 10^{-3}$
6	0.778089599	$0.2 \cdot 10^{-6}$
7	0.778089599	

□

It should be noted that whereas Newton's method converges quadratically near the solution s, it may converge more slowly, even linearly, as long as x_n is not sufficiently close to s. For example, let $F(x) = x^2 - 2 = 0$, $x_0 = 10^6$. Because

$$x_{n+1} = \frac{1}{2}\left(x_n + \frac{2}{x_n}\right) \approx \frac{x_n}{2}$$

we have

$$\epsilon_{n+1} \approx \epsilon_n/2$$

and one must perform about 20 iterations, just to get close to $s = \sqrt{2}$. Thus the convergence starts linearly and turns quadratic only at its final stage.

Let $\{x_n\}$ be a sequence generated by Newton's method and converging to s. Given a tolerance ϵ, one would like to terminate the iterative process when the inequality

$$|x_n - s| < \epsilon$$

first occurs. By using the mean-value theorem we have

$$F(x_n) = F(x_n) - F(s) = (x_n - s)F'(c_n)$$

where c_n is between x_n and s. Thus,

$$|x_n - s| = \left|\frac{F(x_n)}{F'(c_n)}\right| \approx \left|\frac{F(x_n)}{F'(x_n)}\right| = |x_{n+1} - x_n| \tag{3.5.6}$$

provided that x_n is sufficiently close to s.

Example 3.5.5.
Let $F(x) = x^3 - 27 = 0$, $x_0 = 2$. Here $s = 3$, $F'(x) = 3x^2$ and the sequence generated by the NRM is given by

$$x_{n+1} = x_n - \frac{x_n^3 - 27}{3x_n^2} = \frac{2x_n^3 + 27}{3x_n^2}, \quad n \geq 0$$

Table 3.5.4 contains the first four iterations and their associated errors compared with the corresponding $|x_{n+1} - x_n|$, $1 \leq n \leq 5$. It demonstrates the asymptotic relation

Table 3.5.4. Replacing $|x_n - s|$ by $|x_{n+1} - x_n|$ as Error Estimate:
$F(x) = x^3 - 27 = 0$, $x_0 = 2$

n	x_n	$|x_n - s|$	$|x_{n+1} - x_n|$
0	2.00000000	$0.100 \cdot 10^1$	$0.158 \cdot 10^1$
1	3.58333333	$0.583 \cdot 10^0$	$0.494 \cdot 10^0$
2	3.08980830	$0.898 \cdot 10^{-1}$	$0.872 \cdot 10^{-1}$
3	3.00258508	$0.259 \cdot 10^{-2}$	$0.258 \cdot 10^{-2}$
4	3.00000222	$0.222 \cdot 10^{-5}$	

$$\lim_{n \to \infty} \left| \frac{x_{n+1} - x_n}{x_n - s} \right| = 1 \qquad \square$$

As was previously stated the hypothesis $F'(x) \neq 0$, or at least $F'(s) \neq 0$, is a necessary condition for a successful implementation of Newton's method as a quadratically converging scheme for solving Eq. (3.5.1). We shall now discuss the singular case $F'(s) = 0$.

Let us discuss the more general cases:

1. $F(s) = F'(s) = \cdots = F^{(m-1)}(s) = 0, \ F^{(m)}(s) \neq 0$

2. $F'(x) \neq 0, \ x \neq s, \ x$ near s

By defining

$$f(x) = \begin{cases} x - [F(x)/F'(x)], & x \neq s \\ \\ s & , x = s \end{cases}$$

and expanding $F(x), F'(x)$ about $x = s$ using Taylor's theorem, we have

$$f(s + h) = s + h - \frac{F(s + h)}{F'(s + h)}$$

$$= s + h - \frac{(h^m/m!)F^{(m)}(s) + [h^{m+1}/(m+1)!]F^{(m+1)}(c_1)}{[h^{m-1}/(m-1)!]F^{(m)}(s) + (h^m/m!)F^{(m+1)}(c_2)} = s + h - \frac{1}{m}h + O(h^2)$$

where c_1, c_2 are interim points between s and $s + h$. Thus,

$$f(s + h) = f(s) + h\left(1 - \frac{1}{m}\right) + O(h^2)$$

which leads to

$$f'(s) = \lim_{h \to 0} \frac{f(s + h) - f(s)}{h} = 1 - \frac{1}{m}$$

Because $m \neq 1$, the sequence $\{x_n\}$ generated by the NRM: $x_n = f(x_{n-1})$, converges *linearly* to s. However, instead of using $f(x)$, one can define

$$g(x) = \begin{cases} x - [mF(x)/F'(x)] & , x \neq s \\ s & , x = s \end{cases}$$

Now,

$$g(s + h) = s + h - m\left[\frac{1}{m}h + O(h^2)\right] = s + O(h^2)$$
$$= g(s) + O(h^2)$$

and $g'(s) = 0$ (why?). Therefore the sequence x_n, defined as

$$x_n = g(x_{n-1}) = x_n - \frac{mF(x_n)}{F'(x_n)}, \quad n \geq 0$$

converges quadratically to s.

Example 3.5.6.

Consider the equation

$$F(x) = (x-2)^2(x^2+1) = x^4 - 4x^3 + 5x^2 - 4x + 4 = 0$$

Here, $F(2) = F'(2) = 0$, $F''(2) = 10 \neq 0$, $m = 2$. By applying the standard NRM we obtain a sequence $\{x_n\}$ that converges linearly. If, instead, we generate the modified sequence $\{y_n\}$, defined by

$$y_{n+1} = y_n - \frac{2F(y_n)}{F'(y_n)} = y_n - \frac{2(y_n^4 - 4y_n^3 + 5y_n^2 - 4y_n + 4)}{4y_n^3 - 12y_n^2 + 10y_n - 4}$$

$$= \frac{2y_n^4 - 4y_n^3 + 4y_n - 8}{4y_n^3 - 12y_n^2 + 10y_n - 4}$$

we see quadratic convergence to $s = 2$. This is shown in Table 3.5.5 for $x_0 = 3$, $\epsilon = 10^{-8}$.

Table 3.5.5. Modified Newton's Iteration for the Singular Case $F'(s) = 0$

n	x_n	$\|x_n - s\|$	y_n	$\|y_n - s\|$
0	3.00000000	$0.1 \cdot 10^1$	3.00000000	$0.1 \cdot 10^1$
1	2.61538462	$0.6 \cdot 10^0$	2.23076923	$0.2 \cdot 10^0$
2	2.36009826	$0.4 \cdot 10^0$	2.01830166	$0.2 \cdot 10^{-1}$
3	2.20067168	$0.2 \cdot 10^0$	2.00013228	$0.1 \cdot 10^{-3}$

□

In the next section we will discuss various modifications to Newton's method and present other algorithms for solving $F(x) = 0$.

PROBLEMS

1. Let $F(x) = x^k - a = 0$, $0 < a$, $0 < k$. Present Newton's method for solving $F(x) = 0$ and use the scheme for $k = 2.5, 4$; $a = 3$, $x_0 = 1$, $\epsilon = 10^{-6}$.

2. Which of the following cases satisfy all the hypotheses of Theorem 3.5.2?

   (a) $F(x) = x^3 - 7$, $-1 \leq x \leq 2$

   (b) $F(x) = x - e^{-x}$, $1/2 \leq x \leq 2/3$

   (c) $F(x) = x^2 \ln x$, $2/3 \leq x \leq 2$

   (d) $F(x) = (x-1)/\sqrt{x^2+1}$, $-2 \leq x \leq 2$

(e) $F(x) = x^3 - x^2 - x - 1 = 0$, $1.1 \leq x \leq 2$

3. Use the NRM to generate a quadratically converging sequence to the unique solution s of $e^x - 2 - x^2 = 0$, using a starting value $x_0 = 1$. Find a neighborhood around s at which the requirements of Theorem 3.5.2 are fully satisfied.

4. Let $F(x)$ fulfill all the requirements of Theorem 3.5.2 and, in addition,

$$\text{(a) } |F''(x)| \leq M_2, \; a \leq x \leq b \quad \text{(b)} 0 < m_1 \leq |F'(x)|, \; a \leq x \leq b$$

   Denote $\epsilon_n = x_n - s$, $n \geq 0$ where x_n is the sequence generated by the NRM, using a starting value x_0. Find an upper bound for $|\epsilon_n|$.

5. Apply the result of the previous problem to case (b) of problem 2.

6. Let $F(x) = x^3 - 2x^2 = 0$. Choose $x_0 = -1$ and show that Newton's method converges linearly to $s = 0$. Adjust the algorithm to obtain quadratic convergence.

7. The problem of solving the equation $F(x) = 0$ is equivalent to solving $x = x + G(x)F(x)$, provided that $G(x) \neq 0$. Use this concept to derive the Newton–Raphson method.

8. Find, using Newton's method, the roots of $x - \tan x = 0$ near $x_0 = 4.5$, $x_0 = 100$. Discuss and compare the two converging sequences.

9. Compare the NRM for solving $x^2 - a = 0$ with Steffensen's method for solving $x = a/x$. Explain!

10. Show that the equation $F(x) = x + e^{-Kx^2} \cos x = 0$, $K > 0$, has a unique solution s. Choose $x_0 = 0$ and find s for $K = 1, 5, 10, 25,$ and 50. Discuss and explain the behavior of the iteration.

3.6 Related Schemes and Modifications to Newton's Method

We still consider the equation $F(x) = 0$.

3.6.1 The Bisection Method

Let $F(x)$ be a real-valued function defined over $I = [a, b]$ and satisfying the requirements

1. $F(x) \in C[a, b]$

2. $F(a)F(b) < 0$

Then $F(x)$ must have a solution s: $a < s < b$. If we halve the interval $[a, b]$ at $c = (a + b)/2$, then either $F(c) = 0$ or $F(x)$ changes sign on one of the subintervals $[a, c]$, $[c, b]$. Thus, if $F(c) \neq 0$, we have reduced the original interval of search to half its size. This procedure yields the following *bisection method* for solving $F(x) = 0$, given an error tolerance $\epsilon > 0$.

Algorithm 3.6.1.
[Bisection method (BISM)].

Step 1. Denote $c = (a + b)/2$.
Step 2. If $F(c) = 0$ set $s = c$ and stop.
Step 3. If $b - c \leq \epsilon$ set $s = c$ and stop.

Step 4. If $F(a)F(c) < 0$ set $b = c$; otherwise set $a = c$. Go to Step 1.

A subroutine BISECT that incorporates Algorithm 3.6.1 to solve $F(x) = 0$ is found on the attached floppy disk.

Let $[a_n, b_n]$ be the nth subinterval ($a_0 = a$, $b_0 = b$) and define $c_n = (a_n + b_n)/2$. Unless $F(c_n) = 0$ for some finite n, the sequence $\{c_n\}$ must converge to s for which $F(s) = 0$. Indeed, we have

$$b_n - a_n = \frac{1}{2}(b_{n-1} - a_{n-1}), \quad n \geq 1$$

which by induction implies

$$b_n - a_n = \frac{1}{2^n}(b - a), \quad n \geq 0$$

Because $a_n \leq s \leq b_n$ for all n, we can estimate the error of BISM by

$$|c_n - s| \leq \frac{1}{2}(b_n - a_n) = \frac{1}{2^{n+1}}(b - a)$$

Thus $\{c_n\}$ converges to s at least linearly, with a convergence factor $1/2$. For a given error tolerance $\epsilon > 0$, an upper bound for the number of iterations needed for convergence can be concluded by solving

$$\frac{1}{2^{n+1}}(b - a) < \epsilon$$

that is,

$$n \geq \frac{\ln[(b - a)/2\epsilon]}{\ln 2}$$

Example 3.6.1.

Let $F(x) = x^3 - x$, $-3 \leq x \leq 2$. The values of the function at the end points are $F(-3) = -24 < 0$ and $F(2) = 6 > 0$. Because $F(2)F(-3) < 0$ the bisection method can be applied for solving $F(x) = 0$, and its first six iterations are summarized in Table 3.6.1.

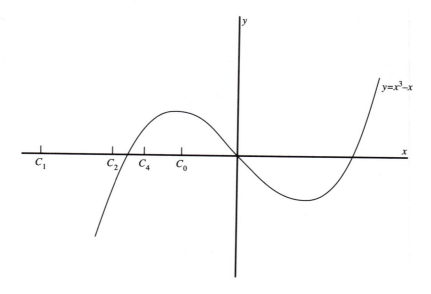

Figure 3.6.1. Bisection method for $x^3 - x = 0$.

Table 3.6.1. Bisection Method Applied to $x^3 - x = 0$

n	a_n	b_n	c_n	$F(c_n)$	$|c_n - s|$
0	−3.0000	2.0000	−0.5000	0.3750	0.5000
1	−3.0000	−0.5000	−1.7500	−3.6094	0.7500
2	−1.7500	−0.5000	−1.1250	−0.2988	0.1250
3	−1.1250	−0.5000	−0.8125	0.2761	0.1875
4	−1.1250	−0.8125	−0.9688	0.0596	0.0313
5	−1.1250	−0.9688	−1.0469	−0.1004	0.0469
6	−1.0469	−0.9688	−1.0078	−0.0158	0.0078

Originally, $F(x)$ has three zeroes: $s_1 = -1$, $s_2 = 0$, $s_3 = 1$ (Fig. 3.6.1). However, the sequence $\{c_n\}$ converges just to one of them, and we have no way of predicting *a priori* to which one (in this case, $s = s_1 = -1$). □

The one advantage of BISM is the guaranteed convergence. Otherwise, it is a slow-converging scheme and rarely used.

The next method for solving $F(x) = 0$ is generally faster than the bisection method and may be interpreted as a modification to Newton's method.

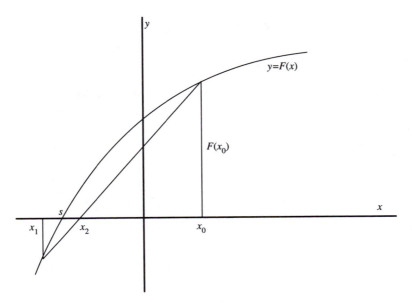

Figure 3.6.2.   The secant method.

3.6.2   The Secant Method

The *secant* method, often referred to as the *regula falsi* method, can be described as follows. Let x_0, x_1 be two initial guesses to the solution s of $F(x) = 0$. The next iteration x_2 is defined as the intersection point of the x axis with the secant that connects $[x_0, F(x_0)]$, $[x_1, F(x_1)]$. This is shown in Fig. 3.6.2.

The secant equation is

$$y - F(x_1) = \frac{F(x_1) - F(x_0)}{x_1 - x_0}(x - x_1)$$

and by substituting $y = 0$ we obtain

$$x_2 = x_1 - \frac{(x_1 - x_0)F(x_1)}{F(x_1) - F(x_0)} = G(x_0, x_1) \qquad (3.6.1)$$

This equation is the basis in the next algorithm for solving $F(x) = 0$, given a tolerance $\epsilon > 0$.

Algorithm 3.6.2.
[Secant method (SECM)].

Step 1. Set two initial guesses to s: x_0, x_1, maximum number of iterations N, a tolerance ϵ, and $n = 2$.
Step 2. If $F(x_{n-1}) = F(x_{n-2})$ output "iteration cannot be completed" and stop.

Step 3. Using Eq. (3.6.1), set $x_n = G(x_{n-2}, x_{n-1})$.

Step 4. If $|x_n - x_{n-1}| \leq \epsilon$ set $s = x_n$ and stop.

Step 5. If $|x_n - x_{n-1}| > \epsilon$ and $n < N$, set $n \leftarrow n + 1$ and go to Step 2. Otherwise, output "maximum number of iterations exceeded" and stop.

A subroutine SECANT, using this algorithm, is found on the attached floppy disk.

Example 3.6.2.

Let $F(x) = x^2 - x = 0$, $x_0 = 3$, $x_1 = 2$. Then

$$x_n = x_{n-1} - \frac{(x_{n-1} - x_{n-2})(x_{n-1}^2 - x_{n-1})}{x_{n-1}^2 - x_{n-1} - (x_{n-2}^2 - x_{n-2})} = \frac{x_{n-1} \, x_{n-2}}{x_{n-1} + x_{n-2} - 1}$$

implementing

$$x_2 = 2/3, \quad x_3 = 6/5, \quad x_4 = 18/17, \quad x_5 = 108/107, \quad \cdots \to 1 \qquad \square$$

Example 3.6.3.

Let $F(x) = x^3 + x - 2 = 0$, $x_0 = 3$, $x_1 = 2$. The sequence constructed by SECM converges to the exact solution $s = 1$. Table 3.6.2 summarizes the first six iterations, using the SECM algorithm, that are necessary to achieve an error tolerance $\epsilon = 10^{-5}$.

Table 3.6.2. Using the Secant Method to Solve $x^3 + x - 2 = 0$

n	x_n	$F(x_n)$	$|x_n - s|$	$|\epsilon_n / \epsilon_{n-1}|$
0	3.0000000	28.0000000	$0.2 \cdot 10^1$	
1	2.0000000	8.0000000	$0.1 \cdot 10^1$	$0.5 \cdot 10^0$
2	1.6000000	3.6960000	$0.6 \cdot 10^0$	$0.6 \cdot 10^0$
3	1.2565056	1.2402844	$0.3 \cdot 10^0$	$0.4 \cdot 10^0$
4	1.0830202	0.3533300	$0.8 \cdot 10^{-1}$	$0.3 \cdot 10^0$
5	1.0139100	0.0562231	$0.1 \cdot 10^{-1}$	$0.2 \cdot 10^0$
6	1.0008319	0.0033297	$0.8 \cdot 10^{-3}$	$0.6 \cdot 10^{-1}$
7	1.0000086	0.0000345	$0.9 \cdot 10^{-5}$	$0.1 \cdot 10^{-1}$

$\square$

The rightmost column of this table seems to converge to zero, which indicates that the rate of convergence of SECM, α, is higher than 1. Before establishing the exact value of α, we will show that the secant method may also be interpreted as a "modification" of Newton's method. Indeed, if x_n, x_{n-1} are two consecutive NRM iterations, then for large n (because $\lim_{n \to \infty} x_n = s$):

$$F'(x_n) \approx \frac{F(x_n) - F(x_{n-1})}{x_n - x_{n-1}}$$

Thus the next NRM iteration, defined as

$$x_{n+1} = x_n - \frac{F(x_n)}{F'(x_n)}$$

can be approximated by

$$\overline{x}_{n+1} = x_n - \frac{(x_n - x_{n-1})(F(x_n)}{F(x_n) - F(x_{n-1})}$$

which is the SECM iteration. If computing $F'(x)$ is particularly costly or impossible, we may prefer SECM over NRM.

If we regard SECM as an approximating scheme to NRM, we may heuristically expect its rate of convergence to be ≈ 2. This apparently is only "heuristically true." If the sequence $\{x_n\}$ originated by SECM converges to s, then the rate of convergence is $(1 + \sqrt{5})/2 \approx 1.62$.

Before going further, we introduce a helpful notation: Let $\{A_n\}$, $\{B_n\}$ be two infinite sequences. We denote $A_n = O(B_n)$ if and only if

$$|A_n| \leq M|B_n|, \quad n > N$$

for some $M > 0$ (independent of n) and N. For example, if $A_n = 2n^2 + 5$, $B_n = n^2$ then $|A_n| \leq 3|B_n|$, $n > 2$. Therefore, $A_n = O(B_n)$.

This concept is not restricted to sequences. In the case of two functions f and g, $f(x) = O[g(x)]$, $x \rightarrow a$ if and only if

$$|f(x)| \leq M|g(x)|, \quad |x - a| \leq \epsilon$$

for some $M > 0$ and $\epsilon > 0$.

The next result establishes the asymptotic behavior of the error, using SECM.

Theorem 3.6.1.

(The secant method for determining error behavior). *Let $F(x)$ be twice continuously differentiable over the interval $[a, b]$ and $F'(x) \neq 0$. For any given x_0, x_1 assume that $\{x_n\}$ generated by*

$$x_{n+1} = x_n - \frac{(x_n - x_{n-1})(F(x_n)}{F(x_n) - F(x_{n-1})}, \quad n \geq 1 \tag{3.6.2}$$

is well defined and converges to s, that is,

$$\lim_{n \to \infty} x_n = s \tag{3.6.3}$$

Then the error sequence $\{\epsilon_n\}$ defined by $\epsilon_n = x_n - s$ satisfies $\lim_{n \to \infty} [\epsilon_n/(\epsilon_{n-1}\epsilon_{n-2})] = M$ (constant), that is, $\epsilon_n \approx M\epsilon_{n-1}\epsilon_{n-2}$ for large n.

Proof.

We will first show that $\epsilon_{n+1} = A_n\epsilon_n$ where $A_n \to 0$ as $n \to \infty$. Indeed, Eq. (3.6.2) yields

$$\epsilon_{n+1} = \epsilon_n - \frac{(x_n - x_{n-1})F(x_n)}{F(x_n) - F(x_{n-1})} = \epsilon_n - \frac{F(s + \epsilon_n)}{F'(c_n)}$$

where c_n lies between x_n and x_{n-1}, that is, $c_n \to s$ as $n \to \infty$.

If in Eq. (3.6.2) we let $n \to \infty$ and use $F'(x) \neq 0$ to maintain a lower bound m for which $0 < m \leq |F'(x)|$, we can easily show that $F(s) = 0$. Thus,

$$\epsilon_{n+1} = \epsilon_n - \frac{F(s + \epsilon_n) - F(s)}{F'(c_n)} = \epsilon_n - \frac{\epsilon_n F'(d_n)}{F'(c_n)}$$

$$= \epsilon_n \left[1 - \frac{F'(d_n)}{F'(c_n)}\right] = A_n\epsilon_n$$

where d_n is between s and $s + \epsilon_n$. Because $\epsilon_n \to 0$ as $n \to \infty$, we have $\lim_{n\to\infty} d_n = s$. Hence $\lim_{n\to\infty} A_n = 0$.

Returning to Eq. (3.6.2) we substitute $x_n = s + \epsilon_n$ and use Taylor's theorem to obtain

$$\epsilon_{n+1} = \epsilon_n - \frac{(\epsilon_n - \epsilon_{n-1})F(s + \epsilon_n)}{F(s + \epsilon_n) - F(s + \epsilon_{n-1})}$$

$$= \epsilon_n - \frac{(\epsilon_n - \epsilon_{n-1})[\epsilon_n F'(s) + O(\epsilon_n^2)]}{(\epsilon_n - \epsilon_{n-1})F'(s + \epsilon_{n-1}) + (1/2)(\epsilon_n - \epsilon_{n-1})^2 F''(\xi_n)}$$

$$= \epsilon_n - \frac{\epsilon_n F'(s) + O(\epsilon_n^2)}{F'(s + \epsilon_{n-1}) + (1/2)(\epsilon_n - \epsilon_{n-1})F''(\xi_n)}$$

where ξ_n lies between x_{n-1} and x_n. Thus

$$\epsilon_{n+1} = \frac{\epsilon_n[F'(s + \epsilon_{n-1}) - F'(s)] + (1/2)\epsilon_n(\epsilon_n - \epsilon_{n-1})F''(\xi_n) + O(\epsilon_n^2)}{F'(s + \epsilon_{n-1}) + (1/2)(\epsilon_n - \epsilon_{n-1})F''(\xi_n)}$$

$$= \frac{\epsilon_n\epsilon_{n-1}F''(\zeta_n) - (1/2)\epsilon_n\epsilon_{n-1}F''(\xi_n) + O(\epsilon_n^2)}{F'(s) + O(\epsilon_{n-1}) + (1/2)(\epsilon_n - \epsilon_{n-1})F''(\xi_n)}$$

where ζ_n lies between s and $s + \epsilon_{n-1}$. Because $\lim_{n\to\infty} \xi_n = \lim_{n\to\infty} \zeta_n = s$ and $\lim_{n\to\infty}(\epsilon_n/\epsilon_{n-1}) = 0$, we finally obtain, after dividing both sides by $\epsilon_n\epsilon_{n-1}$,

$$\lim_{n\to\infty} \frac{\epsilon_{n+1}}{\epsilon_n \epsilon_{n-1}} = \frac{1}{2}\frac{F''(s)}{F'(s)} \qquad \qquad \square$$

Further error analysis leads to

$$\lim_{n\to\infty} \left|\frac{\epsilon_{n+1}}{|\epsilon_n|^\alpha}\right| = K, \qquad \alpha = \frac{1 + \sqrt{5}}{2}, \qquad K = \left|\frac{1}{2}\frac{F''(s)}{F'(s)}\right|^{1/\alpha}$$

Example 3.6.4.

Consider the function $F(x) = x^5 - 15x - 2$, which has a zero $s = 2$. Table 3.6.3 consists of the first six iterations produced using SECM and the initial guesses $x_0 = 3$, $x_1 = 2.5$. The symbol α denotes $(1 + \sqrt{5})/2$, and the exact value of K is ≈ 1.14.

Table 3.6.3. Using the Secant Method to Evaluate
Asymptotic Error Behavior for $F(x) = x^5 - 15x - 2$

n	x_n	$|\epsilon_n|$	$|\epsilon_n/|\epsilon_{n-1}|^\alpha|$
0	3.0000000	$0.1 \cdot 10^1$	
1	2.5000000	$0.5 \cdot 10^0$	0.50
2	2.2890501	$0.3 \cdot 10^0$	0.89
3	2.1123368	$0.1 \cdot 10^0$	0.84
4	2.0308029	$0.3 \cdot 10^{-1}$	1.06
5	2.0038526	$0.4 \cdot 10^{-2}$	1.07
6	2.0001424	$0.1 \cdot 10^{-3}$	1.15

□

3.6.3 The Cubic Newton's Method

If $F(x)$ possesses two easily computed derivatives, then a cubically convergent algorithm for solving $F(x) = 0$, that is,

$$\lim_{n \to \infty} \frac{\epsilon_{n+1}}{\epsilon_n^3} = M \quad \text{(constant)}$$

can be established. Suppose that x_n is the nth iteration that approximates s for which $F(s) = 0$. Denote $s = x_n + h$ and replace $F(x_n + h)$ by its second order Taylor polynomial and remainder about x_n:

$$0 = F(x_n + h) = F(x_n) + hF'(x_n) + \frac{h^2}{2}F''(x_n) + O(h^3)$$

By neglecting the error term $O(h^3)$ one finds

$$h \approx \frac{-F'_n \pm \sqrt{F'^2_n - 2F_n F''_n}}{F''_n}$$

where $F_n = F(x_n)$, $F'_n = F'(x_n)$, $F''_n = F''(x_n)$. The sign of the square root must be equal to the sign of (F'_n) (why?).

The expression on the right-hand side is somewhat awkward. Instead, we first write

$$h \approx \frac{-F'_n + F'_n \left[1 - (2F_n F''_n / F'^2_n)\right]^{\frac{1}{2}}}{F''_n}$$

Now, for small x, we can replace $\sqrt{1-x}$ by its second order Taylor polynomial about $x = 0$, and obtain

$$\sqrt{1-x} = 1 - \frac{x}{2} - \frac{x^2}{8} + O(x^3)$$

Applying this to $x = (2F_n F_n'')/F_n'^2$ (F_n and therefore x approach 0 as $n \to \infty$), we find

$$h \approx -\frac{F_n}{F_n'}\left(1 + \frac{F_n F_n''}{2F_n'^2}\right)$$

which yields the $(n+1)$th iteration as

$$x_{n+1} = x_n - \frac{F_n}{F_n'}\left(1 + \frac{F_n F_n''}{2F_n'^2}\right) \tag{3.6.4}$$

We can now present the following algorithm for solving $F(x) = 0$.

Algorithm 3.6.3.
[Cubic Newton's method (CNM)].

Step 1. Set an initial guess x_0, a tolerance ϵ, maximum number of iterations N, and $n = 0$.

Step 2. If $F'(x_n) = 0$ output "iteration cannot be completed," n, x_n and stop.

Step 3. Calculate x_{n+1}, using Eq. (3.6.4).

Step 4. If $|x_{n+1} - x_n| \le \epsilon$ set $s = x_n$ and stop.

Step 5. If $|x_{n+1} - x_n| > \epsilon$ and $n < N$, set $n \leftarrow n + 1$ and go to step 2. Otherwise, output "maximum number of iterations exceeded," x_n and stop.

The CNM algorithm can be shown to converge cubically. Therefore, after every iteration, provided that we are sufficiently close to the solution, the number of accurate digits is tripled. A subroutine CUBIC, based on Algorithm 3.6.3, is found on the attached floppy disk.

Example 3.6.5.
Let $F(x) = x^2 \ln x$, $x > 0$. The unique solution of $F(x) = 0$ is $s = 1$. The first two derivatives of $F(x)$ are as follow: $F'(x) = 2x \ln x + x$, $F''(x) = 2 \ln x + 3$. For any given $x_0 > 0$, the CNM generates x_{n+1} as

$$\begin{aligned}
x_{n+1} &= x_n - \frac{x_n^2 \ln x_n}{x_n(2 \ln x_n + 1)}\left[1 + \frac{(2 \ln x_n + 3)x_n^2 \ln x_n}{2x_n^2(2 \ln x_n + 1)^2}\right] \\
&= x_n\left\{1 - \frac{\ln x_n}{2 \ln x_n + 1}\left[1 + \frac{(2 \ln x_n + 3) \ln x_n}{2(2 \ln x_n + 1)^2}\right]\right\}
\end{aligned}$$

For $x_0 = 2$, the first four iterations are sufficient to reach an error tolerance $\epsilon = 10^{-8}$ (Table 3.6.4).

Table 3.6.4. Cubic Newton's Method Performance
for $F(x) = x^2 \ln x$, $x_0 = 2$

n	x_n	$|x_n - s|$	$|x_n - x_{n-1}|$
0	2.00000000	$0.1 \cdot 10^1$	
1	1.26397234	$0.3 \cdot 10^0$	$0.7 \cdot 10^0$
2	1.02435780	$0.2 \cdot 10^{-1}$	$0.2 \cdot 10^0$
3	1.00005253	$0.5 \cdot 10^{-4}$	$0.2 \cdot 10^{-1}$
4	1.00000000	$< 10^{-8}$	$0.5 \cdot 10^{-4}$

□

Example 3.6.6.

Let $F(x) = x^3 - x^2 - x - 1 = 0$. Direct differentiation yields $F'(x) = 3x^2 - 2x - 1$, $F''(x) = 6x - 2$. Using the CNM scheme with $x_0 = 2$, only two iterations are needed to reach the exact solution $s = 1.83928675\ldots$ within a tolerance $\epsilon = 10^{-6}$. Table 3.6.5 provides a comparison between the three algorithms NRM, SECM, and CNM. The second initial guess needed to start SECM is computed from x_0 by NRM. The most significant number related to the speed of convergence is the number of function evaluations. If we assume the computations of F_n, F'_n, F''_n to be equally costly, the SECM with five evaluations is preferable to the NRM and CNM (six evaluations each) in this particular case.

Table 3.6.5. Iterations Generated by Newton–Raphson Method, Secant Method, and Cubic Newton's Method for $F(x) = x^3 - x^2 - x - 1$, $x_0 = 2$ *

n	x_n	y_n	z_n
0	2.0000000	2.0000000	2.0000000
1	1.8571429	1.8571429	1.8425656
2	1.8395445	1.8414239	1.8392868
3	1.8392868	1.8393179	
4		1.8392868	

* The iterations are denoted by x_n (NRM), y_n
(SECM), and z_n (CNM).

□

3.6.4 Muller's Method

The secant method is actually obtained by replacing the function $F(x)$ with a straight line passing through the points $[x_{n-1}, F(x_{n-1})]$, $[x_n, F(x_n)]$. The intersection point of this line with the x axis is considered the next iteration, x_{n+1}. This scheme can be generalized. Let x_n, x_{n-1}, ..., x_{n-k} be the previous $(k+1)$ iterations. We construct a kth order interpolating polynomial, $p_k(x)$, to $F(x)$, which coincides with $F(x)$ at these $(k+1)$ points. The intersection point of $p_k(x)$ with the x axis is defined as x_{n+1}. The case $k = 2$ is known as Muller's method and will be given in detail.

Let (x_0, F_0), (x_1, F_1), (x_2, F_2) be three points on the graph of $F(x)$. The parabola $p_2(x)$ through these points is

$$p_2(x) = a(x - x_2)^2 + b(x - x_2) + c$$

where a, b, c satisfy

$$
\begin{aligned}
F_0 &= a(x_0 - x_2)^2 + b(x_0 - x_2) + c \\
F_1 &= a(x_1 - x_2)^2 + b(x_1 - x_2) + c \\
F_2 &= a \cdot 0^2 + b \cdot 0 + c
\end{aligned}
$$

Thus

$$c = F_2 \tag{3.6.5}$$

$$a = \frac{(x_1 - x_2)(F_0 - F_2) - (x_0 - x_2)(F_1 - F_2)}{(x_0 - x_2)(x_1 - x_2)(x_0 - x_1)} \tag{3.6.6}$$

$$b = \frac{(x_0 - x_2)^2(F_1 - F_2) - (x_1 - x_2)^2(F_0 - F_2)}{(x_0 - x_2)(x_1 - x_2)(x_0 - x_1)} \tag{3.6.7}$$

Let the parabola intersect the x axis at $(x_3, 0)$. Then

$$x_3 - x_2 = \frac{-b \pm \sqrt{b^2 - 4ac}}{2a}$$

Only one value of x_3 is agreeable, and it should be chosen so that $|x_3 - x_2|$ is small. If x_2 is already sufficiently close to the solution s of $F(x) = 0$, then c is sufficiently small and $b^2 - 4ac \approx b^2$. Thus one must choose

$$x_3 - x_2 = \frac{-b + \operatorname{sign}(b)\sqrt{b^2 - 4ac}}{2a}$$

The last equation may serve as a case study for round-off error effects. Because the numerator of the right-hand side is the *difference* of two almost identical numbers, we may lose accuracy due to a round-off error effect. A simple way to bypass this difficulty is to rewrite

$$
\begin{aligned}
x_3 - x_2 &= \frac{-b + \operatorname{sign}(b)\sqrt{b^2 - 4ac}}{2a} \frac{-b - \operatorname{sign}(b)\sqrt{b^2 - 4ac}}{-b - \operatorname{sign}(b)\sqrt{b^2 - 4ac}} \\
&= \frac{(-b)^2 - (b^2 - 4ac)}{-2a[b + \operatorname{sign}(b)\sqrt{b^2 - 4ac}]} = -\frac{2c}{b + \operatorname{sign}(b)\sqrt{b^2 - 4ac}}
\end{aligned}
$$

Thus,

$$x_3 = x_2 - \frac{2c}{b + \text{sign}(b)\sqrt{b^2 - 4ac}} \tag{3.6.8}$$

Eqs. (3.6.5) through (3.6.8) are used to formulate the next algorithm for solving $F(x) = 0$.

Algorithm 3.6.4.

[Muller's method (MLR)].

Step 1. Set three initial guesses x_0, x_1, x_2 to the solution s, a tolerance ϵ, maximum number of iterations N, and $n = 2$.

Step 2. Compute F_0, F_1, F_2.

Step 3. Use F_{n-2}, F_{n-1}, F_n and Eqs. (3.6.5) through (3.6.7) to calculate the coefficients a, b, c of the parabola that approximates $F(x)$.

Step 4. Calculate x_{n+1}, using Eq. (3.6.8).

Step 5. If $|x_{n+1} - x_n| \leq \epsilon$ set $s = x_{n+1}$ and stop.

Step 6. If $|x_{n+1} - x_n| > \epsilon$ and $n < N$ set $n \leftarrow n+1$, compute F_n, and go to step 3. Otherwise, output "maximum number of iterations exceeded," x_n and stop.

A subroutine MULLER, based on this algorithm, is found on the attached floppy disk.

Example 3.6.7.

Let $F(x) = x^3 - 3x - 2$. To use the MLR algorithm to approximate the solution $s = 2$ of $F(x) = 0$, we choose the starting values $x_0 = 1.25$, $x_1 = 2.5$, $x_2 = 2.25$. Thus

$$F_0 = -3.796875, \quad F_1 = 6.125, \quad F_2 = 2.640625$$
$$a = 6, \quad b = 12.4375, \quad c = 2.640625$$

leading to $x_3 = 2.0099$. $\qquad \square$

If the sequence generated by MLR converges to a solution s, it can be shown that the order of convergence is $\beta = 1.839\ldots$, which is the only real solution of $x^3 - x^2 - x - 1 = 0$. If one generalizes Muller's approach and approximates $F(x)$, using $(k+1)$ points, it can be shown that the generated sequence converges to s with order of convergence r_k, which is a zero of the polynomial

$$x^{k+1} - x^k - x^{k-1} - \cdots - x - 1 = 0 \tag{3.6.9}$$

and satisfies

$$r_k < 2, \quad \lim_{k \to \infty} r_k = 2 \tag{3.6.10}$$

3.6.5   Which Algorithm to Choose?

So far we have presented several algorithms for solving $F(x) = 0$, all of which converge with order of convergence > 1. None of them is *always* superior. If, for example, calculating F', F'' is particularly time consuming, one should prefer the secant method or Muller's method.

Although the Newton–Raphson method and the cubic Newton method need fewer iterations than the secant and Muller methods, they require two and three function evaluations $(F, F'; F, F', F'')$ per iteration, respectively. If we assume that to compute F, F', F'' is equally costly, we can easily show that, to reach a given tolerance $\epsilon > 0$, Muller's method is the most efficient, provided that *only* function evaluations are counted. From a programmer's point of view, the complexity of Eqs. (3.6.5) through (3.6.8) makes Muller's scheme or its generalization somewhat unattractive. Also, one should bear in mind that all the previously discussed methods with order of convergence > 1 are mostly effective only if the starting value(s) is(are) already within a small neighborhood of the solution s. In that case, Newton's method, for example, may require a single iteration for convergence and is then preferred over the other schemes.

Example 3.6.8.

Let $F(x) = x^2 - x - 1$. The positive solution of $F(x) = 0$ is $s = (1 + \sqrt{5})/2 \approx 1.61803399$. Suppose that $x_0 = 1.62$, $\epsilon = 5 \cdot 10^{-6}$. Newton's method generates $x_1 = 1.61803571$, $x_2 = 1.61803399$ and, because $|x_1 - x_2| = 1.72 \cdot 10^{-6} < \epsilon$, a total of four function evaluations is needed.

If instead we decide to use the secant method, we may, for instance, choose $x_0 = 1.62$, $x_1 = 1.61$. Then $x_2 = 1.61802691$, $x_3 = 1.61803401$, $x_4 = 1.61803399$, and the number of function evaluations needed for convergence is four. In this particular case we naturally prefer Newton's method because of its simplicity. ☐

If a large number of iterations is needed for convergence, then Muller's method would on average perform better. But this is usually not the case, and one should carefully analyze each problem separately before making a decision as to which algorithm to apply.

Finally, when we apply any of the above-mentioned algorithms, the iterative process should stop when the inequality $|x_n - s| < \epsilon$ occurs for the first time. Because s is generally unknown, we may refer only to $|x_n - x_{n-1}|$. However, if the order of convergence is greater than 1, and if one is sufficiently close to s, then $|x_n - x_{n-1}| \approx |s - x_{n-1}|$, that is, $|x_n - s| \approx |x_{n+1} - x_n|$.

PROBLEMS

1. Use the BISM to solve the following equations with an error tolerance $\epsilon = 10^{-3}$.

   (a) The real solution of $x^3 - x^2 - x - 1 = 0$.

   (b) $x - e^{-x/2} = 0$.

   (c) The smallest positive solution of $x - \tan x = 0$.

2. Show that the equation $x = a + b \sin x$ has at least one solution. Use the BISM to find it with an accuracy of $\epsilon = 10^{-3}$ for:

   (a) $a = 1$, $b = 1$

   (b) $a = 10$, $b = 5$

3. Let $F(x)$ be defined over the interval $[-7, 2]$, and $F(2)F(-7) < 0$. What is the maximum number of iterations that may be needed for computing the solution s of $F(x) = 0$, within a tolerance $\epsilon = 10^{-8}$.

4. Find $\sqrt[5]{32}$ within a tolerance $\epsilon = 10^{-3}$, using the BISM. (*Hint*: Consider the equation $x^5 - 32 = 0$.)

5. Using the BISM, find all the solutions of $x^2 - 8\cos x = 0$. (*Hint*: Consider the problem graphically.)

6. Use the BISM to find all the zeroes of the polynomial $x^4 - x^3 - 23x^2 + 3x + 60$ with an accuracy of $\epsilon = 10^{-4}$.

7. Solve the equation $F(x) = x^3 + x^2 - 1 = 0$, using the secant method. Choose $x_0 = 2$, $x_1 = 1$, $\epsilon = 10^{-6}$. How many function evaluations $[F(x)$ or $F'(x)]$ are needed?

8. Solve problem 7 using the NRM with $x_0 = 1.5$, $\epsilon = 10^{-6}$.

9. Let $\{r_n\}$ be the monotonic increasing sequence of all the roots of the equation $xe^{-x} + \cos x = 0$.

   (a) Use the BISM to obtain reasonable starting values (x_0^i, x_1^i) to r_i, $1 \le i \le 4$.

   (b) Insert (x_0^i, x_1^i) into the SECM algorithm and compute r_i, $1 \le i \le 4$ with an accuracy of $\epsilon = 10^{-8}$.

10. Let s be a solution to $F(x) = 0$, and $\{x_n\}$ be a sequence generated by the SECM that converges to s, and $\epsilon_n = x_n - s$. Use the result of Theorem 3.6.1 and show that if

$$\lim_{n \to \infty} \left| \frac{\epsilon_{n+1}}{|\epsilon_n|^\alpha} \right| = K$$

then $\alpha = (1 + \sqrt{5})/2$, $K = |(1/2)[F''(s)/F'(s)]|^{1/\alpha}$.

11. Use the SECM with $x_0 = 0.8$, $x_1 = 0.5$, $\epsilon = 10^{-8}$ to solve $x^{3/2} \ln x = 0$, and validate the results of problem 10.

12. Apply the SECM to solve $x^3 - 3x^2 + 4x - 2 = 0$, using $x_0 = 0.8$, $x_1 = 0.9$, $\epsilon = 10^{-10}$. Study the generated iterations and explain the results.

13. Use the CNM to find a root to the equation $e^{-x} - \tan x = 0$, near $x_0 = 1$, with error tolerance $\epsilon = 10^{-8}$.

14. Use the CNM to find a root to the equation $x^5 - \tan x = 0$, near $x_0 = 1.4$, with tolerance $\epsilon = 10^{-8}$. Repeat the computation using the NRM and compare the performances of both methods.

15. Find the only solution to the equation $\sin x = \cos x$, $0 \le x \le \pi/2$, using the CNM with $x_0 = 0.8$, $\epsilon = 10^{-8}$.

16. Replace $\sin x$, $\cos x$ by their respective third and second order Taylor polynomial approximations $p_3(x)$, $q_2(x)$. Using bisection and then the CNM, find all the roots of the equation $p_3(x) = q_2(x)$ and use one of them as a starting point for solving problem 15.

17. Given $x_0 = 0.9$, $x_1 = 0.8$, $x_2 = 0.7$ use MLR to solve $x = e^{-x}$ with a tolerance $\epsilon = 10^{-6}$.

18. Given $x - e^{-x} = 0$, $x_0 = 0.9$, determine x_1, x_2, using the NRM. Then use MLR with $\epsilon = 10^{-6}$ and compare with the solution of problem 17.

19. Using MLR, solve the equation $x^2 - \sqrt{x+1} = 0$ with $x_0 = 1$, $x_1 = 1.1$, $x_2 = 1.2$ and a tolerance $\epsilon = 10^{-8}$.

20. Show that if $\{x_n\}$, the sequence generated by MLR, converges to a solution s of $F(x) = 0$, then the error $\epsilon_n = x_n - s$ satisfies

$$\lim_{n \to \infty} \frac{\epsilon_{n+1}}{\epsilon_n \epsilon_{n-1} \epsilon_{n-2}} = M \text{ (constant)}$$

3.7   Extensions to Multivariable Systems

In this section we extend the standard iterative method (SIM) and several of its modifications and apply them to multivariable systems. The n-dimensional form of Eq. (3.1.1) is a set of n nonlinear equations with n variables given by

$$
\begin{aligned}
x_1 &= f_1(x_1, \ldots, x_n) \\
x_2 &= f_2(x_1, \ldots, x_n) \\
&\vdots \\
x_n &= f_n(x_1, \ldots, x_n)
\end{aligned}
\tag{3.7.1}
$$

The vector form of Eq. (3.7.1) is $\mathbf{x} = \mathbf{f(x)}$, where

$$
\mathbf{x} \equiv \begin{bmatrix} x_1 \\ \vdots \\ x_n \end{bmatrix}, \quad
\mathbf{f(x)} \equiv \begin{bmatrix} f_1(x_1, \ldots, x_n) \\ \vdots \\ f_n(x_1, \ldots, x_n) \end{bmatrix} = \begin{bmatrix} f_1(\mathbf{x}) \\ \vdots \\ f_n(\mathbf{x}) \end{bmatrix}
\tag{3.7.2}
$$

A solution to Eq. (3.7.1) is any vector

$$
\mathbf{s} = \begin{bmatrix} s_1 \\ \vdots \\ s_n \end{bmatrix} \equiv (s_1, \ldots, s_n)^T
$$

that satisfies

$$\mathbf{s} = \mathbf{f(s)} \tag{3.7.3}$$

For example, the system $x = x^2 - y^2 - 2$, $y = xy - 4$ is solved by $x = 3$, $y = 2$.

Definition 3.7.1. *A sequence of vectors $\{x_m\}$ in the real n-dimensional Euclidean vector space is said to converge to a vector* **x**, *if*

$$\lim_{m \to \infty} \| x_m - x \| = 0 \tag{3.7.4}$$

where the norm $\| x_m - x \|$ is defined as in Section 1.7. We recall Cauchy's necessary and sufficient condition for convergence: a sequence $\{x_m\}$ converges to some **x** *if and only if, for every $\epsilon > 0$, an integer N can be found such that $\| x_p - x_q \| < \epsilon$ for p, q > N. If Eq. (3.7.4) holds, we write $\lim_{m\to\infty} x_m = x$.*

We now extend the Algorithm SIM and apply it to a system of equations [Eq. (3.7.1)].

Algorithm 3.7.1.
[Generalized standard iteration method (GSIM)].

Step 1. Choose a first approximation x_0, a tolerance ϵ, a maximum number of iterations N, and set $m = 1$.

Step 2. Generate x_m, using the vector equation

$$x_m = f(x_{m-1}) \tag{3.7.5}$$

Step 3. If $\| x_m - x_{m-1} \| < \epsilon$ set $s = x_m$ and stop.

Step 4. If $\| x_m - x_{m-1} \| \geq \epsilon$ and $m < N$, set $m \leftarrow m + 1$ and go to step 2. Otherwise, output "maximum number of iterations exceeded," x_m and stop.

Sufficient conditions for a successful implementation of this algorithm are given next.

Theorem 3.7.1.
(Convergence of GSIM). *Let $f_1(x), \ldots, f_n(x)$, be real-valued functions defined over the hyperbox*

$$R \equiv \left\{ x = (x_1, \ldots, x_n)^T \mid a_i \leq x_i \leq b_i, \quad i = 1, \ldots, n \right\}$$

and satisfy the following conditions:

1. $f_i(x) \in C[R]$, $i = 1, \ldots, n$

2. *For any given $x \in R$:* $f(x) = [f_1(x), \ldots, f_n(x)]^T \in R$

3. *There exists a Lipschitz constant $L < 1$ for which the Lipschitz condition*

$$\| f(x_1) - f(x_2) \| \leq L \| x_1 - x_2 \|, \quad x_1, x_2 \in R \tag{3.7.6}$$

holds. Then the system defined by Eq. (3.7.1) has a unique solution $s \in R$, and for any given $x_0 \in R$,

(a) *The sequence $\{x_n\}$ generated by Eq. (3.7.5) converges to* **s**.

(b) *The inequality*

$$\| \mathbf{x}_n - \mathbf{s} \| \leq \frac{L^n}{1 - L} \| \mathbf{x}_1 - \mathbf{x}_0 \|$$

holds for all $n \geq 1$.

The proof is straightforward and is identical to that given for the one-dimensional case.

A mapping $\mathbf{x} \to \mathbf{f(x)}$ governed by Eq. (3.7.6) for $L < 1$ is called a *contraction mapping*. The following theorem suggests a practical way to check whether a given mapping contracts.

Theorem 3.7.2.

(An upper bound for a Lipschitz constant). *Let* $\mathbf{f(x)}$, *a real-valued function defined over R (of Theorem 3.7.1), possess continuous first partial derivatives, that is,*

$$\frac{\partial f_i}{\partial x_j} \in C[R], \quad i, j = 1, \ldots, n$$

Then, Eq. (3.7.6) holds for

$$L = \max_{\mathbf{x} \in R} \left[\sum_{i,j=1}^{n} \left(\frac{\partial f_i}{\partial x_j} \right)^2 \right]^{1/2} \tag{3.7.7}$$

Proof.

We make use of a powerful tool commonly used in functional analysis, the *Cauchy–Schwarz inequality*. For the sake of simplicity, we consider the case $n = 2$ and the functions $f(x, y)$, $g(x, y)$. Let (x_1, y_1), (x_2, y_2) be two given points in R. An arbitrary point on the segment connecting these points is $x(t)$, $y(t)$, where

$$x(t) = x_1 + t(x_2 - x_1), \quad y(t) = y_1 + t(y_2 - y_1), \quad 0 \leq t \leq 1$$

If we denote $f[x(t), y(t)]$, $g[x(t), y(t)]$ by $F(t)$, $G(t)$, respectively, then the well-known *chain rule* yields

$$\frac{dF}{dt} = (x_2 - x_1)f_x + (y_2 - y_1)f_y = hf_x + kf_y$$
$$\frac{dG}{dt} = (x_2 - x_1)g_x + (y_2 - y_1)g_y = hg_x + kg_y$$

where $h = x_2 - x_1$, $k = y_2 - y_1$. Therefore,

$$F(1) - F(0) = \int_0^1 (hf_x + kf_y)\, dt$$

Now, the sum form of the Cauchy–Schwarz inequality yields

$$\left(\sum_{i=1}^{m}\alpha_i\beta_i\right)^2 \le \left(\sum_{i=1}^{m}\alpha_i^2\right)\left(\sum_{i=1}^{m}\beta_i^2\right) \tag{3.7.8}$$

for any real α_i, β_i, $i = 1, \ldots, m$. In particular, $(hf_x + kf_y)^2 \le (h^2 + k^2)(f_x^2 + f_y^2)$. Hence

$$|F(1) - F(0)| = |f(x_2, y_2) - f(x_1, y_1)| \le \sqrt{h^2 + k^2} \int_0^1 \sqrt{f_x^2 + f_y^2} \, dt$$

and by squaring both sides we obtain

$$[f(x_2, y_2) - f(x_1, y_1)]^2 \le (h^2 + k^2)\left[\int_0^1 \sqrt{f_x^2 + f_y^2} \, dt\right]^2 \tag{3.7.9}$$

Now, the integral form of Cauchy–Schwarz inequality, given by

$$\left[\int_a^b u(t)v(t) \, dt\right]^2 \le \int_a^b [u(t)]^2 \, dt \int_a^b [v(t)]^2 \, dt \tag{3.7.10}$$

implies

$$\left(\int_0^1 \sqrt{f_x^2 + f_y^2} \, dt\right)^2 = \left(\int_0^1 1 \cdot \sqrt{f_x^2 + f_y^2} \, dt\right)^2 \le \int_0^1 1^2 \, dt \int_0^1 (f_x^2 + f_y^2) \, dt = \int_0^1 (f_x^2 + f_y^2) \, dt$$

Thus, by substituting in Eq. (3.7.9), we obtain

$$[f(x_2, y_2) - f(x_1, y_1)]^2 \le [(x_2 - x_1)^2 + (y_2 - y_1)^2]\int_0^1 (f_x^2 + f_y^2) \, dt$$

Similarly,

$$[g(x_2, y_2) - g(x_1, y_1)]^2 \le [(x_2 - x_1)^2 + (y_2 - y_1)^2]\int_0^1 (g_x^2 + g_y^2) \, dt$$

By adding the two inequalities we find

$$\| \mathbf{f}(\mathbf{x}_1) - \mathbf{f}(\mathbf{x}_2) \|^2 \le \| \mathbf{x}_1 - \mathbf{x}_2 \|^2 \int_0^1 (f_x^2 + f_y^2 + g_x^2 + g_y^2) \, dt \le \| \mathbf{x}_1 - \mathbf{x}_2 \|^2 L^2 \qquad \Box$$

Example 3.7.1.

Consider the system

$$x = f(x,y) = \frac{1}{4}e^{-xy}, \quad y = g(x,y) = \frac{x^3 + y^3}{6} \text{ for } 0 \le x, y \le 1$$

In this case $f_x^2 + f_y^2 + g_x^2 + g_y^2 = (1/4)[(1/4)(x^2+y^2)e^{-2xy} + x^4 + y^4] \le 5/8$. Consequently, a Lipschitz condition with $L = \sqrt{5/8}$ exists, and hypothesis 3 of Theorem 3.7.1 holds. Also, because

$$0 \le \frac{1}{4}e^{-xy} \le \frac{1}{4}, \quad 0 \le \frac{x^3 + y^3}{6} \le \frac{1}{3}$$

for $0 \le x, y \le 1$, hypothesis 2 is satisfied as well. Therefore a unique solution (s, t) exists. Table 3.7.1 contains the first six iterations generated by the GSIM for the initial values $x_0 = 0.5$, $y_0 = 0.5$. They already provide accuracy to six decimal digits.

Table 3.7.1. Generalized Standard Iteration Method for
$x = (1/4)e^{-xy}$, $y = (x^3 + y^3)/6$, $x_0 = y_0 = 0.5$, $\epsilon = 10^{-6}$

n	x_n	y_n	$f(x_n, y_n)$	$g(x_n, y_n)$
0	0.500000	0.500000	0.194700	0.041667
1	0.194700	0.041667	0.247980	0.001242
2	0.247980	0.001242	0.249923	0.002542
3	0.249923	0.002542	0.249841	0.002602
4	0.249841	0.002602	0.249838	0.002599
5	0.249838	0.002599	0.249838	0.002599
6	0.249838	0.002599		

□

Example 3.7.2.

Let $f(x, y, z) = A_1 \cos(yz)$, $g(x, y, z) = A_2 \cos(xz)$, $h(x, y, z) = A_3 \sin(1 + xy)$ and consider the system $x = f(x, y, z)$, $y = g(x, y, z)$, $z = h(x, y, z)$ defined over the box $R = \{(x, y, z) | -1 \le x, y, z \le 1\}$. The first condition of Theorem 3.7.1 is obviously fulfilled. We now impose $|A_i| \le 1$, $1 \le i \le 3$ to obtain

$$|f(x, y, z)| \le 1, \quad |g(x, y, z)| \le 1, \quad |h(x, y, z)| \le 1$$

and the second requirement is guaranteed. Next, because

$$\begin{aligned}
f_x^2 &+ f_y^2 + f_z^2 + g_x^2 + g_y^2 + g_z^2 + h_x^2 + h_y^2 + h_z^2 \\
&= A_1^2(y^2 + z^2)\sin^2(yz) + A_2^2(x^2 + z^2)\sin^2(xz) + A_3^2(x^2 + y^2)\cos^2(1 + xy) \\
&\le 2(A_1^2 + A_2^2 + A_3^2)^2
\end{aligned}$$

we further impose $A_1^2 + A_2^2 + A_3^2 < 1/2$, and the third condition of Theorem 3.7.1 holds with $L = \sqrt{2}/2$. For example, if $A_i = 1/3$, $1 \le i \le 1/3$, the sequence $\{(x_i, y_i, z_i)\}$ generated by GSIM must converge to the unique solution of the system for any given $(x_0, y_0, z_0) \in R$. Table 3.7.2 consists of the first five iterations, given $x_0 = 0.4$, $y_0 = 0.5$, $z_0 = 0.6$.

Table 3.7.2. Generalized Standard Iteration Method for a Three-Dimensional System

n	x_n	y_n	z_n
0	0.400000	0.500000	0.600000
1	0.318445	0.323779	0.310680
2	0.331648	0.331703	0.297537
3	0.331711	0.331712	0.298568
4	0.331700	0.331700	0.298571
5	0.331700	0.331700	0.298570

$\square$

3.7.1   Error Estimate for Systems

We still consider a two-dimensional system for which the hypotheses of Theorem 3.7.1 hold, and suppose that $f(x,y)$, $g(x,y)$ possess continuous second derivatives as well. Let (s,t) be the unique solution of the system, and for $n = 0, 1, \ldots$ denote

$$\epsilon_n = x_n - s, \quad \delta_n = y_n - t$$

Then, by applying Taylor's general theorem for multivariable functions, we get

$$
\begin{aligned}
\epsilon_{n+1} = x_{n+1} - s &= f(x_n, y_n) - f(s,t) \\
&= f(s + \epsilon_n, t + \delta_n) - f(s,t) \\
&= \epsilon_n f_x(s,t) + \delta_n f_y(s,t) + O_1(\epsilon_n^2 + \delta_n^2)
\end{aligned}
$$

and similarly

$$\delta_{n+1} = \epsilon_n g_x(s,t) + \delta_n g_y(s,t) + O_2(\epsilon_n^2 + \delta_n^2)$$

Thus

$$
\begin{bmatrix} \epsilon_{n+1} \\ \delta_{n+1} \end{bmatrix} = \begin{bmatrix} f_x & f_y \\ g_x & g_y \end{bmatrix}\bigg|_{(s,t)} \begin{bmatrix} \epsilon_n \\ \delta_n \end{bmatrix} + \begin{bmatrix} O_1(\epsilon_n^2 + \delta_n^2) \\ O_2(\epsilon_n^2 + \delta_n^2) \end{bmatrix}
$$

If we denote

$$\mathbf{r}_n = \begin{bmatrix} \epsilon_n \\ \delta_n \end{bmatrix}, \quad \mathbf{J} = \begin{bmatrix} f_x & f_y \\ g_x & g_y \end{bmatrix} \quad \text{(Jacobian matrix of } f, g)$$

then

$$\mathbf{r}_{n+1} = \mathbf{J}(s,t)\mathbf{r}_n + \mathbf{O}\left(\|\mathbf{r}_n\|^2\right) \tag{3.7.11}$$

If $\mathbf{J}(s,t)$ is not identically zero, the convergence is called *linear*. If, however, all four terms of $\mathbf{J}$ vanish at (s,t), the convergence of GSIM is at least quadratic. The next result is a local convergence theorem.

Theorem 3.7.3.

(A sufficient condition for local convergence). *Let $f(x,y)$, $g(x,y) \in C^1[R]$ and let (s,t) be a solution of the system*

$$x = f(x,y), \quad y = g(x,y)$$

for which $\mathbf{J}(s,t) = 0$. Then there exists a circle with radius ϵ, centered at (s,t), such that for any given (x_0, y_0) within this circle the sequence generated by the GSIM converges to (s,t). If, in addition, $f(x,y)$ and $g(x,y)$ possess continuous partial derivatives of the second order, the convergence is (at least) quadratic.

Proof.

Because f_x, f_y, g_x, g_y are continuous in R and equal to zero at (s,t), there exists an $\epsilon > 0$

$$(f_x^2 + f_y^2 + g_x^2 + g_y^2)^{1/2} < 0.5 \quad \text{for} \quad (x-s)^2 + (y-t)^2 < \epsilon^2$$

This guarantees that conditions 2 and 3 of Theorem 3.7.1 are satisfied inside the circle C_ϵ of radius ϵ, centered at (s,t). Therefore

$$\lim_{n\to\infty}(x_n, y_n) = (s,t)$$

for any given $(x_0, y_0) \in C_\epsilon$. Due to Eq. (3.7.11) the convergence must be quadratic. ⬜

3.7.2 Newton's Method for a System

Let the system

$$F(x,y) = 0$$
$$G(x,y) = 0 \tag{3.7.12}$$

possess a solution (s,t). This system can be replaced by

$$\begin{bmatrix} x \\ y \end{bmatrix} = \begin{bmatrix} x \\ y \end{bmatrix} + \begin{bmatrix} h_1(x,y) & h_2(x,y) \\ h_3(x,y) & h_4(x,y) \end{bmatrix} \begin{bmatrix} F(x,y) \\ G(x,y) \end{bmatrix} = \begin{bmatrix} f(x,y) \\ g(x,y) \end{bmatrix} \tag{3.7.13}$$

provided that $(h_1 h_4 - h_2 h_3)_{(s,t)} \neq 0$ (why?).

If we use the GSIM to solve the new set of equations, we should see quadratic convergence provided that $f_x(s,t) = f_y(s,t) = g_x(s,t) = g_y(s,t) = 0$, that is,

$$\begin{aligned}
(x + h_1F + h_2G)_x(s,t) &= (x + h_1F + h_2G)_y(s,t) \\
&= (y + h_3F + h_4G)_x(s,t) \\
&= (y + h_3F + h_4G)_y(s,t) = 0
\end{aligned}$$

We therefore impose the relations

$$(1 + h_{1x}F + h_1F_x + h_{2x}G + h_2G_x)(s,t) = 0$$
$$(h_{1y}F + h_1F_y + h_{2y}G + h_2G_y)(s,t) = 0$$
$$(h_{3x}F + h_3F_x + h_{4x}G + h_4G_x)(s,t) = 0$$
$$(1 + h_{3y}F + h_3F_y + h_{4y}G + h_4G_y)(s,t) = 0$$

But (s,t) is a solution to Eq. (3.7.12). Therefore $F(s,t) = G(s,t) = 0$ and h_i, $1 \le i \le 4$ must satisfy

$$h_1F_x + h_2G_x = -1, \quad h_1F_y + h_2G_y = 0 \tag{3.7.14}$$
$$h_3F_x + h_4G_x = 0, \quad h_3F_y + h_4G_y = -1 \tag{3.7.15}$$

at the solution point (s,t). The solution to Eqs. (3.7.14) and (3.7.15) is given by

$$h_1 = -\frac{G_y}{F_xG_y - F_yG_x}, \quad h_2 = \frac{F_y}{F_xG_y - F_yG_x} \tag{3.7.16}$$

and by

$$h_3 = \frac{G_x}{F_xG_y - F_yG_x}, \quad h_4 = -\frac{F_x}{F_xG_y - F_yG_x} \tag{3.7.17}$$

respectively, computed at (s,t). Because we do not know (s,t) a priori, we simply use Eqs. (3.7.14) through (3.7.17) to define $h_i(x,y)$, $1 \le i \le 4$ everywhere and thus guarantee $\mathbf{J}(s,t) = 0$, and quadratic convergence. When the GSIM is applied to Eq. (3.7.13), it yields

$$x_{n+1} = f(x_n, y_n) = x_n + h_1(x_n, y_n)F(x_n, y_n) + h_2(x_n, y_n)G(x_n, y_n)$$
$$y_{n+1} = g(x_n, y_n) = y_n + h_3(x_n, y_n)F(x_n, y_n) + h_4(x_n, y_n)G(x_n, y_n)$$

or

$$x_{n+1} = x_n + \left(\frac{GF_y - FG_y}{F_xG_y - F_yG_x}\right) \Big|_{(x_n, y_n)}$$
$$y_{n+1} = y_n + \left(\frac{FG_x - GF_x}{F_xG_y - F_yG_x}\right) \Big|_{(x_n, y_n)} \tag{3.7.18}$$

Equation (3.7.18) is Newton's method for two-dimensional systems and generates a quadratically converging sequence $\{(x_n, y_n)\}$, provided that

$$\begin{vmatrix} F_x & G_x \\ F_y & G_y \end{vmatrix} = F_x G_y - F_y G_x \neq 0$$

This is equivalent to the condition $F'(x) \neq 0$ for the one-dimensional case. The algorithm associated with Eq. (3.7.18) is formulated as follows.

Algorithm 3.7.2.

[Generalized Newton's method for systems (GNMS)].

Step 1. Choose a first approximation $\mathbf{x}_0 = (x_0, y_0)$, a tolerance $\epsilon > 0$, a maximum number of iterations N, and set $n = 1$.

Step 2. Generate $\mathbf{x}_n = (x_n, y_n)$, using Eq. (3.7.18).

Step 3. If $\| \mathbf{x}_n - \mathbf{x}_{n-1} \| < \epsilon$ set $\mathbf{s} = \mathbf{x}_n$ and stop.

Step 4. If $\| \mathbf{x}_n - \mathbf{x}_{n-1} \| \geq \epsilon$ and $n < N$, set $n \leftarrow n+1$ and go to step 2. Otherwise, output "maximum number of iterations exceeded," $\mathbf{x}_n$ and stop.

This algorithm can be extended and applied to general systems of any order.

Example 3.7.3.

Consider the two-dimensional system

$$F(x, y) = xy - 1 = 0$$
$$G(x, y) = x^3 - xy^2 + x - 1 = 0$$

which has a solution $(s, t) = (1, 1)$. Here $F_x = y$, $F_y = x$, $G_x = 3x^2 - y^2 + 1$, $G_y = -2xy$, and at the solution point

$$F_x G_y - F_y G_x = -3x^3 - xy^2 - x = -5 \neq 0$$

Newton's sequence is generated by Eq. (3.7.18) and for $x_0 = 0.8$, $y_0 = 1.2$, $\epsilon = 10^{-6}$ we obtain Table 3.7.3, which demonstrates the speed of Newton's method.

Table 3.7.3. Newton's Method for a Two-Dimensional System

n	x_n	y_n	$\| \mathbf{x}_n - \mathbf{x}_{n-1} \|$
0	0.8000000	1.2000000	
1	1.0146789	0.9279817	$0.3 \cdot 10^0$
2	0.9990840	0.9997958	$0.7 \cdot 10^{-1}$
3	1.0000005	0.9999997	$0.9 \cdot 10^{-3}$
4	1.0000000	1.0000000	$0.6 \cdot 10^{-6}$

□

PROBLEMS

1. Show that

$$\mathbf{f}(x,y) = \begin{bmatrix} \sqrt{\dfrac{x^2+y^2}{2}} \\[2mm] \dfrac{xy}{3} \end{bmatrix}, \quad 0 \le x, y \le 1$$

satisfies a Lipschitz condition with a constant $L < 1$.

2. Apply the GSIM algorithm for

$$x = \frac{1}{4} e^{-xy}, \quad y = \frac{x^3+y^3}{6}, \quad 0 \le x,y \le 1$$

for $x_0 = y_0 = 0$ and $\epsilon = 10^{-6}$.

3. Show that the system of equations

$$x = 0.3 \sin x + 0.4 \cos y, \quad -1 \le x,y \le 1$$
$$y = 0.5 \cos x - 0.3 \sin y, \quad -1 \le x,y \le 1$$

satisfies all the conditions of Theorem 3.7.1 and apply the GSIM to obtain its solution for $x_0 = 0, y_0 = 1$, and $\epsilon = 10^{-6}$.

4. The set of equations $x = x^3 - y$, $y = y^3 + x - 1$ has a solution $(s,t) = (1,0)$. Choose $x_0 = 1.1, y_0 = 0.1$ and generate $\{(x_n, y_n)\}$, using the GSIM. Does the sequence converge? Explain.

5. Use the GSIM to solve the system

$$x = \frac{y^2+z^2-1}{6}, \quad y = \frac{z^2+x^3-2}{7}, \quad z = \frac{x^4+2y^2-3}{10}$$

near $x_0 = y_0 = 1.2$, $z_0 = 0.8$. How many iterations are needed to reach a tolerance of $\epsilon = 10^{-6}$?

6. Use the GNMS to find a solution to the system

$$x^3 y - y^3 x - 6 = 0$$
$$x^2 - 4y^3 = 0$$

near $x_0 = 1.8$, $y_0 = 0.8$, with an error tolerance of $\epsilon = 10^{-8}$.

7. Use Newton's method to solve the following systems:

   (a) $x^2 - \ln(x^2 + y^2) = 0$, $x^3 - y = 0$, near $x_0 = 1.25$, $y_0 = 2$

   (b) $x \sin(xy) + 5y^2 = 0$, $x - y^2 - 4 = 0$, near $x_0 = 5$, $y_0 = 1$

   Use an error tolerance of $\epsilon = 10^{-6}$.

8. Consider a system $F(x, y) = 0$, $G(x, y) = 0$. Let (x_0, y_0) be an approximation to the system's solution (s, t) and denote $s = x_0 + h$, $t = y_0 + k$. Derive Newton's method by applying Taylor's theorem to $F(x, y)$ and $G(x, y)$ about (x_0, y_0).

9. Find a solution to the system

$$x - x^2 - y^2 = 0$$
$$y - x^2 + y^2 = 0$$

near $x_0 = 0.8$, $y_0 = 0.4$ for a tolerance $\epsilon = 10^{-6}$.

10. Solve problem 9 with various starting values. Each time increase (decrease) x_0 and/or y_0 by 0.1 until a breakdown occurs.

4

Linear Difference Equations

Many algorithms in numerical analysis are based on finding solutions to *difference equations*. In addition, difference equations are useful in various mathematical subjects such as theory of probability and mathematical economics. There is a great similarity between the theory of difference equations and the theory of differential equations. The fundamental problem in differential equations is solving them (analytically if possible) and studying the asymptotic behavior of the solutions. In difference equations we are doing the same, except that the calculated solution is a *sequence* rather than a *function*.

A sequence $X = \{x_n\}$ is a function defined on a set of integers I. For example, one can define

$$x_n = n - n^2, \quad n \geq 3$$

and obtain the sequence $X = \{-6, -12, -20, \ldots\}$. The set I over which the sequence is defined is $I = \{3, 4, 5, \ldots\}$, and the formula by which x_n is found is of the *explicit* type. Difference equations provide a tool for defining sequences *implicitly*. For example, the relations

$$x_n = 2x_{n-1} + 1, \quad n \geq 5$$
$$x_n = x_{n-1}^2 - x_{n-2}, \quad n \geq 2$$

define implicitly two sequences X_1 and X_2, respectively. The first relation is a *linear difference equation* of the first order and produces X_1 for any *initial value* of x_4. By choosing $x_4 = -2$ we obtain

$$X_1 = \{-2, -3, -5, \ldots\}$$

The second relation is a *nonlinear difference equation* of the second order and determines X_2 uniquely for any given pair of numbers x_0 and x_1. If, for example, we choose $x_0 = 1$, $x_1 = 2$, then

$$X_2 = \{1, 2, 3, 7, \ldots\}$$

Throughout this chapter we will concentrate on the solutions to linear difference equations.

4.1   General Concepts

Definition 4.1.1.   (Difference equations). *The relation*

$$a_{0,n}x_n + a_{1,n}x_{n-1} + \cdots + a_{N,n}x_{n-N} = b_n \tag{4.1.1}$$

forms a linear difference equation (LDE) of order N, for any given sequences $\{a_{0,n}\}$, $\{a_{1,n}\}, \ldots,$ $\{a_{N,n}\}, \{b_n\}$, *provided that* $\{a_{N,n}\}$ *is not identically zero and that* $a_{0,n} \neq 0$ *for all n. If* $b_n = 0$ *for all n, the difference equation is called* homogeneous.

Example 4.1.1.

The relation

$$n^2 x_n + x_{n-1} - n x_{n-2} = 3, \quad n \geq 1$$

defines a linear difference equation of order 2. For the choice $x_{-1} = 0$, $x_0 = 1$ the equation generates the sequence $\{0, 1, 2, 3/4, \ldots\}$. If we cancel the right-hand term, we obtain the homogeneous linear difference equation of order 2

$$n^2 x_n + x_{n-1} - n x_{n-2} = 0$$

and for $x_{-1} = 0$, $x_0 = 1$ it generates the sequence $\{0, 1, -1, 3/4, \ldots\}$.   □

Throughout this chapter we will mainly consider difference equations with constant coefficients, that is,

$$a_{k,n} = a_k, \quad 0 \leq k \leq N, \quad a_0 \neq 0, \quad a_N \neq 0$$

Without any loss of generality we may assume $a_0 = 1$ and thus obtain the form

$$x_n + a_1 x_{n-1} + a_2 x_{n-2} + \cdots + a_n x_{n-N} = b_n, \quad a_N \neq 0 \tag{4.1.2}$$

Let X and Y denote two sequences whose nth terms are $(X)_n = x_n$ and $(Y)_n = y_n$, respectively.

Definition 4.1.2.   *For any given sequences X and Y we define aX (multiplying by a scalar) and X + Y (sum) by*

$$(aX)_n = ax_n, \quad (X+Y)_n = x_n + y_n \tag{4.1.3}$$

If we denote by B and AX the sequences whose nth terms are b_n *and* $x_n + a_1 x_{n-1} + \cdots + a_n x_{n-N}$ *respectively, then Eq. (4.1.2) can be rewritten as*

$$AX = B \tag{4.1.4}$$

In view of Definition 4.1.2 we obtain

$$A(aX + bY) = a(AX) + b(AY) \tag{4.1.5}$$

for any given pair of numbers a and b, that is, the *operator A* is *linear*. We notice the similarity to differential equations: let D denote the differentiation operator. Then $D(af + bg) = a(Df) + b(Dg)$ for any differentiable functions f, g and numbers a, b.

The study of linear differential equations usually starts with a thorough discussion of the homogeneous case. We will follow this course and deal first with homogeneous linear difference equations of order 2.

4.2 Homogeneous Difference Equations of Order 2

Consider the case $N = 2$ in Eq. (4.1.2). The homogeneous equation is

$$x_n + a_1 x_{n-1} + a_2 x_{n-2} = 0 \tag{4.2.1}$$

where a_1, a_2 are constants (real or complex) and $a_2 \neq 0$. The equivalent to Eq. (4.2.1) in linear differential equations is the second order ordinary differential equation

$$y'' + a_1 y' + a_2 y = 0 \tag{4.2.2}$$

which possesses solutions of the form $y = e^{\lambda x}$ (for some specified values of λ). In the case of difference equations x is replaced by n and we may try a solution $x_n = e^{\lambda n} = z^n$, where $z = e^{\lambda} \neq 0$. Thus

$$z^n + a_1 z^{n-1} + a_2 z^{n-2} = z^{n-2}(z^2 + a_1 z + a_2) = 0$$

and because $z \neq 0$ we have

$$z^2 + a_1 z + a_2 = 0 \tag{4.2.3}$$

The polynomial $p_2(z) = z^2 + a_1 z + a_2$ is the *characteristic polynomial* associated with Eq. (4.2.1). It has exactly two roots, z_1 and z_2. If $z_1 \neq z_2$, two solutions to Eq. (4.2.1) in the form of

$$X^{(1)} = \{z_1^n\}, \quad X^{(2)} = \{z_2^n\} \tag{4.2.4}$$

are obtained. In the case $z_1 = z_2$, we have $p'(z_1) = 0$ because z_1 is a multiple root. Whereas the first solution is still $X^{(1)} = \{z_1^n\}$, a second solution is given by $X^{(2)} = \{nz_1^n\}$. Indeed, $p_2'(z_1) = 0$ implies $2z_1 + a_1 = 0$. Therefore

$$nz_1^n + a_1(n-1)z_1^{n-1} + a_2(n-2)z_1^{n-2} = (n-2)(z_1^n + a_1 z_1^{n-1} + a_2 z_1^{n-2}) + 2z_1^n + a_1 z_1^{n-1}$$
$$= (n-2)z_1^{n-2}(z_1^2 + a_1 z_1 + a_2) + z_1^{n-1}(2z_1 + a_1) = 0$$

leading to Theorem 4.2.1.

Theorem 4.2.1.
Consider the linear difference equation of order 2, given by Eq. (4.2.1). If the roots of the associated characteristic polynomial, z_1 and z_2, are distinct, then the sequences $X^{(1)} = \{z_1^n\}$, $X^{(2)} = \{z_2^n\}$ are solutions. If $z_1 = z_2$, the sequences $X^{(1)} = \{z_1^n\}$, $X^{(2)} = \{nz_1^n\}$ solve the given equation.

Example 4.2.1.

The characteristic polynomial of the LDE

$$x_n + x_{n-1} - x_{n-2} = 0$$

is $p_2(z) = z^2 + z - 1$ and has two distinct roots:

$$z_1 = \frac{\sqrt{5} - 1}{2}, \quad z_2 = -\frac{\sqrt{5} + 1}{2}$$

Thus the LDE has two solutions in the form of

$$X^{(1)} = \left\{ \left(\frac{\sqrt{5} - 1}{2} \right)^n \right\}, \quad X^{(2)} = \left\{ \left(-\frac{\sqrt{5} + 1}{2} \right)^n \right\} \qquad \square$$

Example 4.2.2.

Consider the difference equation

$$x_n - 2x_{n-1} + x_{n-2} = 0$$

The characteristic polynomial $p_2(z) = z^2 - 2z + 1$ has the multiple root $z_1 = 1$ and the equation has two solutions, given by

$$X^{(1)} = \{1^n\} = \{1\}, \quad X^{(2)} = \{n \cdot 1^n\} = \{n\} \qquad \square$$

Example 4.2.3.

The difference equation

$$x_n + x_{n-2} = 0$$

provides $p_2(z) = z^2 + 1$ with two complex roots, $z_1 = i, z_2 = -i$. Thus,

$$X^{(1)} = \{i^n\}, \quad X^{(2)} = \{(-i)^n\} \qquad \square$$

4.3 The General Solution to $AX = B$

We will now consider the general LDE of order 2 given by

$$x_n + a_1 x_{n-1} + a_2 x_{n-2} = b_n \qquad (4.3.1)$$

and also written as $AX = B$, where

$$(AX)_n = x_n + a_1 x_{n-1} + a_2 x_{n-2}, \quad (B)_n = b_n \qquad (4.3.2)$$

The process of determining the general solution to Eq. (4.3.1) is identical to that used in differential equations. The first useful result is as follows.

Lemma 4.3.1.

Let $X^{(1)}$ and $X^{(2)}$ be two solutions to the homogeneous LDE [Eq. (4.2.1)]. Then the sequence

$$X = c_1 X^{(1)} + c_2 X^{(2)} \tag{4.3.3}$$

solves Eq. (4.2.1) as well.

Proof.

By virtue of Eq. (4.1.5) we have

$$AX = A(c_1 X^{(1)} + c_2 X^{(2)}) = c_1 AX^{(1)} + c_2 AX^{(2)} = 0$$

where 0 is the null sequence.

The next result is associated with *uniqueness*.

Lemma 4.3.2.

(Initial-value problem: uniqueness). *The homogeneous linear difference equation $AX = 0$, defined over a set I of consecutive integers, has at most one solution that assumes given values α and β at $n = n_0$ and $n = n_0 + 1$.*

Proof.

We still consider an LDE of order 2. Let $X^{(1)}$ and $X^{(2)}$ be two solutions for which

$$[X^{(1)}]_{n_0} = [X^{(2)}]_{n_0} = \alpha, \quad [X^{(1)}]_{n_0+1} = [X^{(2)}]_{n_0+1} = \beta$$

Then the sequence $X = X^{(1)} - X^{(2)} = \{x_n\}$ satisfies $AX = 0$ (in view of the previous lemma), and in addition

$$x_{n_0} = x_{n_0+1} = 0$$

Thus two *consecutive* elements of X vanish. Suppose that X is not the zero sequence. Then a smallest integer $n > n_0 + 1$ such that $x_n \neq 0$ exists, that is,

$$x_n \neq 0, \quad x_{n-1} = x_{n-2} = 0$$

This clearly contradicts $(AX)_n = x_n + a_1 x_{n-1} + a_2 x_{n-2} = 0$ and therefore X must be the zero sequence, that is, $X^{(1)} = X^{(2)}$.

We now define linear dependency between sequences.

Definition 4.3.1. (Linear dependency). *Two sequences of numbers, $X^{(1)}$ and $X^{(2)}$, defined over the same set of integers, are called* linearly dependent *if a two-dimensional constant vector $(c_1, c_2) \neq (0, 0)$ can be found such that*

$$c_1 X^{(1)} + c_2 X^{(2)} = 0$$

that is, if the zero sequence is a nontrivial linear combination of $X^{(1)}$ and $X^{(2)}$. Otherwise the sequences are called linearly independent.

Let $X^{(1)} = \{x_n^{(1)}\}$ and $X^{(2)} = \{x_n^{(2)}\}$ be two solutions to a homogeneous LDE, $AX = 0$. The determinant

$$w_n = \begin{vmatrix} x_n^{(1)} & x_n^{(2)} \\ x_{n-1}^{(1)} & x_{n-1}^{(2)} \end{vmatrix} \tag{4.3.4}$$

is called the *Wronskian* determinant of $X^{(1)}$ and $X^{(2)}$ at point n. The next result associates linear dependency with the Wronskian determinant.

Theorem 4.3.1.

Two solutions $X^{(1)}$ and $X^{(2)}$ to $AX = 0$ are linearly dependent, if and only if $w_n = 0$ for all n.

Proof.

If $X^{(1)}$ and $X^{(2)}$ are linearly dependent, then

$$c_1 X^{(1)} + c_2 X^{(2)} = 0 \tag{4.3.5}$$

for some $(c_1, c_2) \neq (0, 0)$. Thus

$$\begin{aligned} c_1 x_n^{(1)} + c_2 x_n^{(2)} &= 0 \\ c_1 x_{n-1}^{(1)} + c_2 x_{n-1}^{(2)} &= 0 \end{aligned} \tag{4.3.6}$$

for all n. The system of two linear equations given by Eq. (4.3.6) has a nontrivial solution $(c_1, c_2) \neq (0, 0)$. Thus the determinant of its coefficients does not vanish, that is, $w_n \neq 0$ for all n.

The opposite is also true. Let $X^{(1)} = \{x_n^{(1)}\}, X^{(2)} = \{x_n^{(2)}\}$ be solutions to $AX = 0$ and let $w_n = 0$ for all n. We want to determine a vector $(c_1, c_2) \neq (0, 0)$ such that Eq. (4.3.5) holds. Let (c_1, c_2) be a nontrivial solution to the system of Eq. (4.3.6) for some particular choice $n = n_0$. Such a solution exists, because $w_{n_0} = 0$. Therefore the sequence $X = c_1 X^{(1)} + c_2 X^{(2)}$ solves $AX = 0$ and at the same time vanishes at two consecutive points: $(X)_{n_0} = (X)_{n_0-1} = 0$. Thus, in view of Lemma 4.3.2, we must have $X = 0$, that is, the solutions $X^{(1)}$ and $X^{(2)}$ are linearly dependent. □

It is easily seen that if $X^{(1)}$ and $X^{(2)}$ are both solutions to $AX = 0$, then either $w_n = 0$ for all n, or $w_n \neq 0$ for all n (the proof is left for the reader).

We will now obtain the *general solution* to the homogeneous LDE. Let $X^{(1)}$ and $X^{(2)}$ be two linearly independent solutions, that is, $AX^{(1)} = AX^{(2)} = 0$. Let $X = \{x_n\}$ be a third solution. For a particular $n = n_0$, there is a unique vector (c_1, c_2) for which

$$c_1 x_{n_0}^{(1)} + c_2 x_{n_0}^{(2)} = x_{n_0}$$

$$c_1 x_{n_0-1}^{(1)} + c_2 x_{n_0-1}^{(2)} = x_{n_0-1}$$

because $w_{n_0} \neq 0$. The two sequences X and $Y = c_1 X^{(1)} + c_2 X^{(2)}$ are both solutions to $AX = 0$ with two consecutive identical elements: $(X)_{n_0} = (Y)_{n_0}$, $(X)_{n_0-1} = (Y)_{n_0-1}$. By virtue of Lemma 4.3.2 we must have

$$X = Y = c_1 X^{(1)} + c_2 X^{(2)}$$

that is, X is a linear combination of $X^{(1)}$ and $X^{(2)}$. At the same time, every linear combination $c_1 X^{(1)} + c_2 X^{(2)}$ solves $AX = 0$. We have thus proved the following theorem.

Theorem 4.3.2.
Let $X^{(1)}$ and $X^{(2)}$ be two linearly independent solutions to $AX = 0$. Then the general solution is given by $c_1 X^{(1)} + c_2 X^{(2)}$ where c_1 and c_2, are any given numbers.

Example 4.3.1.
Consider the general homogeneous LDE of order 2

$$x_n + a_1 x_{n-1} + a_2 x_{n-2} = 0, \quad a_2 \neq 0$$

Let z_1, z_2 be the roots of the characteristic polynomial and assume that $z_1 \neq z_2$. If $X^{(1)} = \{z_1^n\}$, $X^{(2)} = \{z_2^n\}$ then

$$w_n = \begin{vmatrix} z_1^n & z_2^n \\ z_1^{n-1} & z_2^{n-1} \end{vmatrix} = (z_1 z_2)^{n-1}(z_1 - z_2)$$

In view of $a_2 \neq 0$, we have $z_1, z_2 \neq 0$ and thus $w_n \neq 0$. Therefore the general solution is given by

$$(X)_n = c_1 z_1^n + c_2 z_2^n$$

where c_1, c_2 are arbitrary constants.   □

Example 4.3.2.
In the previous example, let $z_1 = z_2$. Define

$$X^{(1)} = \{z_1^n\}, \quad X^{(2)} = \{n z_1^n\}$$

Then the Wronskian determinant is

$$w_n = \begin{vmatrix} z_1^n & n z_1^n \\ z_1^{n-1} & (n-1)z_1^{n-1} \end{vmatrix} = -z_1^{2n-1} \neq 0$$

and the general solution is

$$(X)_n = c_1 z_1^n + n c_2 z_1^n$$

where c_1, c_2 are any given constants.   □

In view of the previous examples, two linearly independent solutions to the general homogeneous LDE of order 2 can always be found.

We will now prove Theorem 4.3.3.

Theorem 4.3.3.

(Initial-value problem). *The* initial-value problem *defined as finding a solution to the general homogeneous LDE of order 2, given two initial values* $x_0 = \alpha$, $x_1 = \beta$, *has a unique solution for any given constants* α *and* β.

Proof.

Let $X^{(1)}$ and $X^{(2)}$ denote two linearly independent solutions to $x_n + a_1 x_{n-1} + a_2 x_{n-2} = 0$. The system

$$c_1 x_0^{(1)} + c_2 x_0^{(2)} = \alpha$$
$$c_1 x_1^{(1)} + c_2 x_1^{(2)} = \beta$$

of linear equations has a unique solution (c_1, c_2), because

$$w_1 = \begin{vmatrix} x_1^{(1)} & x_1^{(2)} \\ x_0^{(1)} & x_0^{(2)} \end{vmatrix} \neq 0$$

Therefore the sequence $X = c_1 X^{(1)} + c_2 X^{(2)}$ solves the initial-value problem, and by virtue of Lemma 4.3.2 it is also unique.   □

Example 4.3.3.

The LDE $x_n - x_{n-1} - 6 x_{n-2} = 0$ has the characteristic polynomial $p_2(z) = z^2 - z - 6$ with two distinct roots $z_1 = 3$, $z_2 = -2$. Thus the sequences $X^{(1)} = \{3^n\}$ and $X^{(2)} = \{(-2)^n\}$ are two independent solutions. The unique solution $X = c_1 X^{(1)} + c_2 X^{(2)}$ that satisfies the initial conditions $x_0 = 0$, $x_1 = 1$ is determined by solving the two linear equations

$$c_1 \cdot 3^0 + c_2 (-2)^0 = 0$$
$$c_1 \cdot 3 + c_2 (-2) = 1$$

Thus, $c_1 = 1/5$, $c_2 = -1/5$, and

$$x_n = \frac{1}{5} \cdot 3^n - \frac{1}{5}(-2)^n = \frac{1}{5}[3^n - (-2)^n]$$ □

We can now obtain the general solution to the inhomogeneous LDE of order 2, $AX = B$. Let $X^{(1)}$ and $X^{(2)}$ denote two linearly independent solutions to $AX = 0$ and let Y be a particular solution to

$AX = B$. For a general solution Z of the inhomogeneous LDE, we must have

$$A(Z - Y) = B - B = 0$$

Therefore the sequence $Z - Y$ solves the homogeneous LDE and can be represented as

$$Z - Y = c_1 X^{(1)} + c_2 X^{(2)}$$

for some constants c_1 and c_2. Thus

$$Z = Y + c_1 X^{(1)} + c_2 X^{(2)} \tag{4.3.7}$$

The opposite is also true, that is, the right-hand side of Eq. (4.3.7) solves $AX = B$ for any given c_1 and c_2. We have thus proved Theorem 4.3.4 (below).

Theorem 4.3.4.
(General solution to $AX = B$). *The general solution to the inhomogeneous LDE, $AX = B$, is given by* $Y + c_1 X^{(1)} + c_2 X^{(2)}$, *where* $X^{(1)}, X^{(2)}$ *are two independent solutions to $AX = 0$, Y is a particular solution to $AX = B$, and c_1, c_2 are arbitrary constants.*

Example 4.3.4.
To find a particular solution to

$$x_n - x_{n-1} - 2x_{n-2} = 1$$

let us try the constant sequence $Y = \{a, a, \ldots\}$, which implies $a = -1/2$. The characteristic polynomial $p_2(z) = z^2 - z - 2$ has two distinct roots $z_1 = 2$, $z_2 = -1$. Therefore the general solution is

$$x_n = -\frac{1}{2} + c_1(2^n) + c_2(-1)^n$$

If *initial conditions*, say, $x_0 = 0$ and $x_1 = 1$, are imposed, we solve

$$-\frac{1}{2} + c_1 + c_2 = 0$$
$$-\frac{1}{2} + 2c_1 - c_2 = 1$$

to obtain the unique solution

$$x_n = -\frac{1}{2} + \frac{2}{3}(2^n) - \frac{1}{6}(-1)^n \qquad \square$$

In view of the previous discussion, to find the general solution to $AX = B$ one merely needs to obtain one specific solution and add it to the general solution of $AX = 0$. Such a solution can frequently be found by an *intelligent* guess, depending on the form and complexity of $B = \{b_n\}$.

Example 4.3.5.

To find a particular solution of

$$x_n - 2x_{n-2} = n^2$$

we may try $x_n = \alpha n^2 + \beta n + \gamma$, where α, β, and γ are to be determined. Substituting in the LDE we find

$$\alpha n^2 + \beta n + \gamma - 2[\alpha(n-2)^2 + \beta(n-2) + \gamma] = n^2$$

Therefore

$$\alpha - 2\alpha = 1$$
$$\beta + 8\alpha - 2\beta = 0$$
$$\gamma - 8\alpha + 4\beta - 2\gamma = 0$$

that is, $\alpha = -1$, $\beta = -8$, $\gamma = -24$. Thus a particular solution is

$$(Y)_n = -n^2 - 8n - 24$$

and the general solution is

$$(X)_n = -(n^2 + 8n + 24) + c_1(\sqrt{2})^n + c_2(-\sqrt{2})^n \qquad \square$$

Finding a particular solution to the inhomogeneous LDE may not always be as straightforward as in the previous example. In the general case such a solution may be constructed by a procedure that resembles the method of *variation of constants* from differential equations. This is our next result, given here without a proof.

Theorem 4.3.5.

(Finding a particular solution). *Consider the inhomogeneous LDE, $AX = B$, where $B = \{b_n\}$ is defined (for the sake of simplicity) for $n \geq 0$. Let $X^{(1)} = \{x_n^{(1)}\}$ and $X^{(2)} = \{x_n^{(2)}\}$ be two linearly independent solutions to the homogeneous LDE, $AX = 0$, and let $W = \{w_n\}$ denote the sequence of the Wronskian determinants of these solutions. Then a particular solution $Y = \{y_n\}$, $n \geq 0$ to $AX = B$, is given by*

$$y_n = \sum_{m=0}^{n} \frac{\begin{vmatrix} x_n^{(1)} & x_n^{(2)} \\ x_{m-1}^{(1)} & x_{m-1}^{(2)} \end{vmatrix}}{w_m} b_m, \quad n \geq 0 \qquad (4.3.8)$$

Example 4.3.6.

Consider the inhomogeneous LDE

$$x_n - 3x_{n-1} + 2x_{n-2} = 1$$

The two roots of the characteristic polynomial associated with the homogeneous LDE are $z_1 = 1$ and $z_2 = 2$. Thus two linearly independent solutions of $x_n - 3x_{n-1} + 2x_{n-2} = 0$ are

$$X^{(1)} = \{1^n\} = \{1\}, \quad X^{(2)} = \{2^n\}$$

Substituting in Eq. (4.3.8) we obtain

$$y_n = \sum_{m=0}^{n} \frac{\begin{vmatrix} 1 & 2^n \\ 1 & 2^{m-1} \end{vmatrix}}{\begin{vmatrix} 1 & 2^m \\ 1 & 2^{m-1} \end{vmatrix}} = \sum_{m=0}^{n} \frac{2^{m-1} - 2^n}{2^{m-1} - 2^m} = \sum_{m=0}^{n} \frac{1 - 2^{n-m+1}}{1 - 2}$$

$$= \sum_{m=0}^{n} (2^{n-m+1} - 1) = 2^{n+2} - n - 3$$

and one can verify that $Y = \{y_n\}$ is indeed a solution. $\qquad\square$

PROBLEMS

1. Consider the nonlinear difference equation $x_{n+1} = x_n(1 - x_n)$ with an initial value x_0, $0 \le x_0 \le 1$. Show that $\lim_{n \to \infty} x_n = 0$. (*Hint*: Show first that the sequence $\{x_n\}$ is a monotonic decreasing sequence.)

2. Find two linearly independent solutions to the following homogeneous LDEs:
   (a) $x_n - x_{n-1} - 3x_{n-2} = 0$ (b) $x_n + 2x_{n-2} = 0$
   (c) $x_n + 2x_{n-1} + x_{n-2} = 0$ (d) $x_n + ix_{n-1} - x_{n-2} = 0$

3. Consider the LDE $x_n + a_1 x_{n-1} + a_2 x_{n-2} = 0$. Let a_1, a_2 be real and let the roots of the characteristic polynomial be distinct and complex. Thus $z_2 = \bar{z}_1$. The sequences $\{z_1^n\}$, $\{\bar{z}_1^n\}$ are two linearly independent solutions. Construct two *real* sequences $X^{(1)}$ and $X^{(2)}$ that are also linearly independent solutions.

4. Use the result of problem 3 to find two real, linearly independent solutions to $x_n - 2ax_{n-1} + x_{n-2} = 0$, where $|a| < 1$.

5. The solutions to $x_n - x_{n-1} - x_{n-2} = 0$ are called *Fibonacci sequences*. Let $X = \{x_n\}$ denote such a sequence. Show that if $\lim_{n \to \infty} x_n = \infty$, then

$$\lim_{n \to \infty} \frac{x_{n+1}}{x_n} = \frac{1 + \sqrt{5}}{2}$$

Otherwise, show that

$$\lim_{n \to \infty} \frac{x_{n+1}}{x_n} = \frac{1 - \sqrt{5}}{2}$$

6. Solve the following initial-value problems for homogeneous LDEs:

(a) $x_n + x_{n-1} - x_{n-2} = 0, \quad x_0 = 1, \; x_1 = 1$

(b) $x_n - 2x_{n-2} = 0, \quad x_0 = 1, \; x_1 = -1$

(c) $x_n - 2x_{n-1} + x_{n-2} = 0, \quad x_2 = 1, \; x_3 = 2$

7. Obtain particular and general solutions to the following inhomogeneous LDEs:

(a) $x_n + x_{n-1} - 4x_{n-2} = 3$

(b) $x_n - x_{n-1} - x_{n-2} = 2^n$

(c) $x_n - 2x_{n-2} = n$

(d) $x_n + x_{n-1} - 2x_{n-2} = 1$

8. Determine a particular solution to $x_n + a_1 x_{n-1} + a_2 x_{n-2} = A$, where A is any given constant. (*Hint*: One of the choices $x_n = \alpha$, $x_n = \beta n$, or $x_n = \gamma n^2$, where α, β, and γ are constants to be determined, must always work.)

9. Consider the inhomogeneous LDE $x_n - 2x_{n-1} - 3x_{n-2} = 2$. Find a necessary and sufficient relation between x_0 and x_1 such that the sequence $\{x_n\}$ is bounded, that is, $|x_n| \leq C$ for all n, for some constant C. (*Hint*: Present the general solution in terms of x_0 and x_1.)

10. Consider the homogeneous LDE $x_n + 2ax_{n-1} + bx_{n-2} = 0$, where a, b are real. Show that a necessary and sufficient condition for *all* the solutions $X = \{x_n\}$ to satisfy

$$\lim_{n \to \infty} x_n = 0$$

is that the point (a, b) be inside the triangle shown in Fig. 4.3.1.

$$y = 1, \; 2x - y - 1 = 0, \; 2x + y + 1 = 0$$

11. Use Eq. (4.3.8) to determine a specific solution to the inhomogeneous LDE

$$x_n - 3x_{n-1} + 2x_{n-2} = 2^n$$

4.4 Linear Difference Equations of Order N

As in linear differential equations, the theory previously developed for LDEs of order 2, holds for LDEs of order N as well. The general linear difference equation of order N with constant coefficients is given by

$$x_n + a_1 x_{n-1} + \cdots + a_N x_{n-N} = b_n, \quad a_N \neq 0 \tag{4.4.1}$$

where a_i, $1 \leq i \leq N$ and b_n are real or complex numbers. If $b_n = 0$ for all n, the LDE is homogeneous. Equation (4.4.1) can also be written

$$AX = B \tag{4.4.2}$$

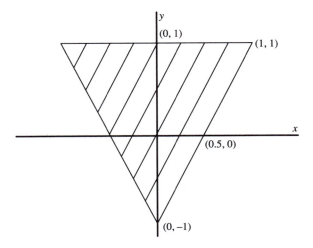

Figure 4.3.1. The triangle of problem 10.

where A is the linear difference operator defined by

$$(AX)_n = x_n + a_1 x_{n-1} + \cdots + a_N x_{n-N} \tag{4.4.3}$$

and $X = \{x_n\}$, $B = \{b_n\}$ are sequences. The homogeneous case is represented by

$$AX = 0 \tag{4.4.4}$$

where 0 is the null sequence, and the inhomogeneous case is represented by

$$AX = B \tag{4.4.5}$$

The following lemmas are the analogs to Lemmas 4.3.1 and 4.3.2.

Lemma 4.4.1.
A linear combination of solutions to $AX = 0$ is also a solution.

Lemma 4.4.2.
The homogeneous LDE, $AX = 0$, defined over a set I of consecutive integers, has at most one solution, which assumes given values $\alpha_1, \alpha_2, \ldots, \alpha_N$ at $n = n_0, n_0 - 1, \ldots, n_0 - N + 1$, respectively.
   Thus a solution to $AX = 0$ that vanishes at N consecutive integers must be identically zero.
   The following definition of linear dependency is extended to any number of sequences.

   Definition 4.4.1.   (Linear dependency). *Let $X^{(1)}, \ldots, X^{(k)}$ be sequences of numbers defined over the same set of integers. If a kth dimensional vector $(c_1, \ldots, c_k) \neq (0, \ldots, 0)$ such that*

$$c_1 X^{(1)} + c_2 X^{(2)} + \cdots + c_k X^{(k)} = 0$$

can be found, the sequences are called linearly dependent. Otherwise they are said to be linearly independent.

Now let $X^{(1)} = \{x_n^{(1)}\}$, $X^{(2)} = \{x_n^{(2)}\}, \ldots, X^{(N)} = \{x_n^{(N)}\}$ be solutions to $AX = 0$. We define the Wronskian determinant of $X^{(1)}, \ldots, X^{(N)}$ at the point n as

$$
w_n = \begin{vmatrix}
x_n^{(1)} & x_n^{(2)} & \cdots & x_n^{(N)} \\
x_{n-1}^{(1)} & x_{n-1}^{(2)} & \cdots & x_{n-1}^{(N)} \\
\vdots & \vdots & & \vdots \\
x_{n-N+1}^{(1)} & x_{n-N+1}^{(2)} & \cdots & x_{n-N+1}^{(N)}
\end{vmatrix}
\tag{4.4.6}
$$

The next result is the extended form of Theorem 4.3.1.

Theorem 4.4.1.

The solutions $X^{(1)}, \ldots, X^{(N)}$ to $AX = 0$ are linearly dependent if and only if $w_n = 0$ for all n.

If a set $X^{(1)}, \ldots, X^{(N)}$ of linearly independent solutions to the homogeneous problem can be found, it can be used as a *basis* to the set of all solutions. This is formulated in the next theorem, an analog to Theorem 4.3.2.

Theorem 4.4.2.

(The general solution to $AX = 0$). *Let $X^{(1)}, \ldots, X^{(N)}$ be N linearly independent solutions to the LDE $AX = 0$ of order N. Then the general solution is given by*

$$
X = \sum_{i=1}^{N} c_i X^{(i)}
\tag{4.4.7}
$$

where c_i, $1 \le i \le N$ are arbitrary constants.

We will now generate such N linearly independent solutions. Let us start by defining the characteristic polynomial associated with the operator A:

$$
p_n(z) = z^n + a_1 z^{n-1} + \cdots + a_{N-1} z + a_N
\tag{4.4.8}
$$

If z_0 is a root of this polynomial, the sequence $X = \{x_n\} = \{z_0^n\}$ is a solution to $AX = 0$. Indeed, because $p_n(z_0) = 0$ we have

$$
\begin{aligned}
(AX)_n &= x_n + a_1 x_{n-1} + \cdots + a_N x_{n-N} \\
&= z^n + a_1 z_0^{n-1} + \cdots + a_N z_0^{n-N} = z_0^{n-N} p_n(z_0) = 0
\end{aligned}
$$

and X is a solution. Let now z_0 be a root of $p_n(z)$ with multiplicity $m > 1$. In this case z_0 is a root of $p_n^{(i)}(z_0)$, $1 \le i \le m - 1$ as well, that is,

$$
p_n(z_0) = p_n'(z_0) = \cdots = p_n^{(m-1)}(z_0) = 0
\tag{4.4.9}
$$

We define the sequences

$$X^{(0)} = \{x_n^{(0)}\} = \{z_0^n\}$$
$$X^{(1)} = \{x_n^{(1)}\} = \{nz_0^{n-1}\}$$
$$X^{(2)} = \{x_n^{(2)}\} = \{n(n-1)z_0^{n-2}\}$$

$$\vdots$$

$$X^{(m-1)} = \{x_n^{(m-1)}\} = \{n(n-1)(n-2)\cdots(n-m+2)z_0^{n-m+1}\} \qquad (4.4.10)$$

and will show that they are solutions to $AX = 0$. The first sequence $X^{(0)}$ is a solution in view of the previous discussion. For any other sequence $X^{(k)}$, $1 \leq k \leq m - 1$ we have

$$
\begin{aligned}
[AX^{(k)}]_n &= x_n^{(k)} + a_1 x_{n-1}^{(k)} + \cdots + a_N x_{n-N}^{(k)} \\
&= n(n-1)\cdots(n-k+1)z_0^{n-k} \\
&\quad + a_1(n-1)(n-2)\cdots(n-k)z_0^{n-k-1} + \cdots \\
&\quad + a_N(n-N)(n-N-1)\cdots(n-N-k+1)z_0^{n-k-N} \\
&= [z^{n-N}p_n(z)]^{(k)}(z_0)
\end{aligned}
$$

We recall the Leibnitz rule for differentiating a product

$$[f(z)g(z)]^{(k)} = \sum_{i=0}^{k} \binom{k}{i} [f(z)]^{(i)} [g(z)]^{(k-i)} \qquad (4.4.11)$$

by which

$$[z^{n-N}p_n(z)]^{(k)} = \sum_{i=0}^{k} \binom{k}{i} (z^{n-N})^{(i)} [p_n(z)]^{(k-i)}$$

Because $1 \leq k \leq m - 1$ and in view of Eq. (4.4.9) we obtain

$$[AX^{(k)}]_n = [z^{n-N}p_n(z)]^{(k)}(z_0) = 0$$

that is, $X^{(k)}$, $0 \leq k \leq m - 1$ solve the homogeneous equation.

Now let $z_1, z_2, \ldots, z_k$ be *all* the distinct roots of $p_n(z)$ with multiplicities $m_1, m_2, \ldots, m_k$, respectively, that is, $\sum_{i=1}^{k} m_i = N$. For each z_i, $1 \leq i \leq k$ we construct the sequences $X^{(i,l)}$, $0 \leq l \leq m_i - 1$ defined as in Eq. (4.4.10). In general, $X^{(i,l)} = \{x_n^{(i,l)}\}$, where

$$x_n^{(i,l)} = n(n-1)\cdots(n-l+1)z_i^l, \quad 1 \leq i \leq k, \quad 0 \leq l \leq m_i - 1 \qquad (4.4.12)$$

We thus obtain a total of N sequences, all of which solve $AX = 0$. In addition, this set of solutions forms a basis for the complete set of solutions.

Theorem 4.4.3.
The N solutions defined by Eq. (4.4.12) to $AX = 0$ are linearly independent.

The proof of this theorem is omitted. Once known to be independent, these solutions form, by virtue of Theorem 4.4.2, a basis for the set of all solutions.

Example 4.4.1.
The characteristic polynomial of the fourth order LDE $x_n - x_{n-2} - 2x_{n-4} = 0$ is $p_4(z) = z^4 - z^2 - 2 = 0$. The four roots are

$$z_1 = \sqrt{2}, \quad z_2 = -\sqrt{2}, \quad z_3 = i, \quad z_4 = -i$$

and are all distinct. The general solution is therefore

$$x_n = c_1(\sqrt{2})^n + c_2(-\sqrt{2})^n + c_3 i^n + c_4(-i)^n$$

where c_i, $1 \le i \le 4$ are arbitrary constants. □

Example 4.4.2.
Consider the LDE given by

$$x_n - 6x_{n-1} + 13x_{n-2} - 12x_{n-3} + 4x_{n-4} = 0$$

The characteristic polynomial

$$p_4(z) = z^4 - 6z^3 + 13z^2 - 12z + 4$$

can be shown to have roots $z_1 = 1$ and $z_2 = 2$, both with multiplicity 2. Thus four solutions that form a basis are

$$X^{(1)} = \{1^n\} = \{1\}$$
$$X^{(2)} = \{n \cdot 1^{n-1}\} = \{n\}$$
$$X^{(3)} = \{2^n\}$$
$$X^{(4)} = \{n \cdot 2^{n-1}\}$$

The general solution is therefore

$$x_n = c_1 + c_2 n + c_3 \cdot 2^n + nc_4 \cdot 2^{n-1}$$ □

If z_0 is a root with multiplicity m, then rather than constructing m linearly independent solutions using Eq. (4.4.10), we may alternatively define

$$X^{(0)} = \{z_0^n\}$$
$$X^{(1)} = \{nz_0^n\}$$
$$X^{(2)} = \{n^2 z_0^n\}$$
$$\vdots$$
$$X^{(m-1)} = \{n^{m-1} z_0^n\} \tag{4.4.13}$$

as m somewhat more convenient linearly independent solutions to $AX = 0$. The proof is left as an exercise for the reader.

Example 4.4.3.

Consider the homogeneous LDE

$$x_n - 3x_{n-1} + 3x_{n-2} - x_{n-3} = 0$$

The characteristic polynomial is

$$p_3(z) = z^3 - 3z^2 + 3z - 1 = (z-1)^3$$

It possesses a single root $z_1 = 1$ with multiplicity 3. The general solution can therefore be expressed as

$$x_n = c_1 + c_2 n + c_3 n^2 \qquad\qquad \square$$

The next result is the analog to the initial-value problem discussed in Theorem 4.3.3.

Theorem 4.4.4.

(Initial-value problem). *A homogeneous LDE of order N has precisely one solution, which assumes given values at the consecutive integers $n = m, m-1, \ldots, m-N+1$ for any given m.*

Once the general solution to the homogeneous LDE of order N is known, one needs to find just one particular solution to the inhomogeneous LDE in order to determine all of them. This is the analog to Theorem 4.3.4 and is our next result.

Theorem 4.4.5.

(General solution to $AX = B$). *Let $X^{(1)}, \ldots, X^{(N)}$ be N linearly independent solutions to $AX = 0$ and let Y solve the inhomogeneous LDE, $AX = B$. Then the general solution to $AX = B$ is given by*

$$Z = Y + \sum_{i=1}^{N} c_i X^{(i)} \tag{4.4.14}$$

where c_i, $1 \le i \le N$ are arbitrary constants.

A particular solution to $AX = B$ may often be guessed if B is not especially complicated. Otherwise, one may use the following result (analog to Theorem 4.3.5) to construct one.

Theorem 4.4.6.

(Constructing a particular solution to $AX = B$). *Let $X^{(i)} = \{x_n^{(i)}\}$, $1 \le i \le N$ be a set of linearly independent solutions to the homogeneous LDE, $AX = 0$, and let $W = \{w_n\}$ be the sequence of their Wronskian determinant. If $B = \{b_n\}$ is a sequence defined over all the nonnegative integers, then $Y = \{y_n\}$, where*

$$y_n = \sum_{m=0}^{n} \frac{b_m}{w_m} \begin{vmatrix} x_n^{(1)} & x_n^{(2)} & x_n^{(N)} \\ x_{m-1}^{(1)} & x_{m-1}^{(2)} & x_{m-1}^{(N)} \\ \vdots & \vdots & \vdots \\ x_{m-N+1}^{(1)} & x_{m-N+1}^{(2)} & x_{m-N+1}^{(N)} \end{vmatrix} \tag{4.4.15}$$

is a specific solution to $AX = B$.

Example 4.4.4.

Consider the inhomogeneous LDE

$$x_n - 3x_{n-1} + 3x_{n-2} - x_{n-3} = 1$$

The characteristic polynomial associated with the homogeneous equation is

$$p_3(z) = z^3 - 3z^2 + 3z - 1 = (z-1)^3$$

It has a single root $z_1 = 1$ with multiplicity 3. Thus a set of three independent solutions to the homogeneous LDE is

$$X^{(1)} = \{1\}, \quad X^{(2)} = \{n\}, \quad X^{(3)} = \{n^2\}$$

The Wronskian determinants are

$$w_n = \begin{vmatrix} 1 & n & n^2 \\ 1 & (n-1) & (n-1)^2 \\ 1 & (n-2) & (n-2)^2 \end{vmatrix} = -2$$

The other determinants that must be computed are

$$\begin{vmatrix} 1 & n & n^2 \\ 1 & m-1 & (m-1)^2 \\ 1 & m-2 & (m-2)^2 \end{vmatrix} = (m-n-1)[(m-2)^2 - n^2] - (m-n-2)[(m-1)^2 - n^2]$$

$$= m(3-2m) - n^2 - (n+1)(m-2)^2 + (n+2)(m-1)^2$$

Substituting in Eq. (4.4.15) yields

$$y_n = \frac{1}{2} \sum_{m=0}^{n} [n^2 + (n+1)(m-2)^2 - m(3-2m) - (n+2)(m-1)^2]$$

$$= \frac{n^3 + 6n^2 + 11n + 6}{6}$$

The general solution is $X = \{x_n\}$, where

$$x_n = c_1 + c_2 n + c_3 n^2 + \frac{n^3 + 6n^2 + 11n + 6}{6} \qquad \Box$$

PROBLEMS

1. Find the general solution to the following homogeneous LDEs:

   (a) $x_n - x_{n-3} = 0$

(b) $x_n - x_{n-1} - x_{n-2} + x_{n-3} = 0$

(c) $x_n + \binom{N}{1}x_{n-1} + \binom{N}{2}x_{n-2} + \cdots + \binom{N}{N-1}x_{n-N+1} + x_{n-N} = 0$

2. Find a homogeneous LDE whose basis of solutions is

(a) $X^{(1)} = \{1\},\ X^{(2)} = \{n\},\ X^{(3)} = \{2^n\}$

(b) $X^{(1)} = \{i^n\},\ X^{(2)} = \{(-i)^n\},\ X^{(3)} = \{2^n\},\ X^{(4)} = \{3^n\}$

(c) $X^{(1)} = \{1\},\ X^{(2)} = \{a^n\},\ X^{(3)} = \{1/a^n\}$

3. Consider the following homogeneous LDE of order N:

$$x_n + x_{n-1} + \cdots + x_{n-N} = 0$$

(a) Find the general solution.

(b) Find the general solution to the inhomogeneous LDE

$$x_n + x_{n-1} + \cdots + x_{n-N} = 1$$

4. Let z_0 be a multiple root of the characteristic polynomial with multiplicity 3. Show that the sequence $\{n^2 z_0^n\}$ is a solution to the homogeneous LDE.

5. Obtain a basis to the solution set of the LDE whose characteristic polynomial is

$$p_5(z) = (z - 1)^3 (z^2 + 1)$$

6. Use Eq. (4.4.15) to determine a particular solution to the inhomogeneous LDE

$$x_n - x_{n-1} - x_{n-2} + x_{n-3} = 1$$

4.5   The Backward Difference Operator

Let $X = \{x_n\}$ be a given sequence defined over a set I of consecutive integers. We define the *backward difference operator* ∇ by

$$(\nabla X)_n = x_n - x_{n-1} \tag{4.5.1}$$

We previously defined the forward difference operator Δ^k by

$$(\Delta X)_n = x_{n+1} - x_n, \quad (\Delta^k X)_n = [\Delta(\Delta^{k-1}X)]_n, \quad k \geq 2$$

The two operators are both linear difference operators and satisfy

$$(\Delta X)_n = (\nabla X)_{n+1} \tag{4.5.2}$$

for all n. We now define

$$(\nabla^k X)_n = [\nabla(\nabla^{k-1} X)]_n, \quad k \geq 2 \tag{4.5.3}$$

For $k = 2$ we have

$$(\nabla^2 X)_n = [\nabla(\nabla X)]_n = (\nabla X)_n - (\nabla X)_{n-1}$$
$$= x_n - x_{n-1} - (x_{n-1} - x_{n-2}) = x_n - 2x_{n-1} + x_{n-2}$$

In general we obtain the following theorem.

Theorem 4.5.1.

For any given sequence $X = \{x_n\}$ and all k we have

$$(\nabla^k X)_n = x_n - \binom{k}{1} x_{n-1} + \binom{k}{2} x_{n-2} - \cdots + (-1)^k x_{n-k} \tag{4.5.4}$$

Proof.

Equation (4.5.4) clearly holds for $k = 2$. We use induction and assume the equation to be valid for arbitrary k. Then

$$\begin{aligned}
(\nabla^{k+1} X)_n &= (\nabla^k X)_n - (\nabla^k X)_{n-1} \\
&= \binom{k}{0} x_n - \binom{k}{1} x_{n-1} + \binom{k}{2} x_{n-2} - \cdots + (-1)^k \binom{k}{k} x_{n-k} \\
&\quad - \left[\binom{k}{0} x_{n-1} - \binom{k}{1} x_{n-2} + \cdots + (-1)^{k-1} \binom{k}{k-1} x_{n-k} + (-1)^k \binom{k}{k} x_{n-k-1} \right] \\
&= \binom{k}{0} x_n - \left[\binom{k}{0} + \binom{k}{1} \right] x_{n-1} + \left[\binom{k}{1} + \binom{k}{2} \right] x_{n-2} - \cdots \\
&\quad + (-1)^k \left[\binom{k}{k-1} + \binom{k}{k} \right] x_{n-k} + (-1)^{k+1} \binom{k}{k} x_{n-k-1} \\
&= x_n - \binom{k+1}{1} x_{n-1} + \binom{k+1}{2} x_{n-2} - \cdots + (-1)^k \binom{k+1}{k} x_{n-k} + (-1)^{k+1} x_{n-k-1} \qquad \square
\end{aligned}$$

Example 4.5.1.

Consider a function $f(x)$ defined for $-\infty < x < \infty$ and let $x_n = nh$, where $h > 0$ is a given constant (step size) and $n \in I$ (the set of all integers). Denote

$$f_n = f(x_n), \quad F = \{f_n\}, \quad \nabla f_n = (\nabla F)_n$$

Table 4.5.1 is called the differences table associated with step h. Each entry is the difference of the two closest entries on its left.

Table 4.5.1. Differences Table

$$
\begin{array}{llll}
\vdots & & & \\
f_{n-2} & \vdots & & \\
 & \nabla f_{n-1} & \vdots & \\
f_{n-1} & & \nabla^2 f_n & \vdots \\
 & \nabla f_n & & \nabla^3 f_{n+1} & \vdots \\
f_n & & \nabla^2 f_{n+1} & & \nabla^4 f_{n+2} & \cdots \\
 & \nabla f_{n+1} & & \nabla^3 f_{n+2} & \vdots \\
f_{n+1} & & \nabla^2 f_{n+2} & \vdots \\
 & \nabla f_{n+2} & \vdots \\
f_{n+2} & \vdots \\
\vdots
\end{array}
$$

The first three columns of differences in the particular case $f(x) = \sin x$, $h = 0.1$, $10 \le n \le 16$ are given in Table 4.5.2. ☐

Table 4.5.2. Differences Table for $f(x) = \sin x$, $h = 0.1$

x_n	f_n	∇f_n	$\nabla^2 f_n$	$\nabla^3 f_n$
1.0	0.841471			
		0.049736		
1.1	0.891207		−0.008905	
		0.040832		−0.000408
1.2	0.932039		−0.009313	
		0.031519		−0.000315
1.3	0.963558		−0.009628	
		0.021892		−0.000219
1.4	0.985450		−0.009846	
		0.012045		−0.000120
1.5	0.997495		−0.009967	
		0.002079		
1.6	0.999574			

Differences tables are particularly applicable in determining the behavior of the derivatives of a given function. In this context the following result is useful.

Theorem 4.5.2.

(Differences of a polynomial function). *Let $f(x)$ be defined for $-\infty < x < \infty$, let k be a positive integer, let $h > 0$, and denote $f_n = f(nh)$. Then*

$$\nabla^k f_n = 0, \quad n \in I \tag{4.5.5}$$

when I is the set of all integers, if and only if $f_n = p(nh)$, where p is a polynomial of degree $< k$.

Proof.

1. Let the sequence $F = \{f_n\}$ satisfy Eq. (4.5.5). In view of Theorem 4.5.1 this sequence is a solution to the homogeneous LDE

$$x_n - \binom{k}{1} x_{n-1} + \binom{k}{2} x_{n-2} - \cdots + (-1)^k x_{n-k} = 0 \tag{4.5.6}$$

The associated characteristic polynomial is

$$p_k(z) = z^k - \binom{k}{1} z^{k-1} + \binom{k}{2} z^{k-2} - \cdots + (-1)^k = (z-1)^k$$

and has a single root $z_1 = 1$ with multiplicity k. Therefore, the sequences $\{1\}, \{n\}, \{n^2\}, \ldots, \{n^{k-1}\}$ form a basis of k linearly independent solutions and a vector $(c_1, c_2, \ldots, c_k)$ such that

$$f_n = c_1 + c_2 n + c_3 n^2 + \cdots + c_k n^{k-1} \tag{4.5.7}$$

exists. Therefore, $f_n = p(nh)$, where

$$p(x) = c_1 + \left(\frac{c_2}{h}\right) x + \left(\frac{c_3}{h^2}\right) x^2 + \cdots + \left(\frac{c_k}{h^{k-1}}\right) x^{k-1} \tag{4.5.8}$$

2. Let $f(x) = p(x)$ at $x = nh$, $n \in I$, where $p(x)$ is a polynomial of degree $< k$, that is,

$$p(x) = a_0 + a_1 x + \cdots + a_{k-1} x^{k-1}$$

Therefore,

$$\begin{aligned}
f_n &= a_0 + a_1(nh) + \cdots + a_{k-1}(nh)^{k-1} \\
&= a_0 + (a_1 h)n + \cdots + (a_{k-1}h^{k-1})n^{k-1}
\end{aligned}$$

By virtue of part 1 (above), the sequences $\{1\}, \{n\}, \ldots, \{n^{k-1}\}$ are solutions of $\nabla^k X = 0$ and so is any of their linear combinations, in particular $F = \{f_n\}$. □

Example 4.5.2.

Let $f(x) = x^4 - x - 1$. In view of the previous theorem, $\nabla^5 f_n = 0$, $n \in I$ as shown in Table 4.5.3.

Table 4.5.3. Differences Table for $f(x) = x^4 - x - 1$, $h = 1$

x_n	f_n	∇f_n	$\nabla^2 f_n$	$\nabla^3 f_n$	$\nabla^4 f_n$	$\nabla^5 f_n$
-3	83					
		-66				
-2	17		50			
		-16		-36		
-1	1		14		24	
		-2		-12		0
0	-1		2		24	
		0		12		0
1	-1		14		24	
		14		36		0
2	13		50		24	
		64		60		0
3	77		110		24	
		174		84		
4	251		194			
		368				
5	619					

□

PROBLEMS

1. Find the general solution to the LDE

$$[(\Delta^2 - \nabla^2)X]_n = 0$$

2. For all $k > 0$ and any given sequence $X = \{x_n\}$, show that

$$(\Delta^k X)_n = (\nabla^k X)_{n+k}$$

and thus deduce

$$(\Delta^k X)_n = x_{n+k} - \binom{k}{1} x_{n+k-1} + \binom{k}{2} x_{n+k-2} - \cdots + (-1)^k x_n$$

3. Using Eq. (4.5.4) and induction, show that for any given sequence $X = \{x_n\}$

$$x_{n-k} = (\nabla^0 X)_n - \binom{k}{1}(\nabla X)_n + \binom{k}{2}(\nabla^2 X)_n - \cdots + (-1)^k(\nabla^k X)_n$$

where $\nabla^0 X = X$.

4. Construct a differences table of order k, $0 \leq k \leq 3$ for the function xe^x. Let $x_n = nh$, where $h = 0.05$ and $0 \leq n \leq 10$.

5. Consider a sequence $F = \{f_n\}$, $-\infty < n < \infty$ for which

$$f_n = \begin{cases} 0 & , \quad n \neq 0 \\ \epsilon > 0, & n = 0 \end{cases}$$

where the zeroes are exact values and ϵ is a computational error that approximates zero. Construct a differences table for F and find the general pattern by which the single error spreads.

4.6 Application: Bernoulli's Method

The extensive discussion in Chapter 3 about solving nonlinear equations assumes that a first good approximation to the solution is given. An algorithm that provides a reasonable initial guess is generally not available. However, if we restrict ourselves to polynomials, such algorithms often exist. In this section we will discuss Bernoulli's method for calculating the *dominant* zeroes of a given polynomial, that is, the zeroes with the largest absolute value. We start with the simplest case of a single dominant zero.

4.6.1 A Single Dominant Zero

Consider a polynomial $p(z)$ of order N

$$p(z) = a_0 z^N + a_1 z^{N-1} + \cdots + a_{N-1} z + a_N, \quad a_N \neq 0, \quad a_0 \neq 0 \tag{4.6.1}$$

with complex coefficients and that has N *distinct* zeroes $z_1, \ldots, z_N$. Thus the LDE associated with this polynomial, that is,

$$a_0 x_n + a_1 x_{n-1} + \cdots + a_N x_{n-N} = 0 \tag{4.6.2}$$

has a general solution of the form

$$x_n = c_1 z_1^n + c_2 z_2^n + \cdots + c_N z_N^n \tag{4.6.3}$$

where $(c_1, \ldots, c_N)^T$ is any given N-dimensional complex vector. Let us assume a single dominant zero, for example,

$$|z_1| > |z_k|, \quad 2 \leq k \leq N \tag{4.6.4}$$

Consider a solution to Eq. (4.6.2) for which $c_1 \neq 0$. Then

$$\begin{aligned} \frac{x_{n+1}}{x_n} &= \frac{c_1 z_1^{n+1} + c_2 z_2^{n+1} + \cdots + c_N z_N^{n+1}}{c_1 z_1^n + c_2 z_2^n + \cdots + c_N z_N^n} \\ &= z_1 \frac{1 + (c_2/c_1)(z_2/z_1)^{n+1} + \cdots + (c_N/c_1)(z_N/z_1)^{n+1}}{1 + (c_2/c_1)(z_2/z_1)^n + \cdots + (c_N/c_1)(z_N/z_1)^n} \end{aligned} \tag{4.6.5}$$

and by virtue of Eq. (4.6.4) one determines

$$\lim_{n \to \infty} \frac{x_{n+1}}{x_n} = z_1 \qquad (4.6.6)$$

We have thus proved the following theorem.

Theorem 4.6.1.

(Bernoulli's method). *If a given polynomial has distinct zeroes and a single dominant zero z_1, and $X = \{x_n\}$ is a solution to Eq. (4.6.2), given by Eq. (4.6.3) where $c_1 \neq 0$, then the sequence q_n defined by*

$$q_n = \frac{x_{n+1}}{x_n} \qquad (4.6.7)$$

converges to z_1 as $n \to \infty$.

To construct X, we choose starting values $x_{-1}, \ldots, x_{-N}$ and define

$$x_n = -\frac{a_1 x_{n-1} + \cdots + a_N x_{n-N}}{a_0}, \quad n \geq 0 \qquad (4.6.8)$$

hoping for $c_1 \neq 0$. We will later adopt a particular choice of starting values that guarantees $c_1 \neq 0$.

Example 4.6.1.

Consider the polynomial

$$p(z) = z^3 - 2z^2 - z + 2 = (z - 1)(z + 1)(z - 2)$$

where $z_1 = 2$ is the only dominant zero. The sequence $\{x_n\}$ is given by

$$x_n = -\frac{-2x_{n-1} - x_{n-2} + 2x_{n-3}}{1} = 2x_{n-1} + x_{n-2} - 2x_{n-3}$$

Table 4.6.1 contains x_n and q_n for the starting values $x_{-3} = 0$, $x_{-2} = -1$, $x_{-1} = 1$ for $0 \leq n \leq 5$. The sequence $\{q_n\}$ converges to $z_1 = 2$, but the convergence is slow.

Table 4.6.1. Bernoulli's Method for
$$p(z) = z^3 - 2z^2 - z + 2$$

n	x_n	q_n
-3	0	
-2	-1	
-1	1	1
0	1	5
1	5	1.8
2	9	2.333
3	21	1.952
4	41	2.073
5	85	
$\vdots$	$\vdots$	$\vdots$
10	2729	2.001
11	5461	

If any of the nondominant zeroes $z_2, \ldots, z_N$ has multiplicity > 1, Eq. (4.6.6) still holds. Indeed, let z_k, $2 \le k \le N$ be a multiple root of $p(z)$ with multiplicity $m_k > 1$. Then the general solution to Eq. (4.6.2), in addition to z_k^n, contains the terms $n^l z_k^n$, $1 \le l \le m_k - 1$. Consequently, the numerator and the denominator of Eq. (4.6.5) include $n^l (z_k/z_1)^{n+1}$ and $n^l (z_k/z_1)^n$, $0 \le l \le m_k - 1$, respectively. Using standard calculus we obtain

$$\lim_{n\to\infty} n^r q^n = 0, \quad -\infty < r < \infty \tag{4.6.9}$$

for all $|q| < 1$. Because $|z_k/z_1| < 1$, $2 \le k \le N$, we still obtain $\lim_{n\to\infty}(x_{n+1}/x_n) = z_1$.

The next case is a dominant zero with multiplicity > 1.

4.6.2 A Multiple Dominant Zero

Let z_1 denote a dominant zero with multiplicity $m > 1$, that is, $z_1 = z_2 = \cdots = z_m$, and let the remaining zeroes $z_{m+1}, \cdots, z_N$ satisfy

$$|z_1| > |z_k|, \quad m + 1 \le k \le N$$

For the sake of simplicity we assume that each z_k, $m + 1 \le k \le N$ is of multiplicity 1. The general solution to Eq. (4.6.2) is therefore

$$x_n = c_1 n^{m-1} z_1^n + c_2 n^{m-2} z_1^n + \cdots + c_m z_1^n + c_{m+1} z_{m+1}^n + \cdots + c_N z_N^n \tag{4.6.10}$$

and can be shown to satisfy

$$\frac{x_{n+1}}{x_n} \approx z_1 \left[1 + \frac{m-1}{n + (c_2/c_1)} \right] \tag{4.6.11}$$

Thus $q_n = x_{n+1}/x_n$ approaches z_1 as $n \to \infty$, but much slower than before. Indeed, q^n (for $|q| < 1$) approaches 0 much faster than $1/n$ as $n \to \infty$.

Example 4.6.2.

The polynomial $p(z) = z^3 - 5z^2 + 8z - 4$ has a dominant zero $z_1 = 2$ of multiplicity 2 and a third zero $z_3 = 1$. The LDE associated with this polynomial is

$$x_n - 5x_{n-1} + 8x_{n-2} - 4x_{n-3} = 0$$

By choosing

$$x_{-3} = x_{-2} = 0, \quad x_{-1} = 1$$

as initial values we obtain Table 4.6.2 for x_n and q_n.   □

Table 4.6.2. Bernoulli's Method in the Case of a
Multiple Single Dominant Zero

n	x_n	q_n
-3	0	
-2	0	
-1	1	
0	5	3.40
1	17	2.88
2	49	2.63
3	129	2.49
4	321	2.40
5	769	2.33
$\vdots$	$\vdots$	$\vdots$
18	$19,922,945$	2.11
19	$41,943,041$	2.10
20	$88,080,385$	

In constructing the sequence $X = \{x_n\}$, whose quotients approach z_1, it is essential to choose initial values that guarantee $c_1 \neq 0$. In the previous example, the choice $x_{-3} = x_{-2} = x_{-1} = 1$ yields $x_n = 1$, $n \geq 0$. It corresponds to $c_1 = c_2 = 0$, and is therefore prohibited.

The following algorithms for choosing $x_{-N}, \ldots, x_{-1}$ guarantee $c_1 \neq 0$.

Algorithm 4.6.1.

The sequence $X = \{x_n\}$ with initial values

$$x_{-N} = \cdots = x_{-2} = 0, \quad x_{-1} = 1$$

guarantees $c_1 \neq 0$ and $\lim_{n \to \infty}(x_{n+1}/x_n) = z_1$.

The second algorithm is more sophisticated and also provides a reasonable speed of convergence.

Algorithm 4.6.2.

For a polynomial given by Eq. (4.6.1), the choice

$$x_{-N} = -\frac{a_1}{a_0}$$

$$x_{-N+1} = -\frac{1}{a_0}(2a_2 + a_1 x_{-N})$$

$$x_{-N+2} = -\frac{1}{a_0}(3a_3 + a_2 x_{-N} + a_1 x_{-N+1})$$

$$\vdots$$

$$x_{-1} = -\frac{1}{a_0}(Na_N + a_{N-1}x_{-N} + a_{N-2}x_{-N+1} + \cdots + a_1 x_{-2}) \tag{4.6.12}$$

for initializing $X = \{x_n\}$ yields $c_1 \neq 0$.

Both proofs use complex variable theory and are omitted. The second algorithm is more attractive in view of the next result, because it increases the speed of convergence [previously given by Eq. (4.6.11)].

Theorem 4.6.2.

Let $z_1, \ldots, z_M$ be the distinct roots of the polynomial given by Eq. (4.6.1) with the multiplicities $m_1, \ldots, m_M$, respectively. Let Bernoulli's sequence $X = \{x_n\}$ be generated by Algorithm 4.6.2. Then

$$x_n = m_1 z_1^{n+1} + m_2 z_2^{n+1} + \cdots + m_M z_M^{n+1} \tag{4.6.13}$$

Again, the proof uses complex variable theory.

Example 4.6.3.

Let $p(z)$ be the polynomial given in the previous example. The initial values determined by Algorithm 4.6.2 are

$$x_{-3} = 5, \quad x_{-2} = 9, \quad x_{-1} = 17$$

and the associated sequences $X = \{x_n\}$, $Q = \{q_n\}$ are given in Table 4.6.3.

A comparison with Table 4.6.2 demonstrates the advantage of using Algorithm 4.6.2 for starting Bernoulli's sequence.   ☐

Table 4.6.3. Bernoulli's Method with
Algorithm 4.6.2 for Initialization

n	x_n	q_n
-3	5	
-2	9	
-1	17	1.941
0	33	1.970
1	65	1.985
2	129	1.992
3	257	1.996
4	513	1.998
5	1025	

We shall now discuss the rate at which Bernoulli's method converges, and follow with a description of an accelerating technique based on Aitken's method.

4.6.3   Accelerating the Convergence

Let z_1 be an only dominant zero to a given polynomial, with multiplicity 1. Furthermore, we assume

$$|z_1| > |z_2| > |z_k|, \quad 2 < k \leq N \tag{4.6.14}$$

and that c_1, c_2 of Eq. (4.6.3) satisfy $c_1, c_2 \neq 0$. Denote

$$d_n = q_n - z_1 \tag{4.6.15}$$

where q_n is given by Eq. (4.6.7). For the sake of simplicity we consider only the case of N distinct zeroes. Thus d_n is given by

$$
\begin{aligned}
d_n &= \frac{c_1 z_1^{n+1} + c_2 z_2^{n+1} + \cdots + c_N z_N^{n+1}}{c_1 z_1^n + c_2 z_2^n + \cdots + c_N z_N^n} - z_1 \\
&= \frac{c_2(z_2 - z_1)z_2^n + \cdots + c_N(z_N - z_1)z_N^n}{c_1 z_1^n + c_2 z_2^n + \cdots + c_N z_N^n} \\
&= \frac{c_2(z_2 - z_1)}{c_1}\left(\frac{z_2}{z_1}\right)^n \frac{1 + \dfrac{c_3(z_3 - z_1)}{c_2(z_2 - z_1)}\left(\dfrac{z_3}{z_2}\right)^n + \cdots + \dfrac{c_N(z_N - z_1)}{c_2(z_2 - z_1)}\left(\dfrac{z_N}{z_2}\right)^n}{1 + \dfrac{c_2}{c_1}\left(\dfrac{z_2}{z_1}\right)^n + \cdots + \dfrac{c_N}{c_1}\left(\dfrac{z_N}{z_1}\right)^n} \\
&= Aq^n(1 + \epsilon_n)
\end{aligned}
$$

where

$$A = \frac{c_2}{c_1}(z_2 - z_1), \quad q = \frac{z_2}{z_1}$$

and

$$1 + \epsilon_n = \frac{1 + \frac{c_3(z_3 - z_1)}{c_2(z_2 - z_1)}\left(\frac{z_3}{z_2}\right)^n + \cdots + \frac{c_N(z_N - z_1)}{c_2(z_2 - z_1)}\left(\frac{z_N}{z_2}\right)^n}{1 + \frac{c_2}{c_1}\left(\frac{z_2}{z_1}\right)^n + \cdots + \frac{c_N}{c_1}\left(\frac{z_N}{z_1}\right)^n}$$

Due to Eq. (4.6.13) we have $|q| < 1$ and $\lim_{n \to \infty} \epsilon_n = 0$. Therefore

$$\frac{d_{n+1}}{d_n} = \frac{Aq^{n+1}(1 + \epsilon_{n+1})}{Aq^n(1 + \epsilon_n)} = q(1 + \delta_n) \tag{4.6.16}$$

where

$$\delta_n = \frac{1 + \epsilon_{n+1}}{1 + \epsilon_n} - 1 = \frac{\epsilon_{n+1} - \epsilon_n}{1 + \epsilon_n} \to 0 \text{ as } n \to \infty$$

We now replace Eq. (4.6.15) by

$$d_{n+1} = (q + \theta_n)d_n \tag{4.6.17}$$

where $\theta_n = q\delta_n$ and satisfies

$$\lim_{n \to \infty} \theta_n = 0 \tag{4.6.18}$$

Equations (4.6.16) and (4.6.17) are identical to Eqs. (3.3.1) and (3.3.2) and suggest the use of Aitken's algorithm to accelerate the convergence of q_n to z_1.

Theorem 4.6.3.

(Accelerating the convergence of Bernoulli's method). *Let the zeroes of a given polynomial $p(z)$, z_1, ..., z_N satisfy Eq. (4.6.13) and let Bernoulli's sequence satisfy $c_1 \neq 0$ and $c_2 \neq 0$. Then, the sequence $Q' = \{q'_n\}$ defined by*

$$q'_n = q_n - \frac{(\Delta q_n)^2}{\Delta^2 q_n} \tag{4.6.19}$$

converges to z_1 faster than $Q = \{q_n\}$, that is,

$$\lim_{n \to \infty} \frac{q'_n - z_1}{q_n - z_1} = 0 \tag{4.6.20}$$

Example 4.6.4.

Let $p(z) = (z - 1)(z - 2)(z - 3) = z^3 - 6z^2 + 11z - 6$. The zeroes of $p(z)$ satisfy Eq. (4.6.13) and $z_1 = 3$. The associated LDE is

$$x_n - 6x_{n-1} + 11x_{n-2} - 6x_{n-3} = 0$$

and the choice $x_{-3} = 0$, $x_{-2} = 0$, $x_{-1} = 1$ provides a Bernoulli's sequence with $c_1 \neq 0$ (Algorithm 4.6.1) that converges to z_1. It can be also shown that $c_2 \neq 0$ as well. Thus the sequence $Q' = \{q'_n\}$

generated by Eq. (4.6.18) outperforms Q in converging to z_1, due to Eq. (4.6.19). The results are given in Table 4.6.4.

Table 4.6.4. Improving Bernoulli's Method by Aitken's Algorithm

n	x_n	q_n	q'_n
-3	0		
-2	0		
-1	1	6	3.346
0	6	4.167	3.135
1	25	3.600	3.058
2	90	3.344	3.026
3	301	3.209	3.012
4	966	3.131	
5	3025	3.084	
6	9330		

□

4.6.4 Two Conjugate Complex Dominant Zeroes

So far we assumed the polynomial $p(z)$ to possess a single dominant zero (possibly multiple). We will now discuss the case of a real polynomial (i.e., with real coefficients) having two conjugate complex dominant zeroes with multiplicity 1. Thus,

$$z_2 = \bar{z}_1, \quad |z_1| > |z_k|, \quad 3 \le k \le N \tag{4.6.21}$$

Let the initial values of $X = \{x_n\}$ be real numbers. Because the coefficients of the polynomial are also real, the whole sequence X must be real. The general term is

$$x_n = c_1 z_1^n + c_2 z_2^n + c_3 z_3^n + \cdots + c_N z_N^n$$

and by virtue of Eq (4.6.20), $c_2 = \bar{c}_1$ (why?). Therefore

$$x_n = c_1 z_1^n + \bar{c}_1 \bar{z}_1^n + c_3 z_3^n + \cdots + c_N z_N^n$$

Let $c_1 = r e^{i\phi}$ and $z_1 = R e^{i\theta}$. Then

$$\begin{aligned} x_n &= 2rR^n \cos(n\theta + \phi) + c_3 z_3^n + \cdots + c_N z_N^n \\ &= 2rR^n[\cos(n\theta + \phi) + \epsilon_n] \end{aligned} \tag{4.6.22}$$

where in view of Eq. (4.6.20)

$$\epsilon_n = \frac{c_3 z_3^n + \cdots + c_N z_N^n}{2rR^n} \to 0 \text{ as } n \to \infty$$

We are interested in finding R and θ. Let us first assume $\epsilon_n = 0$ for $n \geq N_0$, where N_0 is some sufficiently large integer, and restrict the discussion to such n only. We therefore have

$$x_n = c_1 z_1^n + \bar{c}_1 \bar{z}_1^n = 2rR^n \cos(n\theta + \phi) \tag{4.6.23}$$

Now the particular sequence $\{x_n\}$ of Eq. (4.6.22) is a solution to the homogeneous LDE

$$x_n + b x_{n-1} + c x_{n-2} = 0 \tag{4.6.24}$$

where

$$b = -2R \cos\theta, \quad c = R^2 \tag{4.6.25}$$

This can be verified by substituting and using a basic trigonometric identity. To obtain b and c and then R and θ from Eq. (4.6.24), we apply Eq. (4.6.23) consecutively for n and $n+1$ and obtain a system of two linear equations

$$
\begin{aligned}
x_n + b x_{n-1} + c x_{n-2} &= 0 \\
x_{n+1} + b x_n + c x_{n-1} &= 0
\end{aligned}
\tag{4.6.26}
$$

with the two variables b and c. The determinant of the system

$$D_n = \begin{vmatrix} x_{n-1} & x_{n-2} \\ x_n & x_{n-1} \end{vmatrix} \tag{4.6.27}$$

is given by

$$
\begin{aligned}
D_n &= \begin{vmatrix} 2rR^{n-1}\cos[(n-1)\theta + \phi] & 2rR^{n-2}\cos[(n-2)\theta + \phi] \\ 2rR^n \cos(n\theta + \phi) & 2rR^{n-1}\cos[(n-1)\theta + \phi] \end{vmatrix} \\
&= 4r^2 R^{2n-2} \sin^2\theta
\end{aligned}
\tag{4.6.28}
$$

Because z_1 is not real, we must have $\theta \neq 0, \pi$ and therefore $D_n \neq 0$. Thus the system of Eq. (4.6.25) has the unique solution

$$b = \frac{\begin{vmatrix} -x_n & x_{n-2} \\ -x_{n+1} & x_{n-1} \end{vmatrix}}{D_n}, \quad c = \frac{\begin{vmatrix} x_{n-1} & -x_n \\ x_n & -x_{n+1} \end{vmatrix}}{D_n}$$

also written as

$$b = -\frac{D_n'}{D_n}, \quad c = \frac{D_{n+1}}{D_n} \tag{4.6.29}$$

where

$$D_n' = \begin{vmatrix} x_n & x_{n-2} \\ x_{n+1} & x_{n-1} \end{vmatrix} \tag{4.6.30}$$

By virtue of Eq. (4.6.24) we obtain

$$R = \left(\frac{D_{n+1}}{D_n}\right)^{1/2}, \quad \cos\theta = -\frac{b}{2R} = \frac{D'_n}{2(D_nD_{n+1})^{1/2}} \tag{4.6.31}$$

and thus calculate z_1. If, however, $\epsilon_n \neq 0$, one can show that Eq. (4.6.30) is simply replaced by

$$R = \lim_{n\to\infty} \left(\frac{D_{n+1}}{D_n}\right)^{1/2}, \quad \cos\theta = \lim_{n\to\infty} \frac{D'_n}{2(D_nD_{n+1})^{1/2}} \tag{4.6.32}$$

We have thus established the following procedure for calculating a pair of conjugate complex dominant zeroes of a given polynomial

$$p(z) = a_0 z^N + \cdots + a_{N-1}z + a_N, \quad a_0 \neq 0, \quad a_N \neq 0$$

with real coefficients.

Algorithm 4.6.3.

(Calculating a pair of conjugate complex dominant zeroes with multiplicity 1).

Step 1. Construct a solution $X = \{x_n\}$ to the associated LDE

$$a_0 x_n + a_1 x_{n-1} + \cdots + a_N x_{n-N} = 0$$

using Algorithm 4.6.2 for initialization, to guarantee $c_1 \neq 0$.
Step 2. Generate the sequences $\{D_n\}$, $\{D'_n\}$ by Eqs. (4.6.27) and (4.6.30) and calculate the dominant zero $z_1 = Re^{i\theta}$ by using Eq. (4.6.31).

Example 4.6.5.
Consider the polynomial

$$p(z) = (z^2 + 2)(z - 1) = z^3 - z^2 + 2z - 2$$

with conjugate complex dominant zeroes $z_1 = i\sqrt{2}$, $z_2 = -i\sqrt{2}$. The associated LDE is

$$x_n - x_{n-1} + 2x_{n-2} - 2x_{n-3} = 0$$

and by applying Algorithm 4.6.2 for initialization we determine

$$x_{-3} = 1, \quad x_{-2} = -3, \quad x_{-1} = 1; \quad x_n = x_{n-1} - 2x_{n-2} + 2x_{n-3}, \quad n \geq 0$$

Table 4.6.5 demonstrates the convergence of the right-hand sides in Eq. (4.6.31) to $R = \sqrt{2} = 1.4142\ldots$ and to $\cos\theta = 0$ (i.e., $\theta = 90°$ or $\theta = 270°$).

Table 4.6.5. Calculating a Pair of
Conjugate Complex Dominant Zeroes for $p(z) = (z^2 + 2)(z - 1)$

n	x_n	D_n	D'_n	$(D_{n+1}/D_n)^{1/2}$	$D'_n / [2(D_n D_{n+1})^{1/2}]$
-3	1				
-2	-3				
-1	1	8	-12	1.871	-0.401
0	9	28	12	1.690	0.127
1	1	80	24	1.304	0.115
2	-15	136	-24	1.283	-0.069
3	1	224	-48	1.488	-0.072
4	33	496	48	1.481	0.033
5	1	1088	96	1.383	0.032
6	-63	2080	-96	1.381	-0.017
7	1	3968	-192	1.431	-0.017
8	129	8128			

$\square$

Example 4.6.6.

Let $p(z) = z^4 - 6z^3 + 24z^2 + 6z - 25$. The four zeroes of $p(z)$ are

$$z_1 = 3 + 4i, \quad z_2 = 3 - 4i, \quad z_3 = 1, \quad z_4 = -1$$

The associated LDE is

$$x_n - 6x_{n-1} + 24x_{n-2} + 6x_{n-3} - 25x_{n-4} = 0$$

We initialize a particular solution by applying Algorithm 4.6.2 and obtain

$$x_{-4} = 6, \quad x_{-3} = -12, \quad x_{-2} = -234, \quad x_{-1} = -1052$$

The remaining elements are given by

$$x_n = 6x_{n-1} - 24x_{n-2} - 6x_{n-3} + 25x_{n-4}, \quad n \geq 0$$

The first three iterations in the process of calculating z_1 are presented in Table 4.6.6.

Table 4.6.6. Calculating a Pair of Conjugate Complex Dominant Zeroes
for $p(z) = z^4 - 6z^3 + 24z^2 + 6z - 25$

n	x_n	$(D_{n+1}/D_n)^{1/2}$	$D'_n / [2(D_n D_{n+1})^{1/2}]$
-4	6		
-3	-12		
-2	-234	5.2170	0.5646
-1	-1052	4.8616	0.5870
0	-474	5.0061	0.6018
1	23508		

The two rightmost columns converge to $R = 5$ and to $\cos \theta = 0.6$, respectively. The convergence is particularly fast because $|z_3| = 1$ is significantly smaller than $|z_1| = 5$. $\square$

Bernoulli's method has two major disadvantages: treating only dominant zeroes and slow convergence. Let z_1 be a zero of a polynomial $p(z)$. Then, by Theorem 1.4.2, we have

$$p(z) = (z - z_1)q(z)$$

where $q(z)$ is another polynomial whose degree is that of $p(z)$ decreased by one. If z_1 is already computed, we can divide $p(z)$ by $(z - z_1)$, obtain $q(z)$, and apply Bernoulli's method to calculate the dominant zero of $q(z)$. This process is called *deflation* of $p(z)$, and may be applicable if, for example, the zeroes $z_1, \ldots, z_n$ of $p(z)$ are all distinct and satisfy

$$|z_1| > |z_2| > \cdots > |z_N|$$

This process, however, involves round-off error accumulation and we may find the smaller zeroes quite inaccurate. In addition, Bernoulli's method provides linear convergence, at best, and is therefore time consuming. Aitken's algorithm accelerates the process but the convergence is still linear.

It can be shown that Newton's method can be applied to compute complex zeroes of polynomials, provided that a good first approximation is available. Thus one should use Bernoulli's method only to establish a first approximation to the dominant zero and then switch to Newton's method for final computations.

PROBLEMS

1. Use Bernoulli's method to compute a single dominant zero to the following polynomials:
   (a) $p(z) = z^3 - 3z^2 + z - 3$
   (b) $p(z) = z^3 - 4z^2 - z + 4$
   (c) $p(z) = z^4 - 3z^3 + z^2 + 3z - 2$
   Construct the sequence $X = \{x_n\}$ by using Algorithm 4.6.1 and calculate q_n until $|q_n - q_{n-1}| \le 10^{-2}$.

2. The following polynomials all have a multiple single dominant zero. Use Algorithms 4.6.1 and 4.6.2 to construct Bernoulli's sequences and compare the rates of convergence of the associated $Q = \{q_n\}$ to z_1.
   (a) $p(z) = z^3 - 3z^2 + 4$
   (b) $p(z) = z^3 - (1 + 4i)z^2 - (4 - 4i)z + 4$
   (c) $p(z) = z^3 - (5 + 2i)z^2 + (2 + 4i)z + (8 + 6i)$

3. Use Algorithm 4.6.2 to obtain Bernoulli's sequence for the polynomial

$$p(z) = z^3 - z^2 - 10z - 8$$

and then construct the sequences $Q = \{q_n\}$ and $Q' = \{q'_n\}$, which converge to the only dominant zero, $z_1 = 4$. How many elements are needed by either sequence to establish three accurate significant digits?

4. Consider the polynomial

$$p(z) = z^3 - 4z^2 + 2z - 8$$

and apply Algorithm 4.6.1 to construct Bernoulli's sequence. Obtain the sequence $\{Q_n\}$ and use Aitken's algorithm to obtain $\{Q'_n\}$. Have you improved the convergence to $z_1 = 4$? Why?

5. Let $p(z) = (z - 2)^3(z - 1)(z + 1) = z^5 - 6z^4 + 11z^3 - 2z^2 - 12z + 8$.

   (a) Use Algorithms 4.6.1 and 4.6.2 and Bernoulli's method to obtain the multiple dominant zero $z_1 = 2$.

   (b) Do you expect the sequence $\{Q'_n\}$ to converge faster?

6. Approximate a pair of conjugate complex dominant zeroes to three decimal digits in the following cases:

   (a) $p(z) = z^3 + 3z^2 + 7z + 5$

   (b) $p(z) = z^4 + 5z^2 + 6$

   (c) $p(z) = z^4 + 2z^3 + 11z^2 + 2z + 10$

5

Interpolation and Approximation

Rather than approximating numbers, as, for example, the solutions of $F(x) = 0$ (Chapter 3), we will now deal with various techniques for approximating *functions*. The key words are *interpolation* and *approximation*.

Given a function $f(x)$ defined over the interval $[a, b]$, we call $g(x)$ an *approximating function* to $f(x)$, if $|f(x) - g(x)|$ is uniformly *sufficiently small* for all $x \in [a, b]$. For example, the function $\sin x$ can be approximated by $x - (x^3/6)$ for $0 \le x \le 0.1$. The *quality* of the approximation is determined by

$$\max_{0 \le x \le 0.1} \left| \sin x - x + \frac{x^3}{6} \right|$$

The other key word is *interpolation*. It describes the process of evaluating a function whose graph contains a given finite set of points, referred to as a *data set*:

$$S = \{(x_i, \, y_i), \, 1 \le i \le n\} \tag{5.1}$$

This function is called an *interpolating function* or an *interpolator* and is supposed to *approximate* the original unknown function that produced S. The most commonly used interpolators are polynomials, because they are the easiest to manipulate and evaluate. We have already introduced one kind of polynomial approximations, namely, the Taylor polynomials. These, however, are not interpolators.

Interpolation has many applications. The most important are the following:

1. Extension of a given table of a function $f(x) : \{[x_i, f(x_i)], 1 \le i \le n\}$ to include other values of x.

2. Finding a smooth function $f(x)$ that fits a given set of data either exactly or approximately.

3. Replacing a nonpolynomial function by an approximating polynomial for the purpose of *computation*.

4. Replacing a function by a polynomial approximation for the purpose of *numerical integration* and *differentiation*.

We will first discuss the problem of finding an interpolating polynomial to a given data set.

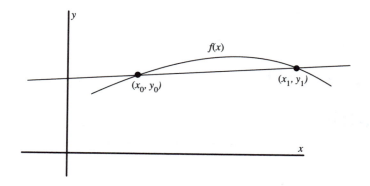

Figure 5.1.1. Two-point interpolation by a straight line.

5.1 Lagrange's Interpolator

Consider the following problem: let $f(x)$ be a real-value function defined on the interval $I = [a, b]$ and let $x_0, x_1, \ldots, x_n$ be $(n + 1)$ distinct points in I:

$$a \le x_i \le b, \quad x_i \ne x_j, \quad 0 \le i, j \le n, \quad i \ne j \tag{5.1.1}$$

Consider the set $S = \{(x_i, y_i), 0 \le i \le n\}$ where

$$y_i = f(x_i), \quad 0 \le i \le n \tag{5.1.2}$$

We are interested in constructing a polynomial $p_n(x)$ that *interpolates* $f(x)$ at $\{x_i\}_{i=0}^n$, that is,

$$p_n(x_i) = y_i, \quad 0 \le i \le n \tag{5.1.3}$$

(We often replace y_i by f_i.)

Example 5.1.1.
Consider the case $n = 1$, that is, $y_0 = f(x_0), y_1 = f(x_1)$. Clearly there is a linear polynomial that interpolates $f(x)$ at $(x_0, y_0), (x_1, y_1)$. It is the straight line, as shown in Fig. 5.1.1:

$$y - y_0 = \frac{y_1 - y_0}{x_1 - x_0}(x - x_0) \tag{5.1.4}$$

Although in Fig. 5.1.1 the linear interpolator seems to "approximate" $f(x)$ at least between x_0 and x_1, the one in Fig. 5.1.2 hardly shows any resemblance to the approximated function. This indicates that an interpolator (not necessarily a linear one) does not have to provide a good approximation to $f(x)$. However, under certain conditions it will, and finding them is the subject of this section. We start by showing the *existence* of an interpolator for a given set of points. ☐

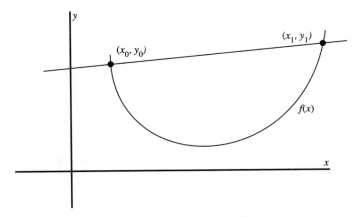

Figure 5.1.2.   A poor approximation by a linear interpolator.

Theorem 5.1.1.

(Existence of an interpolator). *For a given data set defined by Eqs. (5.1) through (5.1.2) there exists an nth order polynomial $p_n(x)$ such that $p_n(x_i) = y_i, 0 \leq i \leq n$.*

Proof.

Define the following nth order polynomials:

$$l_i(x) = \frac{(x - x_0)(x - x_1) \cdots (x - x_{i-1})(x - x_{i+1}) \cdots (x - x_n)}{(x_i - x_0)(x_i - x_1) \cdots (x_i - x_{i-1})(x_i - x_{i+1}) \cdots (x_i - x_n)}, \quad 0 \leq i \leq n$$

or, in short,

$$l_i(x) = \prod_{\substack{j=1 \\ j \neq i}}^{n} \frac{x - x_j}{x_i - x_j}, \quad 0 \leq i \leq n \tag{5.1.5}$$

Because $x_i, 0 \leq i \leq n$ are all distinct points, the denominators in Eq. (5.1.5) cannot vanish and $l_i(x)$ are well-defined nth order polynomials. Their most important feature is satisfying

$$l_i(x_j) = \begin{cases} 0, & i \neq j \\ 1, & i = j \end{cases}, \quad 0 \leq i, j \leq n \tag{5.1.6}$$

Now define

$$p_n(x) = \sum_{i=0}^{n} y_i l_i(x) \tag{5.1.7}$$

Because

$$p_n(x_j) = \sum_{i=0}^{n} y_i l_i(x_j) = y_j \cdot 1 = y_i, \quad 0 \leq j \leq n \tag{5.1.8}$$

$p(x)$ is an nth order *interpolating* polynomial.   ▯

Example 5.1.2.

For $n = 1$, we have

$$p_n(x) = y_0 l_0(x) + y_1 l_1(x) = y_0 \frac{x - x_1}{x_0 - x_1} + y_1 \frac{x - x_0}{x_1 - x_0}$$

which is another form of Eq. (5.1.4).   □

Example 5.1.3.

Consider a four-point data set $S = \{(0,0), (1,1), (2,0), (4,2)\}$ that defines a third order polynomial interpolator. We have

$$l_0(x) = \frac{(x-1)(x-2)(x-4)}{(0-1)(0-2)(0-4)} = -\frac{1}{8}(x^3 - 7x^2 + 14x - 8)$$

$$l_1(x) = \frac{(x-0)(x-2)(x-4)}{(1-0)(1-2)(1-4)} = \frac{1}{3}(x^3 - 6x^2 + 8x)$$

$$l_2(x) = \frac{(x-0)(x-1)(x-4)}{(2-0)(2-1)(2-4)} = -\frac{1}{4}(x^3 - 5x^2 + 4x)$$

$$l_3(x) = \frac{(x-0)(x-1)(x-2)}{(4-0)(4-1)(4-2)} = \frac{1}{24}(x^3 - 3x^2 + 2x)$$

and

$$p_3(x) = 0 \cdot l_0(x) + 1 \cdot l_1(x) + 0 \cdot l_2(x) + 2 \cdot l_3(x) = \frac{1}{12}(5x^3 - 27x^2 + 34x)$$ □

Constructing $p_n(x)$ by using Eq. (5.1.8) is not too efficient, and a more convenient method will be discussed in Section 5.4.

Next, we deal with the issue of uniqueness. Is $p_n(x)$ unique?

Theorem 5.1.2.

(Uniqueness of the interpolator). *The interpolation problem has a unique nth order solution.*

Proof.

Let $p_n(x)$ be the solution given by Eq. (5.1.8) and $q_n(x)$ another nth order polynomial interpolator for the same data set. Consider the difference

$$r_n(x) = p_n(x) - q_n(x) \tag{5.1.9}$$

Because by assumption

$$p_n(x_i) = q_n(x_i) = y_i, \quad 0 \le i \le n \tag{5.1.10}$$

we have

$$r_n(x_i) = y_i - y_i = 0, \quad 0 \le i \le n \tag{5.1.11}$$

Thus an nth order polynomial (at the most!) $r_n(x)$, has $(n+1)$ distinct zeroes. The ultimate conclusion is $r_n(x) = 0$ or

$$q_n(x) = p_n(x) \tag{5.1.12}$$

☐

PROBLEMS

1. Find the Lagrange interpolator for the following data sets:

   (a) $(1,2), (-5,4)$

   (b) $(0,0), (1,1), (3,-1)$

   (c) $(1,2), (2,4), (3,6), (4,8)$

2. What is your conclusion after solving part (c) of problem 1?

3. Construct $p_2(x)$ for $\{(x_i, \sin x_i)\}_{i=1}^{3}$, where $x_1 = 0, x_2 = \pi/4, x_3 = \pi/2$ and compute

$$\left\{ \frac{2}{\pi} \int_0^{\pi/2} \left[p_2(x) - \sin x \right]^2 dx \right\}^{1/2}$$

   as a measure of the quality of the approximation.

4. Find a polynomial $p(x)$ of degree ≤ 3 for which

$$p(x_1) = y_1, \quad p(x_2) = y_2$$
$$p'(x_1) = y_1', \quad p'(x_2) = y_2'$$

   where $x_1 \neq x_2$ and y_1, y_1, y_1', y_2' and given numbers. [*Hint*: Express $p(x)$ by

$$p(x) = y_1 l_1^*(x) + y_1' l_2^*(x) + y_2 l_3^*(x) + y_2' l_4^*(x)$$

   where $l_i^*(x), 1 \leq i \leq 4$ satisfy specified requirements.]

5. Prove that for all n and for all x:

$$l_0(x) + l_1(x) + \cdots + l_n(x) = 1$$

6. Consider the data set $\{(0, 1), (2, 3)\}$ and compute:

   (a) The Lagrange interpolator.

   (b) A function $f(x) = a + be^x$ that interpolates the data [i.e., find a, b such that $f(0) = 1, f(2) = 3$].

   (c) A function $f(x) = a \sin x + b \cos x$ that interpolates the data.

7. Find a function $f(x) = a + b/x + c/x^2$ that interpolates the data $(1, 1), (2, -1), (4, 0)$.

5.2 Error Estimate

Let $p_n(x)$ be the Lagrange interpolator to the function $f(x)$ at the points $x_0, x_1, \ldots, x_n$. We are interested in approximating $f(x)$ by $p_n(x)$ at *all* x, and therefore should first bound the error

$$r_n(x) = f(x) - p_n(x), \quad a \le x \le b \tag{5.2.1}$$

where $[a, b]$ is the interval of definition of the function. The next result provides an error estimate for "sufficiently" smooth functions.

Theorem 5.2.1.

(Error estimate). *Let $f(x) \in C^{n+1}[a, b]$, and $p_n(x)$, the Lagrange interpolator of $f(x)$, be defined by $\{[x_i, f(x_i)], 0 \le i \le n\}$. Then for any given $x \in [a, b]$ there is a point c_x within the smallest interval that contains $x, x_0, x_1, \ldots, x_n$ such that*

$$f(x) - p_n(x) = \frac{1}{(n+1)!} l(x) f^{n+1}(c_x) \tag{5.2.2}$$

where

$$l(x) = (x - x_0)(x - x_1) \cdots (x - x_n) \tag{5.2.3}$$

Proof.

Equation (5.2.2) is clearly valid for $x = x_i, 0 \le i \le n$. Now, consider the function

$$F(t) = f(t) - p_n(t) - Kl(t), \quad a \le t \le b \tag{5.2.4}$$

where K is a constant yet to be determined. Clearly,

$$\begin{aligned} F(x_i) &= f(x_i) - p_n(x_i) - Kl(x_i) \\ &= f(x_i) - f(x_i) - K \cdot 0 = 0, \quad 0 \le i \le n \end{aligned} \tag{5.2.5}$$

Let x be any given point in $[a, b]$ such that $x \ne x_i, 0 \le i \le n$. The constant

$$K = \frac{f(x) - p_n(x)}{l(x)} \tag{5.2.6}$$

is well defined, because $l(x) \ne 0$, and for this particular choice we obtain

$$F(x) = f(x) - p_n(x) - \frac{f(x) - p_n(x)}{l(x)} l(x) = 0 \tag{5.2.7}$$

Therefore $F(t)$ has at least $(n + 2)$ distinct zeroes in $[a, b]$, namely, $x_0, x_1, \ldots, x_n, x$. A multiple use of Rolle's theorem establishes the existence of a point c_x in the smallest interval that contains

$x_0, x_1, \ldots, x_n, x$ for which

$$F^{(n+1)}(c_x) = 0 \tag{5.2.8}$$

The polynomials $p_n(t)$ and $l(t)$ are of degrees n and $n + 1$, respectively, and the leading coefficient of $l(t)$ is 1. Therefore

$$p_n^{(n+1)}(c_x) = 0, \quad l^{(n+1)}(c_x) = (n+1)! \tag{5.2.9}$$

leading to

$$0 = f^{(n+1)}(c_x) - \frac{f(x) - p_n(x)}{l(x)}(n+1)! \tag{}$$

and to Eq. (5.2.2).

Because c_x is not generally known, Eq. (5.2.2) cannot be used to determine the exact value of $f(x) - p_n(x)$.

Example 5.2.1.
Consider the case of one interpolating point, that is, $n = 0$. Here, $p_0(x) = f(x_0)$ and $f(x) - f(x_0) = f'(c_x)(x - x_0)$ where c_x is between x_0 and x. This is, of course, the mean-value theorem.

Example 5.2.2.
For $n = 1$ (two interpolating points) we have

$$p_1(x) = \frac{(x_1 - x)f_0 + (x - x_0)f_1}{x_1 - x_0} \quad \text{(linear interpolation)}$$

with an error of

$$f(x) - p_1(x) = \frac{(x - x_0)(x - x_1)}{2!}f''(c_x) \tag{5.2.10}$$

Suppose that $|f''(x)| \le M_2, a \le x \le b$. Then

$$|f(x) - p_1(x)| \le \frac{|(x - x_0)(x - x_1)|}{2!}M_2 \tag{5.2.11}$$

If $x \in [x_0, x_1]$ then the function $|(x - x_0)(x - x_1)|$ reaches its maximum at $x = (x_0 + x_1)/2$, and

$$|f(x) - p_1(x)| \le \frac{h^2}{8}M_2 \tag{5.2.12}$$

where $h = x_1 - x_0$. Consider, for example, $f(x) = \cos x, x_0 = 0, x_1 = \pi/4$. Here

$$M_2 = \max_{0 \le x \le \pi/4} |-\cos x| = 1$$

$$p_1(x) = \frac{[(\pi/4) - x] \cdot 1 + (x - 0)\cos(\pi/4)}{[(\pi/4) - 0]} \approx 1 - 0.37292x$$

$$|\cos x - p_1(x)| \leq \frac{1}{8}\left(\frac{\pi}{4}\right)^2 \cdot 1 \approx 0.077, \quad 0 \leq x \leq \frac{\pi}{4}$$

If x is outside of $[x_0, x_1]$, the error is likely to increase beyond the bound of Eq. (5.2.11). □

Example 5.2.3.

Consider the case $n = 2$, that is, quadratic interpolation. Let $x_i = x_0 + ih, 0 \leq i \leq 2$ for some $h > 0$, and $|f^{(3)}(x)| \leq M_3$ for all x. Then

$$|f(x) - p_2(x)| \leq \frac{|(x - x_0)(x - x_1)(x - x_2)|}{3!}M_3$$

Suppose that $x_0 \leq x \leq x_2$. To find max $|(x - x_0)(x - x_1)(x - x_2)|$ we should consider only the particular case $x_0 = 0$ (why?) and calculate $L = \max_{0 \leq x \leq 2h} |x(x - h)(x - 2h)|$. By using standard calculus to locate the extrema points of a given function, we find the solutions:

$$x_1 = \frac{3h + \sqrt{3}h}{3}, \quad x_2 = \frac{3h - \sqrt{3}h}{3}$$

Both solutions yield the same maximum $L = (2\sqrt{3}/9)h^3$ and thus we have

$$|f(x) - p_2(x)| \leq \frac{2\sqrt{3}h^3}{9}\frac{M_3}{3!} = \frac{\sqrt{3}}{27}h^3 M_3 \qquad (5.2.13)$$

Let $f(x) = \sin x, x_0 = 0, x_1 = \pi/2, x_2 = \pi$. Clearly $M_3 = 1$. The parabola that interpolates $\sin x$ at $(0,0), (\pi/2, 1), (\pi, 0)$ is $p_2(x) = -(4/\pi^2)x^2 + (4/\pi)x$ and Eq. (5.2.13) provides

$$|\sin x - p_2(x)| \leq \frac{\sqrt{3}}{27}\left(\frac{\pi}{2}\right)^3 \approx 0.25$$

This poor approximation is a direct consequence of choosing a large $h(\approx 1.57)$. □

5.2.1   On the Error Behavior

The error of the Lagrange interpolator given by Eq. (5.2.2) consists of a constant $1/(n + 1)!$, the derivative $f^{(n+1)}$ (computed at some unknown point c_x), and the polynomial $l(x)$, which depends on the interpolation points. Given a function $f(x)$, the first two components of the error are *a priori* determined and the only quantity left for manipulation is $l(x)$. Suppose that $x_0, x_1, \ldots, x_n$ are evenly spaced. Then $l(x)$ provides larger values near both edges of the data set, and smaller values near the middle. Consider, for example, the five-point set $\{x_i = i, -2 \leq i \leq 2\}$. Here

$$l(x) = (x + 2)(x + 1)x(x - 1)(x - 2) = x(x^2 - 1)(x^2 - 4)$$
$$= x^5 - 5x^3 + 4x$$

An extrema point x_0 satisfies $l'(x_0) = 0$, that is, $5x_0^4 - 15x_0^2 + 4 = 0$ and $x_0 \approx \pm 0.544, \pm 1.644$ (Fig. 5.2.1). Thus,

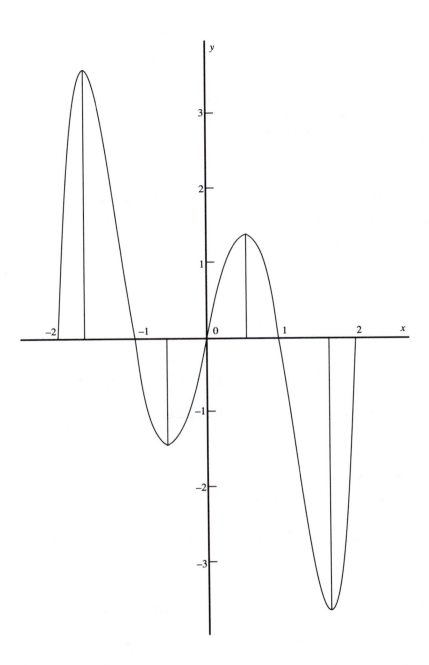

Figure 5.2.1. Behavior of $l(x)$ for $n = 4$.

$$\max_{-1\le x\le 0} |l(x)| = \max_{0\le x\le 1} |l(x)| \approx 1.42$$

$$\max_{-2\le x\le -1} |l(x)| = \max_{1\le x\le 2} |l(x)| \approx 3.63$$

The change in $|l(x)|$ increases as $n \to \infty$. As a result, a high-order interpolator is not often used if the interpolation points are uniformly spaced. We will later show, however, that for a different set of points a high-order polynomial is quite reasonable.

PROBLEMS

1. Let $f(x) = e^x, 0 \le x \le 1$ and $x_0 = 0, x_1 = 1/2, x_2 = 1$. Find a bound for $|e^x - p_2(x)|$.

2. Let $f(x) = \ln x, 1 \le x \le 2$. Find a bound for $|\ln x - p_n(x)|$.

3. Let $p_3(x)$ be a cubic interpolator to $f(x)$ at the points $x_i = x_0 + (i-1)h, 1 \le i \le 4$. Define

$$M_4 = \max_{a\le x\le b} |f^{(4)}(x)|$$

and find a bound to $|f(x) - p_3(x)|$, provided that $x_1 \le x \le x_2$.

4. Apply the result of problem 3 to $f(x) = \sin x, x_0 = 0, h = 0.1$.

5. The function $f(x) = \sqrt{x}$ is tabulated at $x_i, 0 \le i \le n$ where $h = x_{i+1} - x_i, x_0 = a, x_n = b$, and the table is used for linear interpolation. Find an upper bound for h that would guarantee an accuracy of $\epsilon = 10^{-4}$.

6. Implement the result of problem 5 in the following cases:

   (a) $a = 1, b = 2$

   (b) $a = 1, b = 10$

   (c) $a = 10, b = 100$

7. Repeat and solve problem 5 for the case $a = 0$. (*Hint*: Separate the interval $[0, b]$ to $[0, \epsilon]$ and $[\epsilon, b]$.)

8. Let $p_n(x)$ be the Lagrange interpolator of $f(x)$ at $\{x_i\}_{i=0}^n$ and let M_{n+1} be an upper bound of $|f^{(n+1)}(x)|$ over an interval I that contains $x_i, 0 \le i \le n$.

   (a) Prove the inequality

$$|f(x) - p_n(x)| \le \frac{M_{n+1}}{(n+1)!}|l(x)|, \quad x \in I$$

   (b) Find all the cases for which this inequality becomes an equality.

9. If we tabulate a given function $f(x)$ at $\{x_i\}_{i=0}^n$ for the purpose of linear interpolation, we clearly need more points where $f(x)$ changes rapidly and fewer points where it varies slowly, in order to achieve the same accuracy. Apply this approach to tabulate $f(x) = \log_{10} x$ over the interval $[1, 10]$, using three subintervals $[1, a], [a, b], [b, 10]$. Assume evenly spaced points at each subinterval and find upper bounds for the three step sizes that would guarantee an accuracy of $\epsilon = 10^{-6}$.

10. Let $p_n(x)$ interpolate the function $f(x) = e^x$ at $x_i = i, 0 \le i \le n$.

    (a) Find a lower bound for $|e^{n+1} - p_n(n + 1)|$.

    (b) Calculate the exact error $e^{n+1} - p_n(n + 1)$.

11. Let $J_n(x)$ be the *Bessel* function of order n, defined as

$$J_n(x) = \frac{1}{\pi} \int_0^\pi \cos(x \sin t - nt) \, dt$$

and consider a table of $J_n(x)$ with a step size h for interpolation. Find an upper bound for h that would guarantee an acccuracy of $\epsilon = 10^{-4}, 10^{-6}$ using:

    (a) Linear interpolation

    (b) Quadratic interpolation

5.3 Convergence to the Interpolated Function

The most important question related to generating a sequence of interpolators $\{p_n(x)\}_{n=1}^\infty$ to a given function $f(x)$ is whether

$$\lim_{n \to \infty} p_n(x) = f(x), \quad a \le x \le b \tag{5.3.1}$$

for some interval $[a, b]$. A sequence of interpolating polynomials may converge (sometimes uniformly) or not converge to $f(x)$, depending on the *smoothness* of $f(x)$ and on the set of points at which $f(x)$ is interpolated. We start by proving a theorem that is a straightforward result of Eq. (5.2.2).

Theorem 5.3.1.

Let $f(x) \in C^\infty[a, b]$ satisfy the inequalities

$$|f^{(n)}(x)| \le M^n, \quad n = 0, 1, 2, \ldots; \quad a \le x \le b \tag{5.3.2}$$

for some $M > 0$, and let $x_i, i = 0, 1, 2, \ldots$ be an infinite set of distinct points in $[a, b]$. Then the sequence of Lagrange interpolators $\{p_n(x)\}_{n=1}^\infty$ converges uniformly to $f(x)$.

Proof.

The error estimate of Eq. (5.2.2) yields

$$|f(x) - p_n(x)| = \left| \frac{1}{(n + 1)!} l(x) f^{(n+1)}(c_x) \right|$$

and because

$$|l(x)| = |(x - x_0)(x - x_0) \cdots (x - x_n)| \le (b - a)^{n+1}$$

we obtain

$$|f(x) - p_n(x)| \le \frac{1}{(n + 1)!} (b - a)^{n+1} M^{n+1} = \frac{[M(b - a)]^{n+1}}{(n + 1)!}$$

Now, a basic result from calculus is $\lim_{n\to\infty} k^n/n! = 0$ for any given k. Therefore,

$$\lim_{n\to\infty} |f(x) - p_n(x)| = 0 \qquad \qquad \square$$

The hypothesis stated by Eq. (5.3.2) allows $f^{(n)}(x)$ to increase moderately. We may improve this result and replace Eq. (5.3.2) by the weaker inequality

$$|f^{(n)}(x)| \le C\frac{n!}{k^n} \qquad\qquad (5.3.3)$$

for some constants C and k, provided that $b - a < k$ (why?).

Example 5.3.1.

Let $f(x) = \sin(Mx), 0 \le x \le \pi$. Then $f^{(n)}(x) = M^n g(x)$, where $g(x)$ is $\pm\sin(Mx)$ or $\pm\cos(Mx)$. Thus $|f^{(n)}(x)| \le M^n$ and for any set of distinct points $\{x_i\}_{i=0}^{\infty}$ in $[0,\pi]$ we have $\lim_{n\to\infty} p_n(x) = \sin(Mx)$.

$\square$

Example 5.3.2.

Consider the function $f(x) = 1/x, 1/2 \le x \le 1$. Here

$$f'(x) = -\frac{1}{x^2}, \quad f''(x) = \frac{2}{x^3}, \dots, f^n(x) = \frac{(-1)^n n!}{x^{n+1}}$$

and therefore $|f^{(n)}(x)| \le 2^{n+1} n!$. Equation (5.3.3) is thus satisfied with $k = 1/2$, but because $b - a = 1/2 = k$, the convergence of $p_n(x)$ to $1/x$ is not guaranteed. $\square$

Let us now consider a different problem. Suppose that $f(x) \in C^{\infty}(-\infty, \infty)$ and

$$|f^{(n)}(x)| \le M, \quad i = 0, 1, 2, \dots, \quad -\infty < x < \infty$$

[e.g., $f(x) = \cos x$] for some constant M independent of n and x. We wish to approximate $f(x)$ by an interpolator whose interpolating points are evenly spaced along the x axis. Let the interval of approximation be $[0, h]$. Define

$$x_i = -(n-1)h + ih, \quad 0 \le i \le 2n - 1$$

and generate the Lagrange interpolator $p_{2n-1}(x)$ to $f(x)$ at these $2n$ points. Does $p_{2n-2}(x) \to f(x), 0 \le x \le h$ as $n \to \infty$? If so, it would mean that interpolating points, whose distances d_i from the interval of interest $[0, h]$ are such that $\lim_{i\to\infty} d_i = \infty$, maintain some influence within this interval.

Consider the case $f(x) = \sin x, h = \pi$. Here, $\sin(x_i) = 0$ for all i and therefore all the interpolators are identically zero and do not approach $\sin x$ as $n \to \infty$. Thus, if h is *too large*, we cannot expect convergence of $p_{2n-1}(x)$ to $f(x)$. Next we provide an upper bound to h.

Theorem 5.3.2.

[Convergence of $p_{2n-1}(x)$]. *Let a real-valued function $f(x)$ satisfy the hypotheses*

1. $f(x) \in C^{\infty}(-\infty, \infty)$

2. $|f^{(n)}(x)| \le M$

for some M independent of n and x, and let $p_{2n-1}(x)$ interpolate $f(x)$ at $x_i = -(n-1)h + ih$, $0 \le i \le 2n-1$. Then $\lim_{n \to \infty} p_{2n-1}(x) = f(x), 0 \le x \le h$ provided that $h \le 2$.

Proof.

Using Theorem 5.2.1 we obtain

$$|f(x) - p_{2n-1}(x)| = \left| \frac{1}{(2n)!} l(x) f^{(2n)}(c_x) \right|$$

for some unknown c_x, where $l(x) = [x + (n-1)h][x + (n-2)h] \cdots [x - nh]$. The maximum of $|l(x)|$ is maintained at $x = h/2$ (why?) and therefore

$$
\begin{aligned}
|f(x) - p_{2n-1}(x)| &\le \frac{M}{(2n)!} \left| \left[\frac{h}{2} + (n-1)h \right] \left[\frac{h}{2} + (n-2)h \right] \cdots \left[\frac{h}{2} - nh \right] \right| \\
&= \frac{M}{(2n)!} \left[\left(n - \frac{1}{2} \right) h \left(n - \frac{3}{2} \right) h \cdots \frac{3}{2} h \frac{1}{2} h \right]^2 \\
&= \frac{M[(2n-1)(2n-3) \cdots 3 \cdot 1]^2 h^{2n}}{(2n)! 2^{2n}} \\
&= \frac{M h^{2n}}{(2n)! 2^{2n}} \left[\frac{(2n)!}{2^n n!} \right]^2 = \frac{M h^{2n}}{2^{4n}} \frac{(2n)!}{(n!)^2}
\end{aligned}
$$

We now use Stirling's asymptotic formula $n! \approx \sqrt{2\pi n}(n/e)^n$ to obtain

$$|f(x) - p_{2n-1}(x)| \le \frac{M h^{2n}}{2^{4n}} \frac{\sqrt{4\pi n}(2n/e)^{2n}}{2\pi n (n/e)^{2n}} = M \left(\frac{h}{2} \right)^{2n} \frac{1}{\sqrt{\pi n}}$$

and if $|h| \le 2$ the right-hand side approaches zero as $n \to \infty$. ☐

   A problem somewhat *inverse* to that treated by Theorem 5.3.2 is the following: let $f(x)$ be defined over a finite interval $[a, b]$, and let x_i be evenly spaced interpolating points

$$x_i = a + ih, \quad 0 \le i \le n$$

where $h = (b-a)/n$. Does $p_n(x) \to f(x), a \le x \le b$ as $n \to \infty$? Although this is true if $f(x) \in C^\infty[a, b]$ and if $\max_{a \le x \le b} |f^{(n)}(x)|$ does not increase too fast, it can be shown that continuous functions do exists for which convergence of $p_n(x)$ to $f(x)$ does not occur. This is no contradiction of the Weirstrass theorem, which renders a converging sequence of polynomials to any given continuous function. It simply means that for the specific function $f(x)$, Lagrange's interpolators with evenly spaced interpolation points cannot serve as Weirstrass approximating polynomials.

PROBLEMS

1. Calculate $n!$ using Stirling's formula and compare the results with the exact answer.

   (a) $n = 3$

(b) $n = 10$

(c) $n = 20$

2. Let $f(x), \{x_i\}$, and $p_{2n-1}(x)$ be defined as in Theorem 5.3.2 and satisfy hypotheses 1 and 2.

   (a) Explain the difference between the cases $h < 2$ and $h = 2$.

   (b) Can you implement Theorem 5.3.2 to $f(x) = \sin(2x), h = \pi/2$?

3. Let $f(x) \in C^\infty(-\infty, \infty)$ and assume that $|f^{(i)}(x)| \le M^i, i = 0, 1, 2, \ldots$ for some constant M. Let $p_{2n-1}(x)$ be the polynomial that interpolates $f(x)$ at $-n, -n+1, \ldots, -1, 1, \ldots, n-1, n$. Prove that for $M < 2$: $\lim_{n\to\infty} p_{2n-1}(0) = f(0)$.

4. Let $f(x) \in C^\infty(0, \infty)$ satisfy $|f^{(i)}(x)| \le 1, i = 0, 1, 2, \ldots, x \ge 0$ and let $p_n(x)$ be the polynomial that interpolates $f(x)$ at $0, h, 2h, \ldots, nh$ for some given $h > 0$. Give a sufficient condition for which $\lim_{n\to\infty} p_n(x) = f(x), x > 0$.

5. Let $x_i, 0 \le i \le n$ be evenly spaced interpolating points and denote $h = x_{i+1} - x_i$. Find an upper bound for $|l(x)| = |(x - x_0)(x - x_1) \cdots (x - x_n)|$, where $x_0 \le x \le x_n$.

6. In the previous problem let $x_0 = 0, x_n = 1$ and approximate $f(x) = e^x$ by $p_n(x)$. Show that $\lim_{n\to\infty} |e^x - p_n(x)| = 0, 0 \le x \le 1$.

5.4   Divided Differences

Let $f(x)$ be a real-valued function defined at two distinct points x_0, x_1. We define a first order *divided difference* of $f(x)$ as

$$f[x_0, x_1] \equiv \frac{f(x_1) - f(x_0)}{x_1 - x_0} \tag{5.4.1}$$

The divided difference is clearly an approximation to $f'(x)$ near $[x_0, x_1]$ and the first mean-value theorem yields

$$f[x_0, x_1] = f'(c) \tag{5.4.2}$$

where c is some unknown point between x_0 and x_1. Equation (5.4.2) is a formal justification for observing a divided difference as a discrete analog to the derivative of the function. It is, however, relevant only if x_0, x_1 are sufficiently close to each other.

Example 5.4.1.

Let $f(x) = e^x, x_0 = 1.5, x_1 = 1.6$. Here

$$f[x_0, x_1] = \frac{e^{1.6} - e^{1.5}}{1.6 - 1.5} = 4.7134\ldots$$

A good approximation to $f[x_0, x_1]$ is given by $f'[(x_0 + x_1)/2]$. As shown later the difference between these two numbers is $O(|x_1 - x_0|^2)$, that is, extremely small, provided that $x_0 \approx x_1$.   □

Example 5.4.2.

Consider the previous example. The midpoint derivative is $e^{1.55} = 4.7114\ldots$ and

$$\left| f'\left(\frac{x_0 + x_1}{2}\right) - f[x_0.x_1] \right| \le 2 \cdot 10^{-3} \qquad \square$$

Example 5.4.3.

Let $f(x) = \ln x, x_0 = 2, x_1 = 2.5$. The divided difference is

$$\frac{\ln(2.5) - \ln 2}{2.5 - 2} = 0.446\ldots$$

whereas the midpoint derivative is $1/2.25 = 0.444\ldots$. The difference is again bounded by $2 \cdot 10^{-3}$.

$$\square$$

The second divided difference is defined as

$$f[x_0, x_1, x_2] = \frac{f[x_1, x_2] - f[x_0, x_1]}{x_2 - x_0} \qquad (5.4.3)$$

Example 5.4.4.

Let $f(x) = \sin x, x_0 = 1, x_1 = 1.1, x_2 = 1.2$. Then

$$f[x_1, x_2] = \frac{\sin(1.2) - \sin(1.1)}{1.2 - 1.1} \approx 0.408317$$

$$f[x_0, x_1] = \frac{\sin(1.1) - \sin(1)}{1.1 - 1.0} \approx 0.497364$$

$$f[x_0, x_1, x_2] \approx \frac{0.408317 - 0.497364}{1.2 - 1.0} \approx -0.44523 \qquad \square$$

In general, we define a divided difference of order n as

$$f[x_0, x_1, \ldots, x_n] \equiv \frac{f[x_1, \ldots, x_n] - f[x_0, x_1, \ldots, x_{n-1}]}{x_n - x_0} \qquad (5.4.4)$$

provided that $x_0, x_1, \ldots, x_n$ are all distinct points. Divided differences possess certain properties that are useful in various applications. The next result yields an alternative representation for divided differences.

Theorem 5.4.1.

A divided difference of order n defined by Eq. (5.4.4) can also be represented by

$$f[x_0, x_1, \ldots, x_n] = \sum_{i=0}^{n} \frac{f(x_i)}{(x_i - x_0)\ldots(x_i - x_{i-1})(x_i - x_{i+1})\ldots(x_i - x_n)} \qquad (5.4.5)$$

Proof.

We use induction and first verify Eq. (5.4.5) for $n = 1$:

$$f[x_0, x_1] = \frac{f(x_1) - f(x_0)}{x_1 - x_0} = \frac{f(x_0)}{x_0 - x_1} + \frac{f(x_1)}{x_1 - x_0}$$

Next, we assume that Eq. (5.4.5) holds for n. A divided difference of order $(n + 1)$ is given by

$$f[x_0, x_1, \ldots, x_n, x_{n+1}] = \frac{f[x_1, x_2, \ldots, x_n, x_{n+1}] - f[x_0, x_1, \ldots, x_{n-1}, x_n]}{x_{n+1} - x_0} \tag{5.4.6}$$

Each of the divided differences of the right-hand side is of order n. We can therefore apply Eq. (5.4.5) to obtain

$$f[x_0, x_1, \ldots, x_n, x_{n+1}] = \frac{1}{x_{n+1} - x_0} \left[\sum_{i=1}^{n+1} \frac{f(x_i)}{(x_i - x_1) \ldots (x_i - x_{i-1})(x_i - x_{i+1}) \ldots (x_i - x_{n+1})} \right.$$

$$\left. - \sum_{i=0}^{n} \frac{f(x_i)}{(x_i - x_0) \ldots (x_i - x_{i-1})(x_i - x_{i+1}) \ldots (x_i - x_n)} \right]$$

$$= \frac{1}{x_{n+1} - x_0} \left[\frac{f(x_{n+1})}{(x_{n+1} - x_1) \ldots (x_{n+1} - x_n)} - \frac{f(x_0)}{(x_0 - x_1) \ldots (x_0 - x_n)} \right.$$

$$\left. + \sum_{i=1}^{n} \frac{f(x_i)}{(x_i - x_1) \ldots (x_i - x_{i-1})(x_i - x_{i+1}) \ldots (x_i - x_n)} \left(\frac{1}{x_i - x_{n+1}} - \frac{1}{x_i - x_0} \right) \right]$$

$$= \frac{f(x_{n+1})}{(x_{n+1} - x_0) \ldots (x_{n+1} - x_n)} + \frac{f(x_0)}{(x_0 - x_1) \ldots (x_0 - x_{n+1})}$$

$$+ \frac{1}{x_{n+1} - x_0} \sum_{i=1}^{n} \frac{(x_{n+1} - x_0) f(x_i)}{(x_i - x_0) \ldots (x_i - x_{i-1})(x_i - x_{i+1}) \ldots (x_i - x_{n+1})}$$

$$= \sum_{i=0}^{n+1} \frac{f(x_i)}{(x_i - x_0) \ldots (x_i - x_{i-1})(x_i - x_{i+1}) \ldots (x_i - x_{n+1})} \qquad \square$$

Using Theorem 5.4.1 we obtain the following corollary.

Corollary 5.4.1.

Let $(i_0, i_1, \ldots, i_n)$ be a permutation of the numbers $(0, 1, \ldots, n)$. Then

$$f[x_{i_0}, x_{i_1}, \ldots, x_{i_n}] = f[x_0, x_1, \ldots, x_n] \tag{5.4.7}$$

Proof.

Because the right-hand side of Eq. (5.4.5) is invariant to permutations, so is $f[x_0, x_1, \ldots, x_n]$. $\qquad \square$

An important application using divided differences is the construction of a sequence of interpolating polynomials of increasing orders. To calculate Lagrange interpolators from Eq. (5.1.7) is quite inconvenient. Furthermore, given a data set $\{x_0, x_1, \ldots, x_n \ldots\}$ we may be interested in a sequence of interpolators $p_1(x), p_2(x) \ldots, p_n(x), \ldots$. Here, the computation of $p_n(x)$ [Eq. (5.1.7)] is independent of the previously calculated interpolators, and is therefore a costly procedure. By using divided differences we can easily generate a whole sequence of consecutive interpolators to a given function $f(x)$.

Theorem 5.4.2.
(Generating consecutive Lagrange interpolators). *Let $p_j(x), 1 \le j \le n$ be the polynomial that interpolates $f(x)$ at $x_i, 0 \le i \le j$. Then*

$$p_1(x) = f(x_0) + (x - x_0)f[x_0, x_1] \tag{5.4.8}$$
$$p_2(x) = f(x_0) + (x - x_0)f[x_0, x_1] + (x - x_0)(x - x_1)f[x_0, x_1, x_2] \tag{5.4.9}$$

$$\vdots$$

$$p_n(x) = f(x_0) + (x - x_0)f[x_0, x_1] + \cdots$$
$$+ (x - x_0)(x - x_1) \cdots (x - x_{n-1})f[x_0, x_1, \ldots, x_n] \tag{5.4.10}$$

Equation (5.4.10) is *Newton's divided difference formula* for Lagrange interpolators, and leads to the recursive relation

$$p_j(x) = p_{j-1}(x) + (x - x_0)(x - x_1) \cdots (x - x_{j-1})f[x_0, x_1, \ldots, x_j] \tag{5.4.11}$$

Proof.
Using induction, we start with $n = 1$. Denote

$$q_1(x) = f(x_0) + (x - x_0)f[x_0, x_1]$$

Then $q_1(x_1) = f(x_0)$ and

$$q_1(x_1) = f(x_0) + (x_1 - x_0) \left[\frac{f(x_1) - f(x_0)}{x_1 - x_0} \right] = f(x_1)$$

Thus q_1 is linear and interpolates $f(x)$ at x_0, x_1. Due to Theorem 5.1.2 $q_1(x) = p_1(x)$.

We now assume that Eq. (5.4.10) holds for n, and consider the case of $(n + 2)$ interpolating points $x_0, x_1, \ldots, x_n, x_{n+1}$. Define the $(n + 1)$th order polynomial

$$q_{n+1}(x) = f(x_0) + (x - x_0)f[x_0, x_1] + \cdots + (x - x_0) \cdots (x - x_{n-1})f[x_0, x_1, \ldots, x_n]$$
$$+ (x - x_0) \cdots (x - x_n)f[x_0, x_1, \ldots, x_{n+1}]$$

According to our assumption

$$q_{n+1}(x) = p_n(x) + (x - x_0) \cdots (x - x_n)f[x_0, x_1, \ldots, x_{n+1}] \tag{5.4.12}$$

Therefore

$$q_{n+1}(x_i) = p_n(x_i) = f(x_i), \quad 0 \le i \le n$$

We use Eqs. (5.1.7) and (5.4.5) to rewrite $p_n(x)$ and $f[x_0, x_1, \ldots, x_{n+1}]$ in Eq. (5.4.12) and obtain

$$q_{n+1}(x) = \sum_{i=0}^{n} \frac{(x - x_0) \ldots (x - x_{i-1})(x - x_{i+1}) \ldots (x - x_n)}{(x_i - x_0) \ldots (x_i - x_{i-1})(x_i - x_{i+1}) \ldots (x_i - x_n)} f(x_i)$$

$$+ \sum_{i=0}^{n+1} \frac{(x - x_0) \ldots (x - x_n) f(x_i)}{(x_i - x_0) \ldots (x_i - x_{i-1})(x_i - x_{i+1}) \ldots (x_i - x_{n+1})}$$

We now substitute $x = x_{n+1}$. The coefficient of $f(x_{n+1})$ is 1, whereas those of $f(x_i), 0 \le i \le n$ are all 0. Thus, $q_{n+1}(x_{n+1}) = f(x_{n+1})$, that is, $q_{n+1}(x_i) = f(x_i) = p_n(x_i)$, $0 \le i \le n + 1$. We now apply uniqueness and conclude $q_{n+1}(x) = p_{n+1}(x)$.   ☐

Example 5.4.5.

Let $f(x) = \ln(1 + x), x_0 = 0, x_1 = 1, x_2 = 3$. To calculate the interpolators of the first and second order, we first compute

$$f[x_0, x_1] = \frac{\ln 2 - \ln 1}{1 - 0} = \ln 2 \approx 0.693147$$

$$f[x_0, x_1, x_2] = \frac{[(\ln 4 - \ln 2)/(3 - 1)] - \ln 2}{3 - 0} \approx -0.115525$$

Thus

$$p_1(x) \approx \ln 1 + (x - 0) \cdot 0.693147 = 0.693147x$$
$$p_2(x) \approx 0.693147x + (x - 0)(x - 1)(-0.115525)$$
$$= -0.115525x^2 + 0.808672x$$ ☐

Example 5.4.6.

Let $f(x) = \sin x, 0 \le x \le 0.8$. Consider the following interpolating points:

$$x_i = 0.2i, \quad 0 \le i \le 4$$

Table 5.4.1 contains the numerical values of $D_i = f[x_0, x_1, \ldots, x_i], 1 \le i \le 4$ up to six decimal digits.

Table 5.4.1. Divided Differences of $f(x) = \sin x$

i	x_1	$\sin(x_i)$	D_i
0	0	0	0
1	0.2	0.1986693308	$0.993347 \cdot 10^0$
2	0.4	0.3894183423	$-0.990040 \cdot 10^{-1}$
3	0.6	0.5646424734	$-0.158428 \cdot 10^0$
4	0.8	0.7173560909	$0.161179 \cdot 10^{-1}$

The interpolators $p_n(x), 1 \le n \le 4$ are computed as

$$p_n(x) = D_0 + (x - x_0)D_1 + (x - x_0)(x - x_1)D_2 + \cdots$$
$$+ (x - x_0) \cdots (x - x_{n-1})D_n$$

using a simple variation of Horner's algorithm. Table 5.4.2 consists of the interpolator values at $x = 0.3, 0.5, 0.7$ compared with the exact values of $\sin x$.

Table 5.4.2. Interpolating $f(x) = \sin x$

n	$p_n(0.3)$	$p_n(0.5)$	$p_n(0.7)$
1	0.29800	0.49667	0.69534
2	0.29503	0.48182	0.66069
3	0.29551	0.47945	0.64406
4	0.29552	0.47942	0.64423
Exact	0.29552	0.47943	0.64422

We end this section by showing that a divided difference of order n can be interpreted as a discrete analog of $f^{(n)}(x)$. The following theorem states this more precisely.

Theorem 5.4.3.
Let $f(x) \in C^n[a, b]$ and let a set of distinct points $\{x_i, 0 \le i \le n\}$ belong to $[a, b]$. Then

$$f[x_0, x_1, \ldots, x_n] = \frac{1}{n!}f^{(n)}(c) \tag{5.4.13}$$

where c lies in the smallest interval that includes all $x_i, 0 \le i \le n$.

Proof.
The proof is similar to that of Theorem 5.2.1. We define

$$q(x) = p_n(x) - f(x)$$

where $p_n(x)$ is the Lagrange interpolator of $f(x)$ at $x_0, x_1, \ldots, x_n$. Because $q(x_i) = p_n(x_i) - f(x_i) = 0$ for $0 \le i \le n$, there must exist a number c within the smallest interval that includes all $x_i, 0 \le i \le n$

for which $q^{(n)}(c) = 0$, that is, $p_n^{(n)}(c) = f^{(n)}(c)$. Now, Eq. (5.4.10) yields

$$p_n(x) = f[x_0, x_1, \ldots, x_n]x^n + \cdots$$

leading to

$$n! f[x_0, x_1, \ldots, x_n] = f^{(n)}(c)$$

and therefore to Eq. (5.4.13). □

Example 5.4.7.
Let $f(x) = \sqrt{x}, x_0 = 1, x_1 = 1.1, x_2 = 1.2$. Here

$$f[x_0, x_1, x_2] = \frac{1}{1.2 - 1.0}\left(\frac{\sqrt{1.2} - \sqrt{1.1}}{1.2 - 1.1} - \frac{\sqrt{1.1} - \sqrt{1.0}}{1.1 - 1.0}\right) \approx -0.1086$$

Therefore $f''(c) \approx -0.2172$, whereas $f''(x_1) \approx -0.2167$. We will later show that for close, evenly spaced x_0, x_1, x_2 we have $f''(c) \approx f''(x_1)$. □

A subroutine DVDF, which computes divided differences and their associated interpolators, can be found on the attached floppy disk.

PROBLEMS

1. Calculate $f[x_0, x_1]$ and $f[x_0, x_1, x_2]$ in the following cases:
   (a) $f(x) = \sin x, x_0 = 0.5, x_1 = 0.6, x_2 = 0.7$
   (b) $f(x) = e^x \cos x, x_0 = 1.5, x_1 = 1.6, x_2 = 1.8$
   (c) $f(x) = 1/x, x_0 = 0.1, x_1 = 0.2, x_2 = 0.3$
   (d) $f(x) = 3x + 5, x_0 = 0.1, x_2 = 1, x_3 = 2$

2. Compute $f'[(x_0 + x_1)/2]$ and $(1/2)f''(x_1)$ and compare with $f[x_0, x_1]$ and $f[x_1, x_0, x_2]$, respectively, for the following:
   (a) $f(x) = x^3, x_0 = 1, x_1 = 1.2, x_2 = 1.4$
   (b) $f(x) = x \ln x, x_0 = 0.5, x_1 = 0.6, x_2 = 0.7$
   (c) $f(x) = \sqrt{1 - x}, x_0 = 0.8, x_1 = 0.9, x_2 = 1$

3. Use direct calculations and show that for any given $f(x)$ and distinct x_0, x_1, x_2: $f[x_0, x_1, x_2] = f[x_1, x_0, x_2]$.

4. Let $f(x)$ be twice continuously differentiable. Because $\lim_{x_1 \to x_0} f[x_0, x_1] = f'(x_0)$, it is logical to define

$$f[x_0, x_0] = f'(x_0)$$

Obtain meaningful definitions for $f[x_0, x_0, x_2]$ and $f[x_2, x_0, x_0]$.

5. In view of Theorem 5.4.3, obtain a meaningful definition for $f[x_0, x_0, \ldots, x_0]$.

6. Let $f(x)$ be a polynomial of order m. What can be said about $f[x_0,x], f[x_0,x_1,x]$ and the divided difference of order n of $f(x)$.

7. Use DVDF to compute $f[x_0,x_1], f[x_0,x_1,x_2], f[x_0,x_1,x_2,x_3]$, and $f[x_0,x_1,x_2,x_3,x_4]$ in the following cases:

   (a) $f(x) = \sqrt{x^2 + 1}, x_i = 0.2i, 0 \le i \le 4$

   (b) $f(x) = x^4 - 1000x^3, x_i = 0.1i, 0 \le i \le 4$

   (c) $f(x) = e^x, x_0 = 0, x_1 = 2, x_2 = 2.5, x_3 = 3, x_4 = 3.1$

8. Use the divided differences method to calculate the interpolators $p_1(x), p_2(x)$, and $p_3(x)$ to $f(x)$, given the interpolating points x_0, x_1, x_2, x_3 :

   (a) $f(x) = x^3 - x, x_i = i, 0 \le i \le 3$

   (b) $f(x) = \sin(2x), x_i = 0.2i, 0 \le i \le 3$

   (c) $f(x) = \ln(\sin x), x_i = 0.1(i + 1), 0 \le i \le 3$

9. Let $p_n(x)$ be the Lagrange interpolator of $f(x)$ at $x_0, x_1, \ldots, x_n$. By using Eq. (5.4.11) derive the following error formula:

$$f(x) - p_n(x) = (x - x_0)(x - x_1) \cdots (x - x_n) f[x_0, x_1, \ldots, x_n, x]$$

5.5 Interpolation by Splines

Let $f(x)$ be approximated by a Lagrange interpolator $p_n(x)$, which interpolates the function at $\{x_i, 0 \le i \le n\}$. Although this interpolator is clearly a smooth function, it may oscillate too much between the interpolation points, as shown in Fig. 5.1.1. In many cases, however, the structure of the set of points $\{[x_i, f(x_i)], 0 \le i \le n\}$ suggests that the given function does not oscillate at all. We therefore may try to interpolate $f(x)$ differently. The simplest way is to connect the points $P_i = [x_i, f(x_i)]$ using straight segments, replacing the original graph of the unknown $f(x)$ by a polygon. This interpolator is in agreement with the data and does not change drastically between consecutive points. However, it is not smooth at the interpolation points $x_i, 0 \le i \le n$, that is, it does not possess a continuous derivative there. If we use quadratic interpolation rather than linear, we can reduce the number of "corners" by half. This is done by constructing a sequence of parabolas $q_i(x)$, each interpolating $f(x)$ at three consecutive points. The general scheme is

$$P_0, P_1, P_2 \rightarrow q_1(x)$$
$$P_2, P_3, P_4 \rightarrow q_2(x)$$
$$\vdots$$
$$P_{n-2}, P_{n-1}, P_n \rightarrow q_{n/2}(x)$$

and assumes an even n.

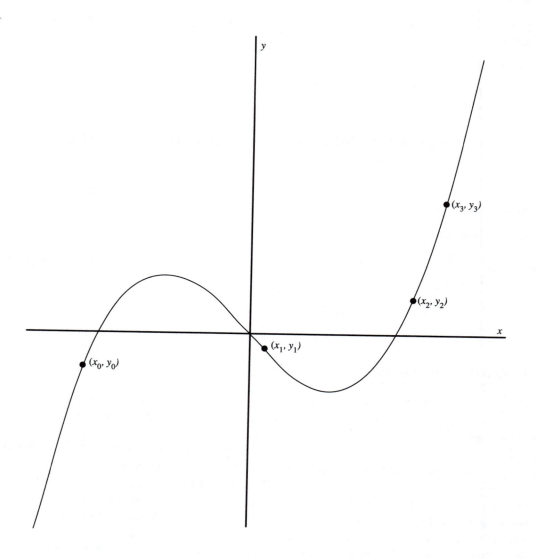

Figure 5.5.1. An oscillating $p_3(x)$.

Example 5.5.1.

Consider the data set of Table 5.5.1.

Table 5.5.1. Interpolation Data for
Linear and Quadratic Interpolators

i	x_i	$f(x_i)$
0	0	0
1	0.5	0.25
2	1	0
3	2	6
4	3.5	3.75
5	4	3
6	6	3

We start with a *piecewise linear interpolation* and replace $f(x)$ by a polygon (Fig. 5.5.2). This polygon has a corner at each P_i except at P_4. The linear interpolator is "accidentally" smooth at P_4 because P_3, P_4, P_5 are on the same straight line.

We now turn to *piecewise quadratic interpolation*. It involves three parabolas:

$$P_0, P_1, P_2 \rightarrow q_1(x) = x - x^2$$
$$P_2, P_3, P_4 \rightarrow q_2(x) = -3x^2 + 15x - 12$$
$$P_4, P_5, P_6 \rightarrow q_3(x) = 0.6x^2 - 6x + 17.4$$

and is shown in Fig. 5.5.3. The quadratic interpolation is smoother than the linear and possesses corners only at P_2 and P_4.   □

So far we have discussed higher order interpolators that are smooth but may oscillate considerably between nodes, and lower order interplators that are only piecewise smooth but do not oscillate much. We will now show how to construct an interpolating function that possesses two continuous derivatives at all points, and does not oscillate much between consecutive nodes.

5.5.1   Spline Interpolation

We start with the following definition.

Definition 5.5.1.   (Cubic spline). *A piecewise cubic polynomial with two continuous derivatives is called a* cubic spline.

Example 5.5.2.

Consider the function

$$s(x) = \begin{cases} s_1(x) = x^3 + 1 & , \ 0 \le x \le 1 \\ \\ s_2(x) = 2x^3 - 3x + 3, & 1 \le x \le 2 \end{cases}$$

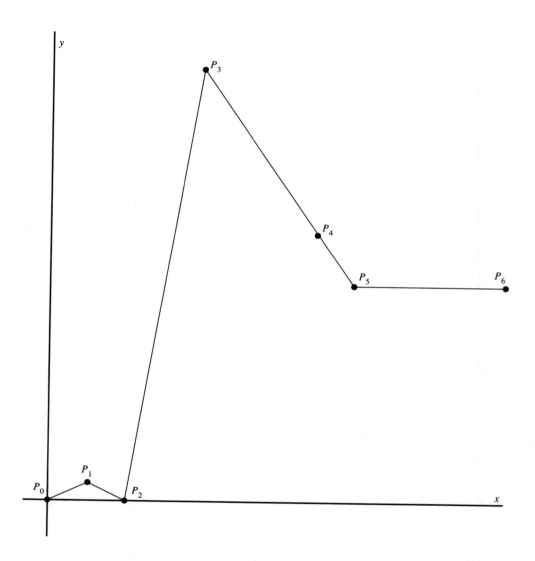

Figure 5.5.2. A piecewise linear interpolator.

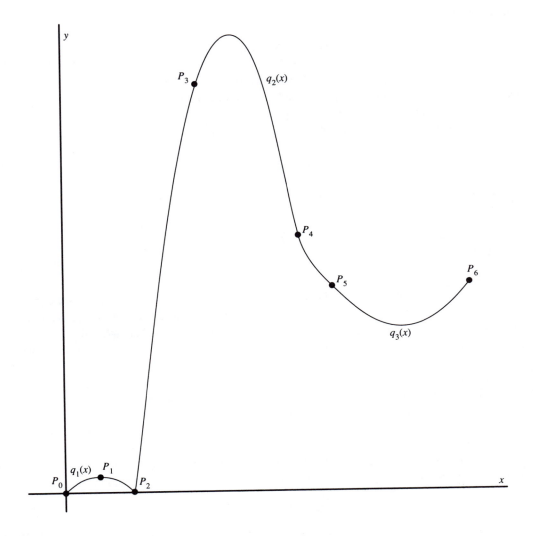

Figure 5.5.3. A piecewise quadratic interpolator.

First, by definition, $s(x)$ is a piecewise cubic polynomial. Now, because

$$s_1(1) = s_2(1) = 2$$
$$s_1'(1) = s_2'(1) = 3$$

we have $s(x) \in C^1[0,2]$ as well. Because

$$s_1''(1) = 6, \quad s_2''(1) = 12$$

it follows that $s(x)$ is not a cubic spline. If we redefine, for example, $s_2(x) = 2x^3 - 3x^2 + 3x$, the function $s(x)$ is a cubic spline. The proof is left for the reader. □

We will now use the cubic spline to interpolate a given set of points.

Consider a data set of $(n+1)$ points $\{(x_i, y_i), 0 \le i \le n\}$ where

$$a = x_0 < x_1 < \cdots < x_n = b \tag{5.5.1}$$

We seek a function $s(x)$ defined over the interval $[a,b]$ that satisfies the following requirements:

1. $s(x_i) = y_i, 0 \le i \le n$
2. $s(x)$ is a cubic polynomial at each $[x_i, x_{i+1}]$
3. $s(x) \in C^2[a,b]$
4. $s''(a) = s''(b) = 0$

The first requirement implies that $s(x)$ interpolates the given data. The second indicates the tendency to avoid higher order interpolators. The desire to replace an unknown function by a smooth interpolator is expressed by the third requirement. Finally, if $s(x)$ does not oscillate too much, $s'(x)$ should not change rapidly, and $|s''(x)|$ is expected to be small. This is enforced by the fourth hypothesis.

Theorem 5.5.1.

(Cubic spline interpolator). *Given a data set $\{(x_i, y_i), 0 \le i \le n\}$ for which Eq. (5.5.1) holds, there is always a unique function that satisfies the four previous requirements.*

Proof.

Let $s(x)$ be *any* piecewise cubic polynomial over the set of intervals $\{[x_0, x_1], \ldots, [x_{n-1}, x_n]\}$, such that $s(x) \in C^2[a,b]$. Denote

$$M_i = s''(x_i), \quad 0 \le i \le n \tag{5.5.2}$$

Because $s(x)$ is a cubic polynomial over each interval $[x_{i-1}, x_i]$, its second derivative must be linear and we can easily determine that

$$s''(x) = \frac{(x_i - x)M_{i-1} + (x - x_{i-1})M_i}{x_i - x_{i-1}}, \quad x_{i-1} \le x \le x_i \tag{5.5.3}$$

By integrating both sides of Eq. (5.5.3) twice and substituting the *boundary conditions*

$$s(x_{i-1}) = y_{i-1}, \quad s(x_i) = y_i \tag{5.5.4}$$

we find

$$s(x) = \frac{(x_i - x)^3 M_{i-1} + (x - x_{i-1})^3 M_i}{6(x_i - x_{i-1})} + \frac{(x_i - x)y_{i-1} + (x - x_{i-1})y_i}{x_i - x_{i-1}}$$

$$- \frac{1}{6}(x_i - x_{i-1})[(x_i - x)M_{i-1} + (x - x_{i-1})M_i] \tag{5.5.5}$$

for $x_{i-1} \le x \le x_i$. Clearly $s(x)$ and $s''(x)$, as represented by Eqs. (5.5.5) and (5.5.3), respectively, are continuous over the whole interval $[a, b]$. So far, no restrictions have been imposed on M_i. However, $s'(x)$ is also continuous.

If we define

$$s_i(x) = s(x), \quad x_{i-1} \le x \le x_i, \quad 1 \le i \le n \tag{5.5.6}$$

then

$$s_i'(x_i) = s_{i+1}'(x_i), \quad 1 \le i \le n - 1 \tag{5.5.7}$$

By differentiating Eq. (5.5.5) and substituting in Eq. (5.5.7) we obtain

$$(x_i - x_{i-1})M_{i-1} + 2(x_{i+1} - x_{i-1})M_i + (x_{i+1} - x_i)M_{i+1}$$

$$+6\left(\frac{y_i - y_{i-1}}{x_i - x_{i-1}} - \frac{y_{i+1} - y_i}{x_{i+1} - x_i}\right) = 0, \quad 1 \le i \le n - 1 \tag{5.5.8}$$

We can relate to Eq. (5.5.8) as a set of $(n - 1)$ linear equations with the $(n + 1)$ variables $M_0, M_1, \ldots, M_n$. Let us impose two additional constraints:

$$M_0 = M_n = 0 \tag{5.5.9}$$

The system defined by Eqs. (5.5.8) and (5.5.9) can be shown to have a unique solution $\{M_0, M_1, M_2, \ldots, M_n\}$ that provides the cubic spline interpolator. $\square$

The spline whose existence and uniqueness are guaranteed by Theorem 5.5.1 is only one type of *cubic spline interpolator* (CSI). Other cubic spline interpolators can be generated by changing the constraints of Eq. (5.5.9).

Example 5.5.3.

Consider the following four-point data set:

$$\{(0,0), (1,1), (2,1), (3,2)\}$$

The CSI related to this set that satisfies Eq. (5.5.9) is determined by the system

$$M_0 + 4M_1 + M_2 + 6 = 0$$
$$M_1 + 4M_2 + M_3 - 6 = 0$$
$$M_0 = 0$$
$$M_3 = 0$$

whose unique solution is $\{M_0, M_1, M_2, M_3\} = \{0, -2, 2, 0\}$. By substituting in Eq. (5.5.5) we get

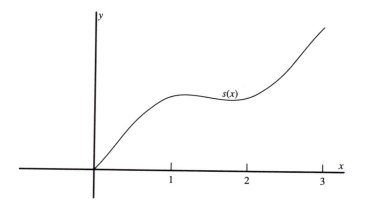

Figure 5.5.4. A cubic spline interpolator for a four-point data set.

$$s(x) = \begin{cases} -\dfrac{1}{3}x^3 + \dfrac{4}{3}x & , \quad 0 \le x \le 1 \\ \dfrac{2}{3}x^3 - 3x^2 + \dfrac{13}{3}x - 1 & , \quad 1 \le x \le 2 \\ -\dfrac{1}{3}x^3 + 3x^2 - \dfrac{23}{3}x + 7, & \quad 2 \le x \le 3 \end{cases} \qquad (5.5.10)$$

and one can show that $s(x), s'(x), s''(x)$ are continuous over the whole interval $[0, 3]$. The CSI defined by Eq. (5.5.10) is shown in Fig. 5.5.4. $\square$

Example 5.5.4.

Let $f(x) = \sin x$ and consider the four-point data set

$$\left\{ (0, \sin 0), \left(\frac{\pi}{3}, \sin\frac{\pi}{3}\right), \left(\frac{2\pi}{3}, \sin\frac{2\pi}{3}\right), (\pi, \sin\pi) \right\} = \left\{ (0, 0), \left(\frac{\pi}{3}, \frac{\sqrt{3}}{2}\right), \left(\frac{2\pi}{3}, \frac{\sqrt{3}}{2}\right), (\pi, 0) \right\}$$

The related CSI is determined by the unique solution to the system

$$M_0 + 4M_1 + M_2 + \frac{27\sqrt{3}}{\pi^2} = 0$$

$$M_1 + 4M_2 + M_3 + \frac{27\sqrt{3}}{\pi^2} = 0$$

$$M_0 = 0$$

$$M_3 = 0$$

which is

$$s(x) = \begin{cases} s_1(x) \approx -0.1508 + 0.9924x & , \quad 0 \le x \le \pi/3 \\ s_2(x) \approx -0.4738x^2 + 1.4886x - 0.1732 & , \quad \pi/3 \le x \le 2\pi/3 \\ s_3(x) \approx 0.1508x^3 - 1.4215x^2 + 3.4734x - 1.5588, & \quad 2\pi/3 \le x \le \pi \end{cases} \qquad (5.5.11)$$

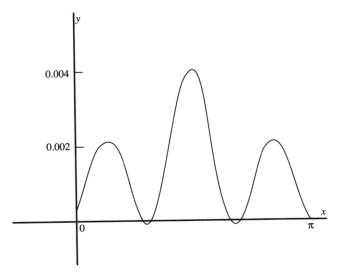

Figure 5.5.5. Sin $x - f(x)$ plotted over $[0, \pi]$.

A comparison between $s(x)$ and $\sin x$ is given in Table 5.5.2 and Fig. 5.5.5.

Table 5.5.2. A Comparison between $\sin x$ and Its Cubic Spline Interpolator

x	$\sin x$	$s(x)$	$\lvert \sin x - s(x) \rvert$
$\pi/9$	0.3420	0.3400	0.0020
$2\pi/9$	0.6428	0.6415	0.0013
$4\pi/9$	0.9848	0.9815	0.0033
$5\pi/9$	0.9848	0.9815	0.0033
$7\pi/9$	0.6428	0.6415	0.0013
$8\pi/9$	0.3420	0.3400	0.0020

5.5.2 Error Estimate

So far, the advantage of constructing a CSI to interpolate a given set of points $\{(x_i, y_i)\}_{i=0}^{n}$ has been the production of a smooth curve $s(x)$ (possessing two continuous derivatives) that is a piecewise polynomial and satisfies

$$s(x_i) = y_i, \quad 0 \le i \le n$$

This curve not only looks right, but may be used for any application that requires the existence of at least two continuous derivatives. If the data set is the *only* input and information about the function, we do not need to be concerned about the accuracy of the spline between the given points. However, if the data set originates from a given $f(x)$, we must consider the error

$$\epsilon(x) = f(x) - s(x) \tag{5.5.12}$$

as a measure of the performance of the CSI.

Let $f(x)$ be a real-valued function defined over the interval $[a, b]$, and let $\{x_i\}_{i=0}^{n}$ be a set of evenly spaced points such that

$$x_i = a + ih, \quad 0 \leq i \leq n \tag{5.5.13}$$

where $h = (b - a)/n$. If $f(x) \in C^4[a, b]$, it can be shown that the previously defined CSI that interpolates $f(x)$ at x_i satisfies

$$|f(s) - s(x)| \leq Kh^2 \tag{5.5.14}$$

where K depends on $f''(x)$ and $f^{(4)}(x)$. Denote this particular CSI by $s_n(x)$. Then Eq. (5.5.14) guarantees

$$\lim_{N \to \infty} s_n(x) = f(x), \quad a \leq x \leq b \tag{5.5.15}$$

The error $O(h^2)$ indicates slow convergence. This is a direct consequence of the constraints imposed by Eq. (5.5.9). Because in general $f''(a) \neq 0$ and $f''(b) \neq 0$, the assumption $M_0 = M_n = 0$ is false and decreases the speed by which $s_n(x)$ converges to $f(x)$. If instead of Eq. (5.5.9) we impose

$$s_n''(x_0) = f''(x_0)$$
$$s_n''(x_n) = f''(x_n) \tag{5.5.16}$$

one can show that

$$|f(x) - s_n(x)| \leq Kh^4 \tag{5.5.17}$$

The particular spline for which Eq. (5.5.16) holds is called the *natural cubic spline interpolator* (NCSI). Another interesting spline is the one for which

$$s_n'(x_0) = f'(x_0)$$
$$s_n'(x_n) = f'(x_n) \tag{5.5.18}$$

Here, too, the rate of convergence is $O(h^4)$.

Example 5.5.5.

Let $f(x) = \sin x$ and consider the four-point data set of Example 5.5.4:

$$\left\{ (0,0), (\pi/3, \sqrt{3}/2), (2\pi/3, \sqrt{3}/2), (\pi, 0) \right\}$$

Because $f''(x) = -\sin x$, the NCSI, $s_4(x)$, must satisfy

$$s_4''(0) = s_4''(\pi) = 0$$

Thus the CSI of Example 5.5.4 is also an NCSI. To appreciate the small error $O(h^4)$ associated with the NCSI, one must choose an h smaller than $\pi/3$. $\square$

Example 5.5.6.

Let $f(x) = \cos x$ and consider the four-point data set

$$\{(0, 1), (\pi/6, \sqrt{3}/2), (\pi/3, 1/2), (\pi/2, 0)\}$$

and the boundary conditions $M_0 = M_3 = 0$. The system of linear equations associated with the corresponding spline is

$$M_0 + 4M_1 + M_2 + \frac{216}{\pi^2}\left(\sqrt{3} - \frac{3}{2}\right) = 0$$

$$M_1 + 4M_2 + M_3 + \frac{216}{\pi^2}\left(1 - \frac{\sqrt{3}}{2}\right) = 0$$

$$M_0 = 0$$

$$M_3 = 0$$

leading to

$$M_0 = 0, \quad M_1 = -1.158799, \quad M_2 = -0.443321, \quad M_3 = 0$$

A comparison between $f(x)$ and $s(x)$ is given in Table 5.5.3.

Table 5.5.3. Cubic Spline Interpolator Performance for $f(x) = \cos x$

x	$\cos x$	$s(x)$	$|\epsilon(x)|$
0.2	0.980067	0.966099	0.013967
0.4	0.921061	0.914494	0.006567
0.6	0.825336	0.827744	0.002408
0.8	0.696707	0.699944	0.003238
1.0	0.540302	0.540901	0.000598
1.2	0.362358	0.361235	0.001122
1.4	0.169967	0.169003	0.000964

The poor accuracy around $x = 0$ is expected, because we impose $s''(0) = 0$, whereas $f''(0) = -1$. On the other hand, the constraint $s''(\pi/2) = 0$ is in agreement with $f''(\pi/2) = 0$. $\square$

Example 5.5.7.

Consider the previous example with the constraints

$$s''(0) = f''(0) = -1, \quad s''(\pi/2) = f''(\pi/2) = 0$$

The system of equations for M_0, M_1, M_2, M_3 is

$$M_0 + 4M_1 + M_2 + \frac{216}{\pi^2}\left(\sqrt{3} - \frac{3}{2}\right) = 0$$

$$M_1 + 4M_2 + M_3 + \frac{216}{\pi^2}\left(1 - \frac{\sqrt{3}}{2}\right) = 0$$

$$M_0 = -1$$

$$M_3 = 0$$

and it is uniquely solved by $M_0 = -1, M_1 = -0.892133, M_2 = -0.509988, M_3 = 0$, which determine the NCSI. A comparison with $f(x) = \cos x$ is given in Table 5.5.4.

Table 5.5.4. Natural Cubic Spline Interpolator Performance for $f(x) = \cos x$

x	$\cos x$	NCSI $(\cos x)$	$|\epsilon(x)|$
0.2	0.980067	0.979577	0.000489
0.4	0.921061	0.920803	0.000258
0.6	0.825336	0.825363	0.000028
0.8	0.696707	0.696634	0.000073
1.0	0.540302	0.540288	0.000014
1.2	0.362358	0.362311	0.000047
1.4	0.169967	0.169891	0.000076

Clearly, the NCSI is significantly closer to $f(x)$ than its CSI.   □

Cubic splines can be used in various applications, such as numerical integration and differentiation, interpolation, and curve-fitting. The conditions imposed at the end points may vary and are problem oriented.

PROBLEMS

1. Determine whether the following functions are cubic splines:

(a) $s(x) = \begin{cases} x^3, & 0 \le x \le 1 \\ x^2, & 1 \le x \le 2 \end{cases}$

(b) $s(x) = \begin{cases} 2x & , \quad 0 \le x \le 1 \\ -\frac{2}{3}x^3 + 2x^2 + \frac{2}{3}, & 1 \le x \le 2 \end{cases}$

(c) $s(x) = \begin{cases} x^3 - x & , \quad 0 \le x \le 1 \\ \frac{4}{3}x^3 - x^2 - \frac{1}{3}, & 1 \le x \le 2 \\ 12x - \frac{53}{3} & , \quad 2 \le x \le 3 \end{cases}$

2. Consider the data set $\{(0,0), (1,2), (2,1)\}$.

(a) Find the piecewise linear interpolator for the set.

(b) Find the quadratic interpolator.

(c) Find the CSI, $s_3(x)$, that satisfies

$$s_3''(0) = s_3''(2) = 0$$

3. Consider the data set $\{(0,0), (1,1), (3,0), (4,1)\}$.

   (a) Find the piecewise linear interpolator for the set.

   (b) Find the cubic polynomial interpolator.

   (c) Find the CSI, $s_4(x)$, that satisfies

$$s_4''(0) = 1, \quad s_4''(4) = 2$$

4. Consider the data set $\{(0,0), (1,1), (2,8), (3,27)\}$. Find the CSI, $s_4(x)$, that satisfies

$$s_4''(0) = 0, \quad s_4''(3) = 18$$

   without solving a system of linear equations.

5. Find a CSI, $s_5(x)$, to the data set

$$\{(-2,4), (-1,1), (0,0), (2,4), (3,9)\}$$

   that satisfies

$$s_5'(-2) = -4, \quad s_5'(2) = 4$$

   Do you have to solve a system of five linear equations to compute $s_5(x)$?

6. Derive the system of linear equations whose solution determines the cubic spline $s_n(x)$ that satisfies Eq. (5.5.18). [*Hint*: Differentiate Eq. (5.5.5) over the first and last subintervals.]

7. Let $f(x) = \sqrt{x}$ and consider its data set:

$$\{(1/4, 1/2), (4/9, 2/3), (16/25, 4/5), (1,1)\}$$

   (a) Find the piecewise linear interpolator, $l(x)$, for these points.

   (b) Find a piecewise quadratic interpolator as follows: construct two quadratic interpolators, $q_1(x)$ and $q_2(x)$, for $\{(x_0, y_0), (x_1, y_1), (x_2, y_2)\}$ and $\{(x_1, y_1), (x_2, y_2), (x_3, y_3)\}$, respectively. A piecewise quadratic interpolator can then be defined as

$$q(x) = \begin{cases} q_1(x), & x_0 \le x \le x_2 \\ q_2(x), & x_2 \le x \le x_3 \end{cases}$$

   (c) Find the NCSI, $s_4(x)$, related to $f(x)$ and the data set.

   (d) Compare $l(x), q(x), s_4(x)$ to $f(x) = \sqrt{x}$ over the interval $[1/4, 1]$.

8. Let $f(x) = \cos x$ and consider the following four-point data set:

$$\{(0, \cos 0), (\pi/6, \cos \pi/6), (\pi/3, \cos \pi/3), (\pi/2, \cos \pi/2)\}$$

(a) Find a CSI, $s_4^{(1)}(x)$, that satisfies

$$s_4^{(1)''}(0) = s_4^{(1)''}(\pi/2) = 1/2$$

(b) Find the NCSI, $s_4^{(2)}(x)$, that satisfies

$$s_4^{(2)''}(0) = f''(0) = -1, \quad s_4^{(2)''}(\pi/3) = f''(\pi/2) = 0$$

(c) Find a CSI, $s_4^{(3)}(x)$, that satisfies

$$s_4^{(3)'}(0) = f'(0) = 0, \quad s_4^{(3)'}(\pi/2) = f'(\pi/2) = -1$$

(d) Compare $f(x) = \cos x$ with $s_4^{(1)}(x), s_4^{(2)}(x), s_4^{(3)}(x)$.

9. Define

$$s_a(x) = \begin{cases} (x-a)^3, & x \geq a \\ 0, & x < a \end{cases}$$

(a) Show that $s_a(x)$ is a cubic spline (using Definition 5.5.1).

(b) Let

$$x_0 < x_1 < \cdots < x_n$$

and define

$$s(x) = \sum_{i=0}^{n} c_i s_{x_i}(x)$$

Show that $s(x)$ is a cubic spline.

(c) What can you say about $s'''(x)$?

10. Use an available code to construct an NCSI that interpolates the function $f(x) = \sin x + \cos x$ over the interval $[0, \pi]$ using 6, 11, and 21 evenly spaced points. Compare the accuracy of the various splines at $x = 0.2, 0.4, 0.6, 0.8, 1.0, 1.2, 1.4$.

5.6 Approximations of Functions

Most functions cannot be evaluated exactly, although they are often treated as if they were completely known. Elementary functions such as e^x, $\sin x$, and $\sqrt{x}$ are most commonly used in various sciences but except for several discrete values of x they can only be *approximated*. Because we are essentially limited to the four basic arithmetic operations, we cannot evaluate *exactly* any function that is not a *rational* function, that is, a function of the form

$$r(x) = p(x)/q(x) \qquad (5.6.1)$$

where $p(x), q(x)$ are polynomials and $q(x) \neq 0$. Any other function can only be approximated either by polynomials or by rational functions. In this section we will concentrate on polynomial approximations of functions. We assume in general that the given function is *sufficiently smooth*, that is, possesses a sufficient number of continuous derivatives.

We previously discussed (Chapter 1) Taylor's polynomials, which are a sequence of polynomial approximations to a given function $f(x)$. If $f(x)$ is expanded about $x = a$, and the numbers

$$f^{(n)}(a), \quad n = 0, 1, 2 \ldots$$

are easily computed, then so are Taylor's polynomials. However, Taylor's approximation is generally recommended only near $x = a$. This is clear from the error formula

$$f(x) - \sum_{n=0}^{N} \frac{f^{(n)}(a)}{n!} (x-a)^n = \frac{(x-a)^{N+1}}{(N+1)!} f^{(N+1)}(c_x) \qquad (5.6.2)$$

where c_x is between a and x. If $|x - a|$ increases, so does the error, and rapidly! Thus we have a high-order polynomial approximation, the computation of which is time consuming; in addition it produces a significant round-off error on one hand and on the other hand may have a large truncation error. Occasionally, the Taylor approximations may be the only alternative. Generally, however, we can replace a Taylor polynomial by another polynomial of lower degree, which provides better accuracy.

We next introduce the concept of the best (polynomial) approximation and later define Chebyshev polynomials, which provide an efficient method, based on interpolation, for constructing polynomial approximations.

5.6.1   Best Approximations

Let $f(x)$ be a real-valued continuous function defined over the inteval $[a, b]$. For each polynomial $p(x)$ we define

$$E[p(x)] = \max_{a \leq x \leq b} |f(x) - p(x)| \qquad (5.6.3)$$

Clearly $E[p(x)]$ is the worst deviation of $p(x)$ from $f(x)$. Now, for any given $n \geq 0$ define

$$\delta_n(f) = \min_{\deg(p) \leq n} E(p) \qquad (5.6.4)$$

A polynomial of degree $\leq n$, for which $\delta_n(f)$ is attained, is called the *best* (minimax) *polynomial approximation* of order n to $f(x)$, and is denoted by $m_n(x)$.

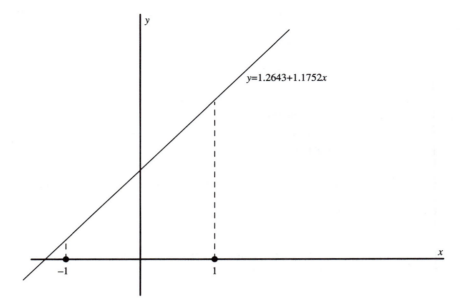

Figure 5.6.1. $m_1(x)$ for $e^x, -1 \le x \le 1$.

Example 5.6.1.

Let $f(x) = e^x$ for $-1 \le x \le 1$. The linear minimax polynomial approximation, $m_1(x)$, is given by

$$m_1(x) = 1.2643 + 1.1752x \tag{5.6.5}$$

It is the straight line that is parallel to the secant connecting $(-1, 1/e)$ and $(1, e)$, and bisects the distance between this secant and its parallel tangent (Fig. 5.6.1).

The linear Taylor polynomial that approximates $f(x)$ about $x = 0$ is

$$t_1(x) = 1 + x \tag{5.6.6}$$

Direct computations or simple geometric observation (Fig. 5.6.2) yields

$$\max_{-1 \le x \le 1} |e^x - t_1(x)| = e^1 - t_1(1) \approx 0.718 \tag{5.6.7}$$

On the other hand the minimax polynomial satisfies (why?)

$$\max_{-1 \le x \le 1} |e^x - m_1(x)| = m_1(c) - e^c \approx 0.279 \tag{5.6.8}$$

where $c = \ln(1.1752)$. The graphs of e^x and its approximations are given in Fig. 5.6.3.   □

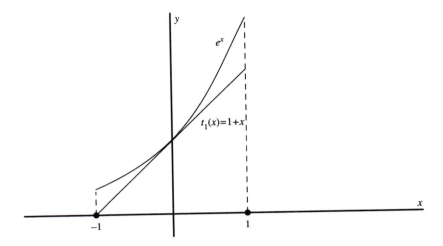

Figure 5.6.2. Determining $\max_{-1 \leq x \leq 1} \left| e^x - t_1(x) \right|$.

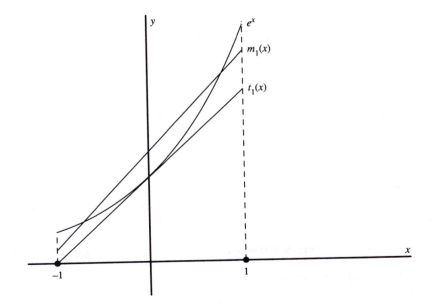

Figure 5.6.3. A comparison between $m_1(x)$ and $t_1(x)$ for $f(x) = e^x$.

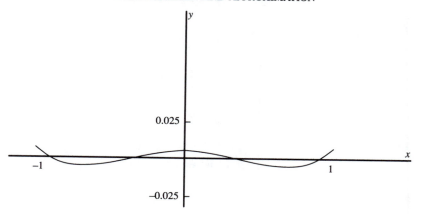

Figure 5.6.4. $e^x - m_3(x)$, $-1 \leq x \leq 1$.

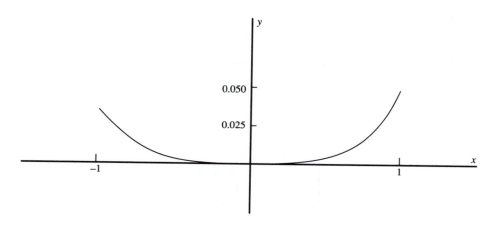

Figure 5.6.5. $e^x - t_3(x)$, $-1 \leq x \leq 1$.

Example 5.6.2.

Let $f(x) = e^x$, $-1 \leq x \leq 1$. It can be shown that the best third order polynomial approximation is given by

$$m_3(x) = 0.994579 + 0.995668x + 0.542973x^2 + 0.179533x^3 \tag{5.6.9}$$

The cubic Taylor approximation about $x = 0$ is

$$t_3(x) = 1 + x + \frac{x^2}{2} + \frac{x^3}{6} \tag{5.6.10}$$

and one can observe that the coefficients of both approximations are "relatively close" to each other. However, if we draw $e^x - m_3(x)$ (Fig. 5.6.4) and $e^x - t_3(x)$ (Fig. 5.6.5) we can find that $m_3(x)$

approximates e^x enormously better than $t_3(x)$. This is also confirmed by Table 5.6.1. The maximum errors for $m_3(x)$ and $t_3(x)$ are 0.00553 and 0.0516, respectively.

Table 5.6.1. A Comparison between $e^x, m_3(x), t_3(x)$ for $-1 \le x \le 1$

x	e^x	$t_3(x)$	$e^x - t_3(x)$	$m_3(x)$	$e^x - m_3(x)$
-1	0.3679	0.3333	0.0345	0.3624	0.0055
-0.8	0.4493	0.4347	0.0147	0.4536	-0.0043
-0.6	0.5488	0.5440	0.0048	0.5539	-0.0051
-0.4	0.6703	0.6693	0.0010	0.6717	-0.0014
-0.2	0.8187	0.8187	0.0001	0.8157	0.0030
0	1.0000	1.0000	0	0.9946	0.0054
0.2	1.2214	1.2213	0.0001	1.2169	0.0045
0.4	1.4918	1.4907	0.0012	1.4912	0.0006
0.6	1.8221	1.8160	0.0061	1.8262	-0.0041
0.8	2.2255	2.2053	0.0202	2.2305	-0.0050
1	2.7183	2.6667	0.0516	2.7128	0.0055

Although the computation of $m_n(x)$ is not an easy and straightforward matter, it is relatively simple to provide a polynomial $p_n(x)$ with the following properties.

1. $p_n(x)$ is "close" to the minimax approximation $m_n(x)$ and thus justifies the name "near minimax approximation."

2. $p_n(x)$ interpolates the approximated function $f(x)$ at $(n+1)$ distinct points. The derivation of $p_n(x)$ is based on *Chebyshev polynomials* and will be introduced in the next two sections. The use of Chebyshev polynomials will also enable us to derive a useful bound for $\delta_n(f)$:

$$\delta_n(f) \le \frac{(b-a)^{n+1}}{2^{2n+1}(n+1)!} \max_{a \le x \le b} |f^{(n+1)}(x)| \tag{5.6.11}$$

PROBLEMS

1. Let $f(x) = \tan^{-1}x$, $-1 \le x \le 1$.
   (a) Explain why $m_1(x)$ should have the form $m_1(x) = ax$ and obtain

$$m_1(x) = 0.833278x$$

   (b) Obtain $t_1(x)$ and compare with $m_1(x)$ and $\tan^{-1}x$ over the set of points $\{x_i = -1 + 0.1i, 0 \le i \le 20\}$.

2. Use Eq. (5.6.11) and obtain bounds for $\delta_n(f), 0 \le n \le 4$ in the following cases:
   (a) $f(x) = e^x, a = 0, b = 1$
   (b) $f(x) = \sin x, a = 0, b = \pi$
   (c) $f(x) = 1/x, a = 1, b = 2$

3. Find an example for which the sequence of bounds given by Eq. (5.6.11) does not converge to zero.

4. Often, to evaluate a given function $f(x)$ within a large interval $[a, b]$, it is only necessary to provide an efficient computational procedure over a much smaller subinterval. Then, by using special properties of $f(x)$, the range of computation of the algorithm is extended to include all of $[a, b]$. The advantage is clear: within a small interval one can replace $f(x)$ by $m_n(x)$, using a relatively small n, and still maintain high accuracy. Use this approach for $f(x) = \cos x$ and construct an algorithm to calculate $\cos x, -\infty < x < \infty$ given by

$$m_3(x) = 0.9986329 + 0.0296140x - 0.6008616x^2 + 0.1125060x^3, \ 0 \le x \le \pi/2$$

to approximate $\cos x$ and present a comparison over the grid $\{x_i = i\pi/20, 0 \le i \le 10\}$ between $\cos x, m_3(x)$, and $t_3(x) = 1 - (x^2/2)$.

5. Based on the approach suggested in problem 4, construct an algorithm to compute $f(x) = e^x, -\infty < x < \infty$.

5.7 Chebyshev Polynomials

The Chebyshev polynomials are useful in many applications. We start by showing several properties of these polynomials and then apply them to derive a near minimax polynomial approximation to a given function $f(x)$.

Definition 5.7.1. (Chebyshev polynomials). *The nth order polynomial*

$$T_n(x) = \cos(n \cos^{-1} x), \quad -1 \le x \le 1, \quad n \ge 0 \tag{5.7.1}$$

is called the Chebyshev polynomial of order n.

The first polynomial is $T_0(x) = 1$, followed by $T_1(x) = x$ and $T_2(x) = 2x^2 - 1$.

The graphs of $T_0(x), T_1(x)$, and $T_2(x)$ are shown in Fig. 5.7.1. Equation (5.7.1) is somewhat awkward and does not even express the polynomial representation of $T_n(x)$. Instead, we may write

$$\theta = \cos^{-1} x, \quad x = \cos \theta, \quad 0 \le \theta \le \pi \tag{5.7.2}$$

leading to

$$T_n(x) = \cos(n\theta), \quad 0 \le \theta \le \pi \tag{5.7.3}$$

Because $\cos(n\theta)$ can be represented as an nth order polynomial in $\cos \theta, T_n(x)$ is clearly an nth order polynomial in x, defined over the interval $-1 \le x \le 1$.

A recursion formula can be used to construct the complete sequence of Chebyshev polynomials.

Theorem 5.7.1.

(Constructing Chebyshev polynomials). *For all $n \ge 2$,*

$$T_n(x) = 2xT_{n-1}(x) - T_{n-2}(x) \tag{5.7.4}$$

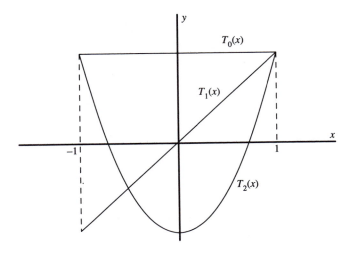

Figure 5.7.1. Graphs of $T_0(x)$, $T_1(x)$, and $T_2(x)$.

The proof is based on elementary trigonometric identities and is left as an exercise for the reader.

Example 5.7.1.

For $n = 3$ we have

$$T_3(x) = 2xT_2(x) - T_1(x)$$
$$= 2x(2x^2 - 1) - x = 4x^3 - 3x \qquad \square$$

Example 5.7.2.

For $n = 4$ we have

$$T_4(x) = 2xT_3(x) - T_2(x)$$
$$= 2x(4x^3 - 3x) - (2x^2 - 1)$$
$$= 8x^4 - 8x^2 + 1 \qquad \square$$

Because $T_n(x) = \cos(n\theta)$, we have

$$|T_n(x)| \leq 1, \quad -1 \leq x \leq 1 \qquad (5.7.5)$$

Also, by directly using Eq. (5.7.4), we obtain

$$T_n(x) = 2^{n-1}x^n + t_{n,n-1}x^{n-1} + \cdots + t_{n,1}x + t_{n,0}, \quad n \geq 1 \qquad (5.7.6)$$

The proof is based on mathematical induction and is left for the reader.

If we define the modified Chebyshev polynomials

$$T_n^* = \frac{1}{2^{n-1}} T_n(x), \quad n \geq 0 \tag{5.7.7}$$

then

$$T_n^*(x) = x^n + \text{lower order terms} \tag{5.7.8}$$

and due to Eq. (5.7.5) we have

$$|T_n^*(x)| \leq \frac{1}{2^{n-1}}, \quad -1 \leq x \leq 1 \tag{5.7.9}$$

Thus, T_n^* is an nth order *monic* (a polynomial whose highest coefficient is 1) that is bounded by $1/2^{n-1}$ for $-1 \leq x \leq 1$. The next theorem emphasizes this fact.

Theorem 5.7.2.

For a given $n \geq 1$, consider $M^{(n)}$, the set of all monics of degree n, defined over the interval $[-1, 1]$. Define

$$\rho_n(p) = \max_{-1 \leq x \leq 1} |p(x)|, \quad p(x) \in M^{(n)} \tag{5.7.10}$$

Then the problem of determining a $p(x)$ for which $\rho_n(p)$ is minimum has a solution

$$p(x) = T_n^*(x) \tag{5.7.11}$$

and the minimum is $1/2^{n-1}$.

Proof.

By definition we have $T_n^*(x) = (1/2^{n-1}) \cos(n\theta)$. The extremae of $\cos(n\theta)$ occur at $n\theta = k\pi$, that is,

$$\theta = k\pi/n, \quad k = 0, 1, \ldots, n$$

Thus there are $(n+1)$ points

$$x_k = \cos(k\pi/n), \quad 0 \leq k \leq n \tag{5.7.12}$$

at which $T_n^*(x)$ achieves its extrema value, which is

$$T_n^*(x_k) = (-1)^k \frac{1}{2^{n-1}}, \quad 0 \leq k \leq n$$

All the extremae share the same absolute value, but oscillate in sign as shown, for example, in Figure 5.7.2, which features $T_5^*(x)$.

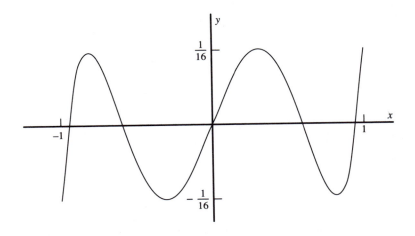

Figure 5.7.2. $T_5^*(x), -1 \le x \le 1$.

Suppose now that a monic polynomial $Q(x) = x^n + \cdots$ for which $\max_{-1 \le x \le 1} |Q(x)| = m < 1/2^{n-1}$ can be found. Denote

$$R(x) = T_n^*(x) - Q(x) \tag{5.7.13}$$

Then, at the $(n+1)$ distinct points defined by Eq. (5.7.12), $R(x)$ must have the same sign as $T_n^*(x)!$. Therefore

$$R(x_k) > 0, \text{ for even } k$$
$$R(x_k) < 0, \text{ for odd } k \tag{5.7.14}$$

Because $x_0 > x_1 > \cdots > x_n$ we conclude that the polynomial $R(x)$ must have *at least* n roots. However, the highest coefficients of $T_n^*(x)$ and $Q(x)$ are 1, they cancel each other, and $R(x)$ is therefore of degree $\le (n-1)$. This contradicts the existence of n roots unless $R(x) \equiv 0$. But then $T_n^*(x) = Q(x)$, which contradicts the assumption

$$m = \max |Q(x)| < 1/2^{n-1}$$

Thus, $\min[\rho(p)] = 1/2^{n-1}$ and it is attained by $T_n^*(x)$.   ⬜

PROBLEMS

1. Use Eq. (5.7.4) to compute $T_5(x)$ and $T_6(x)$.

2. For any given integers $m, n \geq 0$ such that $m \neq n$, show

$$\int_{-1}^{1} \frac{T_n(x) T_m(x)\, dx}{\sqrt{1 - x^2}} = 0$$

3. We define the Chebyshev polynomials of the second kind as

$$S_n(x) = \frac{1}{n+1} T'_{n+1}(x), \quad n \geq 0$$

   (a) Calculate $S_0(x), S_1(x)$.

   (b) Show that

$$S_n(x) = \frac{\sin(n+1)\theta}{\sin \theta}, \quad x = \cos \theta, \quad 0 \leq \theta \leq \pi$$

4. Determine a recursion formula for producing the complete sequence $\{S_n(x)\}_{n=2}^{\infty}$.

5. Let $p(x)$ be a given polynomial of degree $\leq (n-1)$.

   (a) What can be said with respect to $\max_{-1 \leq x \leq 1} |x^n + p(x)|$.

   (b) Find $p(x)$ for which $\max_{-1 \leq x \leq 1} |x^n + p(x)|$ is the smallest possible value.

5.8   Near Minimax Approximation

Consider a function $f(x)$ defined for $-1 \leq x \leq 1$ and interpolated at $x_0, x_1, \ldots, x_n$ by its Lagrange interpolator $p_n(x)$. The error, as previously shown, is

$$f(x) - p_n(x) = \frac{(x - x_0)(x - x_1) \cdots (x - x_n)}{(n+1)!} f^{(n+1)}(c_x)$$

where c_x is within the smallest interval that includes $x_0, x_1, \ldots, x_n, x$. Because we cannot generally locate c_x, the only practical way to reduce $\max_{-1 \leq x \leq 1} |f(x) - p_n(x)|$ is to find a set $\{x_0, x_1, \ldots, x_n\}$ for which

$$\max_{-1 \leq x \leq 1} |l(x)| = \max_{-1 \leq x \leq 1} |(x - x_0)(x - x_1) \cdots (x - x_n)|$$

is minimum.

    Definition 5.8.1.   (A near minimax approximation). *A Lagrange interpolator $p_n(x)$ that interpolates $f(x), -1 \leq x \leq 1$ at $x_0, x_1, \ldots, x_n$, and for which*

$$\max_{-1 \leq x \leq 1} |(x - x_0)(x - x_1) \cdots (x - x_n)|$$

is minimum, is called a near minimax polynomial approximation *to $f(x)$.*

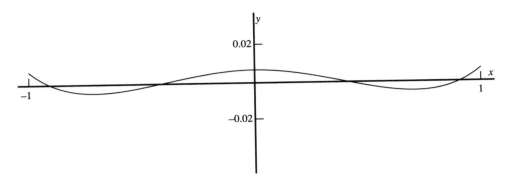

Figure 5.8.1. $e^x - p_3(x)$, plotted for $-1 \le x \le 1$.

As determined in the previous section, the choice

$$l(x) = \frac{1}{2^n} T_{n+1}(x) \tag{5.8.1}$$

provides a near minimax polynomial approximation. The interpolating points $\{x_i, 0 \le i \le n\}$ are the $(n+1)$ distinct roots of $T_{n+1}(x)$. Now, $T_{n+1}(x) = 0$ yields

$$\cos[(n+1)\theta] = 0$$

or

$$(n+1)\theta = \frac{\pi}{2} + k\pi$$

leading to

$$\theta_k = \frac{(\pi/2) + k\pi}{n+1} = \frac{2k+1}{2n+2}\pi, \quad 0 \le k \le n \tag{5.8.2}$$

The actual roots are

$$x_k = \cos\left(\frac{2k+1}{2n+2}\pi\right), \quad 0 \le k \le n \tag{5.8.3}$$

Example 5.8.1.
Let $f(x) = e^x, -1 \le x \le 1$. The interpolation points of the third order near minimax polynomial approximation are

$$x_0 = \cos\frac{\pi}{8}, \quad x_1 = \cos\frac{3\pi}{8}, \quad x_2 = \cos\frac{5\pi}{8}, \quad x_3 = \cos\frac{7\pi}{8}$$

The graph of $e^x - p_3(x)$ is shown in Fig. 5.8.1. The maximum value of $|e^x - p_3(x)|$ is 0.00666, compared with 0.00553 of $|e^x - m_3(x)|$. $\qquad\square$

Consider now the near minimax approximation problem for the particular case in which $f(x)$ is a given monic polynomial of degree $(n+1)$, defined over the interval $[-1, 1]$, that is,

$$f(x) = x^{n+1} + \sum_{i=0}^{n} a_i x^i, \quad -1 \le x \le 1 \tag{5.8.4}$$

Let $p_n(x)$ be the Lagrange interpolator of degree n, which interpolates $f(x)$ at $x_0, x_1, \ldots, x_n$. The error is

$$f(x) - p_n(x) = \frac{(x - x_0)(x - x_1) \cdots (x - x_n)}{(n+1)!} f^{(n+1)}(c_x)$$

and because $f^{(n+1)}(c_x) = (n+1)!$ we have

$$f(x) - p_n(x) = (x - x_0)(x - x_1) \cdots (x - x_n) \tag{5.8.5}$$

By virtue of Theorem 5.7.2, in order to obtain the best approximating interpolator of degree n to $f(x)$, one should choose $x_0, x_1, \ldots, x_n$ as the $(n+1)$ distinct roots of $T_{n+1}(x)$. We then have

$$p_n(x) = f(x) - \frac{1}{2^n} T_{n+1}(x) \tag{5.8.6}$$

Is $p_n(x)$ also the best polynomial approximation of degree n to $f(x)$? Let $q(x)$ be a polynomial of degree n for which

$$m = \max_{-1 \le x \le 1} |f(x) - q(x)| < \max_{-1 \le x \le 1} |f(x) - p_n(x)| = 1/2^n \tag{5.8.7}$$

Then $r(x) = f(x) - q(x)$ is a monic of degree $(n+1)$ and, by using Theorem 5.7.2 once more, we obtain the inequality

$$\max_{-1 \le x \le 1} |r(x)| \ge 1/2^n$$

which contradicts Eq. (5.8.7). We have thus proved the following corollary.

Corollary 5.8.1.

For any given monic polynomial $f(x)$ of degree $(n+1)$ defined over $[-1, 1]$, the best polynomial approximation of order n is given by Eq. (5.8.6).

In other words, if $f(x)$ is an $(n+1)$th order polynomial, the solutions to the minimax and near minimax approximation problems are the same.

Example 5.8.2.

Let $f(x) = x^3 + x + 1, -1 \le x \le 1$. The best polynomial approximation of degree 2 is

$$p_2(x) = x^3 + x + 1 - \frac{1}{2^2} T_3(x) = x^3 + x + 1 - \frac{1}{4}(4x^3 - 3x) = \frac{7x}{4} + 1$$

In this particular case, the solution is even linear.   □

Example 5.8.3.

Let $f(x) = x^3 + 3x^2 + 6x + 6$, $-1 \leq x \leq 1$. The best polynomial approximation of degree 2 to $f(x)$ is given by

$$p_2(x) = x^3 + 3x^2 + 6x + 6 - \frac{1}{4}T_3(x) = 3x^2 + \frac{27}{4}x + 6$$

We notice that $f(x) \approx 6e^x$ (why?). If we now solve the near minimax approximation problem of order 2 for $g(x) = 6e^x$ we obtain $q_2(x) = 3.192251x^2 + 6.778633x + 6$. $\qquad \square$

Corollary 5.8.1 justifies Definition 5.8.1 to a near minimax approximation (explain!).

PROBLEMS

1. Let $f(x)$ be an $(n + 1)$th order polynomial over the interval $[-1, 1]$. We previously showed that the right-hand side of Eq. (5.8.6) is a minimax polynomial approximation of degree $\leq n$ to $f(x)$. Apply some elements in the proof of Theorem 5.7.2 and show that this solution is unique.

2. Find the polynomial of degree $\leq (n - 1)$ that best approximates

$$f(x) = a_n x^n + a_{n-1} x^{n-1} + \cdots + a_1 x + a_0$$

over the interval $[a, b]$ and show that its maximum error is given by

$$\frac{1}{2^{n-1}} \left(\frac{b - a}{2} \right)^n a_n$$

{*Hint*: Define $x = [(a + b)/2 + y(b - a)/2]$, where $-1 \leq y \leq 1$, and use the previous results for the particular interval $[-1, 1]$.}

3. Find the near minimax approximation of order 3 over the interval $[-1, 1]$ to the following functions:
   (a) $\sin x$ (b) $\tan x$ (c) $x^4 + x^3 - 1$ (d) $\ln(2 + x)$

4. Find the best polynomial approximation of degree ≤ 4 to $f(x) = x^5$ over the interval $[0, 2]$.

5. Find the best polynomial approximation of degree ≤ 2 to $f(x) = x^3$ over the interval $[0, 1]$.

6. Approximate x^3 over the interval $[0, 1]$ by a linear polynomial, in two stages:
   (a) Find a best polynomial approximation of degree $\leq 2, p_2(x)$, to $f(x)$.
   (b) Find a best polynomial approximation of degree $\leq 1, p_1(x)$, to $p_2(x)$.

7. Determine the best linear polynomial approximation to $f(x)$ of problem 6. Does it coincide with $p_1(x)$ of the previous problem?

8. Find the best linear polynomial approximation to $f(x) = ax^2 + bx + c$ over the interval $[0, 2]$.

9. Let $p_n(x)$ be the best polynomial approximation of degree $\leq n$ to $f(x)$, over the interval $[a, b]$. Derive the inequality

$$\delta(f) = \max_{a \leq x \leq b} |p_n(x) - f(x)| \leq \frac{(b - a)^{n+1}}{2^{2n+1}(n + 1)!} \max_{a \leq x \leq b} |f^{(n+1)}(x)|$$

which was previously given in Eq. (5.6.11).

5.9   Least-Squares Approximations

Suppose we are given some experimental data related to a function $y = f(x)$. For example, consider the data in Table 5.9.1.

Table 5.9.1. Experimental Data
Related to $y = f(x)$

i	x_i	y_i
1	0	2
2	2	4
3	4	8
4	6	14
5	8	16

To calculate $f(x)$ at nontabulated points, we may replace $f(x)$ by its Lagrange interpolator

$$p_4(x) = -\frac{1}{64}x^4 + \frac{3}{16}x^3 - \frac{7}{16}x^2 + \frac{5}{4}x + 2 \tag{5.9.1}$$

or by some cubic spline. By doing so, we assume that $f(x)$ is a nonlinear function. However, if we sketch the experimental data (Fig. 5.9.1) it seems that $y = f(x)$ is a straight line and the small deviations indicate that the data are not exact. Occasionally we even know *a priori* that the given function is linear and applying nonlinear approximations is simply irrelevant. In either case, we assume a linear relation $y = ax + b$ and find a and b that would best fit the given data. The *discrete least-squares* approach is to find a and b that would minimize the error sum

$$E = \sum_{i=1}^{n}(y_i - ax_i - b)^2 \tag{5.9.2}$$

where n is the number of measurements. The number E measures the total *deviation* of the data from the straight line $y = ax + b$. To find the best a and b, we apply basic calculus and request

$$\frac{\partial E}{\partial a} = 0, \quad \frac{\partial E}{\partial b} = 0 \tag{5.9.3}$$

This leads to

$$-2\sum_{i=1}^{n}(y_i - ax_i - b)x_i = 0, \quad -2\sum_{i=1}^{n}(y_i - ax_i - b) = 0 \tag{5.9.4}$$

and a set of two linear equations with two variables:

$$a\sum_{i=1}^{n}x_i^2 + b\sum_{i=1}^{n}x_i = \sum_{i=1}^{n}x_iy_i$$

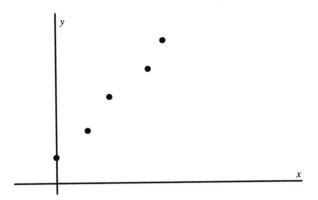

Figure 5.9.1. Experimental data that seem to represent a straight line.

$$a \sum_{i=1}^{n} x_i + bn = \sum_{i=1}^{n} y_i \tag{5.9.5}$$

The unique solution is

$$a = \frac{n \sum_{i=1}^{n} x_i y_i - \sum_{i=1}^{n} x_i \sum_{i=1}^{n} y_i}{\Delta}, \quad b = \frac{\sum_{i=1}^{n} y_i \sum_{i=1}^{n} x_i^2 - \sum_{i=1}^{n} x_i \sum_{i=1}^{n} x_i y_i}{\Delta} \tag{5.9.6}$$

where

$$\Delta = n \sum_{i=1}^{n} x_i^2 - \left(\sum_{i=1}^{n} x_i \right)^2 \tag{5.9.7}$$

Example 5.9.1.
Consider the data given in Table 5.9.1. Here,

$$\sum_{i=1}^{5} x_i = 20, \quad \sum_{i=1}^{5} y_i = 44, \quad \sum_{i=1}^{5} x_i y_i = 252, \quad \sum_{i=1}^{5} x_i^2 = 120, \quad \Delta = 200$$

and therefore $a = 1.9, b = 1.2$. The straight line that best fits the data is $y = 1.9x + 1.2$ and is shown in Fig. 5.9.2. The corresponding deviation is

$$E = \sum_{i=1}^{5} (y_i - 1.9x_i - 1.2)^2 = 4.4 \qquad \square$$

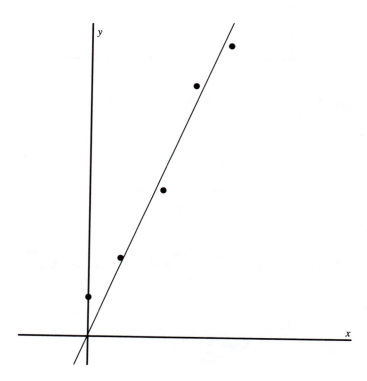

Figure 5.9.2. The least-squares line for the data of Table 5.9.1.

Example 5.9.2.

Consider a set of 10 measurements given by the two leftmost columns of Table 5.9.2. We have

$$a = \frac{(10)(213.585) - (22.5)(72.17)}{206.25} \approx 2.48$$

$$b = \frac{(72.17)(71.25) - (22.5)(213.585)}{206.25} \approx 1.63$$

and the error is $E = \sum_{i=1}^{10}(y_i - \bar{y}_i)^2 = 0.0733$.

Table 5.9.2. A Working Scheme for the Least-Squares Method

x_i	y_i	x_i^2	$x_i y_i$	$\bar{y}_i = 2.48x + 1.63$
0.0	1.62	0.00	0.000	1.63
0.5	2.81	0.25	1.405	2.87
1.0	4.15	1.00	4.150	4.11
1.5	5.27	2.25	7.905	5.35
2.0	6.80	4.00	13.600	6.59
2.5	7.80	6.25	19.500	7.83
3.0	9.08	9.00	27.240	9.07
3.5	10.27	12.25	35.945	10.31
4.0	11.65	16.00	46.600	11.55
4.5	12.72	20.25	57.240	12.79
22.5	72.17	71.25	213.585	

$\square$

If only a single (x_i, y_i) spoils the straight-line structure of the data set, it will have a mild effect on a and b. For example, if in Table 5.9.2 we replace $(2, 6.8)$ by $(2, 8)$, the new computed coefficients of the least-squares line are $a = 2.47$ and $b = 1.78$. If the particular point is completely out of line, it clearly represents a poor measurement and should not participate in the computational procedure.

Sometimes the data should instead be represented by a quadratic or by a higher order polynomial. Let $\{(x_i, y_i)\}_{i=1}^n$ be an experimental data set and we search for a polynomial

$$p_m(x) = a_m x^m + a_{m-1} x^{m-1} + \cdots + a_1 x + a_0 \tag{5.9.8}$$

with unknown coefficients that would best fit the data. The quantity to be minimized is

$$E = \sum_{i=1}^n [y_i - p_m(x_i)]^2 = \sum_{i=1}^n (y_i - a_m x_i^m - \cdots - a_1 x_i - a_0)^2 \tag{5.9.9}$$

As before, we request

$$\frac{\partial E}{\partial a_i} = 0, \quad 0 \leq i \leq m \tag{5.9.10}$$

and thus obtain a system of $(m+1)$ linear equations with the variables $a_0, a_1, \ldots, a_m$:

$$a_0 \sum_{i=1}^{n} x_i^0 + a_1 \sum_{i=1}^{n} x_i^1 + a_2 \sum_{i=1}^{n} x_i^2 + \cdots + a_m \sum_{i=1}^{n} x_i^m = \sum_{i=1}^{n} y_i x_i^0$$

$$a_0 \sum_{i=1}^{n} x_i^1 + a_1 \sum_{i=1}^{n} x_i^2 + a_2 \sum_{i=1}^{n} x_i^3 + \cdots + a_m \sum_{i=1}^{n} x_i^{m+1} = \sum_{i=1}^{n} y_i x_i^1$$

$$\vdots$$

$$a_0 \sum_{i=1}^{n} x_i^m + a_1 \sum_{i=1}^{n} x_i^{m+1} + a_2 \sum_{i=1}^{n} x_i^{m+2} + \cdots + a_m \sum_{i=1}^{n} x_i^{2m} = \sum_{i=1}^{n} y_i x_i^m \qquad (5.9.11)$$

It can be shown that this particular system has a unique solution provided that all $\{x_i\}_{n=1}^{n}$ are distinct.

Example 5.9.3.

Consider the data set given in Table 5.9.3.

Table 5.9.3. A Data Set
Resembling a Parabola

i	x_i	y_i
1	0.0	0.25
2	1.0	1.00
3	1.5	1.20
4	2.0	1.45
5	2.5	1.65

Because the data resemble a parabola rather than a straight line, we may decide to find the best quadratic fit for the given set and thus solve

$$5a_0 + 7a_1 + 13.5a_2 = 5.55$$
$$7a_0 + 13.5a_1 + 28a_2 = 9.825$$
$$13.5a_0 + 28a_1 + 61.125a_2 = 19.8125$$

to obtain $a_0 = 0.259, a_1 = 0.798, a_2 = -0.0988$ (rounded to three significant digits). The best quadratic fit is therefore

$$p_2(x) = -0.0988x^2 + 0.798x + 0.259$$

and the total least-squares error is $E \approx 0.0032$.   □

The least-squares approach is most helpful, if we know *a priori* the exact type of the function and just wish to calculate the values of some unknown parameters. In that case, one applies Eq. (5.9.10) and finally obtains an $n \times n$ system of linear or nonlinear equations. If the function is some unknown polynomial (with a given degree), the system is always linear. Otherwise we must deal with a system of nonlinear equations.

Example 5.9.4.

Let the experimental data behave like

$$y = ae^{\alpha x} \tag{5.9.12}$$

Table 5.9.4 contains a four-point data set related to this exponential function.

Table 5.9.4. An Experimental Data Set
for $y = ae^{\alpha x}$

x_i	y_i	$\bar{y}_i = ae^{\alpha x_i}$
0.0	0.48	0.497
0.5	0.31	0.292
1.0	0.17	0.172
1.5	0.10	0.101

To obtain a best fit we should minimize the total error

$$E = \sum_{i=1}^{4} (y_i - ae^{\alpha x_i})^2 \tag{5.9.13}$$

with respect to a and α. Thus

$$\frac{\partial E}{\partial a} = 2 \sum_{i=1}^{4} e^{\alpha x_i}(y_i - ae^{\alpha x_i}) = 0$$

$$\frac{\partial E}{\partial \alpha} = 2 \sum_{i=1}^{4} ax_i e^{\alpha x_i}(y_i - ae^{\alpha x_i}) = 0$$

leading to

$$a \sum_{i=1}^{4} e^{2\alpha x_i} - \sum_{i=1}^{4} y_i e^{\alpha x_i} = 0$$

$$a \sum_{i=1}^{4} x_i e^{2\alpha x_i} - \sum_{i=1}^{4} x_i y_i e^{\alpha x_i} = 0 \tag{5.9.14}$$

We may eliminate a from the first equation, substitute in the second, and find

$$F(\alpha) = \sum_{i=1}^{4} y_i e^{\alpha x_i} \sum_{i=1}^{4} x_i e^{2\alpha x_i} - \sum_{i=1}^{4} e^{2\alpha x_i} \sum_{i=1}^{4} x_i y_i e^{\alpha x_i} = 0 \tag{5.9.15}$$

Equation (5.9.15) is nonlinear and may be solved by Newton's method. However, an implementation of Newton's method requires computation of $F'(\alpha)$ and an intelligent initial guess α_0 as well.

Another approach is to rewrite Eq. (5.9.12) as

$$\ln y = \ln a + \alpha x \tag{5.9.16}$$

and find numbers a and α that would minimize

$$E' = \sum_{i=1}^{4} (\ln a + \alpha x_i - \ln y_i)^2 \tag{5.9.17}$$

The new problem is clearly different. The values of a, α that minimize E of Eq. (5.9.13) generally differ from those that minimize E'. But the new system of equations is *linear* in the variables $\ln a$ and α and is therefore easily solved. We have

$$\alpha \sum_{i=1}^{4} x_i^2 + (\ln a) \sum_{i=1}^{4} x_i = \sum_{i=1}^{4} x_i \ln(y_i)$$

$$\alpha \sum_{i=1}^{4} x_i + 4 \ln a = \sum_{i=1}^{4} \ln(y_i)$$

and in our case

$$3.5\alpha + 3 \ln a = -5.811$$
$$3\alpha + 4 \ln a = -5.980$$

with the unique solution $\alpha = -1.061, \ln a = -0.699$. The best exponential fit to the data, using one exponent and one coefficient, is thus given by

$$\bar{y} = ae^{\alpha x} = 0.497e^{-1.061x}$$

and produces the rightmost column of Table 5.9.4. If we choose to solve Eq. (5.9.15), which represents the original least-squares problem, we obtain

$$\bar{y}* = 0.486e^{-1.008x}$$

as the best exponential fit. The graphs of $\bar{y}, \bar{y}*$ and the data set are given in Fig. 5.9.3.

$\square$

A subroutine LSQR that, given a general data set, provides the user with linear and/or quadratic least-squares approximations is found on the attached floppy disk.

PROBLEMS

1. Find the linear least-squares approximations to the following data sets:
   (a) (1.5, –0.93), (2.2, 0.81), (3.4, 3.72), (4.0, 5.63), (6.4, 11.38)
   (b) (0, 0.30), (1, 1.42), (2, 2.55), (4, 4.47)

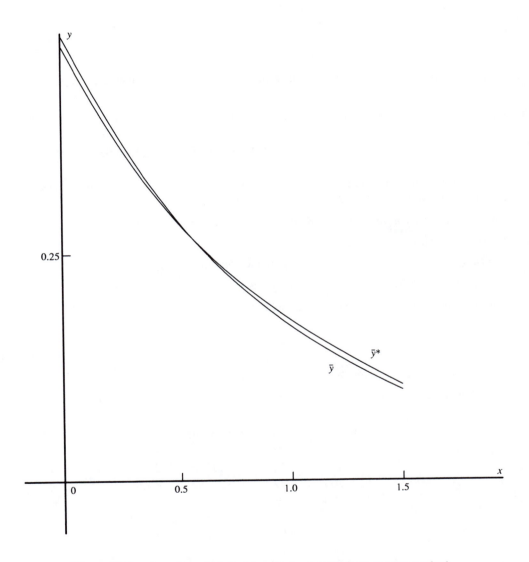

Figure 5.9.3. An exponential fit determined using the least-squares method.

(c) $(0, -3)$, $(2, 2)$, $(3, 4)$, $(5, 9)$, $(7, 17)$

and calculate the respective errors.

2. Find the quadratic least-squares approximations to the following data sets:

   (a) $(0, 0)$, $(1, 2)$, $(2, 6)$, $(3, 15)$, $(4, 20)$

   (b) $(0, 0)$, $(1, 1)$, $(4, 2)$, $(9, 3)$

   Sketch the two sets of points and comment.

3. Let $p(x) = 0.0001x^2 - 2x + 3$ be the quadratic least-squares approximation for a given data set. What can be said about this particular set?

4. Find the one-exponent exponential *least-squares* approximation for the data set

$$(0, 1.95), (0.25, 4.21), (0.50, 8.98), (0.75, 18.99), (1, 40.03)$$

by deriving an equation similar to Eq. (5.9.15) and solving it using the secant method. Choose starting values $\alpha_0 = 2.7, \alpha_1 = 2.8$.

5. Solve problem 4 by minimizing the related error E' given by Eq. (5.9.17).

6. Use basic calculus to show that

   (a) The system given by Eq. (5.9.5) has a unique solution.

   (b) The solution of Eq. (5.9.5) indeed minimizes the error E given by Eq. (5.9.2).

7. There is a reason to believe that some given data set $\{(x_i, y_i)\}_{i=1}^n$ is represented by the relation

$$y = ae^{\alpha x} + be^{\beta x}$$

Derive a least-squares scheme based on this assumption.

8. The following table consists of the homework and final grades of 20 students.

Homework	Final	Homework	Final
210	85	200	98
163	51	175	42
265	72	265	88
300	93	247	73
193	80	225	62
220	65	172	45
237	75	215	80
290	56	262	87
245	100	230	70
285	78	270	81

   (a) Obtain the least-squares line for this data.

   (b) Use this line to predict the homework grade necessary for a final grade of 75.

   (c) Discuss the answer to part (b).

10   Other Approximations

In this section we will discuss several other approaches for approximating a given *function* (not a data set). We start by extending the ideas of the previous section.

5.10.1   Approximations: Using the Least-Squares Technique and Orthogonality

Let $f(x) \in [a, b]$ be approximated by polynomials. The polynomial $p_n(x)$ of degree $\leq n$ that minimizes the error

$$\int_a^b [f(x) - p_n(x)]^2 \, dx \tag{5.10.1}$$

is called the *least-squares polynomial approximation* of degree $\leq n$ to $f(x)$. This is an extension to the discrete least-squares scheme discussed in Section 5.9. Consider a general polynomial of degree $\leq n$

$$p_n(x) = a_n x^n + a_{n-1} x^{n-1} + \cdots + a_1 x + a_0$$

and define

$$E = \int_a^b \left[f(x) - \sum_{i=0}^n a_i x^i \right]^2 dx \tag{5.10.2}$$

To minimize E we must solve a system of $(n+1)$ equations, namely

$$\frac{\partial E}{\partial a_i} = 0, \quad j = 0, 1, \ldots, n \tag{5.10.3}$$

However,

$$\frac{\partial E}{\partial a_j} = 2 \int_a^b \left[f(x) - \sum_{i=0}^n a_i x^i \right] (-x^j) \, dx$$

leading to

$$\sum_{i=0}^n a_i \int_a^b x^{i+j} \, dx = \int_a^b x^j f(x), \quad j = 0, 1, \ldots, n \tag{5.10.4}$$

It can be shown that there exists a unique solution to Eq. (5.10.4) and that it minimizes E, provided that $f \in C[a, b]$ and $a \neq b$. [The continuity of $f(x)$ is a sufficient condition for the existence of the right-hand side of Eq. (5.10.4).]

Example 5.10.1.
Let $f(x) = x^3, 0 \leq x \leq 1$ and we seek the least-squares quadratic approximation to $f(x)$. Here, $p_2(x) = a_2 x^2 + a_1 x + a_0$ and the system to be solved is

$$a_0 \int_0^1 1 \, dx + a_1 \int_0^1 x \, dx + a_2 \int_0^1 x^2 \, dx = \int_0^1 x^3 \, dx$$

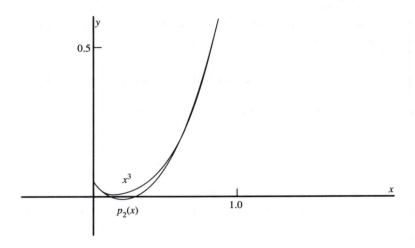

Figure 5.10.1. Least-squares quadratic approximation for x^3, $0 \le x \le 1$.

$$a_0 \int_0^1 x \, dx + a_1 \int_0^1 x^2 \, dx + a_2 \int_0^1 x^3 \, dx = \int_0^1 x^4 \, dx$$

$$a_0 \int_0^1 x^2 \, dx + a_1 \int_0^1 x^3 \, dx + a_2 \int_0^1 x^4 \, dx = \int_0^1 x^5 \, dx$$

or

$$a_0 + \frac{1}{2}a_1 + \frac{1}{3}a_2 = \frac{1}{4}$$

$$\frac{1}{2}a_0 + \frac{1}{3}a_1 + \frac{1}{4}a_2 = \frac{1}{5}$$

$$\frac{1}{3}a_0 + \frac{1}{4}a_1 + \frac{1}{5}a_2 = \frac{1}{6}$$

with the unique solution $a_0 = 1/20, a_1 = -3/5, a_2 = 3/2$. Thus the least-squares quadratic approximation to x^3 over the interval $[0, 1]$ is

$$p_2(x) = \frac{3}{2}x^2 - \frac{3}{5}x + \frac{1}{20}$$

A comparison between x^3 and $p_2(x)$ is given in Fig. 5.10.1 and Table 5.10.1.

Table 5.10.1. A Comparison between x^3 and $p_2(x)$ at Selected Points

x	x^3	$p_2(x)$	$|x^3 - p_2(x)|$
0.0	0.000	0.050	0.050
0.2	0.008	−0.010	0.018
0.4	0.064	0.050	0.014
0.6	0.216	0.230	0.014
0.8	0.512	0.530	0.018
1.0	1.000	0.950	0.050

The total error given by Eq. (5.10.2) is $E \approx 0.000357$. $\square$

The system given by Eq. (5.10.4) has two disadvantages. First, one must solve a set of $(n + 1)$ linear equations, which, for large n, is time consuming. Second, the system whose coefficients are given by

$$\int_a^b x^{i+j} \, dx = \frac{b^{i+j+1} - a^{i+j+1}}{i+j+1}$$

is represented by a Hilbert matrix, which is numerically ill conditioned. We can bypass these difficulties by replacing the general polynomial $a_n x^n + \cdots + a_1 x + a_0$, which is a *linear combination* of the functions $1, x, x^2, \ldots, x^n$, with a linear combination of *orthogonal functions*.

We start by introducing the concepts of *linear independence* and *orthogonality* between functions.

Definition 5.10.1.   (Linear independence). *A set of $(n+1)$ functions, $\{\phi_0(x), \phi_1(x), \ldots, \phi_n(x)\}$, defined over the interval $[a, b]$, is said to be* linearly independent *on $[a, b]$ if the relation*

$$\sum_{i=0}^n c_i \phi_i = c_0 \phi_0 + c_1 \phi_1 + \cdots + c_n \phi_n = 0, \quad a \le x \le b \qquad (5.10.5)$$

implies $c_i = 0, 0 \le i \le n$. Otherwise, the set is linearly dependent. *If $\{\phi_i\}_{i=0}^n$ is linearly dependent, then for some set of constants $\{c_i\}_{i=0}^n$, which are not all zero, Eq. (5.10.5) holds for* all $x \in [a, b]$.

Example 5.10.2.
The set $\{\sin^2 x, \cos^2 x, 1\}$ is linearly dependent for $-\infty < x < \infty$, because

$$1 \cdot \sin^2 x + 1 \cdot \cos^2 x - 1 \cdot 1 = 0, \quad -\infty < x < \infty$$

On the other hand, $\{\sin^2 x, \cos^2 x\}$ is linearly independent. Let

$$c_0 \sin^2 x + c_1 \cos^2 x = 0$$

and assume $c_0 \ne 0$. Then $\tan^2 x = -c_1/c_0$, which is true only for a *discrete* set of points. $\square$

Theorem 5.10.1.
(Linear independence theorem for polynomials). *Let $\phi_i, 0 \le i \le n$ be a polynomial of degree i. Then the set $\{\phi_0, \phi_1, \ldots, \phi_n\}$ is linearly independent over any interval.*

Proof.

Consider the interval $[a, b]$, and let

$$c_0\phi_0 + c_1\phi_1 + \cdots + c_n\phi_n = 0, \quad a \le x \le b$$

Because an nth order polynomial has exactly n zeroes, we obtain $c_n = 0$, then $c_{n-1} = 0$, and so on.

□

The next result follows immediately.

Corollary 5.10.1.

The polynomials $\{1, x, x^2, \ldots, x^n\}$ are linearly independent.

Another basic notation is the weight function.

Definition 5.10.2. *A function $w(x)$, defined over the interval $[a, b]$, is called a* weight function *if it satisfies the following requirements.*

1. $w(x)$ *is integrable over* $[a, b]$.

2. $w(x) \ge 0, a \le x \le b$.

3. $w(x) \ne 0$ *on* any *subinterval of* $[a, b]$.

For example, $w(x) = x^2, -1 \le x \le 1$ is a weight function. On the other hand, the function

$$w(x) = \begin{cases} 0, & 0 \le x \le 1 \\ x^2, & 1 < x \le 3 \end{cases}$$

is not a weight function.

Definition 5.10.3. (Orthogonality). *Let $w(x)$ be a weight function over the interval $I = [a, b]$ and let $f(x), g(x) \in C[a, b]$. If*

$$\int_a^b w(x)f(x)g(x)\, dx = 0 \tag{5.10.6}$$

then f, g are said to be orthogonal *on $[a, b]$ with respect to w.*

Example 5.10.3.

The functions $f(x) = x$ and $f(x) = x^2$ are orthogonal on $[-1, 1]$ with respect to $w(x) = 1$, because

$$\int_{-1}^1 1 \cdot x \cdot x^2\, dx = \left.\frac{x^4}{4}\right|_{-1}^1 = 0 \qquad \qquad \square$$

Example 5.10.4.

Let $w(x) = 1/\sqrt{1-x^2}, f(x) = 1, g(x) = x$ for $-1 \le x \le 1$. At $x = \pm 1$ define $w = 1$. Then

$$\int_{-1}^{1} w(x)f(x)g(x)\,dx = \int_{-1}^{1} \frac{x}{\sqrt{1-x^2}}\,dx = 0$$

Thus 1 and x are orthogonal with respect to $1/\sqrt{1-x^2}$.   □

We will now extend the concept of least-squares approximations, using a general set of linearly independent functions rather than the set $\{1, x, \ldots, x^n\}$.

Definition 5.10.4.   (General least-squares approximations). *Let $\{\phi_i\}_{i=0}^{n}$ be a set of linearly independent continuous functions, defined over the interval $[a, b]$, let w be a weight function on $[a, b]$, and let $f \in C[a, b]$. A linear combination*

$$p_n(x) = c_0\phi_0 + c_1\phi_1 + \cdots + c_n\phi_n = \sum_{i=0}^{n} c_i\phi_i \tag{5.10.7}$$

that minimizes the error

$$E(c_0, c_1, \ldots, c_n) = \int_{a}^{b} w(x)\left[f(x) - \sum_{i=0}^{n} c_i\phi_i\right]^2 dx \tag{5.10.8}$$

is called a least-squares approximation of $f(x)$, with respect to $\{\phi_i\}_{i=0}^{n}$ and w.

This particular $p_n(x)$ is the best approximation to $f(x)$ within the set of all the linear combinations of $\{\phi_i\}_{i=0}^{n}$, which is denoted by

$$\text{span}\{\phi_0, \phi_1, \ldots, \phi_n\} = \left\{\sum_{i=0}^{n} c_i\phi_i \mid -\infty < c_i < \infty, \quad 0 \le i \le n\right\} \tag{5.10.9}$$

To find a least-squares approximation we solve

$$\frac{\partial E}{\partial c_i} = 0, \quad 0 \le i \le n$$

and obtain the system

$$\sum_{i=0}^{n} c_i \int_{a}^{b} w(x)\phi_i(x)\phi_j(x)\,dx = \int_{a}^{b} w(x)f(x)\phi_j(x), \quad 0 \le i \le n \tag{5.10.10}$$

which yields a unique solution for $(c_0, c_1, \ldots, c_n)$.

As previously mentioned, to calculate the c values we must solve a set of $(n+1)$ linear equations.

An exception to the rule is the particular case when

$$\int_a^b w(x)\phi_i(x)\phi_j(x)\,dx = \begin{cases} 0\,, & i \neq j \\[2mm] A_i, & i = j \end{cases} \tag{5.10.11}$$

If Eq. (5.10.11) holds, then, because $A_i \neq 0, 0 \leq i \leq n$ (why?), we have

$$c_i = \frac{1}{A_i}\int_a^b w(x)f(x)\phi_i(x)\,dx, \quad 0 \leq i \leq n \tag{5.10.12}$$

Definition 5.10.5. (Orthogonality of a set of functions). *If the set $\{\phi_0, \phi_1, \ldots, \phi_n\}$ satisfies Eq. (5.10.11), it is said to be orthogonal over the interval $[a, b]$ with respect to w.*

The extension of this definition to an infinite set is straightforward.

Theorem 5.10.2.
If $\{\phi_0, \phi_1, \ldots, \phi_n\}$ is orthogonal, it is also linearly independent.

Proof.
Let

$$c_0\phi_0 + c_1\phi_1 + \cdots + c_n\phi_n = 0, \quad a \leq x \leq b$$

If we multiply both sides by $w\phi_i$ and integrate over $[a, b]$, we obtain

$$c_i \int_a^b w(x)\phi_i^2(x)\,dx = c_i A_i = 0$$

Because $A_i \neq 0$, we must have $c_i = 0, 0 \leq i \leq n$. ⬜

Example 5.10.5.
Consider the set $\{1, \cos(mx), \sin(mx)\}_{m=1}^n$ over the interval $[-\pi, \pi]$ with $w = 1$. We can easily verify

$$\int_{-\pi}^{\pi} \cos(mx)\cos(kx)\,dx = 0, \quad m \neq k$$

$$\int_{-\pi}^{\pi} \sin(mx)\sin(kx)\,dx = 0, \quad m \neq k$$

$$\int_{-\pi}^{\pi} \sin(mx)\cos(kx)\,dx = 0$$

$$\int_{-\pi}^{\pi} \cos^2(mx)\,dx = \int_0^{2\pi} \sin^2(mx)\,dx = \pi, \quad \int_0^{2\pi} 1^2 dx = 2\pi$$

and thus the set is orthogonal. ☐

The next result is concluded directly from Eq. (5.10.12).

Corollary 5.10.2.

Let $\{\phi_0, \phi_1, \ldots, \phi_n\}$ be an orthogonal set with respect to w, over the interval $[a, b]$. Then the least-squares approximation of $f(x) \in C[a, b]$, related to $\{\phi_i\}_{i=0}^n$ and w, is given by $\sum_{i=0}^n c_i \phi_i$, where

$$c_i = \frac{\int_a^b w(x) f(x) \phi_i(x)\, dx}{\int_a^b w(x) \phi_i^2(x)\, dx} \tag{5.10.13}$$

Definition 5.10.6.   (Orthonormality). *If the requirements of Definition 5.10.5 are fulfilled, and if in addition*

$$\int_a^b w(x) \phi_i^2(x)\, dx = 1, \quad 0 \le i \le n$$

then the set $\{\phi_i\}_{i=0}^n$ is called orthonormal over $[a, b]$ with respect to w.

Example 5.10.6.

Define $\{\phi_0, \phi_1, \ldots, \phi_{2n}\}$ over $[-\pi, \pi]$ as

$$\phi_0(x) = \frac{1}{\sqrt{2\pi}}, \quad \phi_{2k-1}(x) = \frac{1}{\sqrt{\pi}} \cos(kx), \quad \phi_{2k}(x) = \frac{1}{\sqrt{\pi}} \sin(kx), \quad 1 \le k \le n$$

By virtue of Example 5.10.5 this set is orthonormal. The least-squares approximation to a given function $f(x)$, using this particular set, is called the *trigonometric polynomial* of degree n that approximates $f(x)$. $\qquad\square$

Using orthogonal or orthonormal sets in a search for a least-squares approximation renders another advantage. Suppose that $\{\phi_0, \phi_1, \ldots, \phi_n, \ldots\}$ is an infinite orthogonal set over $[a, b]$ with respect to w. Let

$$p_n(x) = \sum_{i=0}^n c_i \phi_i(x)$$

be the least-squares approximation to $f(x)$ with respect to $\{\phi_i\}_{i=0}^n$. Then the next approximation $p_{n+1}(x)$ is given by

$$p_{n+1}(x) = p_n(x) + c_{n+1} \phi_{n+1}(x) \tag{5.10.14}$$

where

$$c_{n+1} = \frac{\int_a^b w(x) f(x) \phi_{n+1}(x)\, dx}{\int_a^b w(x) \phi_{n+1}^2(x)\, dx}$$

Thus we do not have to recalculate all the coefficients, as we must when solving the set given by Eq. (5.10.4).

Example 5.10.7.

Let $f(x) = x$, $-\pi \leq x \leq \pi$. The trigonometric polynomial of degree 1 is given by

$$S_1(x) = c_0 \frac{1}{\sqrt{2\pi}} + \frac{c_1}{\sqrt{\pi}} \cos x + \frac{c_2}{\sqrt{\pi}} \sin x$$

where

$$c_0 = \frac{1}{\sqrt{2\pi}} \int_{-\pi}^{\pi} x \, dx = 0, \quad c_1 = \frac{1}{\sqrt{\pi}} \int_{-\pi}^{\pi} x \cos x \, dx = 0, \quad c_2 = \frac{1}{\sqrt{\pi}} \int_{-\pi}^{\pi} x \sin x \, dx = 2\sqrt{\pi}$$

Thus

$$S_1(x) = 2 \sin x$$

is the trigonometric polynomial of degree 1 for $f(x) = x$, on $[-\pi, \pi]$. To calculate the trigonometric polynomial of degree 2, we calculate only the additional integrals

$$c_3 = \frac{1}{\sqrt{\pi}} \int_{\pi}^{\pi} x \cos(2x) \, dx = 0, \quad c_4 = \frac{1}{\sqrt{\pi}} \int_{-\pi}^{\pi} x \sin(2x) \, dx = -\sqrt{\pi}$$

and obtain

$$S_2(x) = S_1(x) + c_3 \frac{1}{\sqrt{\pi}} \cos(2x) + c_4 \frac{1}{\sqrt{\pi}} \sin(2x) = 2 \sin x - \sin(2x)$$

The function and its least-squares approximation $S_2(x)$ are illustrated in Fig. 5.10.2.   □
The trigonometric polynomial of degree n, associated with $f(x)$, is

$$S_n(x) = \sum_{m=0}^{n} [a_m \cos(mx) + b_m \sin(mx)] \tag{5.10.15}$$

where

$$a_m = \frac{1}{\pi} \int_{-\pi}^{\pi} f(x) \cos(mx) \, dx, \quad m > 0$$

$$b_m = \frac{1}{\pi} \int_{-\pi}^{\pi} f(x) \sin(mx) \, dx, \quad m > 0 \tag{5.10.16}$$

and

$$a_0 = \frac{1}{2\pi} \int_{-\pi}^{\pi} f(x) \, dx \tag{5.10.17}$$

$S_n(x)$ is the best approximation [in the sense of minimizing E of Eq. (5.10.8)] to $f(x)$ within the set of all linear combinations of the functions 1 and $\{\sin(mx), \cos(mx)\}_{m=1}^{n}$. If we let $n \to \infty$, we obtain the *Fourier series* of $f(x)$, which is a common notation in ordinary and partial differential equations.

Two types of functions that are particularly important and strongly related to Fourier series theory are even and odd functions.

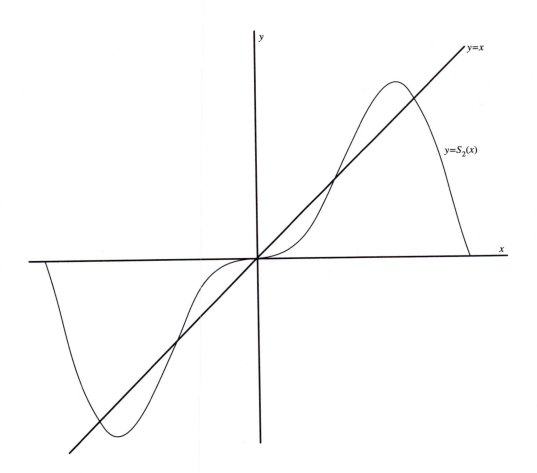

Figure 5.10.2. Trigonometric polynomial of $f(x) = x$, $-\pi \leq x \leq \pi$.

Definition 5.10.7. (Even and odd functions). *A function $f(x), -a \leq x \leq a$ is said to be* even *or* odd *if*

$$f(-x) = f(x), \quad 0 \leq x \leq a \text{ (even)} \tag{5.10.18}$$

or

$$f(-x) = -f(x), \quad 0 \leq x \leq a \text{ (odd)} \tag{5.10.19}$$

respectively.

For example, the functions x, x^3, and $\sin x$ are odd, whereas $x^2, 1 + x^4$, and $\cos(3x)$ are even.

Corollary 5.10.3.

Let a_m, b_m be the Fourier coefficients of $f(x)$ as given by Eqs. (5.10.16) and (5.10.17). If $f(x)$ is even or odd, then $b_m = 0, m > 0$ or $a_m = 0, m \geq 0$, respectively. The proof is left to the reader.

Example 5.10.8.

Let $f(x) = x, -\pi \leq x \leq \pi$. We have

$$a_m = 0, \quad m \geq 0; \quad b_m = \frac{1}{\pi} \int_{-\pi}^{\pi} x \sin(mx)\, dx = -\frac{2}{m}(-1)^m, \quad m > 0$$

and the Fourier series of $f(x)$ is

$$S(x) = 2 \sin x - \sin(2x) + \frac{2}{3} \sin(3x) - \frac{1}{2} \sin(4x) + \cdots \qquad \square$$

Example 5.10.9.

Let $f(x) = x^2, -\pi \leq x \leq \pi$. Here $b_m = 0, m > 0$, and

$$a_0 = \frac{1}{2\pi} \int_{-\pi}^{\pi} x^2\, dx = \frac{\pi^2}{3}, \quad a_m = \frac{1}{\pi} \int_{-\pi}^{\pi} x^2 \cos(mx)\, dx = (-1)^m \frac{4}{m^2}$$

The Fourier series is therefore

$$S(x) = \frac{\pi^2}{3} + \sum_{i=1}^{\infty} (-1)^m \frac{4}{m^2} \cos(mx)$$

and it converges for all x, because

$$\left| (-1)^m \frac{4}{m^2} \cos(mx) \right| \leq \frac{4}{m^2}$$

and $\sum_{m=1}^{\infty} 4/m^2$ converges. $\qquad \square$

For a given set of functions, C, let span (C) denote the set of all *finite* linear combinations of functions in C. For example,

$$\text{span}\{1, x, \sin x\} = \{c_0 + c_1 x + c_2 \sin x | -\infty < c_i < \infty, \quad 0 \leq i \leq 2\}$$

$$\text{span}\{1, x, x^2, \ldots\} = \left\{ \sum_{i=0}^{n} c_i x^i \mid -\infty < c_i < \infty, \quad 0 \leq n < \infty \right\}$$

We will now show that given a linearly independent set of functions $A = \{\phi_0, \phi_1, \ldots, \phi_n\}$ and a weight function w, defined over an interval $[a, b]$, we can construct another set of functions $B = \{\theta_0, \theta_1, \ldots, \theta_n\}$ such that

1. B is orthogonal over $[a, b]$ with respect to w.

2. span (A) = span (B).

Thus, instead of minimizing E of Eq. (5.10.8) by solving each time a set of linear equations, we may calculate $\{\theta_0, \theta_1, \ldots, \theta_n\}$ just *once*, and implement Eq. (5.10.13) for any given $f(x)$.

5.10.2   The Gram–Schmidt Process

Let $\{\phi_0, \phi_1, \ldots, \phi_n\}$ be a set of linearly independent functions defined over an interval $[a, b]$, along with a weight function $w(x)$. Suppose that $\{\theta_0, \theta_1, \ldots, \theta_k\}$ is already orthogonal over $[a, b]$ with respect to $w(x)$, for some $k \geq 0$, and that

$$\text{span}\{\phi_0, \phi_1, \ldots, \phi_k\} = \text{span}\{\theta_0, \theta_1, \ldots, \theta_k\} \tag{5.10.20}$$

Define

$$\theta_{k+1} = \phi_{k+1} + \sum_{i=0}^{k} \alpha_i \theta_i \tag{5.10.21}$$

where $\alpha_i, 0 \leq i \leq k$ are still underdetermined. If we multiply both sides of Eq. (5.10.21) by $w\theta_j, 0 \leq j \leq k$ and integrate over $[a, b]$, we obtain

$$\int_a^b w\theta_{k+1}\theta_j \, dx = \int_a^b w\phi_{k+1}\theta_j \, dx + \alpha_j \int_a^b w\theta_j^2 \, dx, \quad 0 \leq j \leq k \tag{5.10.22}$$

Let us choose α_j so that the left-hand side of Eq. (5.10.22) vanishes. Consequently,

$$\alpha_j = -\frac{\int_a^b w\phi_{k+1}\theta_j \, dx}{\int_a^b w\theta_j^2 \, dx}, \quad 0 \leq j \leq k \tag{5.10.23}$$

and θ_{k+1} is orthogonal to $\{\theta_0, \theta_1, \ldots, \theta_k\}$. But we also have

$$\int_a^b w\theta_{k+1}^2 \, dx > 0$$

otherwise $\theta_{k+1} = 0$ and Eqs. (5.10.20) and (5.10.21) force a contradiction (why?). Thus the whole set $\{\theta_0, \theta_1, \ldots, \theta_k, \theta_{k+1}\}$ is orthogonal with respect to w over $[a, b]$ and

$$\text{span}\{\phi_0, \phi_1, \ldots, \phi_k, \phi_{k+1}\} = \text{span}\{\theta_0, \theta_1, \ldots, \theta_k, \theta_{k+1}\}$$

To start the set $\{\theta_0, \theta_1, \ldots, \theta_n\}$, we define $\theta_0 = \phi_0$ and then apply Eq. (5.10.21) n times.

Example 5.10.10.

Let $\phi_0 = 1, \phi_2 = e^x, w = x$ for $0 \le x \le 1$. We have

$$\theta_0 = 1, \quad \theta_2 = e^x - \frac{\int_0^1 x e^x \cdot 1 \, dx}{\int_0^1 x \cdot 1^2 \, dx} \cdot 1 = e^x - 2$$

and clearly span $\{1, e^x\} = \text{span}\{1, e^x - 2\}$. $\qquad\square$

Example 5.10.11.

Consider the linearly independent set

$$\{\phi_0, \phi_1, \ldots, \phi_n\} = \{1, x, \ldots, x^n\}, \quad -1 \le x \le 1$$

and the weight function $w = 1$. Here

$$\theta_0 = 1, \quad \theta_2 = x - \frac{\int_{-1}^1 1 \cdot x \cdot 1 \, dx}{\int_{-1}^1 1 \cdot 1^2 \, dx} \cdot 1 = x,$$

$$\theta_3 = x^2 - \frac{\int_{-1}^1 1 \cdot x^2 \cdot 1 \, dx}{\int_{-1}^1 1 \cdot 1^2 \, dx} \cdot 1 - \frac{\int_{-1}^1 1 \cdot x^2 \cdot x \, dx}{\int_{-1}^1 1 \cdot x^2 \, dx} x = x^2 - \frac{1}{3} \text{ and so on}$$

The function θ_k is clearly a kth order polynomial and is called the *Legendre polynomial* of degree k. $\qquad\square$

Example 5.10.12.

The least-squares quadratic with $w = 1$, which approximates $f(x) = x^3, -1 \le x \le 1$, is given by

$$c_0 + c_1 x + c_2 \left(x^2 - \frac{1}{3} \right)$$

where

$$c_0 = \frac{\int_{-1}^1 x^3 \, dx}{\int_{-1}^1 1^2 \, dx} = 0, \quad c_1 = \frac{\int_{-1}^1 x^3 \cdot x \, dx}{\int_{-1}^1 x^2 \, dx} = \frac{3}{5}, \quad c_2 = \frac{\int_{-1}^1 x^3 \left[x^2 - (1/3) \right] dx}{\int_{-1}^1 \left[x^2 - (1/3) \right]^2 dx} = 0$$

The least-squares quadratic approximation is therefore $(3/5)x$.

5.10.3 Approximation by Rational Functions

The advantages of using polynomials to approximate functions have been previously discussed. The main disadvantage, however, is the tendency of the polynomial to oscillate, which often causes a large maximum error, that is, a large error bound. To decrease this error bound we introduce a new class of approximating functions, namely, the *rational functions*

$$r(x) = \frac{p(x)}{q(x)} \tag{5.10.24}$$

where $p(x)$ and $q(x)$ are polynomials.

Definition 5.10.8.   (Degree of a rational function). *Let $p(x)$ and $q(x)$ denote polynomials of degrees n and m, respectively. Then the degree of $r(x)$ as given by Eq. (5.10.24) is defined as $N = m + n$.*

For example, the degree of $(2x + 5)/(x^2 - 1)$ is 3.

In the particular case $q(x) = 1$, $r(x)$ is a polynomial. Thus, by approximating a function using all rational functions of degree $\leq n$, we can only decrease the error bound associated with polynomial approximations of the same degree. Occasionally the use of rational functions instead of polynomials is a necessity. For example, consider the function

$$f(x) = \frac{1}{x + x^2\sqrt{x}}, \quad \frac{1}{10^6} \leq x \leq 1$$

Because $f(x) \approx 1/x$ near $x = 1/10^6$, $f(x)$ is better approximated by a rational function than by a polynomial.

Let $f(x)$ by defined over the interval $[a, b]$ that, for the sake of simplicity, contains the point $x = 0$. Let

$$r(x) = \frac{p(x)}{q(x)} = \frac{\sum_{i=0}^{n} a_i x^i}{\sum_{j=0}^{m} b_j x^j}, \quad b_0 \neq 0 \tag{5.10.25}$$

denote an approximation to $f(x)$. We can normalize the right-hand side of Eq. (5.10.25) and choose $b_0 = 1$. The degree of $r(x)$ is $N = n + m$ and the number of free parameters is $(n + m + 1)$. They are generally uniquely determined if we impose, for example,

$$f^{(k)}(0) = r^{(k)}(0), \quad 0 \leq k \leq N \tag{5.10.26}$$

Indeed, let us consider the error

$$f(x) - r(x) = f(x) - \frac{p(x)}{q(x)} = \frac{f(x)q(x) - p(x)}{1 + b_1 x + \cdots + b_m x^m}$$

Assume that $f(x)$ can be replaced by its Maclaurin series, that is,

$$f(x) = \sum_{l=0}^{\infty} c_l x^l$$

Then, to satisfy Eq. (5.10.26), we should choose $\{a_i\}_{i=0}^{n}$ and $\{b_j\}_{j=1}^{m}$ such that the powers $\{x^k, 0 \leq k \leq N\}$ will not appear in the numerator $f(x)q(x) - p(x)$ (why?). The result is a system of $(N + 1)$

linear equations

$$\sum_{i=0}^{j} c_i b_{j-i} - a_j = 0, \quad 0 \le j \le N$$

with the $(N+1)$ unknowns $a_0, a_1, \ldots, a_n$ and $b_1, \ldots, b_m$. This scheme for determining $p(x)$ and $q(x)$ is known as Padé approximation, and it is generally recommended to choose $m \approx n$.

Example 5.10.13.

Let $f(x) = \ln(1+x), 0 \le x \le 1$. For $m = n = 2$ we find

$$f(x)q(x) - p(x) = (x - \frac{x^2}{2} + \frac{x^3}{3} - \frac{x^4}{4} + \cdots)(1 + b_1 x + b_2 x^2) - a_0 - a_1 x - a_2 x^2$$

and the system of linear equations is

$$x^0 : -a_0 = 0$$
$$x^1 : 1 - a_1 = 0$$
$$x^2 : -\frac{1}{2} + b_1 - a_2 = 0$$
$$x^3 : \frac{1}{3} - \frac{b_1}{2} + b_2 = 0$$
$$x^4 : -\frac{1}{4} + \frac{b_1}{3} - \frac{b_2}{2} = 0$$

The unique solution of the system is $a_0 = 0, a_1 = 1, a_2 = 0.5, b_1 = 1, b_2 = 1/6$, that is,

$$f(x) = \frac{x + (x^2/2)}{1 + x + (x^2/6)}$$

Table 5.10.2. Padé and Taylor Approximations for $\ln(1+x)$

x	$\ln(1+x)$	$r(x)$	$|r(x) - \ln(1+x)|$	$p_4(x)$	$|p_4(x) - \ln(1+x)|$
0.2	0.18232	0.18232	$0.11 \cdot 10^{-5}$	0.18227	$0.55 \cdot 10^{-4}$
0.4	0.33647	0.33645	$0.24 \cdot 10^{-4}$	0.33493	$0.15 \cdot 10^{-2}$
0.6	0.47000	0.46988	$0.12 \cdot 10^{-3}$	0.45960	$0.10 \cdot 10^{-1}$
0.8	0.58779	0.58741	$0.37 \cdot 10^{-3}$	0.54827	$0.40 \cdot 10^{-1}$
1.0	0.69315	0.69231	$0.84 \cdot 10^{-3}$	0.58333	$0.11 \cdot 10^{0}$

and a comparison between $r(x)$ and the fourth order Taylor approximation

$$p_4(x) = x - \frac{x^2}{2} + \frac{x^3}{3} - \frac{x^4}{4}$$

is given in Table 5.10.2, where the numbers are rounded to five significant digits. The Padé approximation is superior to the Taylor approximation over the whole interval. Both errors are small near $x = 0$ but increase with x. This is expected because the requirements in either case are at $x = 0$. However, the error of Padé approximation increases at a slower rate.   □

The next example demonstrates the advantage of Padé approximation when an infinite discontinuity occurs outside but "sufficiently close" to the interval of approximation.

Example 5.10.14.

Let

$$f(x) = \frac{1}{\sin x + (1/1000)}, \quad 0 \le x \le 1$$

Here, $f(0) = 1000, f'(0) = -10^6, f''(0) = 2 \cdot 10^9$. The Taylor approximation of order 2 is

$$p_2(x) = 10^3 - 10^6 x + 10^9 x^2$$

The Padé approximation of order 2 with $m = n = 1$ is generated by computing b_1, a_0, a_1, for which the first three coefficients of

$$(10^3 - 10^6 x + 10^9 x^2 + \cdots)(1 + b_1 x) - (a_0 + a_1 x)$$

are zero. The system of equations is

$$10^3 - a_0 = 0$$
$$-10^6 + 10^3 b_1 - a_1 = 0$$
$$10^9 - 10^6 b_1 = 0$$

and it yields the unique solution $a_0 = 10^3, b_1 = 10^3, a_1 = 0$. Thus

$$r(x) = \frac{10^3}{1 + 10^3 x}, \quad 0 \le x \le 1$$

Obviously p_2 does not approximate $f(x)$ for large x. But even for x as small as $1/1000$ we obtain

$$f\left(\frac{1}{1000}\right) = 500, \quad p_2\left(\frac{1}{1000}\right) = 1000, \quad r\left(\frac{1}{1000}\right) = 500$$

Padé approximation is therefore superior to Taylor approximation near $x = 0$. For $x = 0.2, r(x)$ still yields two accurate digits: $f(0.2) = 5.01, r(0.2) = 4.98$.   □

PROBLEMS

1. Find the least-squares linear, quadratic, and cubic approximations to the following functions:
   (a) $f(x) = x^4, 0 \le x \le 1$
   (b) $f(x) = \sqrt{x+1}, 0 \le x \le 1$

(c) $f(x) = \sin x, 0 \le x \le \pi$

2. Compare $\sin x$ with its three approximations of problem 1 at $x = i(\pi/8), 0 \le i \le 8$.

3. Check whether the following sets are linearly independent.

 (a) $\sin x, \cos x, \sin(2x)$

 (b) $\ln(1 + x), \ln(1 - x), 1$

 (c) $x, \sin x, \tan x$

 (d) $\cos(2x), \sin^2 x, 1$

4. Find the trigonometric polynomial of degree ≤ 2 in the following cases:

 (a) $f(x) = x^2 + 1, -\pi \le x \le \pi$

 (b) $f(x) = e^x, -\pi \le x \le \pi$

 (c) $f(x) = \sin(3x), -\pi \le x \le \pi$

5. Find the Fourier series for $f(x) = |x|, -\pi \le x \le \pi$ and show its convergence.

6. Show that any given function defined over the interval $[-a, a]$ can be expressed as the sum of an even and an odd function.

7. Use the Gram–Schmidt process and replace the set $\{1, x, x^2\}$ by an orthogonal set $\{\theta_0, \theta_1, \theta_2\}$ with respect to $w = 1$, over the interval $[0, 1]$.

8. Repeat and solve problem 7 for $\{1, x, \sin x\}$ and $w(x) = \cos x$. Express each function as a linear combination of the orthogonal set.

9. Find the Padé approximation of order 2, with $m = n = 1$ in the following cases:

 (a) $f(x) = \sin x + \cos x, 0 \le x \le 1$

 (b) $f(x) = \sqrt{1 + x}, 0 \le x \le 1$

 (c) $f(x) = e^x, 0 \le x \le 1$

10. Find the Padé approximation of order 3, with $m = 2, n = 1$ for

 (a) $f(x) = 1/(0.1 + x^2 + x^3), 0 \le x \le 1$

 (b) $f(x) = 1/[0.1 + \ln(1 + x)], 0 \le x \le 1$

Numerical Integration and Differentiation

6.1 Numerical Integration: The Trapezoidal and Simpson Rules

The definite integral

$$I(f; a, b) = \int_a^b f(x)\, dx \tag{6.1.1}$$

of a given function $f(x)$ is defined as the limit of the Riemann sums of $f(x)$. Let $f(x)$ be a real-valued function defined over the interval $[a, b]$, and let

$$a = x_0 < x_1 < \cdots < x_n = b \tag{6.1.2}$$

be some partition of $[a, b]$ to subintervals. A sum

$$R_n = \sum_{i=1}^n f(c_i)\Delta x_i = \sum_{i=1}^n f(c_i)(x_i - x_{i-1}) \tag{6.1.3}$$

where $x_{i-1} \le c_i \le x_i$, $i = 1, \ldots, n$, is called a *Riemann sum*. If

$$\lim_{\max(\Delta x_i) \to 0} R_n \tag{6.1.4}$$

exists, $f(x)$ has a *Riemann integral* and we define

$$\int_a^b f(x)\, dx = \lim_{\max(\Delta x_i) \to 0} R_n \tag{6.1.5}$$

It can be shown that

$$\int_a^b f(x)\, dx = F(b) - F(a) \tag{6.1.6}$$

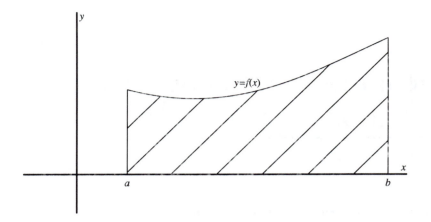

Figure 6.1.1. Geometric interpretation of $I(f,a,b)$.

provided that

$$F'(x) = f(x), \quad a \le x \le b \tag{6.1.7}$$

This is known as the *fundamental theorem* of differential calculus.

Geometrically, $I(f; a, b)$ represents the "area" that is bounded between the x axis, the straight lines $x = a$ and $x = b$, and the function $y = f(x)$ (Fig. 6.6.1).
It should be emphasized that an area below the x axis is considered negative. This follows directly from Eq. (6.1.3). Thus, for example (Fig. 6.1.2),

$$\int_{-1}^{1} x \, dx = 0.$$

If an integral can be calculated by using Eq. (6.1.6) we say that $f(x)$ can be *analytically integrated*. This is true for all polynomials, $\sin x$, $\cos x$, e^x, and so on. In most cases, however, an *antiderivative* $F(x)$ of the *integrand* $f(x)$ cannot be found and expressed in terms of *elementary functions*. Simple examples are

$$\int_{1}^{2} \frac{e^x}{x} \, dx, \quad \int_{0}^{1} \sqrt{1 + x^5} \, dx$$

This demonstrates the necessity to provide *numerical procedures* to *approximate* $I(f; a, b)$. In this section we present the trapezoidal and Simpson rules, which basically consist of *replacing $f(x)$ by approximating polynomials*, which are then integrated analytically.

6.1.1 The Trapezoidal Rule

Let $f(x)$ be a real-valued function defined over the interval $[a, b]$ and consider the subintervals

$$[x_0, x_1], [x_1, x_2], \ldots, [x_{n-1}, x_n]$$

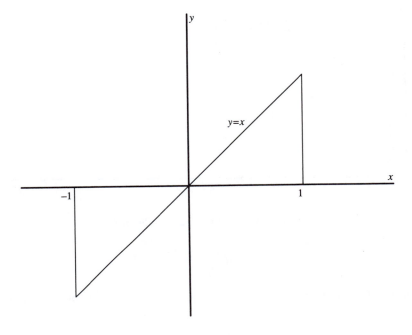

Figure 6.1.2.   Area below the x axis is negative.

where $x_0 = a$ and $x_n = b$. Throughout the interval $[x_i, x_{i+1}], f(x)$ is approximated by the linear polynomial

$$y = y_i + \frac{y_{i+1} - y_i}{x_{i+1} - x_i}(x - x_i) \tag{6.1.8}$$

where $y_i = f(x_i), 0 \leq i \leq n$. It represents the secant connecting (x_i, y_i) and (x_{i+1}, y_{i+1}) (Fig. 6.1.3).

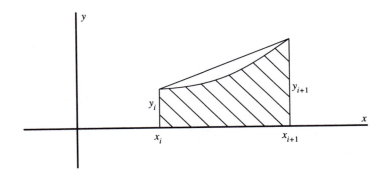

Figure 6.1.3.   Illustration of the trapezoidal rule.

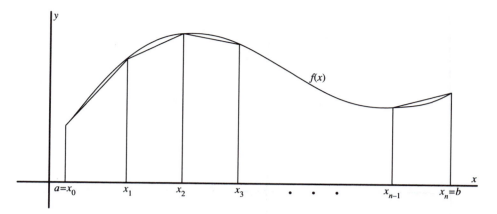

Figure 6.1.4. The trapezoidal rule for $f(x)$.

The integral of $f(x)$ between x_i and x_{i+1} is approximated by the area of a trapezoid, that is,

$$\int_{x_i}^{x_{i+1}} f(x)\, dx \approx \frac{f(x_i)+f(x_{i+1})}{2}(x_{i+1}-x_i) \tag{6.1.9}$$

If all the trapezoids are added (Fig. 6.1.4), we obtain the *trapezoidal rule*, given by

$$\int_a^b f(x)\, dx \approx \frac{1}{2}\sum_{i=1}^{n}[f(x_i)+f(x_{i-1})]\Delta x_i \tag{6.1.10}$$

A special case of Eq. (6.1.10) is

$$\Delta x_i = h, \quad 1 \leq i \leq n \tag{6.1.11}$$

for which

$$\int_a^b f(x)\, dx \approx h\left[\frac{f(x_0)}{2}+f(x_1)+f(x_2)+\cdots+f(x_{n-1})+\frac{f(x_n)}{2}\right] \tag{6.1.12}$$

Example 6.1.1.

Let $f(x) = e^x$ for $0 \leq x \leq 2$ and consider the subintervals $[0,1], [1,1.5], [1.5,2]$. Using the trapezoidal rule, we find

$$\int_0^2 e^x\, dx \approx \frac{1}{2}[(e^0+e^1)\cdot 1 + (e^1+e^{1.5})\cdot 0.5 + (e^{1.5}+e^2)\cdot 0.5] \approx 6.627$$

whereas the exact value is $6.389\ldots$.   □

Example 6.1.2.

Let $f(x) = x \ln x$, $1 \le x \le 2$. Consider the eight equal subintervals defined by

$$x_n = 1 + (n-1)h, \quad 1 \le n \le 9$$

where $h = 0.125$. The approximation to $\int_1^2 x \ln x \, dx$, using the trapezoidal rule and rounding to four digits, is

$$0.125 \left[\frac{1 \cdot \ln(1)}{2} + 1.125 \cdot \ln(1.125) + \cdots + 1.875 \cdot \ln(1.875) + \frac{2 \cdot \ln(2)}{2} \right] = 0.6372$$

The exact value of the integral is 0.6363. □

Let us denote by $T_n(f; a, b)$ the right-hand side of Eq. (6.1.12), that is, the approximation to $I(f; a, b)$, using the trapezoidal rule with n equal subintervals. We naturally expect $T_n(f; a, b)$ to satisfy

$$\lim_{n \to \infty} T_n(f; a, b) = \int_a^b f(x) \, dx \tag{6.1.13}$$

The next theorem provides an error estimate and sufficient conditions for which Eq. (6.1.13) holds.

Theorem 6.1.1.

(Error estimate for the trapezoidal rule). *Let $f(x)$ be a real-valued function possessing two continuous derivatives over the interval $[a, b]$, and let $\{x_i\}_{i=0}^n$ be evenly spaced points that satisfy $a = x_0 < x_1 < \cdots < x_n = b$, $h = (b-a)/n$. Then the error associated with the trapezoidal rule is given by*

$$I(f; a, b) - T_n(f; a, b) = -\frac{h^2(b-a)}{12} f''(c) \tag{6.1.14}$$

where $a < c < b$.

Proof.

The contribution of each subinterval $[x_i, x_{i+1}]$ to $T_n(f; a, b)$ is

$$\int_{x_i}^{x_{i+1}} p_1(x; x_i, x_{i+1}) \, dx \tag{6.1.15}$$

where $p_1(x; x_i, x_{i+1})$ is the linear Lagrange interpolator to $f(x)$ with interpolating points x_i, x_{i+1}. By using Eq. (5.2.2) we obtain

$$f(x) - p_1(x; x_i, x_{i+1}) = \frac{(x-x_i)(x-x_{i+1})}{2!} f''(c_{ix}) \tag{6.1.16}$$

where c_{ix} is between x_i and x_{i+1}. Then, by integrating both sides of Eq. (6.1.16) and applying the continuity of $f''(x)$, we have

$$\int_{x_i}^{x_{i+1}} f(x)\,dx - \int_{x_i}^{x_{i+1}} p_1(x; x_i, x_{i+1})\,dx = \int_{x_i}^{x_{i+1}} \frac{(x-x_i)(x-x_{i+1})}{2!} f''(c_{ix})\,dx$$

$$= f''(c_i) \int_{x_i}^{x_{i+1}} \frac{(x-x_i)(x-x_{i+1})}{2!}\,dx = -\frac{h^3}{12} f''(c_i)$$

where $x_i < c_i < x_{i+1}$ (why?). Therefore

$$I(f; a, b) - T_n(f; a, b) = -\frac{h^3}{12} \sum_{i=1}^{n} f''(c_i) \tag{6.1.17}$$

Let m and M be the minimum and the maximum of $f''(x)$ within $[a, b]$. Then

$$nm \leq f''(c_1) + \cdots + f''(c_n) \leq nM$$

or $m \leq \frac{1}{n} \sum_{i=1}^{n} f''(c_i) \leq M$.

Because $f''(x)$ is continuous, there is a point c between a and b for which

$$f''(c) = \frac{1}{n} \sum_{i=1}^{n} f''(c_i)$$

Therefore

$$I(f; a, b) - T_n(f; a, b) = -\frac{nh^3}{12} f''(c) = -\frac{h^2(b-a)}{12} f''(c) \qquad \square$$

An immediate result of Theorem 6.1.1 is the following corollary.

Corollary 6.1.1.

(Error bound for the trapezoidal rule). *Let $f(x)$ satisfy the requirements of Theorem 6.1.1 and denote $M_2 = \max_{a \leq x \leq b} |f''(x)|$. Then*

$$|I(f; a, b) - T_n(f; a, b)| \leq \frac{h^2(b-a)}{12} M_2 \tag{6.1.18}$$

Example 6.1.3.

Consider the function $\sin x$ over $[0, \pi]$ and let $n = 6, h = \pi/6$. $T_n(f; a, b)$, rounded to six significant digits, is given by

$$T_6(\sin x; 0, \pi) = \frac{\pi}{6}\left[\frac{\sin 0}{2} + \sin\left(\frac{\pi}{6}\right) + \cdots + \sin\left(\frac{5\pi}{6}\right) + \frac{\sin \pi}{2}\right] \approx 1.95410$$

whereas $I(\sin x; 0, \pi) = 2$. The right-hand side of Eq. (6.1.18) is $(1/12)(\pi/6)^2(\pi - 0) \cdot 1 \approx 0.0718$, whereas $|2 - 1.95410| = 0.0459 < 0.0718$, as expected. $\square$

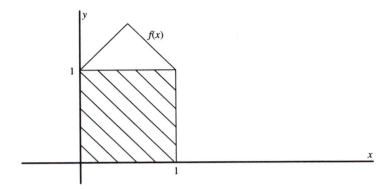

Figure 6.1.5.   The trapezoidal rule for a nonsmooth $f(x)$.

Example 6.1.4.

Let $f(x) = e^x$ for $0 \leq x \leq 2, n = 10$, and $h = 0.2$. Here we have

$$T_{10}(e^x; 0, 2) = 0.2 \left(\frac{e^0}{2} + e^{0.2} + \cdots + e^{1.8} + \frac{e^2}{2} \right) \approx 6.41034$$

The exact value of $\int_0^2 e^x \, dx$ is ≈ 6.38906 and by using Eq. (6.1.14) we find

$$6.38906 - 6.41034 \approx -\frac{0.2^2 \cdot (2-0)}{12} e^c$$

Therefore, $e^c \approx 3.192$ and $c \approx 1.161$. $\square$

Example 6.1.5.

Consider the function

$$f(x) = \begin{cases} 1 + x \ , & 0 \leq x \leq 0.5 \\[2mm] -x + 2 \ , & 0.5 \leq x \leq 1 \end{cases}$$

and let $x_0 = 0, x_1 = 1, n = 1, h = 1$ (Fig. 6.1.5).

For this case $T_1(f; 0, 1) = 1, I(f; 0, 1) = 1.25$. The error is 0.25 while the right-hand side of Eq. (6.1.14) is $-f''(c)/12$. Because $f''(x) = 0$ except at $x = 0.5$, Eq. (6.1.14) does not hold. This is expected, because $f(x)$ does not have a continuous derivative at $x = 0.5$ as requested by Theorem 6.1.1. This is a simple but a powerful example demonstrating the risk of applying a theorem without prechecking whether all the requirements are satisfied. $\square$

By virtue of Theorem 6.1.1

$$I(f; a, b) - T_n(f; a, b) = O(h^2) \tag{6.1.19}$$

In general, we are unable to calculate the interim point c in Eq. (6.1.14), or M_2 in Eq. (6.1.18). Instead, given a tolerance $\epsilon > 0$, we generate a sequence $T_{n_i}(f; a, b), i = 0, 1, \ldots$ that converges to $I(f; a, b)$, and quit when $|T_{n_{i+1}} - T_{n_i}| \le \epsilon$ for the first time. This can be done efficiently by choosing $n_k = 2n_{k-1}, k = 1, 2, \ldots$. Thus, to compute $T_{n_k}(f; a, b)$, we use the values of $f(x)$, previously calculated at $n_{k-1} + 1$ points, and calculate the function only at the n_{k-1} additional points.

Example 6.1.6.

Let $f(x) = \ln x, 1 \le x \le 4.2$. The exact value of $I(\ln x; 1, 4.2)$ is

$$\int_1^{4.2} \ln x \, dx = x \ln x \Big|_1^{4.2} - \int_1^{4.2} dx \approx 2.82736$$

Table 6.1.1 presents values of $T_{2^n}(\ln x; 1, 4.2)$ computed for $0 \le n \le 5$ and rounded to six digits, and their associated errors.

Table 6.1.1. A Converging Sequence of $T_{n_i}(\ln x; 1, 4.2), n_i = 2^i, 0 \le i \le 5$

n	Subintervals	$T_{2^n}(\ln x; 1, 4.2)$	Error
0	1	2.29614	0.53122
1	2	2.67689	0.15047
2	4	2.78769	0.03967
3	8	2.81726	0.01010
4	16	2.82482	0.00254
5	32	2.82672	0.00064

☐

If we carefully observe the error column of Table 6.1.1, we notice that the error at each step is asymptotically reduced by a factor of 4. This is true in general, as follows from the next result.

Theorem 6.1.2.

(An error estimate for the trapezoidal rule). *Let $f(x)$ and $\{x_i\}_{i=0}^n$ satisfy the requirements of Theorem 6.1.1. Then*

$$\lim_{h \to 0} \frac{I(f; a, b) - T_n(f; a, b)}{h^2} = -\frac{1}{12} \int_a^b f''(x) \, dx \tag{6.1.20}$$

that is, a reduction of h by a factor of two reduces the error by a factor of four as $h \to 0$.

Proof.

We first divide Eq. (6.1.17) by h^2 and obtain

$$\frac{I(f; a, b) - T_n(f; a, b)}{h^2} = -\frac{1}{12} h \sum_{i=1}^n f''(c_i) \tag{6.1.21}$$

Because $f''(x)$ is continuous, $I(f''; a, b) = \int_a^b f''(x) \, dx$ exists and the right-hand side of Eq. (6.1.21) approaches $-\frac{1}{12} I(f''; a, b)$ as $h \to 0$, in agreement with Eq. (6.1.20). ☐

Corollary 6.1.2.
Let $f(x)$ and $\{x_i\}_{i=0}^n$ satisfy the requirements of Theorem 6.1.1. Then the asymptotic relation

$$I(f; a, b) - T_n(f; a, b) = -\frac{h^2}{12}[f'(b) - f'(a)], \quad h \to 0 \qquad (6.1.22)$$

holds. The proof is left for the reader.

If $f'(x)$ is easily computed, Eq. (6.1.22) provides a practical error estimate for the trapezoidal rule.

Example 6.1.7.
Consider the function $x \sin x$ over $[0, \pi]$. If we approximate $\int_0^x x \sin x \, dx$ by T_4 and T_8 we find $T_4 \approx 2.97842$ and $T_8 \approx 3.10112$, whereas the exact value is $I = \pi = 3.14159\ldots$. Without further computations we may now determine an approximation that is superior to either T_4 or T_8. Because

$$\frac{I - 2.97842}{(\pi/4)^2} \approx -\frac{1}{12} \int_0^\pi (x \sin x)'' \, dx$$

$$\frac{I - 3.10112}{(\pi/8)^2} \approx -\frac{1}{12} \int_0^\pi (x \sin x)'' \, dx$$

we easily obtain

$$I \approx \frac{4 \cdot 3.10112 - 2.97842}{3} = 3.14202 \qquad \square$$

The approach applied in Example 6.1.7 for better approximation without additional computations is *Richardson's extrapolation method*. The general scheme will be discussed in Section 6.1.3.

6.1.2   Simpson's Rule

The trapezoidal rule basically suggests replacing $f(x)$ between two consecutive points x_i, x_{i+1} by its linear interpolator. To obtain a better approximation, we consider three consecutive points x_{i-1}, x_i, x_{i+1} such that $x_i - x_{i-1} = x_{i+1} - x_i = h$ and replace $f(x)$ by its second order Lagrange interpolator, given by

$$p_2(x) = \frac{(x - x_i)(x - x_{i+1})}{(x_{i-1} - x_i)(x_{i-1} - x_{i+1})} f(x_{i-1}) + \frac{(x - x_{i-1})(x - x_{i+1})}{(x_i - x_{i-1})(x_i - x_{i+1})} f(x_i)$$
$$+ \frac{(x - x_{i-1})(x - x_i)}{(x_{i+1} - x_{i-1})(x_{i+1} - x_i)} f(x_{i+1})$$

The analytical integration of $p_2(x)$ yields

$$I(f; x_{i-1}, x_{i+1}) \approx \int_{x_{i-1}}^{x_{i+1}} p_2(x) \, dx = \frac{h}{3}[f(x_{i-1}) + 4f(x_i) + f(x_{i+1})] \qquad (6.1.23)$$

The scheme presented by Eq. (6.1.23) is illustrated in Fig. 6.1.6.

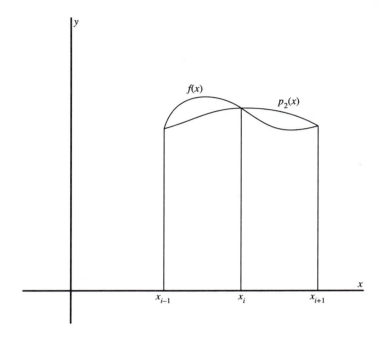

Figure 6.1.6.   Simpson's rule for two subintervals.

To use Eq. (6.1.23) and approximate $I(f; a, b)$, we divide $[a, b]$ to an even number of equal subintervals. Thus $x_i = a + ih, 0 \leq i \leq n$, where $h = (b - a)/n$ and $n = 2m$. If we add up the contributions of each pair of adjacent subintervals, we find

$$I(f; a, b) \approx S_n(f; a, b) = \frac{h}{3}[f(x_0) + 4f(x_1) + 2f(x_2) + 4f(x_3) + 2f(x_4) + \cdots$$
$$+ 2f(x_{n-2}) + 4f(x_{n-1}) + f(x_n)] \qquad (6.1.24)$$

which is *Simpson's rule* for approximating $I(f; a, b)$. This is a popular integration scheme, extremely accurate, and easy to program and manipulate.

Example 6.1.8.

Let $f(x) = x \cos x, 0 \leq x \leq \pi/2$. The exact value of the integral is

$$I\left(x \cos x; 0, \frac{\pi}{2}\right) = \int_0^{\pi/2} x \cos x \, dx = \frac{\pi}{2} - 1 = 0.5707 \ldots$$

Using Simpson's rule, we approximate the integral by

$$S_2\left(x \cos x; 0, \frac{\pi}{2}\right) = \frac{\pi}{12}\left[0 \cdot \cos 0 + 4\left(\frac{\pi}{4}\right)\cos\left(\frac{\pi}{4}\right) + \frac{\pi}{2}\cos\left(\frac{\pi}{2}\right)\right]$$

$$= \frac{\pi^2}{12} \frac{\sqrt{2}}{2} \approx 0.5816$$

Thus two subintervals already provide a reasonable approximation.   □

Example 6.1.9.

Consider the function $f(x) = x^3 - 2x$ defined over $[0, 1]$. The exact value of $I(f; 0, 1)$ is -0.75. By applying the trapezoidal and Simpson rules with four subintervals we determine that

$$T_4(f; 0, 1) = -0.734\ldots, \quad S_4(f; 0, 1) = -0.75$$

Simpson's rule not only provides a better approximation, but for this particular case yields the exact value. The reason will become clear in the section on error analysis (Theorem 6.1.3, below).   □

No integration scheme is applicable to *all* problems; the next example presents a case in which both the trapezoidal and Simpson rules perform poorly.

Example 6.1.10.

Define

$$f(x) = \begin{cases} 1/\sqrt{x}, & 0 < x \le 1 \\ 0, & x = 0 \end{cases}$$

$f(x)$ is integrable over $[0, 1]$ and

$$\int_0^1 f(x)\, dx = \int_0^1 \frac{dx}{\sqrt{x}} = 2\sqrt{x}\big|_0^1 = 2$$

Table 6.1.2 presents computed values of $T_n(f; 0, 1)$ and $S_n(f; 0, 1)$ for various values of n and evenly spaced integration nodes.

Table 6.1.2. Slow Convergence of T_n and S_n
as $n \to \infty$ for $f(x) = 1/\sqrt{x}, 0 < x \le 1$

n	T_n	S_n
2	0.957107	1.109476
4	1.267229	1.370602
8	1.483036	1.554972
16	1.634749	1.685319

The table shows that both the trapezoidal and Simpson rules fail to approximate the exact value of the integral for a "reasonable" n. Although Simpson's rule is consistently better than the trapezoidal rule, it still yields a poor approximation even for $n = 16$ (a relative error of approximately 15%!). The slow convergence of T_n and S_n to the exact value, 2, is connected with the singularity of $f(x)$ at $x = 0$. This singularity cannot be removed by redefining $f(x)$ at $x = 0$. One approach to treat such a singularity is to modify Simpson's rule and use it with a *variable step size*.   □

Theorem 6.1.3.

(Error estimate for Simpson's rule). *Let $f(x) \in C^4[a, b]$ and let $\{x_i\}_{i=0}^n$ divide $[a, b]$ evenly, so that $x_n = a + ih, 0 \leq i \leq n$, where $h = (b-a)/n$ and n is even. Then*

$$I(f; a, b) - S_n(f; a, b) = -\frac{h^4(b-a)}{180} f^{(4)}(c) \tag{6.1.25}$$

where c is an interim point between a and b.

Proof.

The error associated with Simpson's rule, confined to two adjacent subintervals, is given by

$$\int_{x_{i-1}}^{x_{i+1}} f(x)\, dx - \int_{x_{i-1}}^{x_{i+1}} p_2(x; x_{i-1}, x_i, x_{i+1})\, dx = \int_{x_{i-1}}^{x_{i+1}} \frac{(x - x_{i-1})(x - x_i)(x - x_{i+1})}{3!} f^{(3)}(c_x) \tag{6.1.26}$$

where $p_2(x; x_{i-1}, x_i, x_{i+1})$ is the Lagrangian quadratic that interpolates $f(x)$ at x_{i-1}, x_i, x_{i+1}, and c_x is between x_{i-1} and x_{i+1}. Because the term $(x - x_{i-1})(x - x_i)(x - x_{i+1})$ in Eq. (6.1.26) changes its sign between x_{i-1} and x_{i+1}, one cannot use the mean-value theorem for integrals, and replace the whole right-hand side by

$$\frac{1}{3!} f^{(3)}(c) \int_{x_{i-1}}^{x_{i+1}} (x - x_{i-1})(x - x_i)(x - x_{i+1})\, dx$$

(as in the proof of Theorem 6.1.1). Instead, we try another approach and interpolate $f(x)$ by a cubic polynomial at $x_{i-1}, x_i - \epsilon, x_i + \epsilon, x_{i+1}$, denoted by $p_3(x; x_{i-1}, x_i - \epsilon, x_i + \epsilon, x_{i+1})$. For the sake of simplicity, we assume $x_{i-1} = -h$ and obtain

$$\int_{-h}^{h} f(x)\, dx - \int_{-h}^{h} p_3(x; -h, -\epsilon, \epsilon, h)\, dx = \int_{-h}^{h} \frac{(x + h)(x + \epsilon)(x - \epsilon)(x - h)}{4!} f^{(4)}(\bar{c}_x)\, dx$$

$$= \int_{-h}^{-\epsilon} \frac{(x + h)(x + \epsilon)(x - \epsilon)(x - h)}{4!} f^{(4)}(\bar{c}_x)\, dx$$

$$+ \int_{\epsilon}^{h} \frac{(x + h)(x + \epsilon)(x - \epsilon)(x - h)}{4!} f^{(4)}(\bar{c}_x)\, dx$$

$$+ \int_{-\epsilon}^{\epsilon} \frac{(x + h)(x + \epsilon)(x - \epsilon)(x - h)}{4!} f^{(4)}(\bar{c}_x)\, dx \tag{6.1.27}$$

As $\epsilon \to 0$, the third integral at the right-hand side of Eq. (6.1.27) approaches zero. Denote the first two integrals by I_1, I_2. Because the fourth order polynomials that are the integrands of I_1, I_2 do not change signs over the respective domains of integration, we can apply the mean-value theorem for integrals and conclude

$$I_1 = \frac{f^{(4)}(c_{1\epsilon})}{4!} \int_{-h}^{-\epsilon} (x + h)(x + \epsilon)(x - \epsilon)(x - h)\, dx$$

$$I_2 = \frac{f^{(4)}(c_{2\epsilon})}{4!} \int_\epsilon^h (x+h)(x+\epsilon)(x-\epsilon)(x-h)\, dx$$

where $-h < c_{1\epsilon} < h, -h < c_{2\epsilon} < h$. If we now let $\epsilon \to 0$, the right-hand side of Eq. (6.1.27) approaches

$$\frac{f^{(4)}(c_1)}{4!} \int_{-h}^0 (x+h)x^2(x-h)\, dx + \frac{f^{(4)}(c_2)}{4!} \int_0^h (x+h)x^2(x-h)\, dx = -\frac{h^5}{180}[f^{(4)}(c_1) + f^{(4)}(c_2)]$$

where $-h \le c_1 \le h, -h \le c_2 \le h$.

It may be seen that $f^{(4)}(c_1) + f^{(4)}(c_2)$ can be replaced by $2f^{(4)}(\bar c)$, where $-h \le \bar c \le h$. Therefore

$$\lim_{\epsilon \to 0} \left[\int_{-h}^h f(x)\, dx - \int_{-h}^h p_3(x; -h, -\epsilon, \epsilon, h)\, dx \right] = -\frac{h^5}{90} f^{(4)}(\bar c) \tag{6.1.28}$$

We now write

$$p_3(x; -h, -\epsilon, \epsilon, h) = \frac{(x+\epsilon)(x-\epsilon)(x-h)}{(-h+\epsilon)(-h-\epsilon)(-h-h)} f(-h) + \frac{(x+h)(x-\epsilon)(x-h)}{(-\epsilon+h)(-\epsilon-\epsilon)(-\epsilon-h)} f(-\epsilon)$$
$$+ \frac{(x+h)(x+\epsilon)(x-h)}{(\epsilon+h)(\epsilon+\epsilon)(\epsilon-h)} f(\epsilon) + \frac{(x+h)(x+\epsilon)(x-\epsilon)}{(h+h)(h+\epsilon)(h-\epsilon)} f(h)$$

integrate each term at the right-hand side separately, and let $\epsilon \to 0$. The first and fourth terms contribute $(h/3)f(-h)$ and $(h/3)f(h)$, respectively, and the two middle terms contribute $-(4h/3)f(0)$. Thus

$$\int_{-h}^h f(x)\, dx - \frac{h}{3}[f(-h) + 4f(0) + f(h)] = -\frac{h^5}{90} f^{(4)}(\bar c) \tag{6.1.29}$$

where $\bar c$ is between $-h$ and h. If we implement Eq. (6.1.29) to the whole interval $[a, b]$, we obtain

$$I(f; a, b) - S_n(f; a, b) = -\frac{h^5}{90}\left[f^{(4)}(\bar c_1) + \cdots + f^{(4)}(\bar c_{n/2}) \right]$$
$$= -\frac{h^5}{90}\left(\frac{n}{2}\right) f^{(4)}(c) = -\frac{h^4(b-a)}{180} f^{(4)}(c) \tag{6.1.30}$$

where $a \le c \le b$.

By using Eq. (6.1.30) we can easily prove the next result.

Corollary 6.1.3.

If $f(x)$ and $\{x_i\}_{i=0}^n$ satisfy all the requirements of Theorem 6.1.3, then

$$\lim_{h \to 0} \frac{\int_a^b f(x)\, dx - S_n(f; a, b)}{h^4} = -\frac{1}{180} \int_a^b f^{(4)}(x)\, dx = -\frac{1}{180}\left[f^{(3)}(b) - f^{(3)}(a) \right]$$

The proof is left as an exercise for the reader.

Thus, if $f^{(3)}(x)$ is easily computed and if h is sufficiently small, an accurate estimate for the error associated with Simpson's rule can be calculated.

Example 6.1.11.

The exact value of $\int_0^\pi \sin x \, dx$ is 2. If we apply Simpson's rule, using eight equal subintervals, we have

$$S_8(\sin x; 0, \pi) \approx 2.000269$$

and the error is thus ≈ -0.000269. By using Corollary 6.1.3 the approximate error is

$$-\frac{h^4}{180}[-\cos(\pi) + \cos(0)] = -\frac{\pi^4 \cdot 2}{8^4 \cdot 180} \approx -0.000264$$

and is in good agreement with the computed error.   □

Example 6.1.12.

Consider the problem of calculating $\int_0^2 e^x \, dx$, using Simpson's rule, with a guaranteed error tolerance of $\epsilon = 10^{-6}$. Here we must choose h that satisfies

$$\frac{h^4(2-0)}{180} \max_{0 \le x \le 2}(e^x) \le 10^{-6}$$

that is, $h \le 0.059$.   □

Simpson's rule as given by Eq. (6.1.24) can also be written as

$$S_n(f; a, b) = \frac{h}{3}\left[f(a) + f(b) + 4S_n^{(1)} + 2S_n^{(2)}\right] \tag{6.1.31}$$

where $S_n^{(1)}$ and $S_n^{(2)}$ are the sums of the values of the integrand at the even and odd integration nodes, respectively. By doubling n we obtain

$$S_{2n}(f; a, b) = \frac{h}{6}\left[f(a) + f(b) + 4S_{2n}^{(1)} + 2S_{2n}^{(2)}\right]$$

where $S_{2n}^{(2)} = S_n^{(1)} + S_n^{(2)}$. The only new term that must be calculated to obtain $S_{2n}(f; a, b)$ is $S_{2n}^{(1)}$, the sum of the values of the integrand at the n new additional nodes. Because the error, using Simpson's rule, is of the $O(h^4)$ type, one reduces the error by a factor of 16 by doubling the computing time.

Given a function $f(x)$, an interval $[a, b]$, an even integer n_0, an error tolerance ϵ, and a maximum number of partitions N, the next algorithm generates $S_{n_0}(f; a, b), S_{2n_0}(f; a, b), S_{4n_0}(f; a, b), \dots$. The last computed number is considered the approximate value of $\int_a^b f(x) \, dx$.

Algorithm 6.1.1.

[Simpson's rule for integration (SIMP)].

Step 1. Set $m = 0, h = (b-a)/n_0, A = f(a) + f(b)$. Compute $B = S_{n_0}^{(1)}, C = S_{n_0}^{(2)}$ and set $S_1 = (h/3)(A + 4B + 2C)$.

Step 2. Set $h \leftarrow h/2, n_0 \leftarrow 2n_0, C \leftarrow B + C, m \leftarrow m + 1$; compute $B = S_{n_0}^{(1)}$ and set $S_2 = (h/3)(A + 4B + 2C)$.

Step 3. If $|(S_2 - S_1)/S_2| \leq \epsilon$ set $I(f; a, b) = S_2$ and stop.

Step 4. If $|(S_2 - S_1)/S_2| > \epsilon$ and $m < N$ set $S_1 \leftarrow S_2$ and go to Step 2. Otherwise output "maximum number of partitions exceeded," S_2 and stop.

SIMP applies Simpson's rule with evenly spaced nodes. This is undesirable if the function and its derivatives increase or decrease rapidly. To modify SIMP so it can efficiently treat problems such as Example 6.1.10, one should design a procedure that will automatically adjust the density of the nodes to the local behavior of $f(x)$.

The subroutine SIMP, which implements Algorithm 6.1.1, is found on the attached floppy disk.

We end this section by introducing a useful technique that enables us to improve the approximations to a given integral without performing additional function evaluations.

6.1.3 Richardson's Extrapolation

Both the trapezoidal and Simpson rules yield an error estimate formula given by

$$I - I_c \approx K h^n \tag{6.1.32}$$

where I and I_c are the exact and computed values of the integral, respectively, h is the step size, and n is the order of convergence (2 and 4 for the trapezoidal and Simpson rules, respectively). Let us apply the integration scheme twice, for h and $h/2$. The computed values, I_{c1} and I_{c2}, respectively, satisfy

$$I - I_{c1} \approx K h^n$$
$$I - I_{c2} \approx K \left(\frac{h}{2}\right)^n \tag{6.1.33}$$

leading to

$$I \approx \frac{2^n I_{c2} - I_{c1}}{2^n - 1} \tag{6.1.34}$$

The right-hand side of Eq. (6.1.34) will generally approximate I better than both I_{c1} and I_{c2}, providing an accuracy of $O(h^{n+1})$.

PROBLEMS

1. Apply the trapezoidal rule to calculate
   (a) $\int_0^2 [1/(1 + x^4)] dx$
   (b) $\int_1^2 (e^x/x) dx$
   (c) $\int_0^{1.57} \tan x \, dx$
   Apply two, four, and eight equal subintervals. Which of the results would you expect to be most inaccurate?

2. Apply the trapezoidal rule to calculate

   (a) $\int_0^3 e^x \, dx$

   (b) $\int_1^2 (x^2/2)\left[\ln x - (1/2)\right] \, dx$

   using nine evenly spaced integration nodes, and bound the error in either case.

3. Consider the proof of Theorem 6.1.1, in which it is assumed that the right-hand side of Eq. (6.1.16), that is,

$$\frac{(x - x_i)(x - x_{i+1})}{2!} f''(c_{ix})$$

   is integrable over the interval $[x_i, x_{i+1}]$. A sufficient condition for this integrability is the continuity of $f''(c_{ix})$ as a function of x. Show that $f''(c_{ix})$ is indeed continuous. [*Hint*: Write $f''(c_{ix})$ in terms of other quantities.]

4. Approximate the integrals

   (a) $\int_0^5 e^{-x} \, dx$

   (b) $\int_0^\pi [1/(1 - \cos x)] \, dx$

   (c) $\int_{0.001}^{1.5707} (\ln x)(\tan x) \, dx$

   using Simpson's rule with $n = 2, 4, 8,$ and 16 equal subintervals. Which of the integrals may cause difficulty and why?

5. (a) Approximate the integral

$$I = \int_0^2 \left(\sin x + \frac{x^5}{2} + x^5 \sqrt{x} \right) dx$$

   using Simpson's rule with 2, 4, 8, and 16 equal subintervals.

   (b) Evaluate the exact value of I and compare with S_8 and S_{16}.

6. (a) Calculate the integral of

$$f(x) = \begin{cases} 1/(\sqrt{x} + x^3), & 0 < x \le 1 \\ \\ 10 & , & x = 0 \end{cases}$$

   over the interval $[0, 1]$ using Simpson's rule with 10 equal subintervals.

   (b) Improve the result of part (a) by dividing $[0, 1]$ to $[0, x_0]$ and $[x_0, 1]$, using Simpson's rule with a different step size for each subinterval.

7. Let $f(x) = e^x(2 + x - x^2)$. Evaluate $I = \int_0^2 f(x) \, dx$ and compare with S_8 and S_{16}. Use Richardson's method to obtain a better approximation without further computations.

8. For a given integration scheme using a constant step size h, the difference between the exact and approximate integrals is

$$I(\text{exact}) - I_c = Kh^p$$

   (a) Obtain a scheme for determining p, if I is known.

(b) Determine I, if p and K are unknown.

9. Validate Corollary 6.1.3 in the following cases:

   (a) $\int_1^2 xe^x \, dx$, using eight equal subintervals

   (b) $\int_0^{1.57} \tan x \, dx$, using 16 equal subintervals

10. The integral $\int_0^1 (1/\sqrt{x + x^4}) \, dx$ cannot be calculated analytically. To approximate it by the trapezoidal or Simpson rule, with equal step size, is inefficient. Suggest an alternative (other than the trapezoidal and Simpson rules with a variable step size) that combines an analytical approach and the trapezoidal (or Simpson) rule with equal step size.

11. Let $f(x)$ be a periodic function with period T, that is,

$$f(x) = f(x + T), \quad -\infty < x < \infty$$

   and let this relation not hold for any $T' < T$. For example, $\sin(2x + 1)$ is periodic with period π. Let $f(x) \in C^4[a, b]$, and let $(b - a)$ be an integer multiple of T. Show that the errors associated with the trapezoidal and Simpson rules converge to zero faster than in the case when $(b - a)$ is not an integer multiple of T.

6.2 Gaussian Integration

The integration schemes previously discussed consisted of integrating linear and quadratic interpolators to the integrand $f(x)$. Both the trapezoidal and Simpson rules approximate the integral as

$$I(f; a, b) \approx \sum_{i=0}^{n} c_i f(x_i)$$

where $\{x_i\}_{i=0}^n$ are (generally) evenly spaced nodes. Each rule is characterized only by the set of coefficients $\{c_i\}_{i=0}^n$.

   In this section we introduce a different approach. No subdivision is performed. Instead, we regard the nodes and coefficients (weights) as parameters and search for the *optimal* nodes $\{x_i\}_{i=1}^n$ and weights $\{w_i\}_{i=1}^n$ to approximate the integral as

$$I(f; a, b) \approx \sum_{i=1}^{n} w_i f(x_i) \tag{6.2.1}$$

where $a \leq x_i \leq b, 1 \leq i \leq n$. One way of doing so is to prefix n and find nodes and weights for which the right-hand side of Eq. (6.2.1) is the exact value of the integral, provided that $f(x)$ is a polynomial of degree $\leq k$. The weights and the locations of the nodes depend only on n [not on $f(x)$!] and for each n we try to determine the maximum value of k.

   Let us discuss the motivation for this approach. Consider a function $f(x) \in C^{(k+1)}[a, b]$, and let $m_k(x)$ denote the minimax polynomial approximation of degree $\leq k$ to $f(x)$. As previously shown (Chapter 5)

$$\rho_k(f) = \max_{a \le x \le b} |f(x) - m_k(x)| \le \frac{(b-a)^{k+1}}{(k+1)!2^{2k+1}} M_{k+1} \tag{6.2.2}$$

where

$$M_{k+1} = \max_{a \le x \le b} |f^{(k+1)}(x)| \tag{6.2.3}$$

Let $\{x_i\}_{i=1}^n, \{w_i\}_{i=1}^n$ be the nodes and the weights that guarantee an exact integration formula for all the polynomials of degree $\le k$ [but not $(k+1)$], over the interval $[a, b]$. Then

$$\int_a^b f(x)\,dx - \sum_{i=1}^n w_i f(x_i) = \int_a^b [f(x) - m_k(x)]\,dx + \int_a^b m_k(x)\,dx$$

$$- \sum_{i=1}^n w_i m_k(x_i) - \sum_{i=1}^n w_i [f(x_i) - m_k(x_i)]\,dx$$

$$= \int_a^b [f(x) - m_k(x)]\,dx - \sum_{i=1}^n w_i [f(x_i) - m_k(x_i)]$$

It can be further shown that $w_i > 0, 1 \le i \le n$ and that

$$\sum_{i=1}^n w_i = (b-a) \tag{6.2.4}$$

Thus

$$\left| \int_a^b f(x)\,dx - \sum_{i=1}^n w_i f(x_i) \right| \le \int_a^b |f(x) - m_k(x)|\,dx + \sum_{i=1}^n |w_i [f(x_i) - m_k(x_i)]|$$

$$\le (b-a)\rho_k(f) + \rho_k(f) \sum_{i=1}^n w_i = 2(b-a)\rho_k(f)$$

or

$$\left| \int_a^b f(x)\,dx - \sum_{i=1}^n w_i f(x_i) \right| \le \frac{(b-a)^{k+2}}{(k+1)!2^{2k}} M_{k+1} \tag{6.2.5}$$

The scheme given by Eq. (6.2.1) is therefore effective whenever the right-hand side of Eq. (6.2.5) converges rapidly to zero (which is usually the case). Equation (6.2.1) is the *Gaussian integration method* of order n, and $\{x_i\}_{i=1}^n, \{w_i\}_{i=1}^n$ are the Gaussian integration nodes and weights, respectively.

We will now discuss in detail the Gaussian method for $n = 1, 2$ and then outline the general case. For the sake of simplicity, we consider the domain of integration to be the interval $[-1, 1]$.

Case 1. $n = 1$: We must determine two parameters: a single integration node x_1 and its associated weight w_1. The general polynomial of degree ≤ 1, $p_1(x) = a_0 + a_1 x$ also has two parameters, and it is reasonable to assume that x_1 and w_1, for which the relation

$$\int_{-1}^{1} f(x)\,dx \approx w_1 f(x_1) \tag{6.2.6}$$

is exact, provided that $f(x) = a_0 + a_1 x$ can be found. To find such x_1 and w_1 we replace $f(x)$ in Eq. (6.2.6) by $a_0 + a_1 x$ and, after integrating it, equate the coefficients of a_0 and a_1 on both sides. Alternatively, we can substitute $f(x) = 1$ and $f(x) = x$ in Eq. (6.2.6), obtain two equations with the variables x_1 and w_1, and solve them. Then, if Eq. (6.2.6) is exact for the particular integrands 1 and x, it must be exact for a general $p_1(x)$ as well (why?). For $f(x) = 1$, we have

$$2 = \int_{-1}^{1} 1\,dx = w_1 \cdot 1 \tag{6.2.7}$$

and for $f(x) = x$

$$0 = \int_{-1}^{1} x\,dx = w_1 x_1 \tag{6.2.8}$$

The unique solution of Eqs. (6.2.7) and (6.2.8) is $x_1 = 0, w_1 = 2$, and the Gaussian integration formula of order 1 is therefore

$$\int_{-1}^{1} f(x)\,dx \approx G_1(f; -1, 1) = 2f(0) \tag{6.2.9}$$

Example 6.2.1.
Let $f(x) = -17x + 5, -1 \le x \le 1$. Then

$$I(f; -1, 1) = \int_{-1}^{1} (-17x + 5)\,dx = -\frac{17}{2}x^2 + 5x\big|_{-1}^{1} = 10$$

By using the Gaussian integration method of order 1 we find that

$$G_1(f; -1, 1) = 2(-17x + 5)_{x=0} = 10 \qquad \square$$

Example 6.2.2.
Let $f(x) = x^2, -1 \le x \le 1$. Clearly Eq. (6.2.9) is not exact for $f(x)$, because $I(f; -1, 1) = 2/3$ whereas $2f(0) = 0$. $\qquad \square$

Case 2. $n = 2$: We need to determine four parameters: x_1, x_2, w_1, and w_2. The general polynomial of degree ≤ 3, that is, $p_3(x) = a_0 + a_1 x + a_2 x^2 + a_3 x^3$, has four parameters as well. If we confine ourselves to such polynomials it is sufficient to maintain a Gaussian formula that will integrate exactly the particular polynomials $1, x, x^2$, and x^3. We thus obtain

$$f(x) = 1 : 2 = \int_{-1}^{1} dx = w_1 \cdot 1 + w_2 \cdot 1$$

$$f(x) = x : 0 = \int_{-1}^{1} x \, dx = w_1 x_1 + w_2 x_2$$

$$f(x) = x^2 : \frac{2}{3} = \int_{-1}^{1} x^2 \, dx = w_1 x_1^2 + w_2 x_2^2$$

$$f(x) = x^3 : 0 = \int_{-1}^{1} x^3 \, dx = w_1 x_1^3 + w_2 x_2^3 \qquad (6.2.10)$$

that is, a nonlinear system of four equations that can be shown to possess the unique solution

$$x_1 = -\sqrt{3}/3, \quad x_2 = \sqrt{3}/3, \quad w_1 = 1, \quad w_2 = 1$$

The Gaussian integration method of order 2 is therefore

$$\int_{-1}^{1} f(x) \, dx \approx G_2(f; -1, 1) = f\left(-\frac{\sqrt{3}}{3}\right) + f\left(\frac{\sqrt{3}}{3}\right) \qquad (6.2.11)$$

and if $f(x)$ is a polynomial of degree ≤ 3, Eq. (6.2.11) provides the exact value of the integral.

Example 6.2.3.

Let $f(x) = x^3 - 2x, -1 \leq x \leq 1$. The Gaussian integration method of order 2 yields

$$G_2(f; -1, 1) = \left[\left(-\frac{\sqrt{3}}{3}\right)^3 - 2\left(-\frac{\sqrt{3}}{3}\right)\right] + \left[\left(\frac{\sqrt{3}}{3}\right)^3 - 2\left(\frac{\sqrt{3}}{3}\right)\right] = 0$$

which is the exact value of $I(f; -1, 1)$. □

Example 6.2.4.

Let $f(x) = x^4, -1 \leq x \leq 1$. The exact value of $I(f; -1, 1)$ is $2/5$ whereas, by using the Gaussian formula, we obtain

$$G_2(x^4; -1, 1) = \left(-\sqrt{3}/3\right)^4 + \left(\sqrt{3}/3\right)^4 = 2/9 \neq 2/5$$

Therefore the Gaussian integration formula of order 2 is not exact for a general polynomial of degree 4. □

Example 6.2.5.

Let $f(x) = \cos x, -1 \leq x \leq 1$. By applying Eq. (6.2.11) we have

$$\int_{-1}^{1} \cos x \, dx \approx G_2(\cos x; -1, 1) = \cos\left(-\sqrt{3}/3\right) + \cos\left(\sqrt{3}/3\right) \approx 1.6758$$

The exact integral is $2\sin(1) \approx 1.6829$ and the error

$$E = \int_{-1}^{1} \cos x \, dx - G_2(\cos x; -1, 1) \approx 0.0071$$

If, alternatively, we apply Eq. (6.2.5) to determine an error bound, we have

$$|E| \leq \frac{(b-a)^{k+2}}{(k+1)!\,2^{2k}} M_{k+1} = \frac{2^5}{(4!)2^6} = \frac{1}{48} \approx 0.0208 \qquad \Box$$

6.2.1 The General Gaussian Formula

We want to obtain $\{x_i\}_{i=1}^{n}$ and $\{w_i\}_{i=1}^{n}$ for which Eq. (6.2.1) yields the exact integral for all polynomials of degree $\leq k$, and for the largest possible k. Because the number of parameters is $2n$, we confine ourselves to polynomials of degree $\leq 2n - 1$, whose number of coefficients is also $2n$. We can verify that it is sufficient to consider the monomials $1, x, x^2, \ldots, x^{2n-1}$, for which we obtain

$$
\begin{aligned}
f(x) &= 1: \ 2 = w_1 + w_2 + \cdots + w_n \\
f(x) &= x: \ 0 = w_1 x_1 + w_2 x_2 + \cdots + w_n x_n \\
f(x) &= x^2: \ 2/3 = w_1 x_1^2 + w_2 x_2^2 + \cdots + w_n x_n^2 \\
&\quad \vdots \\
f(x) &= x^{2n-2}: \ 2/(2n-1) = w_1 x_1^{2n-2} + w_2 x_2^{2n-2} + \cdots + w_n x_n^{2n-2} \\
f(x) &= x^{2n-1}: \ 0 = w_1 x_1^{2n-1} + w_2 x_2^{2n-1} + \cdots + w_n x_n^{2n-1}
\end{aligned}
\qquad (6.2.12)
$$

Equation (6.2.12) represents a system of $2n$ nonlinear equations that can be shown to possess a unique solution. Tables for $\{x_i\}_{i=1}^{n}$ and $\{w_i\}_{i=1}^{n}$ exist. Clearly, the most commonly used values of n are the lowest. The solutions of the system given by Eq. (6.2.12) for $1 \leq n \leq 8$ are given, rounded to 10 significant digits, in Table 6.2.1.

Table 6.2.1. Gaussian Nodes and Weights for $1 \leq n \leq 8$

n	x_i	w_i
1	0	2
2	± 0.5773502692	1
3	± 0.7745966692	0.5555555556
	0	0.8888888889
4	± 0.8611363116	0.3478548451
	± 0.3399810436	0.6521451549
5	± 0.9061798459	0.2369268851
	± 0.5384693101	0.4786286705
	0	0.5688888889
6	± 0.9324695142	0.1713244924
	± 0.6612093865	0.3607615730
	± 0.2386191861	0.4679139346
7	± 0.9491079123	0.1294849662
	± 0.7415311856	0.2797053915
	± 0.4058451514	0.3818300505
	0	0.4179591837
8	± 0.9602898565	0.1012285363
	± 0.7966664774	0.2223810345
	± 0.5255324099	0.3137066459
	± 0.1834346425	0.3626837834

Example 6.2.6.

Consider the function $f(x) = e^x, -1 \leq x \leq 1$, whose integral is computed by the trapezoidal, Simpson, and Gaussian procedures. If we use four equal subintervals for the trapezoidal and Simpson rules and five nodes for the Gaussian method, we perform five function evaluations per computation. The results are $T_4(e^x; -1, 1) = 2.3991, S_4(e^x; -1, 1) = 2.3511, G_5(e^x; -1, 1) = 2.35040238$ whereas the exact value is $I(e^x; -1, 1) = 2.350402387$. The Gaussian quadrature scheme yields here a far better approximation while consuming the same computing time. $\qquad\square$

A detailed error analysis for the Gaussian method yields $\int_{-1}^{1} f(x)\, dx = \sum_{i=1}^{n} w_i f(x_i) + R_n$, where

$$R_n = \frac{2^{2n+1}(n!)^4}{(2n+1)[(2n)!]^3} f^{(2n)}(\xi), \quad -1 < \xi < 1. \tag{6.2.13}$$

Thus

$$\left| \int_{-1}^{1} f(x)\, dx - \sum_{i=1}^{n} w_i f(x_i) \right| \leq \frac{2^{2n+1}(n!)^4}{(2n+1)[(2n)!]^3} M_{2n} \tag{6.2.14}$$

where $M_{2n} = \max_{-1 \leq x \leq 1} |f^{(2n)}(x)|$.

By using Stirling's formula, we can show that the error bound given by Eq. (6.2.14) is better than the one previously computed [Eq. (6.2.5)].

To extend the Gaussian scheme for an arbitrary interval $[a, b]$, we introduce a new variable:

$$y = \frac{a+b}{2} + \frac{b-a}{2}x, \quad -1 \le x \le 1 \tag{6.2.15}$$

Thus

$$\int_a^b f(y)\, dy = \frac{(b-a)}{2} \int_{-1}^1 f\left(\frac{a+b}{2} + \frac{b-a}{2}x\right) dx \tag{6.2.16}$$

By using the formula for the interval $[-1, 1]$ we obtain

$$\int_a^b f(y)\, dy = \frac{(b-a)}{2}\left[\sum_{i=1}^n w_i f\left(\frac{a+b}{2} + \frac{b-a}{2}x_i\right) + R_n'\right]$$

where

$$R_n' = \left(\frac{b-a}{2}\right)^{2n} \frac{2^{n+1}(n!)^4}{(2n+1)[(2n)!]^3} f^{(2n)}(\xi), \quad a \le \xi \le b \tag{6.2.17}$$

Thus if $\{x_i\}, \{w_i\}$ denote the Gaussian nodes and weights for the interval $[-1, 1]$, respectively, the general Gaussian scheme for the interval $[a, b]$ is

$$\int_a^b f(x)\, dx = \sum_{i=1}^n w_i' f(x_i') + R_n$$

where

$$x_i' = \frac{a+b}{2} + \frac{b-a}{2}x_i, \quad w_i' = \frac{b-a}{2}w_i \tag{6.2.18}$$

and

$$R_n = \frac{(b-a)^{2n+1}(n!)^4}{(2n+1)[(2n)!]^3} f^{(2n)}(\xi), \quad a \le \xi \le b \tag{6.2.19}$$

Equation (6.2.19) yields the error bound

$$\left| \int_a^b f(x)\, dx - \sum_{i=1}^n w_i' f(x_i') \right| \le \frac{(b-a)^{2n+1}(n!)^4}{(2n+1)[(2n)!]^3} M_{2n} \tag{6.2.20}$$

where $M_{2n} = \max_{a \le x \le b} |f^{(2n)}(x)|$.

The error, as determined by a Gaussian integration scheme, is generally significantly smaller than the one produced by either the trapezoidal or Simpson rule. This is easily verified by comparing the errors given by Eq. (6.1.22), (6.1.30), and (6.2.19), using Stirling's formula to estimate $(n!)$ and $(2n)!$. However, one should note that if the values of the integrand at the integration nodes are not exact, but approximated, a Gaussian procedure may not be the better choice. A Gaussian quadrature formula for $f(x)$ is powerful, provided that at the Gaussian nodes $f(x)$ can be approximated with great accuracy and that $f(x)$ possesses a sufficient number of continuous derivatives. Otherwise,

Eq. (6.2.20) may not be applicable and other numerical integration schemes should be considered instead.

Another case in which a Gaussian scheme may not be recommended is when integrating a periodic function. Here the trapezoidal and Simpson rules present, as previously shown, an extremely small error and are often preferable.

Example 6.2.7.

Consider the function $f(x) = xe^x, -1 \le x \le 1$. The exact value of $I(f; -1, 1)$ is $0.73575888\ldots$. A comparison between the trapezoidal (T), Simpson (S), and Gaussian (G) schemes is given in Table 6.2.2 for $n = 2,\ldots,5$ function evaluations. The trapezoidal and Simpson rules are used with evenly-spaced nodes and all the results are rounded to six significant digits.

Table 6.2.2. A Comparison between the Different Integration Schemes for $f(x) = xe^x, -1 \le x \le 1$

n	T	$\|\text{Error } (T)\|$	S	$\|\text{Error } (S)\|$	G	$\|\text{Error } (G)\|$
2	2.350402	$0.16 \cdot 10^1$	—	—	0.704326	$0.31 \cdot 10^{-1}$
3	1.175201	$0.44 \cdot 10^0$	0.783467	$0.48 \cdot 10^{-1}$	0.735362	$0.40 \cdot 10^{-3}$
4	0.934374	$0.20 \cdot 10^0$	—	—	0.735757	$0.24 \cdot 10^{-5}$
5	0.848148	$0.11 \cdot 10^0$	0.739131	$0.34 \cdot 10^{-2}$	0.735759	$< 10^{-8}$

The Gaussian quadrature clearly provides the best results and five function evaluations are sufficient to produce six accurate digits. To achieve the same degree of accuracy using the Simpson rule, as much as six times more computing time is consumed. □

The next example is of a periodic function for which the Gaussian method performs poorly.

Example 6.2.8.

Consider the periodic function

$$f(x) = \frac{1}{2 + \cos x}, \quad 0 \le x \le 2\pi$$

for which $I(f; 0, 2\pi) = 2\pi/\sqrt{3} \approx 3.6275987$. A comparison between the approximate values computed by the trapezoidal, Simpson, and Gaussian schemes (denoted by T, S, and G respectively) is definitely in favor of the trapezoidal method, as demonstrated in Table 6.2.3. The number of function evaluations is denoted by N. All the results are rounded to six significant digits. □

Table 6.2.3. Performance of the Various Integration Schemes for the Periodic Function $f(x) = (2 + \cos x)^{-1}, 0 \le x \le 2\pi$

n	T	$\|\text{Error } (T)\|$	S	$\|\text{Error } (S)\|$	G	$\|\text{Error } (G)\|$
2	2.09440	$0.15 \cdot 10^1$	—	—	2.80422	$0.82 \cdot 10^0$
3	4.18879	$0.56 \cdot 10^0$	4.88692	$0.13 \cdot 10^1$	4.05745	$0.43 \cdot 10^0$
4	3.49066	$0.14 \cdot 10^0$	—	—	3.45099	$0.18 \cdot 10^0$
5	3.66519	$0.38 \cdot 10^{-1}$	3.49066	$0.14 \cdot 10^0$	3.70884	$0.81 \cdot 10^{-1}$
6	3.61759	$0.10 \cdot 10^{-1}$	—	—	3.59214	$0.35 \cdot 10^{-1}$
7	3.63028	$0.27 \cdot 10^{-2}$	3.67683	$0.49 \cdot 10^{-1}$	3.64343	$0.16 \cdot 10^{-1}$

Although the trapezoidal rule is preferable to the Gaussian quadrature formula in Example 6.2.8, we should realize that this is not always the case when integrating periodic functions. Whether the Gaussian method should be used depends *only* on the right-hand side of Eq. (6.2.20), that is, on the number $M_{2n} = \max_{a \leq x \leq b} |f^{(2n)}(x)|$.

The next example presents another periodic function whose integral is best approximated by a Gaussian scheme.

Example 6.2.9.
Let $f(x) = \sqrt{1 + \cos(\pi x)}, -1 \leq x \leq 1$. The exact value of the integral $I[f(x); -1, 1]$ is $4\sqrt{2}/\pi \approx 1.80063263$. The approximate values computed by the trapezoidal, Simpson, and Gaussian schemes (denoted by T, S, and G, respectively), and rounded to six significant digits, are given in Table 6.2.4, using $n = 2, \ldots, 7$ function evaluations.

Table 6.2.4. Another Periodic Function Integrated by the
Trapezoidal, Simpson and Gaussian Schemes

n	T	$\lvert$Error $(T)\rvert$	S	$\lvert$Error $(S)\rvert$	G	$\lvert$Error $(G)\rvert$
2	0.00000	$0.18 \cdot 10^{1}$	—	—	1.74285	$0.58 \cdot 10^{-1}$
3	1.41421	$0.39 \cdot 10^{0}$	1.88562	$0.85 \cdot 10^{-1}$	1.80188	$0.13 \cdot 10^{-2}$
4	1.63299	$0.17 \cdot 10^{0}$	—	—	1.80062	$0.14 \cdot 10^{-4}$
5	1.70711	$0.94 \cdot 10^{-1}$	1.80474	$0.41 \cdot 10^{-2}$	1.80063	$0.12 \cdot 10^{-6}$
6	1.74100	$0.60 \cdot 10^{-1}$	—	—		
7	1.75931	$0.41 \cdot 10^{-1}$	1.80141	$0.78 \cdot 10^{-3}$		

Here, the Gaussian quadrature formula is clearly superior, and only five evaluations of the integrand are needed to achieve an accuracy of 10^{-7}. $\square$

6.2.2 Infinite Discontinuities

If the integrated function $f(x)$ possesses a jump discontinuity at $x = c, a < c < b$ (Fig. 6.2.1), we express $I(f; a, b)$ as

$$I(f; a, b) = I(f; a, c) + I(f; c, b)$$

and integrate separately over each subinterval. The case of an infinite discontinuity is more complicated. Consider, for example, the function

$$f(x) = \begin{cases} 1/(\sqrt{x} + x) , & 0 < x \leq 1 \\ 0 , & x = 0 \end{cases}$$

The integral $I(f; 0, 1)$ is defined as

$$I(f; 0, 1) = \lim_{\epsilon \to 0} I(f; \epsilon, 1) = 2 \ln 2 \approx 1.38629436$$

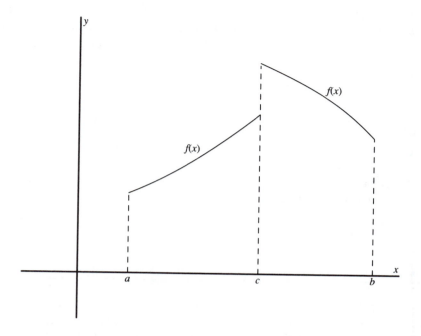

Figure 6.2.1. An integrand with a jump discontinuity.

However, if we apply any of the previous methods we find the convergence to $I(f; 0, 1)$ quite slow. Simpson's rule, for example, with 16 subintervals, yields 1.102, a poor approximation (off by $\sim 20\%$). Even the generally more accurate Gaussian quadrature scheme does not provide a significan' improvement. The use of eight nodes and weights produces 1.284 (off by $\sim 7\%$).

The reason for the poor performance of the various schemes is the infinite discontinuity of $f(x)$ at $x = 0$, which produces unbounded derivatives and large errors. To bypass it we apply an indirect approach. We notice that the integrand satisfies

$$\frac{1}{\sqrt{x} + x} \approx \frac{1}{\sqrt{x}} \text{ near } x = 0$$

and $1/\sqrt{x}$ can be integrated *analytically* [so is $1/(\sqrt{x} + x)$, which is integrated numerically only for the sake of demonstration]. We therefore rewrite

$$\int_0^1 \frac{dx}{\sqrt{x} + x} = \int_0^1 \left(\frac{1}{\sqrt{x} + x} - \frac{1}{\sqrt{x}} \right) dx + \int_0^1 \frac{dx}{\sqrt{x}} = \int_0^1 g(x)\, dx + 2$$

where

$$g(x) = \frac{1}{\sqrt{x} + x} - \frac{1}{\sqrt{x}} = -\frac{1}{1 + \sqrt{x}}$$

is continuous over the whole interval $[0, 1]$. Although $g(x)$ does not possess continuous derivatives at $x = 0$, it can already be integrated successfully by any of the previous schemes. If, for example, we apply a Gaussian formula with two nodes, we obtain $\int_0^1 g(x)\,dx + 2 \approx 1.393$ (off by ~0.5%).

PROBLEMS

1. Show that the system of equations that is generated in the case of Gaussian quadrature with two nodes [Eq. (6.2.10)] possesses a unique solution.

2. Compute $\int_0^2 x^2 \sin x\,dx$ using Gaussian quadrature with two, three, and four nodes. Compare with the exact value of the integral.

3. (a) Use the trapezoidal and Simpson rules to evaluate

$$\int_0^1 e^x \cos x\,dx$$

   using two, four, and six equal subintervals.

   (b) Apply the Gaussian scheme to the same integral, using the same number of function evaluations, and compare the results.

4. Show that any integration formula given by Eq. (6.2.1), which is exact for the integrands $1, x, x^2, \ldots, x^m$, must also be exact for any polynomial of degree $\leq m$.

5. Let $f(x)$ be "sufficiently smooth" over $[0, 1]$ and consider the integral

$$IL[f(x); 0, 1] = \int_0^1 f(x) \ln x\,dx$$

   Apply a Gaussian approach and

   (a) Find x_1 and w_1 such that

$$IL[f(x); 0, 1] = w_1 f(x_1)$$

   for all $f(x)$ that are polynomials of degree ≤ 1.

   (b) Find a nonlinear system with the unknowns x_1, x_2 and w_1, w_2, whose solution satisfies

$$IL[f(x); 0, 1] = w_1 f(x_1) + w_2 f(x_2)$$

   for all $f(x)$ that are polynomials of degree ≤ 3. *Hint:* Integrate by parts and show

$$\int_0^1 x^m \ln x\,dx = -\frac{1}{(m+1)^2}, \quad m \geq 0$$

   (c) Apply the result of part (a) to approximate

$$\int_0^1 x^2 \ln x\,dx$$

   and compare with the exact value.

6. Apply an indirect Gaussian scheme, using two and three nodes, to calculate the following integrals:

(a) $\int_0^1 [1/(\sqrt{x} + x^3)]\, dx$

(b) $\int_0^1 [1/(\sqrt[4]{x} + \sqrt{x})]\, dx$

(c) $\int_0^2 [1/(\sqrt{x} + x)]\, e^{-x}\, dx$

7. Construct an integrand whose integral over the interval $[0, 1]$ is better approximated by the trapezoidal rule than by Gaussian quadrature with the same number of function evaluations.

6.3   The Romberg Method

The trapezoidal rule discussed in Section 6.1 is easy to implement but converges relatively slowly as $h = x_i - x_{i-1} \to 0$. This is demonstrated, for example, in Table 6.1.1, where at each step the number of subintervals is doubled and the error is asymptotically reduced by the modest factor of 4. If, however, we combine the trapezoidal rule with the Richardson extrapolation method, we can significantly speed up the convergence at almost no additional cost.

Let $f(x)$ be a real-valued function defined over $[a, b]$ and consider the following set of partitions of this interval by evenly spaced points:

$$
\begin{aligned}
P_1 &: \quad x_i = a + (i-1)h_1, \quad 0 \le i \le n_1 \\
P_2 &: \quad x_i = a + (i-1)h_2, \quad 0 \le i \le n_2 \\
&\qquad\qquad \vdots \\
P_k &: \quad x_i = a + (i-1)h_k, \quad 0 \le i \le n_k \\
&\qquad\qquad \vdots
\end{aligned}
\tag{6.3.1}
$$

where

$$
h_1 = b - a; \quad h_k = \frac{h_{k-1}}{2}, \quad k \ge 2
\tag{6.3.2}
$$

$$
n_1 = 1; \quad n_k = 2n_{k-1}, \quad k \ge 2
\tag{6.3.3}
$$

that is, $h_k = (b-a)2^{-k+1}$ and $n_k = 2^{k-1}$ for all $k \ge 1$. By virtue of Eq. (6.1.20) we obtain

$$
\int_a^b f(x)\, dx = \frac{h_k}{2} \left[f(a) + f(b) + 2 \sum_{i=1}^{n_k - 1} f(a + ih_k) \right] + E_k
\tag{6.3.4}
$$

where

$$
E_k \approx -\frac{h_k^2}{12} [f'(b) - f'(a)] \text{ as } h_k \to 0
\tag{6.3.5}
$$

If $f(x) \in C^4[a, b]$ it can further be shown that

$$
E_k = -\frac{h_k^2}{12} [f'(b) - f'(a)] + \frac{(b-a)h_k^4}{720} f^{(4)}(\xi_k)
\tag{6.3.6}
$$

where ξ_k is some interim point between a and b.

The *Romberg procedure* first generates

$$R_{1,1} = \frac{h_1}{2}[f(a) + f(b)]$$

$$R_{2,1} = \frac{h_2}{2}[f(a) + f(b) + 2f(a + h_2)]$$

$$\vdots$$

$$R_{k,1} = \frac{h_k}{2}[f(a) + f(b) + 2\sum_{i=1}^{n_k-1} f(a + ih_k)]$$

$$\vdots$$

which are the sequence of approximations to $I(f; a, b)$ based on the trapezoidal rule and the partitions $P_1, P_2, \ldots, P_k, \ldots$, respectively. An efficient way to calculate these numbers is by using the relation

$$R_{k,1} = \frac{1}{2}\left\{R_{k-1,1} + h_{k-1}\sum_{i=1}^{n_k-1} f\left[a + \left(i - \frac{1}{2}\right)h_{k-1}\right]\right\} \tag{6.3.7}$$

Now, in view of Eq. (6.3.6) we have

$$\int_a^b f(x)\, dx = R_{k-1,1} - \frac{h_{k-1}^2}{12}[f'(b) - f'(a)] + \frac{(b-a)h_{k-1}^4}{720}f^{(4)}(\xi_{k-1})$$

$$= R_{k,1} - \frac{h_k^2}{12}[f'(b) - f'(a)] + \frac{(b-a)h_k^4}{720}f^{(4)}(\xi_k)$$

$$= R_{k,1} - \frac{h_{k-1}^2}{48}[f'(b) - f'(a)] + \frac{(b-a)h_k^4}{720}f^{(4)}(\xi_k)$$

for all $k \geq 2$ and the Richardson extrapolation method yields

$$\int_a^b f(x)\, dx = \frac{4R_{k,1} - R_{k-1,1}}{3} + O(h_k^4), \quad k \geq 2 \tag{6.3.8}$$

The next step of the Romberg scheme is to define

$$R_{k,2} = \frac{4R_{k,1} - R_{k-1,1}}{3}, \quad k \geq 2 \tag{6.3.9}$$

which approximates $I(f; a, b)$ better than $R_{k,1}$, $k \geq 1$ and converges faster as $k \to \infty$. In general, the method generates

$$R_{k,i} = \frac{4^{i-1}R_{k,i-1} - R_{k-1,i-1}}{4^{i-1} - 1}, \quad k = 2, 3, \ldots, \quad i = 2, \ldots, k \tag{6.3.10}$$

which can be shown to approximate $I(f; a, b)$ with a truncation error of $O(h_k^{2i})$. The construction of

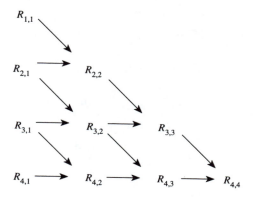

Figure 6.3.1. Generating the Romberg scheme.

$R_{k,i}$ for $k \leq 4$ is shown in Fig. 6.3.1.

If some general conditions are fulfilled the diagonal terms $R_{k,k}$ converge to $I(f; a, b)$ as $k \to \infty$ and the convergence is much faster than that of $R_{k,1}$.

Example 6.3.1.

Consider the integral

$$I(f; a, b) = \int_0^\pi x \sin x \, dx$$

whose exact value is $\pi \approx 3.14159265$. Using the trapezoidal rule for $n_k = 1, 2, 4,$ and 8 generates the approximations

$$R_{1,1} = 0, \quad R_{2,1} = 2.4670110, \quad R_{3,1} = 2.97841660, \quad R_{4,1} = 3.10111575$$

which are not particularly impressive. However, these numbers initialize the Romberg scheme given in Table 6.3.1 and the diagonal terms satisfy

$$|R_{2,2} - \pi| \approx 0.15 \cdot 10^0, \quad |R_{3,3} - \pi| \approx 0.22 \cdot 10^{-2}, \quad |R_{4,4} - \pi| \approx 0.87 \cdot 10^{-5}$$

Table 6.3.1. $R_{k,i}$, $1 \leq k \leq 4$ Calculated for $\int_0^{\pi} x \sin x \, dx$

k	$i = 1$	$i = 2$	$i = 3$	$i = 4$
1	0			
2	2.46740110	3.28986813		
3	2.97841660	3.14875510	3.13934756	
4	3.10111575	3.14201546	3.14156616	3.14160137

Two important features characterize the Romberg procedure. First, it is clear that function evaluations are needed only to generate $R_{k,1}$. The rest of the numbers are calculated using Eq. (6.3.10). A second and most useful property is the fact that the Romberg scheme, as shown in Fig. 6.3.1, can be generated *row by row*. Thus if $R_{5,5}$, for example, is sufficiently accurate one need perform only the function evaluations needed to calculate $R_{k,1}$, $1 \leq k \leq 5$, that is, 17 evaluations.

The complete Romberg algorithm, including the last feature, is given next. The generation of $R_{k,i}$ ends when $|R_{k,k} - R_{k,k-1}|$ is less than some given tolerance ϵ.

Algorithm 6.3.1.

(Romberg algorithm). This algorithm approximates the integral

$$I(f; a, b) = \int_a^b f(x) \, dx$$

Input: End points a, b; maximum number of rows N; tolerance ϵ.
Output: An array $R_{k,i}$, $1 \leq k \leq n$, $1 \leq i \leq k$ for some $n \leq N$. The array is generated row by row and only the last two rows are saved.
Step 1. Set $h = b - a$; $R_{1,1} = (h/2)[f(a) + f(b)]$; $i = 1$.
Step 2. Output $R_{1,1}$.
Step 3. Set

$$R_{2,1} = \frac{1}{2}\left\{R_{1,1} + h \sum_{j=1}^{2^{i-1}} f\left[a + \left(j - \frac{1}{2}\right)h\right]\right\}$$

Step 4. For $l = 2, \ldots, (i+1)$ set

$$R_{2,l} = \frac{4^{l-1}R_{2,l-1} - R_{1,l-1}}{4^{l-1} - 1}$$

Step 5. Set $i \leftarrow i + 1$ and output $R_{2,l}$, $1 \leq l \leq i$.
Step 6. If $|R_{2,i} - R_{2,i-1}| > \epsilon$ and $i < N$ set $h \leftarrow h/2$ and $R_{1,l} = R_{2,l}$, $1 \leq l \leq i$, and go to Step 3. Otherwise go to Step 7.
Step 7. If $|R_{2,i} - R_{2,i-1}| > \epsilon$ and $i \geq N$ go to Step 9. Otherwise output i (number of iterations) and go to Step 8.
Step 8. Stop (procedure completed successfully).
Step 9. Output "maximum number of iterations exceeded," $R_{2,i}$ and stop.

Example 6.3.2.

Consider the integral

$$I(f; a, b) = \int_0^2 e^x \, dx$$

The exact value is $e^2 - 1 = 6.3890560989\ldots$ The implementation of Algorithm 6.3.1 with $\epsilon = 10^{-6}$ and $N = 8$ provides the approximation 6.38905610 (rounded to nine significant figures) after five iterations. ☐

In the next example the integrand is not bounded.

Example 6.3.3.

Consider the integral

$$I(f; a, b) = \int_0^1 \frac{dx}{\sqrt{x}} = 2$$

where $f(x)$ is defined as zero at $x = 0$. Using the Romberg method with $\epsilon = 10^{-4}$, $N = 8$ determines a state of "convergence" after six iterations, as observed from Table 6.3.2. This makes little sense because

$$|R_{6,6} - 2| = |1.78549 - 2| \approx 0.215$$

The reason for this considerable deviation is that $f(x)$ is singular at $x = 0$, and the Romberg algorithm performs well only if $f(x)$ is sufficiently smooth.

Table 6.3.2. $R_{k,i}$, $1 \leq k \leq 6$ Calculated for $\int_0^1 (dx/\sqrt{x})$

k	i = 1	i = 2	i = 3	i = 4	i = 5	i = 6
1	0.50000					
2	0.95711	1.10948				
3	1.26723	1.37060	1.38801			
4	1.48304	1.55497	1.56726	1.57011		
5	1.63475	1.68532	1.69401	1.69602	1.69651	
6	1.74180	1.77749	1.78363	1.78505	1.78540	1.78549

Should we wish to apply the Romberg method to integrals whose integrands possess singularities, these singularities must be removed and dealt with prior to the computational process. ☐

PROBLEMS

1. Use Algorithm 6.3.1 to calculate $R_{4,4}$ for the following integrals:

(a) $\int_0^\pi \sin x \, dx$

(b) $\int_0^2 x^4 \, dx$

(c) $\int_0^1 xe^x \, dx$

and compare with the exact values.

2. Use Algorithm 6.3.1 to determine the number of iterations needed for convergence in the following cases and compare the approximate and exact integrals.

(a) $\int_0^\pi \sin x \, dx$, $\quad \epsilon = 10^{-5}$

(b) $\int_0^1 (1/\sqrt{x}) \, dx$, $\quad \epsilon = 10^{-5}$

(c) $\int_0^1 \ln x \, dx$, $\quad \epsilon = 10^{-4}$

In cases (b) and (c) we define the integrand at $x = 0$ as zero.

3. Compare the number of function evaluations needed for calculating

$$\int_0^\pi x \sin x \, dx$$

(a) Using the Romberg algorithm with $\epsilon = 10^{-5}$.

(b) Using the standard trapezoidal rule until

$$|R_{k+1,1} - R_{k,1}| < 10^{-5}$$

4. Use the Romberg algorithm with $\epsilon = 10^{-5}$ to approximate

$$I(f; a, b) = \int_0^2 \sqrt{x} \, dx$$

5. (a) Use the Romberg algorithm with $\epsilon = 10^{-5}$ to approximate

$$I(f; a, b) = \int_0^1 x^2 e^x \, dx$$

and compare the result with the exact value.

(b) How many function evaluations are needed to obtain the same accuracy using Simpson's rule?

6. Apply the Romberg algorithm with $\epsilon = 10^{-5}$ to approximate

$$I(f; a, b) = \int_0^1 \frac{dx}{\sqrt{e^x - 1} + \sqrt[3]{x}}$$

(a) Directly.

(b) By first treating the singularity analytically.

6.4 Multiple Integrals

The techniques presented in the previous sections for approximating the one-dimensional integral

$$I(f; a, b) = \int_a^b f(x)\, dx$$

can be extended for multiple integrals

$$I(f; R) = \iint_R f(x, y)\, dx dy \tag{6.4.1}$$

where R is a general two-dimensional region. We shall first consider a rectangular region

$$R = \{(x, y) \mid a \le x \le b,\ c \le y \le d\} \tag{6.4.2}$$

and construct a Simpson-type scheme to approximate $I(f; R)$.

6.4.1 A Two-Dimensional Simpson's Rule for a Rectangle

Let us divide the intervals $[a, b]$ and $[c, d]$ into $2n$ and $2m$ equal subintervals, respectively. Denote

$$h = \frac{b - a}{2n}, \quad k = \frac{d - c}{2m} \tag{6.4.3}$$

and define

$$x_i = a + ih, \quad 0 \le i \le 2n; \quad y_j = c + jk, \quad 0 \le j \le 2m \tag{6.4.4}$$

The multiple integral can be presented in several forms, for example,

$$\iint_R f(x, y)\, dx dy = \int_a^b \left[\int_c^d f(x, y)\, dy \right] dx \tag{6.4.5}$$

We first use the one-dimensional Simpson rule to approximate

$$G(x) = \int_c^d f(x, y)\, dy \tag{6.4.6}$$

for a given constant x and then repeat it to calculate

$$I(f; R) = \int_a^b G(x)\, dx \tag{6.4.7}$$

Given a constant x we have

$$\int_c^d f(x,y)\, dy = \frac{k}{3}\left[f(x,y_0) + 4\sum_{j=1}^{m} f(x,y_{2j-1}) + 2\sum_{j=1}^{m-1} f(x,y_{2j}) + f(x,y_{2m}) \right]$$
$$-\frac{(d-c)k^4}{180}\frac{\partial^4 f(x,\eta)}{\partial y^4}, \quad \eta = \eta(x) \tag{6.4.8}$$

where $c < \eta < d$. To integrate the right-hand side of Eq. (6.4.8) we first write

$$\int_a^b f(x,y_l)\, dx = \frac{h}{3}\left[f(x_0,y_l) + 4\sum_{i=1}^{n} f(x_{2i-1},y_l) + 2\sum_{i=1}^{n-1} f(x_{2i},y_l) + f(x_{2n},y_l) \right]$$
$$-\frac{(b-a)h^4}{180}\frac{\partial^4 f(\xi_l,y_l)}{\partial x^4} \tag{6.4.9}$$

where $a < \xi_l < b$, $0 \le l \le 2n$. By substituting Eq. (6.4.9) in Eq. (6.4.8) we obtain

$$\iint_R f(x,y)\, dxdy = \int_a^b \left[\int_c^d f(x,y)\, dy \right] dx$$

$$= \frac{hk}{9}\left[f(x_0,y_0) + 4\sum_{i=1}^{n} f(x_{2i-1},y_0) + 2\sum_{i=1}^{n-1} f(x_{2i},y_0) + f(x_{2n},y_0) \right.$$

$$+4\sum_{j=1}^{m} f(x_0,y_{2j-1}) + 16\sum_{j=1}^{m}\sum_{i=1}^{n} f(x_{2i-1},y_{2j-1}) + 8\sum_{j=1}^{m}\sum_{i=1}^{n-1} f(x_{2i},y_{2j-1})$$

$$+4\sum_{j=1}^{m} f(x_{2n},y_{2j-1}) + 2\sum_{j=1}^{m-1} f(x_0,y_{2j}) + 8\sum_{j=1}^{m-1}\sum_{i=1}^{n} f(x_{2i-1},y_{2j})$$

$$+4\sum_{j=1}^{m-1}\sum_{i=1}^{n-1} f(x_{2i},y_{2j}) + 2\sum_{j=1}^{m-1} f(x_{2n},y_{2j}) + f(x_0,y_{2m})$$

$$\left. +4\sum_{i=1}^{n} f(x_{2i-1},y_{2m}) + 2\sum_{i=1}^{n-1} f(x_{2i},y_{2m}) + f(x_{2n},y_{2m}) \right]$$

$$+O(h^4 + k^4) \tag{6.4.10}$$

The error E can be shown to be given by

$$E = -\frac{S}{180}\left[h^4\frac{\partial^4 f(\alpha,\beta)}{\partial x^4} + k^4\frac{\partial^4 f(\gamma,\delta)}{\partial y^4} \right] \tag{6.4.11}$$

where $S = (d-c)(b-a)$ and $(\alpha,\beta), (\gamma,\delta) \in R$.

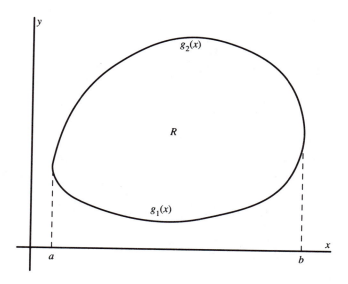

Figure 6.4.1. A nonrectangular region [Eq. (6.4.12)].

Example 6.4.1.

Consider the integral

$$\int_0^\pi \int_0^\pi (\sin x)(\sin y)\, dxdy = \int_0^\pi (\sin x)\, dx \int_0^\pi (\sin y)\, dy = 4$$

Implementing the two-dimensional Simpson rule given by Eq. (6.4.10) provides the results given in Table 6.4.1. The approximate values to the exact integral are denoted by $S(f; n, m)$ and are in good agreement with the relation $E = O(h^4 + k^4)$.

Table 6.4.1. Two-Dimensional Simpson's Rule
Applied to Several Meshes

h	k	$S(f; n, m)$
$\pi/4$	$\pi/4$	4.01826
$\pi/8$	$\pi/8$	4.00108
$\pi/16$	$\pi/16$	4.00007

□

Simpson's rule can be modified and applied to two-dimensional regions that are not necessarily rectangular. Consider a region R whose boundary is given by the curves $g_1(x)$, $a \le x \le b$ and $g_2(x)$, $a \le x \le b$ (Fig. 6.4.1) where

$$g_1(x) \le g_2(x), \quad a \le x \le b \tag{6.4.12}$$

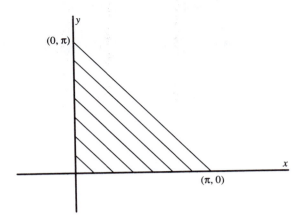

Figure 6.4.2. A triangular region (Example 6.4.2).

Because

$$\iint_R f(x,y)\, dxdy = \int_a^b \left[\int_{g_1(x)}^{g_2(x)} f(x,y)\, dy \right] dx \tag{6.4.13}$$

a two-dimensional Simpson integration scheme is again available by twice implementing the one-dimensional Simpson rule on the right-hand side of Eq. (6.4.13).

Example 6.4.2.

Consider the multiple integral

$$I(f;R) = \iint_R (\sin x)(\sin y)\, dxdy$$

where R is the triangle with vertices $(0,0), (0,\pi)$ and $(\pi,0)$ (Fig. 6.4.2). Following Eq. (6.4.13) we obtain

$$I(f;R) = \int_0^\pi \left[\int_0^{\pi-x} (\sin x)(\sin y)\, dy \right] dx$$

The exact value of this integral can be found easily and is equal to 2. To approximate it via a two-dimensional Simpson integration scheme we first write

$$I(f;R) = \int_0^\pi G(x)\, dx$$

where

$$G(x) = \int_0^{\pi-x} (\sin x)(\sin y)\, dy$$

Implementing the one-dimensional Simpson's rule with two subintervals yields

$$I(f;R) \approx \frac{\pi}{6} \left[G(0) + 4G\left(\frac{\pi}{2}\right) + G(\pi) \right]$$

Because $\sin(0) = \sin(\pi) = 0$ we have $G(0) = G(\pi) = 0$. Therefore

$$I(f;R) \approx \frac{2\pi}{3} G\left(\frac{\pi}{2}\right) = \frac{2\pi}{3} \int_0^{\pi/2} (\sin y) \, dy$$

We now reapply the one-dimensional Simpson's rule and obtain

$$\int_0^{\pi/2} (\sin y) \, dy \approx \frac{\pi}{12} \left[\sin(0) + 4\sin\left(\frac{\pi}{4}\right) + \sin\left(\frac{\pi}{2}\right) \right] = \frac{\pi}{12} \left(2\sqrt{2} + 1 \right)$$

Thus $I(f;R)$ is approximated by $(2\pi/3)(\pi/12)(2\sqrt{2} + 1) \approx 2.099$, which yields 5% accuracy. □

Simpson's rule is only one example of several methods that can be modified to approximate multiple integrals. Another example is the Gaussian integration method. We will again simplify the problem and assume a rectangular region.

6.4.2 A Two-Dimensional Gaussian Integration Scheme

Let $f(x, y)$ be integrable over

$$R = \{(x, y) \mid a \leq x \leq b, \quad c \leq y \leq d\}$$

Let us again express the multiple integral as

$$I(f;R) = \int_a^b \left[\int_c^d f(x, y) \, dy \right] dx$$

and perform Gaussian integration of order n and m in the coordinates x and y, respectively. Denote by $\{x_i, w_i, 1 \leq i \leq n\}$ the Gaussian points and weights associated with the interval $[a, b]$ and let $\{y_j, w'_j, 1 \leq j \leq m\}$ denote the points and weights of $[c, d]$. The implementation of the one-dimensional Gaussian integration scheme in the x coordinate yields $I(f;R) \approx \sum_{i=1}^n w_i G(x_i)$, where $G(x) = \int_c^d f(x, y) \, dy$. A second implementation leads to $G(x_i) \approx \sum_{j=1}^m w'_j f(x_i, y_j)$. Thus

$$I(f;R) \approx \sum_{i=1}^n \sum_{j=1}^m w_i w'_j f(x_i, y_j) \tag{6.4.14}$$

The computation of the right-hand side of Eq. (6.4.14) involves mn function evaluations. If $f(x, y)$ is a polynomial of order $(2n - 1)$ in x and of order $(2m - 1)$ in y, the approximate value is exact.

Example 6.4.3.

Consider the multiple integral

$$I(f;R) = \int_0^\pi \int_0^\pi (\sin x)(\sin y) \, dxdy$$

Assuming Gaussian integration of order 2 in each coordinate, we have [Eq. (6.2.18)]

$$w_i = w_j' = \pi/2, \quad 1 \le i,j \le 2$$

$$x_1 = y_1 = \frac{\pi}{2}\left(1 - \frac{\sqrt{3}}{3}\right), \quad x_2 = y_2 = \frac{\pi}{2}\left(1 + \frac{\sqrt{3}}{3}\right)$$

Thus

$$I(f;R) = \frac{\pi^2}{4} \sum_{i=1}^{2} \sum_{j=1}^{2} \sin(x_i) \sin(y_j) = 3.747\ldots$$

The exact value as previously stated (Example 6.4.1) is 4.   □

Example 6.4.4.

The multiple integral

$$I(f;R) = \int_0^2 \left(\int_0^1 x^2 y^3 \, dy \right) dx$$

can be calculated analytically and equals 2/3. Consider a two-dimensional Gaussian integration scheme of the second order in both coordinates x and y. By virtue of Eq. (6.2.18) and Table 6.2.1 we have

$$x_1 = 0.4226497308, \quad x_2 = 1.5773502692, \quad w_1 = w_2 = 1$$
$$y_1 = 0.2113248654, \quad y_2 = 0.7886751346, \quad w_1' = w_2' = 0.5$$

Substituting in Eq. (6.4.14) we find

$$I(f;R) = \frac{1}{2} \sum_{i=1}^{2} \sum_{j=1}^{2} x_i^2 y_j^3 = \frac{2}{3}$$

because the Gaussian procedure is exact for the particular integrand $x^2 y^3$.   □

The subroutines SIMP2D and GAUS2D, which calculate multiple integrals over a rectangular region by using Simpson's rule and a Gaussian integration method, respectively, are found on the attached floppy disk.

PROBLEMS

1. Let $f(x,y) = e^x \cos y$ and $R = \{(x,y) \mid 0 \le x \le 2, \ 0 \le y \le \pi/2\}$. Apply the two-dimensional

Simpson integration technique to approximate $I(f; R)$ using

$$(a) \ \ n = 2, m = 3 \quad (b) \ \ n = 4, m = 6 \quad (c) \ \ n = 8, m = 12$$

and compare with the exact value of the multiple integral.

2. Let

$$f(x, y) = \begin{cases} 1/\sqrt{xy}, \ x > 0, \quad \quad y > 0 \\ 0 \quad \quad , \ x = 0 \ \ \text{or} \ \ y = 0 \end{cases}$$

Apply two-dimensional Simpson integration technique to approximate

$$I(f; R) = \int_0^1 \int_0^1 f(x, y) \, dxdy$$

using

$$(a) \ \ n = m = 2 \quad (b) \ \ n = m = 4 \quad (c) \ \ n = m = 8$$

and compare to the exact value 4.

3. (a) Apply a two-dimensional Gaussian integration scheme with $n = m = 2$ to approximate

$$I(f; R) = \int_0^1 \int_0^1 f(x, y) \, dxdy$$

where

$$f(x, y) = \begin{cases} 1/(\sqrt{xy} + \sqrt[3]{xy}) , \quad xy \neq 0 \\ 0 \quad \quad \quad \quad , \quad xy = 0 \end{cases}$$

(b) Repeat the computation by first treating the singularity at $(0, 0)$ analytically.

4. Generate a two-dimensional Gaussian integration scheme with $n = m = 3$ to approximate

$$I(f; R) = \int_0^1 \left(\int_0^{1-x} e^{x-y} \, dy \right) dx$$

and compare with the exact value of the integral.

5. Design a two-dimensional trapezoidal rule to approximate

$$I(f; R) = \iint_R f(x, y) \, dxdy$$

for the rectangle

$$R = \{(x, y) \mid a \leq x \leq b, \quad c \leq y \leq d\}$$

6. Apply the two-dimensional trapezoidal rule to approximate

$$\int_0^1 \int_0^1 e^x \sin y \, dx dy$$

using four equal subintervals to integrate in the direction of each coordinate.

7. Write a two-dimensional Simpson algorithm to approximate

$$I(f;R) = \iint_R f(x,y) \, dx dy$$

where R is bounded by the curves

$$y_1 = g_1(x), \quad y_2 = g_2(x), \quad a \le x \le b$$

and the straight lines $x = a$, $x = b$ [assume $g_1(x) \le g_2(x)$].

8. Apply the algorithm of problem 7 with $n = 3$, $m = 2$ to approximate

$$\iint_R \sqrt{xy + x^2 y^2} \, dx dy$$

where R is the region bounded between the curves

$$y_2 = \sqrt{x}, \quad y_1 = x^2, \quad 0 \le x \le 1$$

Approximate the integral using a two-dimensional Gaussian integration scheme with three nodes per each coordinate.

6.5 Numerical Differentiation

To approximate the derivative $f'(x)$ of a given function $f(x)$ we use the basic relation

$$f'(x) = \lim_{h \to 0} \frac{f(x+h) - f(x)}{h}$$

which suggests the approximation

$$f'(x) \approx \frac{f(x+h) - f(x)}{h} = D(f;x,h) \tag{6.5.1}$$

provided that the step h is sufficiently small. The right-hand side of Eq. (6.5.1) is simply the slope of the secant between the points $[x, f(x)]$ and $[x+h, f(x+h)]$ (Fig. 6.5.1).

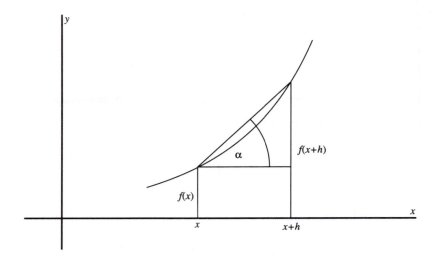

Figure 6.5.1. Approximating $f'(x)$ by $\tan \alpha$.

Example 6.5.1.
Let $f(x) = \tan x$. The derivative $f'(x) = 1/\cos^2 x$ equals 2 at $x = \pi/4$. Table 6.5.1 presents a set of approximations calculated using Eq. (6.5.1).

Table 6.5.1. Approximating $f'(x)$ Using $D(f; x, h)$

h	$D(\tan x; \pi/4, h)$	|Error|
0.20000	2.54249	0.54249
0.10000	2.23049	0.23049
0.05000	2.10711	0.10711
0.02500	2.05172	0.05172
0.01250	2.02542	0.02542
0.00625	2.01260	0.01260

□

The fact that the error in $D(f; x, h)$ is proportional to h as $h \to 0$ is well demonstrated by the previous example. This is true in general: by virtue of Taylor's theorem we have

$$f(x + h) = f(x) + hf'(x) + \frac{h^2}{2}f''(\xi), \quad x < \xi < x + h$$

or

$$\frac{f(x + h) - f(x)}{h} = f'(x) + \frac{h}{2}f''(\xi)$$

which implies

$$\lim_{h \to 0} \frac{D(f; x, h) - f'(x)}{h} = \frac{1}{2}f''(x) \tag{6.5.2}$$

We have thus proved the following result.

Theorem 6.5.1.
Let $f(x) \in C^2[a, b]$. Then Eq. (6.5.2) holds for all $x \in [a, b]$.

6.5.1 Differentiating the Lagrange Interpolator

Let $p_n(x)$ be the polynomial of degree n that interpolates $f(x)$ at the $(n + 1)$ points $x_i, 0 \le i \le n$. To approximate $f'(x)$ we may replace $f(x)$ by $p_n(x)$ and write

$$f'(x) \approx p'_n(x) \tag{6.5.3}$$

For each particular set of points we obtain a different approximation for $f'(x)$. A popular case is $n = 2$ with evenly spaced points:

$$x_1 = x_0 + h, \quad x_2 = x_0 + 2h$$

Lagrange's interpolator is

$$p_2(x) = \frac{(x - x_1)(x - x_2)}{2h^2} f(x_0) + \frac{(x - x_0)(x - x_2)}{(-h^2)} f(x_1) + \frac{(x - x_0)(x - x_1)}{2h^2} f(x_2)$$

and its derivative at $x = x_1$ is

$$
\begin{aligned}
p'_2(x_1) &= \frac{(x_1 - x_2)}{2h^2} f(x_0) + \frac{(x_1 - x_2)}{(-h^2)} f(x_1) + \frac{(x_1 - x_0)}{(-h^2)} f(x_1) + \frac{(x_1 - x_0)}{2h^2} f(x_2) \\
&= \frac{f(x_2) - f(x_0)}{2h}
\end{aligned} \tag{6.5.4}
$$

Substituting in Eq. (6.5.3) we have

$$f'(x) \approx \frac{f(x + h) - f(x - h)}{2h} \tag{6.5.5}$$

The right-hand side of Eq. (6.5.5) is called a *central difference*. It approximates $f'(x)$ better than $D(f; x, h)$, as shown next.

Theorem 6.5.2.
Let $f(x) \in C^3[a, b]$ and $x - h, x, x + h \in [a, b]$. Then

$$\frac{f(x + h) - f(x - h)}{2h} - f'(x) \approx \alpha h^2, \quad h \to 0$$

for some constant α.

Proof.

By applying Taylor's theorem we obtain

$$f(x + h) = f(x) + \frac{h}{1!}f'(x) + \frac{h^2}{2!}f''(x) + \frac{h^3}{3!}f'''(\xi_1), \quad x < \xi_1 < x + h$$

$$f(x - h) = f(x) - \frac{h}{1!}f'(x) + \frac{h^2}{2!}f''(x) - \frac{h^3}{3!}f'''(\xi_2), \quad x - h < \xi_2 < x$$

leading to

$$f(x + h) - f(x - h) = 2hf'(x) + \frac{h^3}{3!}[f'''(\xi_1) + f'''(\xi_2)]$$

Thus

$$\lim_{h \to 0} \frac{\{[f(x + h) - f(x - h)]/2h\} - f'(x)}{h^2} = \frac{f'''(x)}{6} \tag{6.5.6}$$

□

The right-hand side of Eq. (6.5.5) will be denoted by $D_c(f; x, h)$.

Example 6.5.2.

Table 6.5.2 presents a comparison between $D(f; x, h)$ and $D_c(f; x, h)$ for $f(x) = \tan x$, $x = \pi/4$.

Table 6.5.2. $D(f; x, h)$ vs. $D_c(f; x, h)$

h	$D(\tan x; \pi/4, h)$	$D_c(\tan x; \pi/4, h)$
0.20000	2.54249	2.11397
0.10000	2.23049	2.02710
0.05000	2.10711	2.00669
0.02500	2.05172	2.00167
0.01250	2.02542	2.00042
0.00625	2.01260	2.00010

□

A general error estimate associated with Eq. (6.5.3) is given next without proof.

Theorem 6.5.3.

Let $f(x)$ possess $(n + 2)$ continuous derivatives over the interval $[a, b]$ and let $p_n(x)$ denote the Lagrange interpolator that interpolates $f(x)$ at $x_0, x_1, \ldots, x_n \in [a, b]$. Then

$$f'(x) - p'_n(x) = l_n(x)\frac{f^{(n+2)}(\xi_1)}{(n + 2)!} + l'_n(x)\frac{f^{(n+1)}(\xi_2)}{(n + 1)!} \tag{6.5.7}$$

where $a < \xi_1, \xi_2 < b$ and

$$l_n(x) = (x - x_0)(x - x_1) \cdots (x - x_n) \tag{6.5.8}$$

In the particular case $n = 2$, $x_1 = x_0 + h$, $x_2 = x_0 + 2h$ we have

$$l_2(x) = (x - x_0)(x - x_1)(x - x_2)$$
$$l_2(x_1) = 0$$
$$l_2'(x_1) = -h^2$$

and thus

$$f'(x_1) - p_2'(x_1) = -h^2 \frac{f'''(\xi_2)}{6}$$

in agreement with Eq. (6.5.6).

To perform numerical differentiation of order higher than 1, we introduce the following technique.

6.5.2 Undetermined Coefficients

To calculate $f^{(k)}(x)$ using only given values of $f(t)$ at the evenly spaced nodes

$$\ldots, x - 2h, x - h, x, x + h, x + 2h, \ldots$$

we approximate $f^{(k)}(x)$ by a finite linear combination of these values with *undetermined coefficients*. We then replace each value of $f(t)$ by its Taylor expansion about x and determine the coefficients that minimize the difference between $f^{(k)}(x)$ and its approximation. For example, consider approximating $f''(x)$ using only $f(x - h), f(x), f(x + h)$, that is,

$$f''(x) \approx \alpha f(x - h) + \beta f(x) + \gamma f(x + h) \tag{6.5.9}$$

We replace $f(x - h)$ and $f(x + h)$ by their Taylor expansions and obtain

$$f''(x) \approx \alpha \left[f(x) - hf'(x) + \frac{h^2}{2!}f''(x) - \frac{h^3}{3!}f^{(3)}(x) + \frac{h^4}{4!}f^{(4)}(\xi_1) \right]$$
$$+ \beta f(x) + \gamma \left[f(x) + hf'(x) + \frac{h^2}{2!}f''(x) + \frac{h^3}{3!}f^{(3)}(x) + \frac{h^4}{4!}f^{(4)}(\xi_2) \right] \tag{6.5.10}$$

where $x - h < \xi_1 < x$, $x < \xi_2 < x + h$.

The next step is to determine α, β, and γ such that the coefficients of $f^{(j)}(x)$, $j = 0, 1, \ldots$ on both sides of Eq. (6.5.10) are identical. This request leads to a system of linear equations:

$$
\begin{aligned}
j = 0: \quad & \alpha + \beta + \gamma = 0 \\
j = 1: \quad & (-\alpha + \gamma)h = 0 \\
j = 2: \quad & (\alpha + \gamma)\frac{h^2}{2!} = 1 \\
j = 3: \quad & (-\alpha + \gamma)\frac{h^3}{3!} = 0
\end{aligned}
\tag{6.5.11}
$$

The equations for $j = 1$ and $j = 3$ are practically the same and the system yields the unique solution

$$\alpha = \gamma = 1/h^2, \quad \beta = -2/h^2$$

Thus

$$f''(x) \approx \frac{f(x-h) - 2f(x) + f(x+h)}{h^2} \qquad (6.5.12)$$

The error associated with this approximation is

$$f''(x) - \frac{f(x-h) - 2f(x) + f(x+h)}{h^2} = -\frac{h^2}{4!}[f^{(4)}(\xi_1) + f^{(4)}(\xi_2)]$$

and if $f(t) \in C^4[a, b]$ we obtain

$$f''(x) - \frac{f(x-h) - 2f(x) + f(x+h)}{h^2} = -\frac{h^2}{12} f^{(4)}(\xi) \qquad (6.5.13)$$

for some ξ between $x - h$ and $x + h$.

Example 6.5.3.

For $f(x) = \tan x$ we have $f'' = 2\sin x / \cos^3 x$. Thus, $f''(\pi/4) = 4$. Table 6.5.3 provides a set of approximations denoted by $D^{(2)}(f; x, h)$ to $f''(\pi/4)$, using Eq. (6.5.12).

Table 6.5.3. Approximating $f''(x)$ by Using Eq. (6.5.12)

h	$D^{(2)}(\tan x; \pi/4, h)$	$|\text{Error}|$
0.20000	4.28522	0.28522
0.10000	4.06777	0.06777
0.05000	4.01673	0.01673
0.02500	4.00417	0.00417
0.01250	4.00104	0.00104
0.00625	4.00026	0.00026

□

Numerical differentiation of a higher order must clearly involve a greater number of nodes.

So far, the error analysis associated with numerical differentiation has not included the effect of using inaccurate function values. Two major sources for such inaccuracies are as follow:

1. Round-off errors that occur in the process of computing $f(x)$.

2. Obtaining $f(x)$ empirically.

In either case this error must be taken in account. For example, let $\epsilon_{-1}, \epsilon_0, \epsilon_1$ denote the errors in $f(x - h), f(x), f(x + h)$, respectively. Using Eq. (6.5.12) we find

$$f''(x) - \frac{f(x-h) + \epsilon_{-1} - 2[f(x) + \epsilon_0] + f(x+h) + \epsilon_1}{h^2} = -\frac{h^2}{12} f^{(4)}(\xi)$$

or

$$f''(x) - \frac{f(x-h) - 2f(x) + f(x+h)}{h^2} = -\frac{h^2}{12} f^{(4)}(\xi) + \frac{\epsilon_{-1} - 2\epsilon_0 + \epsilon_1}{h^2}$$

If ϵ_i is random and does not exceed some given ϵ in absolute value, we have

$$\left| f''(x) - \frac{f(x-h) - 2f(x) + f(x+h)}{h^2} \right| \leq \frac{M_4 h^2}{12} + \frac{4\epsilon}{h^2} \tag{6.5.14}$$

where

$$M_4 = \max_{x-h \leq t \leq x+h} |f^{(4)}(t)|$$

Thus numerical differentiation is not effective if h is too small. The right-hand side of Eq. (6.5.14) increases indefinitely and Eq. (6.5.12) is not valid. There is, however, some h_0 given by

$$h_0 = \left(\frac{48\epsilon}{M_4} \right)^{1/4} \tag{6.5.15}$$

which minimizes

$$E(h) = \frac{M_4 h^2}{12} + \frac{4\epsilon}{h^2} \tag{6.5.16}$$

The constant M_4 may not be known, but can be replaced by $f^{(4)}(x)$, because h is presumably small.

Example 6.5.4.

For $f(x) = \sin x$ we have

$$f''(x) = -\sin x, \quad f^{(4)}(x) = \sin x$$

A set of approximations to $f''(\pi/4)$, whose exact value is $-\sqrt{2}/2 = -0.70710678\ldots$, is presented in Table 6.5.4.

Table 6.5.4. Round-Off Effect in Numerical Differentiation

h	$D^{(2)}(\sin x; \pi/4, h)$	$|\text{Error}|$
10^{-1}	-0.70651772	0.00058906
10^{-2}	-0.70710089	0.00000589
10^{-3}	-0.70710672	0.00000006
10^{-4}	-0.70710655	0.00000023
10^{-5}	-0.70713213	0.00002535
10^{-6}	-0.71054274	0.00343595

The computations were subject to $\epsilon \approx 10^{-14}$. Because $M_4 \approx \sqrt{2}/2$ we obtain $h_0 \approx 0.0009$, in agreement with the results. □

If each value of $f(x)$ is given empirically and yields a considerable error, then instead of numerically differentiating the data it is often recommended that the function first be replaced by an interpolating cubic spline $s(x)$, followed by substituting $s'(x)$ for $f'(x)$.

PROBLEMS

1. Calculate and compare $D[f(x); x, h]$ and $D_c[f(x); x, h]$ for
   (a) $f(x) = \ln(x);\ x = 1,\ h = 0.1, 0.05, 0.025, 0.0125$
   (b) $f(x) = e^x;\ x = 2,\ h = 0.1, 0.05, 0.025, 0.0125$
   (c) $f(x) = (1+x)/(1+x^2);\ x = 1,\ h = 0.4, 0.2, 0.1, 0.05, 0.25$

2. Calculate $D_c[f(x); x, h]$ and validate Eq. (6.5.5) in the following cases:
   (a) $f(x) = \sin(x^2);\ x = 1,\ h = 0.1, 0.05, 0.025, 0.0125$
   (b) $f(x) = x^3;\ x = 1,\ h = 0.1, 0.05, 0.025$
   (c) $f(x) = x^2 e^{-x};\ x = 2,\ h = 0.1, 0.01, 0.001$

3. Calculate $D^{(2)}[f(x); x, h]$ and validate Eq. (6.5.13) in the following cases:
   (a) $f(x) = \cos(x)\ ;\ x = \pi/3,\ h = 0.1, 0.05, 0.025, 0.0125$
   (b) $f(x) = \sqrt{x}\ ;\ x = 1,\ h = 0.1, 0.05, 0.025, 0.0125$
   (c) $f(x) = \ln x\ ;\ x = 2,\ h = 0.1, 0.01, 0.001$

4. (a) Determine the Lagrange interpolator $p_4(x)$ of $\sin(x)$ associated with the five nodes

$$x_i = i\frac{\pi}{4}, \quad 0 \le i \le 4$$

   (b) Compare $[\sin(x)]'$ and $p_4'(x)$ at $x = \pi/4$ and validate Eq. (6.5.7).

5. Let $f(x)$ be approximated by its Lagrange interpolator $p_n(x)$. Use Eq. (6.5.7) to bound $|f'(x) - p_n'(x)|$.

6. Use $f(x)$, $f(x+h)$, $f(x+2h)$ and the method of undetermined coefficients to approximate $f'(x)$. Estimate the error.

7. Repeat problem 6, assuming a random error ϵ in each function value.

8. Use $f(x - k)$, $f(x)$, $f(x + h)$ and the method of undetermined coefficients to approximate $f'(x)$ and estimate the error.

9. (a) Show that h_0 of Eq. (6.5.15) minimizes $E(h)$ of Eq. (6.5.16).

   (b) Let $f(x) = x^4 + \alpha x^3 + \beta x^2 + \gamma x + \delta$, where α, β, γ, and δ are unknown. For each x, $f(x)$ is given empirically with a random error $|\epsilon| = 10^{-5}$. Find the optimal mesh size h for approximating $f''(x)$, using Eq. (6.5.12).

7

Linear Equations

Systems of simultaneous linear equations play a major role in numerical analysis. They occur in many areas such as mathematics, physics, statistics, engineering, and social sciences. When they occur, they are part of the mathematical models that are supposed to deal with real problems. There are various algorithms for solving systems of linear equations. It is the *structure* of the system matrix that indicates which algorithm should be implemented. At this point, the particular area and problem that produced the system are of no significance. In other words, *similar* systems with different background are treated by the same algorithm.

We start this chapter by defining linear equations, using matrix notation. Then, although it is assumed that the reader is familiar with basics such as matrices and determinants, we present a short review of elementary matrix algebra.

7.1 System of Linear Equations: Matrices

Definition 7.1.1. (System of simultaneous linear equations). *A set of equations*

$$a_{11}x_1 + a_{12}x_2 + \cdots + a_{1n}x_n = b_1$$
$$a_{21}x_1 + a_{22}x_2 + \cdots + a_{2n}x_n = b_2$$
$$\vdots$$
$$a_{m1}x_1 + a_{m2}x_2 + \cdots + a_{mn}x_n = b_m \qquad (7.1.1)$$

with constant coefficients $\{a_{ij}, b_i | 1 \leq i \leq m, 1 \leq j \leq n\}$ *and variables* $\{x_j | 1 \leq j \leq n\}$ *is called a system of linear equations with m equations and n variables.*

Unless otherwise stated, it is assumed that the numbers of equations and variables are equal, that is, $m = n$. In most practical applications this is indeed the case and we then refer to n as the *order* of the system.

Any set of numbers $\{x_i\}_{i=1}^{n}$ that satisfies Eq. (7.1.1) is a *solution* of the given system. All the solutions compose the *solution set* of the system.

Example 7.1.1.

The system of three linear equations

$$2x_1 - x_2 + x_3 = 1$$
$$x_1 + x_2 = 2$$
$$-x_1 + x_3 = -1$$

yields the unique solution $x_1 = x_2 = 1, x_3 = 0$.   □

Example 7.1.2.

The system of two linear equations

$$x_1 + x_2 = 1$$
$$2x_1 + 2x_2 = 3$$

has no solutions (why?). However, if the second equation is replaced by

$$2x_1 + 2x_2 = 2$$

the system has an infinite number of solutions, all of which can be represented in the parametric form

$$\{x_1 = t, x_2 = 1 - t \mid -\infty < t < \infty\}$$ □

Definition 7.1.2. (Equivalence). *Two systems of linear equations, S_1 and S_2, are said to be equivalent if they have the same solutions.*

Example 7.1.3.

Consider the systems

$$S_1 \begin{cases} x_1 + x_2 = 3 \\ x_1 - x_2 = 1 \end{cases}, \qquad S_2 \begin{cases} x_1 + 2x_2 = 4 \\ 3x_1 - x_2 = 5 \end{cases}$$

Because both yield the unique solution $x_1 = 2, x_2 = 1$, they are equivalent.   □

All the algorithms for solving systems of linear equations work by replacing the original set of equations with equivalent systems before and during the computations. Let E_i denote the ith equation of a given system. The following *elementary* operations on E_i provide equivalent systems, that is, leave the solution set of the original system unchanged:

1. Replacing E_i by kE_i, where k is any given nonzero constant

2. Interchanging E_i and E_j, where $i \neq j$

3. Replacing E_j by $E_j + kE_i$, where k is any given constant

Example 7.1.4.

Consider the system

$$E_1: x_1 - 2x_2 + x_3 + 5x_4 = 1$$
$$E_2: 2x_1 + x_2 - x_3 - x_4 = -3$$
$$E_3: x_1 - x_2 + 2x_3 + 3x_4 = 7$$
$$E_4: x_1 - 4x_2 + x_3 + x_4 = 0$$

By implementing the elementary operations $E_1 \leftarrow 2E_1, E_3 \leftarrow E_3 + 2E_2, E_4 \leftarrow E_4 + E_2$ we obtain the system

$$E_1: 2x_1 - 4x_2 + 2x_3 + 10x_4 = 2$$
$$E_2: 2x_1 + x_2 - x_3 - x_4 = -3$$
$$E_3: 5x_1 + x_2 + x_4 = 1$$
$$E_4: 3x_1 - 3x_2 = -3$$

with an identical solution set. $\square$

Example 7.1.5.

The systems

$$S_1 \begin{cases} x_1 + x_2 = 1 \\ 2x_1 + 2x_2 = 2 \end{cases}, \qquad S_2 \begin{cases} x_1 + 2x_2 = 2 \\ 2x_1 + 4x_2 = 4 \end{cases}$$

have the solution sets

$$\{x_1 = t, x_2 = 1 - t | -\infty < t < \infty\} \text{ for } S_1$$
$$\{x_1 = t, x_2 = (2 - t)/2 | -\infty < t < \infty\} \text{ for } S_2$$

Because the particular pair $(x_1, x_2) = (0.5, 0.5)$ belongs only to the first set, the two systems are not equivalent. Consequently, for example, we cannot obtain S_2 from S_1 by using a finite number of elementary operations. $\square$

A general system of linear equations given by Eq. (7.1.1) can be written as

$$Ax = b \qquad (7.1.2)$$

where

$$A = (a_{ij}) = \begin{bmatrix} a_{11} & a_{12} & \cdots & a_{1n} \\ a_{21} & a_{22} & \cdots & a_{2n} \\ \vdots & \vdots & & \vdots \\ a_{m1} & a_{m2} & \cdots & a_{mn} \end{bmatrix} \qquad (7.1.3)$$

is an $m \times n$ matrix and x, b are the $n \times 1$ and $m \times 1$ matrices

$$x = \begin{bmatrix} x_1 \\ x_2 \\ \vdots \\ x_n \end{bmatrix}, \qquad b = \begin{bmatrix} b_1 \\ b_2 \\ \vdots \\ b_n \end{bmatrix} \qquad (7.1.4)$$

The $m \times (n+1)$ matrix

$$B = (A, b) = \begin{bmatrix} a_{11} & a_{12} & \cdots & a_{1n} & b_1 \\ a_{21} & a_{22} & \cdots & a_{2n} & b_2 \\ \vdots & \vdots & & \vdots & \vdots \\ a_{m1} & a_{m2} & \cdots & a_{mn} & b_m \end{bmatrix} \tag{7.1.5}$$

is called the *augmented* matrix of the given system. It is the structure of A and B that determines whether a set of linear equations has a solution. Before introducing several algorithms for solving Eq. (7.1.2) we present a short review of elementary matrix algebra, which is essential to the formulation and understanding of these algorithms.

7.1.1 Matrices: Definition and Arithmetic

Definition 7.1.3.   ($m \times n$ matrix). *A rectangular array of numbers*

$$A = \begin{bmatrix} a_{11} & a_{12} & \cdots & a_{1n} \\ a_{21} & a_{22} & \cdots & a_{2n} \\ \vdots & \vdots & & \vdots \\ a_{m1} & a_{m2} & \cdots & a_{mn} \end{bmatrix} \tag{7.1.6}$$

is called a matrix. *The matrix A is said to have m rows and n columns. Each number a_{ij} is called an* entry *or* element.

For each pair of $m \times n$ matrices $A = (a_{ij})$ and $B = (b_{ij})$ we define the $m \times n$ sum matrix $C = (c_{ij})$ by

$$c_{ij} = a_{ij} + b_{ij}; \quad 1 \leq i \leq m, \quad 1 \leq j \leq n$$

and write $C = A + B$. For arbitrary number α and an $m \times n$ matrix A, we define

$$\alpha A = (\alpha a_{ij})$$

Let A and B be $m \times r$ and $r \times n$ matrices, respectively. The $m \times n$ matrix $C = (c_{ij})$, where

$$c_{ij} = \sum_{k=1}^{r} a_{ik} b_{kj}; \quad 1 \leq i \leq m, \quad 1 \leq j \leq n \tag{7.1.7}$$

is called the *product* of A and B (in this order) and we write $C = AB$.

Example 7.1.6.
For the matrices

$$A = \begin{bmatrix} 1 & -1 & 2 \\ 0 & 1 & 3 \end{bmatrix}, \quad B = \begin{bmatrix} 0 & 2 & 0 \\ 3 & 5 & -1 \end{bmatrix}, \quad C = \begin{bmatrix} 1 & 2 \\ -1 & 6 \\ 1 & 0 \end{bmatrix}$$

we obtain

$$A + B = \begin{bmatrix} 1 & 1 & 2 \\ 3 & 6 & 2 \end{bmatrix}, \quad AC = \begin{bmatrix} 4 & -4 \\ 2 & 6 \end{bmatrix}$$

and

$$(-2)B = \begin{bmatrix} 0 & -4 & 0 \\ -6 & -10 & 2 \end{bmatrix} \qquad \qquad \square$$

For every $m \times n$ matrix A we define the $n \times m$ transpose matrix

$$A^T = \begin{bmatrix} a_{11} & a_{21} & \cdots & a_{m1} \\ a_{12} & a_{22} & \cdots & a_{m2} \\ \vdots & \vdots & & \vdots \\ a_{1n} & a_{2n} & \cdots & a_{mn} \end{bmatrix} \tag{7.1.8}$$

An $m \times n$ matrix A for which $m = n$ is called a *square* matrix of order n. A square matrix of order n for which

$$a_{ij} = 0, \quad i \neq j$$
$$a_{ii} = 1, \quad 1 \leq i \leq n$$

is called an *identity* matrix of order n and is denoted by I_n. For an arbitrary square matrix A of order n, we have

$$AI_n = I_nA = A \tag{7.1.9}$$

A matrix $A = (a_{ij})$ for which $a_{ij} = 0; 1 \leq i \leq m, 1 \leq j \leq n$ is called a *zero* matrix and is denoted by 0. We clearly have

$$A + 0 = 0 + A = A$$

for arbitrary A.

7.1.2 Determinants

To each square matrix A we attach a number, the *determinant* of A. This number plays a major role in determining whether or not a linear system $Ax = b$ has a unique solution. It is denoted by $|A|$ or by $\det(A)$ and is defined next.

Definition 7.1.4. [Row representation of $\det(A)$]. *For arbitrary square matrix A of order n we define*

1. $\det(A) = a_{11}, n = 1$
2. $\det(A) = \sum_{j=1}^{n}(-1)^{i+j}a_{ij}\det(A_{ij}), n > 1$

where A_{ij} is the $(n-1) \times (n-1)$ matrix obtained by removing the ith row and jth column from A.

A basic result from linear algebra implies that the definition of $\det(A)$ is independent of i. Also, it can be shown that a similar *column representation* of $\det(A)$, that is,

$$\det(A) = \sum_{i=1}^{n} (-1)^{i+j} a_{ij} \det(A_{ij}), \quad n > 1 \tag{7.1.10}$$

produces the same number. Another important result is the following theorem.

Theorem 7.1.1.

(Elementary operations on rows and columns). *For arbitrary square matrix A:*

1. *Interchanging two rows (columns) of A changes the sign of $\det(A)$.*

2. *Replacing a row (column) E_i by $E_i + kE_j$, where k is an arbitrary number and E_j is another row (column) such that $i \neq j$, does not change $\det(A)$.*

3. *Replacing a row (column) E_i by kE_i changes $\det(A)$ to $k \cdot \det(A)$.*

The proof of Theorem 7.1.1 can be found in any textbook of linear algebra.

Example 7.1.7.

1. $\begin{vmatrix} a & b \\ c & d \end{vmatrix} = a|d| - b|c| = ad - bc$

2. $\begin{vmatrix} 3 & 1 & -2 \\ 0 & 2 & 1 \\ 1 & 4 & 2 \end{vmatrix} = 3 \begin{vmatrix} 2 & 1 \\ 4 & 2 \end{vmatrix} - \begin{vmatrix} 0 & 1 \\ 1 & 2 \end{vmatrix} - 2 \begin{vmatrix} 0 & 2 \\ 1 & 4 \end{vmatrix} = 3 \cdot 0 + 1 + 2 \cdot 2 = 5$ □

Definition 7.1.5.   (Inverse matrix). *Given a square matrix A, if a matrix B such that*

$$AB = BA = I \tag{7.1.11}$$

exists, B is called the inverse *matrix of A and is denoted by A^{-1}.*

There is similarity between arithmetic operations on numbers and arithmetic operations on matrices. However, the analogy is not complete. The rules that are valid for matrices are listed below. Let A, B, and C denote general matrices (not necessarily square). Then

$$(A + B) + C = A + (B + C) \tag{7.1.12}$$
$$(AB)C = A(BC) \tag{7.1.13}$$
$$A + B = B + A \tag{7.1.14}$$
$$A(B + C) = AB + AC \tag{7.1.15}$$
$$(A + B)C = AC + BC \tag{7.1.16}$$
$$(AB)^T = B^T A^T \tag{7.1.17}$$

$$(A + B)^T = A^T + B^T \tag{7.1.18}$$

The next set of rules, which are related to inverse matrices and determinants, apply only to square matrices.

$$(AB)^{-1} = B^{-1}A^{-1} \tag{7.1.19}$$

$$(cA)^{-1} = \frac{1}{c}A^{-1}, \quad c \neq 0 \tag{7.1.20}$$

$$\det(AB) = \det(A)\det(B) \tag{7.1.21}$$

$$\det(A^T) = \det(A) \tag{7.1.22}$$

$$\det(cA) = c^n\det(A), \quad n = \text{order}(A) \tag{7.1.23}$$

It should be emphasized that Eqs. (7.1.11) through (7.1.23) hold, provided that their left-hand sides are well defined. For example, Eq. (7.1.13) is valid only for A, B, and C of orders $m \times n$, $n \times r$, and $r \times s$, respectively.

We now resume our discussion of linear equations. In most practical applications the matrix A of Eq. (7.1.2) is a square matrix, and the next result holds.

Theorem 7.1.2.
(Solving linear equations with a square matrix). *The following statements are equivalent:*

1. The inverse matrix A^{-1} exists.

2. $\det(A) \neq 0$.

3. The linear system given by Eq. (7.1.2) has a unique solution.

4. $r(A) = n$, where $r(A)$ is the rank of A (defined below) and n is the order of A.

If *any* of the statements hold, A is called a *nonsingular matrix*. In this case we multiply both sides of Eq. (7.1.2) by the inverse matrix A^{-1} and express the unique solution of the system as

$$x = A^{-1}b \tag{7.1.24}$$

If $\det(A) = 0$, A is a *singular* matrix and Eq. (7.1.24) does not hold. It is also not applicable if the number of equations and the number of variables do not coincide, that is, if $m \neq n$. In these cases the solution is not unique, or does not exist. Let us denote by r_1 and r_2 the maximum numbers of linearly independent rows and columns in A. A well-known result from linear algebra is

$$r_1 = r_2 \tag{7.1.25}$$

and the number, denoted by $r(A)$, is called the *rank* of A.

Theorem 7.1.3.
(Existence of a solution to a general linear system). *A solution to the linear system given by Eq. (7.1.2) exists if and only if*

$$r(A) = r(B) \tag{7.1.26}$$

where B is the augmented matrix defined by Eq. (7.1.5).

The proof is left as an exercise for the reader.

Example 7.1.8.

Consider the system of linear equations

$$Ax = \begin{bmatrix} 2 & 1 & -3 \\ 0 & 3 & 4 \\ 2 & 4 & 1 \end{bmatrix} \begin{bmatrix} x_1 \\ x_2 \\ x_3 \end{bmatrix} = \begin{bmatrix} 1 \\ 0 \\ 2 \end{bmatrix} = b$$

Because $\det(A) = 0$ there is no unique solution. Whether or not a solution exists depends only on b. It is easily seen that $r(B) = 3$ and thus $r(A) \neq r(B)$. Therefore the given system does not have a solution.   □

Example 7.1.9.

Consider a set of two equations and four variables, given by

$$x_1 - x_2 + x_3 + 2x_4 = 3$$
$$x_1 + x_2 - x_3 - x_4 = 1$$

that is,

$$A = \begin{bmatrix} 1 & -1 & 1 & 2 \\ 1 & 1 & -1 & -1 \end{bmatrix}, \quad B = (A, b)$$

where

$$b = \begin{bmatrix} 3 \\ 1 \end{bmatrix}$$

The first two columns of A are linearly independent because

$$\begin{vmatrix} 1 & -1 \\ 1 & 1 \end{vmatrix} = 2 \neq 0$$

Thus $r(A) = 2$. Also, $r(A) \leq r(B) \leq 2$ (why?). Therefore $r(A) = r(B)$ and the linear system has a solution.   □

We end this section by pointing out an important arithmetic rule that works for numbers but not for matrices. Given a pair of numbers a and b, the commutative law

$$ab = ba \tag{7.1.27}$$

holds. The analog $AB = BA$ for matrices is generally false. First, A and B must have the same order n. But even then we generally have $AB \neq BA$.

Example 7.1.10.
Consider the matrices

$$A = \begin{bmatrix} 1 & 2 & 0 \\ -1 & 0 & 1 \\ 0 & 3 & 0 \end{bmatrix}, \quad B = \begin{bmatrix} 1 & 0 & 0 \\ 0 & 0 & 1 \\ 0 & 1 & 0 \end{bmatrix}$$

Then

$$AB = \begin{bmatrix} 1 & 0 & 2 \\ -1 & 1 & 0 \\ 0 & 0 & 3 \end{bmatrix}, \quad BA = \begin{bmatrix} 1 & 2 & 0 \\ 0 & 3 & 0 \\ -1 & 0 & 1 \end{bmatrix}$$

and $AB \neq BA$. $\qquad\square$

PROBLEMS

1. For a square matrix A, define

$$A^1 = A$$
$$A^{n+1} = A^n A, \quad n \geq 1$$

Calculate A^2, A^3 for the following matrices:

(a) $\begin{bmatrix} 1 & 0 \\ 2 & 1 \end{bmatrix}$

(b) $\begin{bmatrix} 1 & -1 & 0 \\ 1 & 2 & 0 \\ -1 & 0 & 3 \end{bmatrix}$

(c) $\begin{bmatrix} 0 & 1 & 0 \\ 1 & 0 & 0 \\ 0 & 0 & 1 \end{bmatrix}$

2. (a) Find the values of a for which the linear system

$$a^2 x_1 = 1$$
$$x_1 + a x_2 + x_3 = 0$$
$$a x_1 - x_2 + x_3 = 2$$

has a unique solution.

(b) Denote the matrix of the system by A and solve $\det[A(a)] = 0$.

(c) Let $\det[A(a)] = 0$ for some particular a_0. What can be said about the solutions of the system for $a = a_0$?

3. A matrix A is said to be *symmetric* if $A^T = A$ and antisymmetric if $A^T = -A$. Show that

   (a) A symmetric matrix must be square.

   (b) Every square matrix A can be expressed as $A = B + C$, where B and C are symmetric and antisymmetric matrices, respectively.

4. A *scalar* matrix is defined to be a square matrix $A = (a_{ij})$ that satisfies

$$a_{ii} = c(\text{constant}), \quad 1 \le i \le n$$
$$a_{ij} = 0, \quad i \ne j$$

   or simply $A = cI$. Show that

   (a) If A is a scalar matrix, then $AB = BA$ for any given square matrix B.

   (b) If $AB = BA$ for all B, then A is scalar.

5. What conditions must be satisfied by a, b, c, and d so that the matrix

$$A = \begin{bmatrix} a & b \\ c & d \end{bmatrix}$$

   will commute with

$$B = \begin{bmatrix} 1 & -1 \\ 1 & 2 \end{bmatrix}$$

6. A square matrix $A = (a_{ij})$ of order n, which satisfies

$$a_{ij} = 0, \quad i > j$$

   is called *upper triangular*. Show that the sum and the product of such matrices are also upper triangular.

7. Let B be an inverse of A, that is,

$$AB = BA = I$$

   Show that B is unique. [*Hint*: Assume two inverses B_1 and B_2 and use Eq. (7.1.11) to obtain $B_1 = B_2$.]

8. Consider the system $Ax = 0$, where A is square and $\det(A) = 0$. By using Theorem 7.1.2 we must have more than one solution. Therefore the given system has a nontrivial solution $x \ne 0$. Find such an x for the system

$$x_1 - 2x_2 + 2x_3 = 0$$
$$x_1 + x_2 + 5x_3 = 0$$
$$2x_1 - x_2 + 7x_3 = 0$$

9. Let

$$C = \begin{bmatrix} c_1 \\ c_2 \\ \vdots \\ c_n \end{bmatrix}$$

be a column matrix for which $C^T C = 1$. Define

$$A = CC^T$$

and show that A is a square matrix such that $A^2 = A$.

10. Solve the equation

$$\det(\lambda I - A) = 0$$

for

$$A = \begin{bmatrix} 1 & 2 \\ 2 & 1 \end{bmatrix}$$

For each λ use the discussion of problem 8 and find a vector $x \neq 0$ such that

$$(\lambda I - A)x = 0$$

The number λ is called an *eigenvalue* of A, and x is an *eigenvector* of A associated with λ.

7.2 Gaussian Elimination

If a given matrix of a system of linear equations has a particular structure such as being upper triangular or symmetric, a special algorithm for solving the system may be designed. However, for a general system the most useful and efficient approach is the Gaussian elimination method. We will demonstrate this technique by solving a specific system of three linear equations. Consider the set

$$\begin{aligned} E_1 &: -x_1 + x_2 - 2x_3 = 1 \\ E_2 &: 3x_1 + 4x_3 = 0 \\ E_3 &: 2x_1 - x_2 + x_3 = 3 \end{aligned} \tag{7.2.1}$$

Step 1. Eliminate x_1 from equations E_2 and E_3 by adding $3(E_1)$ to E_2 and $2(E_1)$ to E_3, that is,

$$E_2 \leftarrow E_2 + 3E_1, \quad E_3 \leftarrow E_3 + 2E_1 \tag{7.2.2}$$

The new set of equations

$$E_1: -x_1 + x_2 - 2x_3 = 1$$
$$E_2: 3x_2 - 2x_3 = 3$$
$$E_3: x_2 - 3x_3 = 5 \qquad\qquad (7.2.3)$$

is equivalent to the original set, that is, both have the same solution(s).

Step 2. Eliminate x_2 from E_3 by adding $(-1/3)(E_2)$ to E_3, that is,

$$E_3 \leftarrow E_3 + (-1/3)(E_2) \qquad\qquad (7.2.4)$$

The new equivalent system is given by

$$E_1: -x_1 + x_2 - 2x_3 = 1$$
$$E_2: 3x_2 - 2x_3 = 3$$
$$E_3: -(7/3)x_3 = 4 \qquad\qquad (7.2.5)$$

Step 3. Perform *back substitution* on the final set given by Eq. (7.2.5) to obtain x_3, x_2, x_1 (in this order):

$$x_3 = \frac{4}{(-7/3)} = -\frac{12}{7}$$
$$x_2 = \frac{3 + 2x_3}{3} = \frac{3 + 2(-12/7)}{3} = -1/7$$
$$x_1 = x_2 - 2x_3 - 1 = \left(-\frac{1}{7}\right) - 2\left(-\frac{12}{7}\right) - 1 = \frac{16}{7} \qquad (7.2.6)$$

The first two steps are the *elimination* steps and the entire scheme is called *Gaussian elimination*.

We will now outline the algorithm to a system of n linear equations with n variables, provided that the matrix A of the system is nonsingular. Consider the system

$$E_1: a_{11}^{(1)}x_1 + a_{12}^{(1)}x_2 + \cdots + a_{1n}^{(1)}x_n = b_1^{(1)}$$
$$E_2: a_{21}^{(1)}x_1 + a_{22}^{(1)}x_2 + \cdots + a_{2n}^{(1)}x_n = b_2^{(1)}$$
$$\vdots$$
$$E_n: a_{n1}^{(1)}x_1 + a_{n2}^{(1)}x_2 + \cdots + a_{nn}^{(1)}x_n = b_n^{(1)} \qquad (7.2.7)$$

The upper index 1 at each coefficient is introduced for the sake of uniformity in the presentation. We now assume $a_{11}^{(1)} \neq 0$. Otherwise we search through $\{a_{i1}^{(1)}\}_{i=1}^{n}$ until we locate the first i_0 for which $a_{i_0 1}^{(1)} \neq 0$. We then interchange E_1 and E_{i_0}. A failure to find such an i_0 contradicts the nonsingularity of A. Now perform the first step: eliminate x_1 from E_i, $2 \leq i \leq n$ by adding $[-a_{i1}^{(1)}/a_{11}^{(1)}](E_1)$ to E_i, that is,

$$E_i \leftarrow E_i + \left[-\frac{a_{i1}^{(1)}}{a_{11}^{(1)}}\right]E_1 \qquad\qquad (7.2.8)$$

The result is an equivalent set of equations

$$E_1: \quad a_{11}^{(1)}x_1 + a_{12}^{(1)}x_2 + \cdots + a_{1n}^{(1)}x_n = b_1^{(1)}$$

$$E_2: \qquad\qquad a_{22}^{(2)}x_2 + \cdots + a_{2n}^{(2)}x_n = b_2^{(2)}$$

$$\vdots$$

$$E_n: \qquad\qquad a_{n2}^{(2)}x_2 + \cdots + a_{nn}^{(2)}x_n = b_n^{(2)} \qquad (7.2.9)$$

After $(k-1)$ steps $(k \leq 2)$ we find

$$E_1: \quad a_{11}^{(1)}x_1 + a_{12}^{(1)}x_2 + \cdots + a_{1n}^{(1)}x_n = b_1^{(1)}$$

$$E_2: \qquad\qquad a_{22}^{(2)}x_2 + \cdots + a_{2n}^{(2)}x_n = b_2^{(2)}$$

$$\vdots$$

$$E_k: \qquad\qquad a_{kk}^{(k)}x_k + \cdots + a_{kn}^{(k)}x_n = b_k^{(k)}$$

$$\vdots$$

$$E_n: \qquad\qquad a_{nk}^{(k)}x_k + \cdots + a_{nn}^{(k)}x_n = b_n^{(k)} \qquad (7.2.10)$$

which is a new set of equations equivalent to the original set given by Eq. (7.2.7). Because the new matrix is also nonsingular, at least one coefficient from the set $\{a_{ik}^{(k)}, k \leq i \leq n\}$ does not vanish, and we may assume, for example, $a_{kk}^{(k)} \neq 0$.

Step k. Eliminate x_k from $E_i, k+1 \leq i \leq n$ by defining the multipliers

$$m_{ik} = a_{ik}^{(k)}/a_{kk}^{(k)}, \quad k+1 \leq i \leq n \qquad (7.2.11)$$

and replacing

$$E_i \leftarrow E_i - m_{ik}(E_k), \quad k+1 \leq i \leq n \qquad (7.2.12)$$

After exactly $(n-1)$ steps we have a set of equations in an upper triangular form:

$$E_1: \quad a_{11}^{(1)} x_1 + a_{12}^{(1)} x_2 + \cdots + a_{1n}^{(1)} x_n = b_1^{(1)}$$

$$E_2: \qquad\qquad a_{22}^{(2)} x_2 + \cdots + a_{2n}^{(2)} x_n = b_2^{(2)}$$

$$\vdots$$

$$E_n: \qquad\qquad\qquad\qquad\qquad a_{nn}^{(n)} x_n = b_n^{(n)} \qquad (7.2.13)$$

which satisfies $a_{kk}^{(k)} \neq 0, 1 \leq k \leq n$.

Step n. We obtain $x_n, x_{n-1}, \ldots, x_1$ by back substitution:

$$x_n = b_n^{(n)} / a_{nn}^{(n)}$$

$$x_{n-1} = \frac{b_{n-1}^{(n-1)} - a_{n-1,n}^{(n-1)} x_n}{a_{n-1,n-1}^{(n-1)}}$$

$$\vdots$$

$$x_1 = \frac{b_1^{(1)} - \sum_{k=2}^{n} a_{1k}^{(1)} x_k}{a_{11}^{(1)}} \qquad (7.2.14)$$

Throughout the Gaussian elimination, the coefficients change while the unknowns remain the same and serve only as position pointers. Thus we can carry out the previous process using just the coefficients a_{ij} and b_i.

Example 7.2.1.

Consider the system

$$E_1: \; x_2 - 2x_3 + x_4 = 1$$
$$E_2: \; x_1 + x_2 - x_4 = 0$$
$$E_3: \; 2x_1 - 2x_2 + x_3 + 2x_4 = 3$$
$$E_4: \; x_2 - x_3 = 4$$

The augmented matrix on which the Gaussian elimination process is first applied is

$$B = (A, b) = \begin{bmatrix} 0 & 1 & -2 & 1 & 1 \\ 1 & 1 & 0 & -1 & 0 \\ 2 & -2 & 1 & 2 & 3 \\ 0 & 1 & -1 & 0 & 4 \end{bmatrix}$$

Because $a_{11}^{(1)} = 0$, we interchange rows 1 and 2 to obtain

$$B^1 = \begin{bmatrix} 1 & 1 & 0 & -1 & 0 \\ 0 & 1 & -2 & 1 & 1 \\ 2 & -2 & 1 & 2 & 3 \\ 0 & 1 & -1 & 0 & 4 \end{bmatrix}$$

Next we form the first set of multipliers $m_{i1}, 2 \leq i \leq 4$, and eliminate the unknown x_1 from $E_i, 2 \leq i \leq 4$:

$$\begin{matrix} m_{21} = 0 \\ m_{31} = 2 \\ m_{41} = 0 \end{matrix} \begin{bmatrix} 1 & 1 & 0 & -1 & 0 \\ 0 & 1 & -2 & 1 & 1 \\ 2 & -2 & 1 & 2 & 3 \\ 0 & 1 & -1 & 0 & 4 \end{bmatrix} \rightarrow \begin{bmatrix} 1 & 1 & 0 & -1 & 0 \\ 0 & 1 & -2 & 1 & 1 \\ 0 & -4 & 1 & 4 & 3 \\ 0 & 1 & -1 & 0 & 4 \end{bmatrix}$$

The elimination of x_2 from $E_i, 3 \leq i \leq 4$ is given by

$$\begin{matrix} \\ \\ m_{32} = -4 \\ m_{42} = 1 \end{matrix} \begin{bmatrix} 1 & 1 & 0 & -1 & 0 \\ 0 & 1 & -2 & 1 & 1 \\ 0 & -4 & 1 & 4 & 3 \\ 0 & 1 & -1 & 0 & 4 \end{bmatrix} \rightarrow \begin{bmatrix} 1 & 1 & 0 & -1 & 0 \\ 0 & 1 & -2 & 1 & 1 \\ 0 & 0 & -7 & 8 & 7 \\ 0 & 0 & 1 & -1 & 3 \end{bmatrix}$$

followed by

$$\begin{matrix} \\ \\ \\ m_{43} = -1/7 \end{matrix} \begin{bmatrix} 1 & 1 & 0 & -1 & 0 \\ 0 & 1 & -2 & 1 & 1 \\ 0 & 0 & -7 & 8 & 7 \\ 0 & 0 & 1 & -1 & 3 \end{bmatrix} \rightarrow \begin{bmatrix} 1 & 1 & 0 & -1 & 0 \\ 0 & 1 & -2 & 1 & 1 \\ 0 & 0 & -7 & 8 & 7 \\ 0 & 0 & 0 & 1/7 & 4 \end{bmatrix}$$

Finally, we perform back substitution to obtain the numbers

$$x_4 = \frac{4}{(1/7)} = 28$$
$$x_3 = \frac{7 - 8x_4}{(-7)} = 31$$
$$x_2 = 1 - x_4 - (-2)x_3 = 35$$
$$x_1 = -(-1)x_4 - x_2 = -7$$

that uniquely solve the original set of equations as well as the upper diagonal system. $\qquad\square$

At each step of the Gaussian elimination we impose the condition $a_{kk}^{(k)} \neq 0$. Otherwise we interchange rows. Apparently, due to round-off errors, this is not always efficient and the original process must by modified.

7.2.1   Pivoting

Consider the following set of linear equations:

$$x_1 + (1/3)x_2 - x_3 = 1$$
$$2x_1 + (2/3)x_2 + x_3 = 3$$
$$x_1 - x_3 = 0$$

The exact solution is given by

$$x_1 = 1/3, \quad x_2 = 3, \quad x_3 = 1/3$$

Let us use a rounding decimal computer with floating point representation and a four-digit mantissa. The Gaussian elimination process yields

$$
\begin{aligned}
m_{21} &= 2.000 \\
m_{31} &= 1.000
\end{aligned}
\qquad
\begin{bmatrix}
1.000 & 0.3333 & -1.000 & 1.000 \\
2.000 & 0.6667 & 1.000 & 3.000 \\
1.000 & 0 & -1.000 & 0
\end{bmatrix}
$$

$$
\longrightarrow
\begin{bmatrix}
1.000 & 0.3333 & -1.000 & 1.000 \\
0 & 0.0001 & 3.000 & 1.000 \\
0 & -0.3333 & 0 & -1.000
\end{bmatrix}
$$

$$
\longrightarrow
\begin{bmatrix}
1.000 & 0.3333 & -1.000 & 1.000 \\
0 & 0.0001 & 3.000 & 1.000 \\
0 & 0 & 9999 & 3332
\end{bmatrix}
$$

and by using back substitution we find

$$x_3 = 0.3332, \quad x_2 = 4.000, \quad x_1 = 0$$

which is clearly a poor approximation. The reason is that the element in position $(2,2)$ of the second augmented matrix is, due to rounding error, 0.0001 rather than 0. Thus it has an infinite relative error that is carried into every computation that involves this element. To bypass this difficulty we interchange rows 2 and 3 in the second matrix, before eliminating the unknown x_2:

$$
m_{32} = 0.0003
\qquad
\begin{bmatrix}
1.000 & 0.3333 & -1.000 & 1.000 \\
0 & -0.3333 & 0 & -1.000 \\
0 & 0.0001 & 3.000 & 1.000
\end{bmatrix}
$$

$$
\longrightarrow
\begin{bmatrix}
1.000 & 0.3333 & -1.000 & 1.000 \\
0 & -0.3333 & 0 & -1.000 \\
0 & 0 & 3.000 & 0.9997
\end{bmatrix}
$$

Now we find

$$x_3 = 0.3332, \quad x_2 = 3.000, \quad x_1 = 0.3331$$

which is in excellent agreement with the exact solution.

The strategy to avoid obtaining $a_{kk}^{(k)} \approx 0$ is, at each step k, to calculate

$$p = \max_{k \leq i \leq n} \left| a_{ik}^{(k)} \right| = \left| a_{i(k),k}^{(k)} \right| \tag{7.2.15}$$

and then interchange rows k and $i(k)$. Thus the final value at position (k, k) is the largest coefficient in absolute value among $\{a_{kk}^{(k)}, a_{k+1,k}^{(k)}, \ldots, a_{nk}^{(k)}\}$. This element is called the *pivot element* at step k of the Gaussian elimination. The process itself is called *partial pivoting* or simply *pivoting*. It generally prevents having a very small $a_{kk}^{(k)}$ and, because now $|m_{ik}| \leq 1$, the loss of significant digits throughout the computations is reduced.

In some cases, the lack of pivoting may lead to an extreme deviation from the exact solution, as demonstrated in the previous example. Generally, partial pivoting simply *improves* the accuracy of the solution.

We will now estimate the computing cost of the Gaussian elimination method, by performing an operations count.

7.2.2 Operations Count

Consider an elimination step $k, 1 \leq k \leq n - 1$. Let AS_k denote the number of additions and subtractions and MD_k denote the number of multiplications and divisions that are performed during this step. If we ignore the search for a pivot, the computation is composed of $(n - k)(n - k + 1)$ multiplications, $(n - k)$ divisions, and $(n - k)(n - k + 1)$ additions and subtractions. Thus,

$$(\text{AS})_k = (n - k)(n - k + 1), \quad 1 \leq k \leq n - 1$$
$$(\text{MD})_k = (n - k + 2)(n - k), \quad 1 \leq k \leq n - 1 \tag{7.2.16}$$

Let A' denote the upper triangular matrix to which A is converted and b' denote the final right-hand column that satisfies $A'x = b'$. Then the operations count for $(A, b) \rightarrow (A', b')$ is

$$\text{AS} = \sum_{k=1}^{n-1} \text{AS}_k = \sum_{k=1}^{n-1} (n - k)(n - k + 1) = \frac{n(n^2 - 1)}{3}$$

$$\text{MD} = \sum_{k=1}^{n-1} \text{MD}_k = \sum_{k=1}^{n-1} (n - k + 2)(n - k) = \frac{n(n - 1)(2n + 5)}{6} \tag{7.2.17}$$

The back substitution step consists of $[n(n - 1)]/2$ additions and subtractions and $[n(n + 1)]/2$ multiplications and divisions. Thus the total number of operations is

$$\text{AS}(x) = \text{AS} + \frac{n(n - 1)}{2} = \frac{n(n - 1)(2n + 5)}{6}$$

$$\text{MD}(x) = \text{MD} + \frac{n(n+1)}{2} = \frac{n(n^2 + 3n - 1)}{3} \tag{7.2.18}$$

and for large n

$$\text{AS}(x) \approx n^3/3, \quad \text{MD}(x) \approx n^3/3 \tag{7.2.19}$$

Most of the computations are needed to obtain A' and b', because

$$\text{AS} \approx n^3/3, \quad \text{MD} \approx n^3/3, \quad n \to \infty$$

as well.

A subroutine GELIM, which solves a system $Ax = b$ by using Gaussian elimination with partial pivoting, is found on the attached floppy disk.

7.2.3 The Inverse Matrix

When solving a system $Ax = b$ it is sometimes necessary to calculate the inverse matrix A^{-1} as well. This can be done straightforwardly by using Gaussian elimination. Consider the system

$$Ax = b$$

and let Y denote the inverse matrix of A, that is,

$$AY = YA = I_n$$

Denote the columns of Y by $y_1, \ldots, y_n$ and those of I_n by $e_1, \ldots, e_n$. Thus

$$y_1 = \begin{bmatrix} y_{11} \\ \vdots \\ y_{n1} \end{bmatrix}, \ldots, \quad y_n = \begin{bmatrix} y_{1n} \\ \vdots \\ y_{nn} \end{bmatrix}$$

and

$$e_1 = \begin{bmatrix} 1 \\ 0 \\ \vdots \\ 0 \end{bmatrix}, \ldots, \quad e_n = \begin{bmatrix} 0 \\ \vdots \\ 0 \\ 1 \end{bmatrix}$$

and

$$\begin{bmatrix} a_{11} & \cdots & a_{1n} \\ \vdots & & \vdots \\ a_{n1} & \cdots & a_{nn} \end{bmatrix} \begin{bmatrix} y_{11} & \cdots & y_{1n} \\ \vdots & & \vdots \\ y_{n1} & \cdots & y_{nn} \end{bmatrix} = \begin{bmatrix} 1 & 0 & \cdot & \cdot & 0 \\ 0 & 1 & 0 & \cdot & \cdot \\ \cdot & 0 & \cdot & \cdot & \cdot \\ \cdot & \cdot & \cdot & \cdot & 0 \\ 0 & \cdot & \cdot & 0 & 1 \end{bmatrix} \tag{7.2.20}$$

The system given by Eq. (7.2.20) is composed of n sets of n linear equations each, namely

$$Ay_1 = e_1, \quad Ay_2 = e_2, \ldots, \quad Ay_n = e_n$$

These sets share the same matrix A and can be efficiently solved for $y_1, \ldots, y_n$ by Gaussian elimination, provided that all the right-hand columns are treated simultaneously.

Example 7.2.2.

Consider the matrix

$$A = \begin{bmatrix} 1 & -1 & 2 \\ 3 & 1 & 1 \\ -1 & 3 & 2 \end{bmatrix}$$

To find A^{-1} we implement Gaussian elimination on the augmented matrix

$$(A, I_3) = \begin{bmatrix} 1 & -1 & 2 & | & 1 & 0 & 0 \\ 3 & 1 & 1 & | & 0 & 1 & 0 \\ -1 & 3 & 2 & | & 0 & 0 & 1 \end{bmatrix}$$

Step 1.a. Partial pivoting:

$$\begin{bmatrix} 1 & -1 & 2 & | & 1 & 0 & 0 \\ 3 & 1 & 1 & | & 0 & 1 & 0 \\ -1 & 3 & 2 & | & 0 & 0 & 1 \end{bmatrix} \rightarrow \begin{bmatrix} 3 & 1 & 1 & | & 0 & 1 & 0 \\ 1 & -1 & 2 & | & 1 & 0 & 0 \\ -1 & 3 & 2 & | & 0 & 0 & 1 \end{bmatrix}$$

Step 1.b. Elimination:

$$\begin{matrix} \\ m_{21} = 1/3 \\ m_{31} = -1/3 \end{matrix} \begin{bmatrix} 3 & 1 & 1 & | & 0 & 1 & 0 \\ 1 & -1 & 2 & | & 1 & 0 & 0 \\ -1 & 3 & 2 & | & 0 & 0 & 1 \end{bmatrix} \rightarrow \begin{bmatrix} 3 & 1 & 1 & | & 0 & 1 & 0 \\ 0 & -4/3 & 5/3 & | & 1 & -1/3 & 0 \\ 0 & 10/3 & 7/3 & | & 0 & 1/3 & 1 \end{bmatrix}$$

Step 2.a. Partial pivoting:

$$\begin{bmatrix} 3 & 1 & 1 & | & 0 & 1 & 0 \\ 0 & -4/3 & 5/3 & | & 1 & -1/3 & 0 \\ 0 & 10/3 & 7/3 & | & 0 & 1/3 & 1 \end{bmatrix} \rightarrow \begin{bmatrix} 3 & 1 & 1 & | & 0 & 1 & 0 \\ 0 & 10/3 & 7/3 & | & 0 & 1/3 & 1 \\ 0 & -4/3 & 5/3 & | & 1 & -1/3 & 0 \end{bmatrix}$$

Step 2.b. Elimination:

$$\begin{matrix} \\ \\ m_{32} = -2/5 \end{matrix} \begin{bmatrix} 3 & 1 & 1 & | & 0 & 1 & 0 \\ 0 & 10/3 & 7/3 & | & 0 & 1/3 & 1 \\ 0 & -4/3 & 5/3 & | & 1 & -1/3 & 0 \end{bmatrix} \rightarrow \begin{bmatrix} 3 & 1 & 1 & | & 0 & 1 & 0 \\ 0 & 10/3 & 7/3 & | & 0 & 1/3 & 1 \\ 0 & 0 & 13/5 & | & 1 & -1/5 & 2/5 \end{bmatrix}$$

Step 3. Back substitution: the three upper triangular systems and their corresponding solutions are

1.

$$\begin{bmatrix} 3 & 1 & 1 \\ 0 & 10/3 & 7/3 \\ 0 & 0 & 13/5 \end{bmatrix} \begin{bmatrix} y_{11} \\ y_{21} \\ y_{31} \end{bmatrix} = \begin{bmatrix} 0 \\ 0 \\ 1 \end{bmatrix} \quad y_{31} = 5/13, \quad y_{21} = -7/26, \quad y_{11} = -1/26$$

2.

$$\begin{bmatrix} 3 & 1 & 1 \\ 0 & 10/3 & 7/3 \\ 0 & 0 & 13/5 \end{bmatrix} \begin{bmatrix} y_{12} \\ y_{22} \\ y_{32} \end{bmatrix} = \begin{bmatrix} 1 \\ 1/3 \\ -1/5 \end{bmatrix} \quad y_{32} = -1/13, \quad y_{22} = 2/13, \quad y_{12} = 4/13$$

3.

$$\begin{bmatrix} 3 & 1 & 1 \\ 0 & 10/3 & 7/3 \\ 0 & 0 & 13/5 \end{bmatrix} \begin{bmatrix} y_{13} \\ y_{23} \\ y_{33} \end{bmatrix} = \begin{bmatrix} 0 \\ 1 \\ 2/5 \end{bmatrix} \quad y_{33} = 2/13, \quad y_{23} = 5/26, \quad y_{13} = -3/26$$

Therefore the inverse matrix of A is

$$A^{-1} = \begin{bmatrix} -1/26 & 4/13 & -3/26 \\ -7/26 & 2/13 & 5/26 \\ 5/13 & -1/13 & 2/13 \end{bmatrix} = 1/26 \begin{bmatrix} -1 & 8 & -3 \\ -7 & 4 & 5 \\ 10 & -2 & 4 \end{bmatrix} \qquad \square$$

PROBLEMS

1. Solve the following linear systems using Gaussian elimination without pivoting:

(a)

$$\begin{bmatrix} 1 & 0 & 4 & 1 \\ 3 & 2 & -1 & 0 \\ -1 & 3 & 0 & 2 \\ 1 & 1 & 1 & -1 \end{bmatrix} \begin{bmatrix} x_1 \\ x_2 \\ x_3 \\ x_4 \end{bmatrix} = \begin{bmatrix} -1 \\ 0 \\ 1 \\ 1 \end{bmatrix}$$

(b)

$$\begin{bmatrix} 1 & 2 & 3 \\ 1 & -1 & 0 \\ 3 & 0 & 3 \end{bmatrix} \begin{bmatrix} x_1 \\ x_2 \\ x_3 \end{bmatrix} = \begin{bmatrix} 2 \\ 0 \\ 1 \end{bmatrix}$$

(c)

$$\begin{bmatrix} 1 & 0 & 1 & 0 \\ 0 & 1 & 0 & 1 \\ -1 & 2 & -1 & 3 \\ 1 & 1 & -1 & 0 \end{bmatrix} \begin{bmatrix} x_1 \\ x_2 \\ x_3 \\ x_4 \end{bmatrix} = \begin{bmatrix} 1 \\ 1 \\ 0 \\ 0 \end{bmatrix}$$

2. Solve the following systems using Gaussian elimination, with and without partial pivoting. Assume a rounding machine with four-digit floating-point arithmetic.

(a)

$$0.101x_1 - x_2 + x_3 = 1.5$$
$$2x_1 + x_2 - 0.17x_3 = 0$$
$$17.53x_1 - 4x_2 + 3x_3 = 1$$

(b)

$$0.3x_1 + 0.8x_2 - 0.3x_3 = 1$$
$$11.3x_1 + x_3 = 2$$
$$0.05x_1 + x_2 - 14x_3 = 0$$

3. At each elimination step k of the Gaussian elimination method, we eliminate x_k from equations 1 to $(k - 1)$ as well as from equations $(k + 1)$ to n. The result is the *Gauss–Jordan* reduction. The final set of equations is of the form

$$a_{11}^{(1)}x_1 = b_1'$$
$$a_{22}^{(2)}x_2 = b_2'$$
$$\vdots$$
$$a_{nn}^{(n)}x_n = b_n' \qquad\qquad (7.2.21)$$

with the unique solution $x_i = b_i'/a_{ii}^{(i)}, 1 \le i \le n$. Find the total number of operations needed to obtain the solution and compare with that of the Gaussian elimination procedure.

4. Let A, B be nonsingular $n \times n$ matrices. Show that

$$(AB)^{-1} = B^{-1}A^{-1}$$

5. Consider the two nonsingular matrices

$$A = \begin{bmatrix} 1 & 0 & 1 \\ 2 & -1 & 5 \\ 0 & 1 & 1 \end{bmatrix}, \quad B = \begin{bmatrix} 7 & 1 & 1 \\ -1 & 0 & 2 \\ 6 & 1 & 4 \end{bmatrix}$$

(a) Calculate A^{-1} and B^{-1}.

(b) Calculate $(AB)^{-1}$ directly and also by using the result of problem 4.

6. Solve the linear system given by

$$A = \begin{bmatrix} 1 & 0 & -1 & 2 \\ 3 & 0.5 & 1 & 1 \\ -1 & 3 & 0 & 1 \\ 2 & 1 & 1 & 0 \end{bmatrix}, \quad b = \begin{bmatrix} -1 \\ -1 \\ 1 \\ 1 \end{bmatrix}$$

and calculate A^{-1} as well, using Gaussian elimination.

7. Find the inverse matrix in the following cases:
   (a)

$$\begin{bmatrix} a & b \\ c & d \end{bmatrix}$$

   (b)

$$\begin{bmatrix} a & b & -1 \\ b & c & 0 \\ 1 & 0 & 1 \end{bmatrix}$$

7.3 *LU* Factorization

Consider a set of linear equations $Ax = b$. By using Gaussian elimination, the system is reduced to an upper triangular form given by

$$Ux = f \tag{7.3.1}$$

where

$$U = \begin{bmatrix} u_{11} & & \cdots & u_{1n} \\ 0 & & & \\ \vdots & \ddots & \ddots & \vdots \\ 0 & \cdots & 0 & u_{nn} \end{bmatrix}, \quad f = \begin{bmatrix} f_1 \\ \vdots \\ f_n \end{bmatrix}$$

are given by

$$u_{ij} = a_{ij}^{(i)}, \quad f_i = b_i^{(i)} \tag{7.3.2}$$

Suppose that during the elimination process pivoting is not considered and interchanging rows is not necessary. Define the lower triangular matrix as

$$L = \begin{bmatrix} 1 & 0 & \cdots & 0 \\ m_{21} & 1 & \ddots & \vdots \\ \vdots & & \ddots & 0 \\ m_{n1} & \cdots & m_{n,n-1} & 1 \end{bmatrix} \tag{7.3.3}$$

Then the relation between matrices U and L and matrix A is given by the following theorem.

Theorem 7.3.1.

(Factorization of a nonsingular matrix). *Let A be a nonsingular matrix, L and U be defined by Eqs. (7.3.2) and (7.3.3), and U be produced without interchanging rows. Then*

$$LU = A \qquad (7.3.4)$$

Equation (7.3.4) is called the LU factorization of A.
    The proof is left as an exercise for the reader.

Example 7.3.1.

Consider the linear system given in Example 7.2.1. The upper triangular matrix is

$$U = \begin{bmatrix} 1 & 1 & 0 & -1 \\ 0 & 1 & -2 & 1 \\ 0 & 0 & -7 & 8 \\ 0 & 0 & 0 & 1/7 \end{bmatrix}$$

the lower triangular matrix is

$$L = \begin{bmatrix} 1 & 0 & 0 & 0 \\ 0 & 1 & 0 & 0 \\ 2 & -4 & 1 & 0 \\ 0 & 1 & -1/7 & 1 \end{bmatrix}$$

and their multiplication is

$$LU = \begin{bmatrix} 1 & 0 & 0 & 0 \\ 0 & 1 & 0 & 0 \\ 2 & -4 & 1 & 0 \\ 0 & 1 & -1/7 & 1 \end{bmatrix} \begin{bmatrix} 1 & 1 & 0 & -1 \\ 0 & 1 & -2 & 1 \\ 0 & 0 & -7 & 8 \\ 0 & 0 & 0 & 1/7 \end{bmatrix} = \begin{bmatrix} 1 & 1 & 0 & -1 \\ 0 & 1 & -2 & 1 \\ 2 & -2 & 1 & 2 \\ 0 & 1 & -1 & 0 \end{bmatrix}$$

which is the original coefficient matrix with rows 1 and 2 interchanged.   □
From Eq. (7.3.4) it is clear that solving the system $Ax = b$ is equivalent to solving

$$LUx = b \qquad (7.3.5)$$

or to solving the two triangular systems

$$Ly = b, \quad Ux = y \qquad (7.3.6)$$

The second system, $Ux = y$, is an upper triangular system and is solved by back substitution. The first is a lower triangular system given by

$$
\begin{array}{llll}
y_1 & & & = b_1 \\
m_{21}y_1 + & y_2 & & = b_2 \\
\vdots & & & \vdots \\
m_{n1}y_1 + m_{n2}y_2 + \cdots + m_{n,n-1}y_{n-1} + y_n & = b_n
\end{array}
\tag{7.3.7}
$$

and can be solved by forward substitution. Thus, once the factorization of A is known, solving $Ax = b$ is equivalent to solving two triangular systems.

Example 7.3.2.

Let U and L be the triangular matrices of Example 7.2.1. The first system is

$$
\begin{bmatrix}
1 & 0 & 0 & 0 \\
0 & 1 & 0 & 0 \\
2 & -4 & 1 & 0 \\
0 & 1 & -1/7 & 1
\end{bmatrix}
\begin{bmatrix}
y_1 \\
y_2 \\
y_3 \\
y_4
\end{bmatrix}
=
\begin{bmatrix}
0 \\
1 \\
3 \\
4
\end{bmatrix}
$$

solved by the vector

$$
\begin{bmatrix}
y_1 \\
y_2 \\
y_3 \\
y_4
\end{bmatrix}
=
\begin{bmatrix}
0 \\
1 \\
7 \\
4
\end{bmatrix}
$$

The second system is

$$
\begin{bmatrix}
1 & 1 & 0 & -1 \\
0 & 1 & -2 & 1 \\
0 & 0 & -7 & 8 \\
0 & 0 & 0 & 1/7
\end{bmatrix}
\begin{bmatrix}
x_1 \\
x_2 \\
x_3 \\
x_4
\end{bmatrix}
=
\begin{bmatrix}
0 \\
1 \\
7 \\
4
\end{bmatrix}
$$

solved by

$$
\begin{bmatrix}
x_1 \\
x_2 \\
x_3 \\
x_4
\end{bmatrix}
=
\begin{bmatrix}
-7 \\
35 \\
31 \\
28
\end{bmatrix}
$$

in agreement with the results of the original example. $\qquad\square$

7.3.1 Doolittle's Method

Rather than constructing L and U during the Gaussian elimination process, we may try and compute them directly. Let us consider the general case for $n = 3$, provided that pivoting is not necessary.

The factorization $A = LU$ is given by

$$\begin{bmatrix} a_{11} & a_{12} & a_{13} \\ a_{21} & a_{22} & a_{23} \\ a_{31} & a_{32} & a_{33} \end{bmatrix} = \begin{bmatrix} 1 & 0 & 0 \\ m_{21} & 1 & 0 \\ m_{31} & m_{32} & 1 \end{bmatrix} \begin{bmatrix} u_{11} & u_{12} & u_{13} \\ 0 & u_{22} & u_{23} \\ 0 & 0 & u_{33} \end{bmatrix} \qquad (7.3.8)$$

We use Eq. (7.3.8) to obtain the rows of U and the columns of L alternately. We start by multiplying the first row of L by the three columns of U and matching a_{11}, a_{12}, and a_{13}:

$$a_{11} = u_{11}, \quad a_{12} = u_{12}, \quad a_{13} = u_{13} \qquad (7.3.9)$$

Next we match a_{21} and a_{31} and calculate the first column of L:

$$a_{21} = m_{21}u_{11}, \quad a_{31} = m_{31}u_{11} \qquad (7.3.10)$$

We now write

$$a_{22} = m_{21}u_{12} + u_{22}, \quad a_{23} = m_{21}u_{13} + u_{23} \qquad (7.3.11)$$

and determine u_{22} and u_{23}, that is, the second row of U. Next we calculate m_{32} from

$$a_{32} = m_{31}u_{12} + m_{32}u_{22} \qquad (7.3.12)$$

and finally u_{33} from

$$a_{33} = m_{31}u_{13} + m_{32}u_{23} + u_{33} \qquad (7.3.13)$$

The procedure given by Eqs. (7.3.9) through (7.3.13) is called *Doolittle's method*. Because pivoting is not necessary, a division by zero does not occur.

Example 7.3.3.

Consider the matrix

$$\begin{bmatrix} 1 & -3 & 0 \\ 2 & 1 & -1 \\ 1 & 4 & 2 \end{bmatrix}$$

From Eqs. (7.3.9) and (7.3.10) we obtain

$$u_{11} = 1, \quad u_{12} = -3, \quad u_{13} = 0, \quad m_{21} = a_{21}/u_{11} = 2, \quad m_{31} = a_{31}/u_{11} = 1$$

Substituting in Eq. (7.3.11) we obtain

$$u_{22} = a_{22} - m_{21}u_{12} = 7, \quad u_{23} = a_{23} - m_{21}u_{13} = -1$$

and finally, from Eqs. (7.3.12) and (7.3.13),

$$m_{32} = \frac{a_{32} - m_{31}u_{12}}{u_{22}} = 1, \quad u_{33} = a_{33} - m_{31}u_{13} - m_{32}u_{23} = 3$$

Thus the factorization of A is given by

$$\begin{bmatrix} 1 & -3 & 0 \\ 2 & 1 & -1 \\ 1 & 4 & 2 \end{bmatrix} = \begin{bmatrix} 1 & 0 & 0 \\ 2 & 1 & 0 \\ 1 & 1 & 1 \end{bmatrix} \begin{bmatrix} 1 & -3 & 0 \\ 0 & 7 & -1 \\ 0 & 0 & 3 \end{bmatrix}$$

$\square$

A successful direct computation of L and U requests that the computed u_{ii} satisfy

$$u_{ii} \neq 0, \quad 1 \leq i \leq n - 1$$

This is indeed the case if, during the Gaussian elimination, interchanging rows is not necessary. A special class of matrices in this category is given below.

Definition 7.3.1. *An $n \times n$ matrix A is said to be strictly diagonally dominant if*

$$|a_{ii}| > \sum_{\substack{j=1 \\ j \neq i}}^{n} |a_{ij}|, \quad 1 \leq i \leq n \tag{7.3.14}$$

Example 7.3.4.
The matrix

$$A = \begin{bmatrix} 7 & 3 & -3 \\ 3 & 4 & -1 \\ 1 & 2 & 10 \end{bmatrix}$$

is not strictly diagonally dominant, because

$$a_{22} = 4 = |a_{21}| + |a_{23}| = 4$$

in contrast to Eq. (7.3.14). On the other hand, the matrix

$$B = \begin{bmatrix} 1 & 0 & 0 & 0 \\ -1 & 3 & 0 & 1 \\ 7 & 0 & 8 & 0 \\ 1 & 2 & 3 & 7 \end{bmatrix}$$

satisfies Eq. (7.3.14) and is strictly diagonally dominant. $\square$

Theorem 7.3.2.
(Factorization of a strictly diagonally dominant matrix). *Let A be an $n \times n$ strictly diagonally dominant matrix. Then A is nonsingular and can be factorized using Doolittle's method.*

Proof.

If A is singular, the system $Ax = 0$ must have some nonzero solution x, given by $x^T = (x_1, x_2, \ldots, x_n)$. Therefore,

$$\sum_{j=1}^{n} a_{ij} x_j = 0, \quad 1 \leq i \leq n \tag{7.3.15}$$

Let k satisfy

$$|x_k| = \max |x_j|, \quad 1 \leq j \leq n \tag{7.3.16}$$

By using Eq. (7.3.15) we have in particular

$$\sum_{j=1}^{n} a_{kj} x_j = 0$$

which leads to

$$a_{kk} x_k = -\sum_{\substack{j=1 \\ j \neq k}}^{n} a_{kj} x_j \tag{7.3.17}$$

Because $x_k \neq 0$ we can divide both sides of Eq. (7.3.17) by x_k, implement Eq. (7.3.16) and the inequality of the triangle and obtain

$$|a_{kk}| \leq \sum_{\substack{j=1 \\ j \neq k}}^{n} |a_{kj}| \left| \frac{x_j}{x_k} \right| \leq \sum_{\substack{j=1 \\ j \neq k}}^{n} |a_{kj}| \tag{7.3.18}$$

Because Eq. (7.3.18) contradicts the strict diagonal dominance of A, we conclude that A is nonsingular. To show that interchanging rows during the Gaussian elimination is not necessary, we first apply

$$|a_{11}| > \sum_{i=2}^{n} |a_{1i}| \geq 0$$

to conclude that interchanging rows is not needed during the first step of the Gaussian elimination. Next, consider the matrix

$$A^{(2)} = \begin{bmatrix} a_{22}^{(2)} & a_{23}^{(2)} & \cdots & a_{2n}^{(2)} \\ \vdots & & & \vdots \\ a_{n2}^{(2)} & a_{n3}^{(2)} & \cdots & a_{nn}^{(2)} \end{bmatrix}$$

If $A^{(2)}$ can be shown to also be strictly diagonally dominant, then interchanging rows is not needed during the next elimination step, and by using induction we conclude that this holds throughout the

Gaussian elimination process. Because

$$a_{ij}^{(2)} = a_{ij} - \frac{a_{i1}}{a_{11}} a_{1j}, \quad 2 \le i, j \le n \tag{7.3.19}$$

we need to show that

$$\left| a_{ii} - \frac{a_{i1}}{a_{11}} a_{1i} \right| > \sum_{\substack{j=2 \\ j \ne i}}^{n} \left| a_{ij} - \frac{a_{i1}}{a_{11}} a_{1j} \right|, \quad 2 \le i \le n \tag{7.3.20}$$

The right-hand side of Eq. (7.3.20) satisfies

$$\sum_{\substack{j=2 \\ j \ne i}}^{n} \left| a_{ij} - \frac{a_{i1}}{a_{11}} a_{1j} \right| \le \sum_{\substack{j=2 \\ j \ne i}}^{n} |a_{ij}| + \sum_{\substack{j=2 \\ j \ne i}}^{n} |a_{i1}| \left| \frac{a_{1j}}{a_{11}} \right|$$

$$= \sum_{\substack{j=2 \\ j \ne i}}^{n} |a_{ij}| + |a_{i1}| \sum_{\substack{j=2 \\ j \ne i}}^{n} \left| \frac{a_{1j}}{a_{11}} \right| \le \sum_{\substack{j=2 \\ j \ne i}}^{n} |a_{ij}| + |a_{i1}| \left(1 - \left| \frac{a_{1i}}{a_{11}} \right| \right)$$

$$= \sum_{\substack{j=1 \\ j \ne i}}^{n} |a_{ij}| - |a_{i1}| \left| \frac{a_{1i}}{a_{11}} \right| < |a_{ii}| - |a_{i1}| \left| \frac{a_{1i}}{a_{11}} \right|$$

$$\le \left| a_{ii} - a_{i1} \frac{a_{1i}}{a_{11}} \right| = \left| a_{ii}^{(2)} \right|$$

which concludes the strict diagonal dominance of $A^{(2)}$.   □

Example 7.3.5.

Consider the matrix

$$A = \begin{bmatrix} 4 & 0 & 1 & -1 \\ 2 & 6 & 2 & 1 \\ -1 & 1 & 3 & 0 \\ 0 & 1 & 0 & 2 \end{bmatrix}$$

which is clearly strictly diagonally dominant. A direct computation of L and U is feasible, and the factorization

$$\begin{bmatrix} 4 & 0 & 1 & -1 \\ 2 & 6 & 2 & 1 \\ -1 & 1 & 3 & 0 \\ 0 & 1 & 0 & 2 \end{bmatrix} = \begin{bmatrix} 1 & 0 & 0 & 0 \\ m_{21} & 1 & 0 & 0 \\ m_{31} & m_{32} & 1 & 0 \\ m_{41} & m_{42} & m_{43} & 1 \end{bmatrix} \begin{bmatrix} u_{11} & u_{12} & u_{13} & u_{14} \\ 0 & u_{22} & u_{23} & u_{24} \\ 0 & 0 & u_{33} & u_{34} \\ 0 & 0 & 0 & u_{44} \end{bmatrix}$$

is carried out by calculating alternately the rows of U and the columns of L. The first row of U is given by

$$u_{11} = 4, \quad u_{12} = 0, \quad u_{13} = 1, \quad u_{14} = -1$$

and the first column of L is given by

$$m_{21} = a_{21}/u_{11} = 1/2, \quad m_{31} = a_{31}/u_{11} = -1/4, \quad m_{41} = a_{41}/u_{11} = 0$$

Next we determine the second row of U by solving

$$6 = m_{21}u_{12} + u_{22}, \quad 2 = m_{21}u_{13} + u_{23}, \quad 1 = m_{21}u_{14} + u_{24}$$

Thus

$$u_{22} = 6, \quad u_{23} = 3/2, \quad u_{24} = 3/2$$

The second column of L is calculated from

$$1 = m_{31}u_{12} + m_{32}u_{22}, \quad 1 = m_{41}u_{12} + m_{42}u_{22}$$

that is,

$$m_{32} = 1/6, \quad m_{42} = 1/6$$

Next we solve

$$3 = m_{31}u_{13} + m_{32}u_{23} + u_{33}, \quad 0 = m_{31}u_{14} + m_{32}u_{24} + u_{34}$$

to obtain the third row of U:

$$u_{33} = 3, \quad u_{34} = -1/2$$

The third column of L is determined by computing the single number m_{43}, which satisfies

$$0 = m_{41}u_{13} + m_{42}u_{23} + m_{43}u_{33}$$

that is, $m_{43} = -1/12$. Finally, we solve

$$2 = m_{41}u_{14} + m_{42}u_{24} + m_{43}u_{34} + u_{44}$$

and find that $u_{44} = 41/24$. The complete factorization is given by

$$\begin{bmatrix} 4 & 0 & 1 & -1 \\ 2 & 6 & 2 & 1 \\ -1 & 1 & 3 & 0 \\ 0 & 1 & 0 & 2 \end{bmatrix} = \begin{bmatrix} 1 & 0 & 0 & 0 \\ 1/2 & 1 & 0 & 0 \\ -1/4 & 1/6 & 1 & 0 \\ 0 & 1/6 & -1/12 & 1 \end{bmatrix} \begin{bmatrix} 4 & 0 & 1 & -1 \\ 0 & 6 & 3/2 & 3/2 \\ 0 & 0 & 3 & -1/2 \\ 0 & 0 & 0 & 41/24 \end{bmatrix}$$

Let us now solve the linear system $Ax = b$ for the particular case $b^T = (1, 1, 1, 2)$. We first determine the unique vector y for which $Ly = b$, that is,

$$
\begin{bmatrix}
1 & 0 & 0 & 0 \\
1/2 & 1 & 0 & 0 \\
-1/4 & 1/6 & 1 & 0 \\
0 & 1/6 & -1/12 & 1
\end{bmatrix}
\begin{bmatrix}
y_1 \\ y_2 \\ y_3 \\ y_4
\end{bmatrix}
=
\begin{bmatrix}
1 \\ 1 \\ 1 \\ 2
\end{bmatrix}
$$

The computed components of y are

$$
y_1 = 1, \quad y_2 = 1/2, \quad y_3 = 7/6, \quad y_4 = 145/72
$$

Next we solve $Ux = y$, that is,

$$
\begin{bmatrix}
4 & 0 & 1 & -1 \\
0 & 6 & 3/2 & 3/2 \\
0 & 0 & 3 & -1/2 \\
0 & 0 & 0 & 41/24
\end{bmatrix}
\begin{bmatrix}
x_1 \\ x_2 \\ x_3 \\ x_4
\end{bmatrix}
=
\begin{bmatrix}
1 \\ 1/2 \\ 7/6 \\ 145/72
\end{bmatrix}
$$

and obtain the unique solution of $Ax = b$,

$$
x_4 = 145/123, \quad x_3 = 24/41, \quad x_2 = -44/123, \quad x_1 = 49/123 \qquad \square
$$

The two procedures for computing L and U, that is, Gaussian elimination and the Doolittle method, are equivalent and need the same amount of arithmetic. However, the Doolittle method, slightly modified, significantly reduces the number of round-off errors throughout the computation process. For large n, the formulas for m_{ij} and u_{ij} involve many sums of many multiplications. Assume that all the numbers are stored in single-precision format, but that multiplications and their sum are computed in double-precision arithmetic. The sum is then rounded to a single-precision number. Consequently, this sum involves a single round-off error instead of many, and the total of $O(n^3)$ round-off errors produced by a standard Gaussian elimination is reduced to $O(n^2)$ round-off errors with the Doolittle method. For large values of n, the reduction is significant.

Another important feature of the Doolittle method is that while L and U are computed, new storage is not needed. The numbers u_{ij}, $j \geq i$ are stored instead of a_{ij}, $j \geq i$, and the numbers m_{ij}, $i > j$ are stored instead of a_{ij}, $i > j$.

7.3.2 Tridiagonal Systems

Definition 7.3.2. *A linear system $Ax = b$ is said to be tridiagonal if the coefficient matrix*

A is given by

$$A = \begin{bmatrix} b_1 & c_1 & 0 & 0 & \cdots & 0 \\ a_2 & b_2 & c_2 & 0 & \cdots & 0 \\ 0 & a_3 & b_3 & c_3 & \cdots & 0 \\ \vdots & & \ddots & \ddots & \ddots & \vdots \\ 0 & \vdots & & a_{n-1} & b_{n-1} & c_{n-1} \\ 0 & \vdots & & 0 & a_n & b_n \end{bmatrix} \qquad (7.3.21)$$

A is called a tridiagonal matrix. For example, the matrices

$$A = \begin{bmatrix} 2 & 1 & 0 & 0 \\ -1 & 0 & 2 & 0 \\ 0 & 0 & 1 & -1 \\ 0 & 0 & 2 & 3 \end{bmatrix}, \quad B = \begin{bmatrix} 1 & 3 \\ 2 & -1 \end{bmatrix}$$

are tridiagonal, whereas

$$C = \begin{bmatrix} 1 & 1 & 0 \\ -1 & 1 & 2 \\ 1 & 0 & 1 \end{bmatrix}$$

is not. Tridiagonal systems occur frequently, particularly when mathematical models involve differential equations. Because most of the A entries vanish, a similar behavior is expected from L and U, provided that an *LU* factorization is feasible.

Theorem 7.3.3.

(Factorization of a tridiagonal system). *Let A be a tridiagonal matrix and assume that the Gaussian elimination process, applied to A, does not require interchanging rows. Then $A = LU$, where*

$$L = \begin{bmatrix} 1 & 0 & \cdots & & 0 \\ \alpha_2 & 1 & & & \\ 0 & \alpha_3 & 1 & & \\ \vdots & & & & \vdots \\ & & \ddots & \ddots & \ddots & 0 \\ 0 & \cdots & & 0 & \alpha_n & 1 \end{bmatrix}, \quad U = \begin{bmatrix} \beta_1 & c_1 & 0 & \cdots & 0 \\ 0 & \beta_2 & c_2 & 0 & 0 \\ \vdots & \ddots & \ddots & \ddots & \vdots \\ & & & & c_{n-1} \\ 0 & & \cdots & & \beta_n \end{bmatrix} \qquad (7.3.22)$$

and

$$\begin{aligned} \beta_1 &= b_1 \\ \alpha_j \beta_{j-1} &= a_j, \quad \alpha_j c_{j-1} + \beta_j = b_j, \quad 2 \leq j \leq n \end{aligned} \qquad (7.3.23)$$

The proof is left as an exercise for the reader.

The solution of the system described by Eq. (7.3.23) is given by

$$
\begin{aligned}
\beta_1 &= b_1 \\
\alpha_j &= \frac{a_j}{\beta_{j-1}}, \ 2 \le j \le n \\
\beta_j &= b_j - \alpha_j c_{j-1}, \ 2 \le j \le n
\end{aligned}
\tag{7.3.24}
$$

Thus α_j and β_j are calculated alternately. To solve a tridiagonal system $Ax = f$, we consider a pair of systems

$$
Ly = f, \quad Ux = y
$$

The first system, $Ly = f$, is solved using forward substitution to obtain

$$
\begin{aligned}
y_1 &= f_1 \\
y_j &= f_j - \alpha_j y_{j-1}, \quad 2 \le j \le n
\end{aligned}
\tag{7.3.25}
$$

The system $Ux = y$ is solved by back substitution, which in this particular case yields

$$
\begin{aligned}
x_n &= y_n / \beta_n \\
x_j &= \frac{y_j - c_j x_{j+1}}{\beta_j}, \quad j = n-1, n-2, \ldots, 1
\end{aligned}
\tag{7.3.26}
$$

The entire scheme given by Eqs. (7.3.24) through (7.3.26) for solving tridiagonal systems is extremely rapid and has an operations count of $O(n)$ multiplications and divisions. In addition, the storage space is needed for only three diagonals of the matrix A. As previously stated (Theorem 7.3.3), the factorization of A is feasible, provided that throughout the Gaussian elimination interchanging rows is not necessary. The next result provides sufficient practical conditions for the factorization of a tridiagonal matrix A.

Theorem 7.3.4.

Let A be a tridiagonal matrix given by Eq. (7.3.21) such that

1. $|b_1| > |c_1|$
2. $|b_j| \ge |a_j| + |c_j|$, $\quad a_j \ne 0$, $\quad b_j \ne 0$, $\quad c_j \ne 0$, $\quad 2 \le j \le n-1$
3. $|b_n| \ge |a_n|$, $\quad b_n \ne 0$

Then the scheme given by Eqs. (7.3.24) through (7.3.26) is well defined, that is, $\beta_j \ne 0, 1 \le j \le n$.

Proof.

We first apply mathematical induction to show that $\beta_j \ne 0$, $\quad |c_j / \beta_j| < 1$ for $1 \le j \le n-1$:

1. $j = 1$. Because $\beta_1 = b_1$ and $|b_1| > |c_1|$, we obtain $\beta_1 \ne 0$ and $|c_1 / \beta_1| < 1$.
2. Assume $\beta_j \ne 0$, $\quad |c_j / \beta_j| < 1$ for some $j \ge 1$. Then

$$
\alpha_{j+1} = \frac{a_{j+1}}{\beta_j}, \quad \beta_{j+1} = b_{j+1} - \alpha_{j+1} c_j
$$

Thus α_{j+1} is well defined, and

$$\beta_{j+1} = b_{j+1} - \frac{a_{j+1}}{\beta_j} c_j$$

Because $|b_{j+1}| \geq |a_{j+1}| + |c_{j+1}|$, $|a_{j+1}| > 0$, $|c_{j+1}| > 0$, and $|c_j/\beta_j| < 1$, we also have

$$|\beta_{j+1}| > |b_{j+1}| - |a_{j+1}| \geq |c_{j+1}| > 0$$

that is,

$$\beta_{j+1} \neq 0, \quad \left| \frac{c_{j+1}}{\beta_{j+1}} \right| < 1$$

Now, because $\beta_n = b_n - (a_n/\beta_{n-1})c_{n-1}$, $|b_n| \geq |a_n|$, and $|c_{n-1}/\beta_{n-1}| < 1$, we obtain

$$|\beta_n| \geq |b_n| - \frac{|b_n|}{|\beta_{n-1}|}|c_{n-1}| = |b_n| \left(1 - \left| \frac{c_{n-1}}{\beta_{n-1}} \right| \right) > 0$$

that is, $\beta_n \neq 0$, which concludes the proof.   ☐

Example 7.3.6.

The matrix

$$A = \begin{bmatrix} 2 & 1 & 0 & 0 & 0 \\ 1 & 2 & 1 & 0 & 0 \\ 0 & 2 & 3 & 1 & 0 \\ 0 & 0 & 1 & 4 & 1 \\ 0 & 0 & 0 & 2 & 2 \end{bmatrix}$$

satisfies all the requirements of Theorem 7.3.4 and thus can be factorized using Eqs. (7.3.23) and (7.3.24). Substituting in Eq. (7.3.24) yields

$$\begin{aligned} & & \beta_1 &= b_1 = 2 \\ \alpha_2 &= a_2/\beta_1 = 1/2, & \beta_2 &= b_2 - \alpha_2 c_1 = 3/2 \\ \alpha_3 &= a_3/\beta_2 = 4/3, & \beta_3 &= b_3 - \alpha_3 c_2 = 5/3 \\ \alpha_4 &= a_4/\beta_3 = 3/5, & \beta_4 &= b_4 - \alpha_4 c_3 = 17/5 \\ \alpha_5 &= a_5/\beta_4 = 10/17, & \beta_5 &= b_5 - \alpha_5 c_4 = 24/17 \end{aligned}$$

and the factorization $A = LU$ is maintained by

$$L = \begin{bmatrix} 1 & 0 & 0 & 0 & 0 \\ 1/2 & 1 & 0 & 0 & 0 \\ 0 & 4/3 & 1 & 0 & 0 \\ 0 & 0 & 3/5 & 1 & 0 \\ 0 & 0 & 0 & 10/17 & 1 \end{bmatrix}, \quad U = \begin{bmatrix} 2 & 1 & 0 & 0 & 0 \\ 0 & 3/2 & 1 & 0 & 0 \\ 0 & 0 & 5/3 & 1 & 0 \\ 0 & 0 & 0 & 17/5 & 1 \\ 0 & 0 & 0 & 0 & 24/17 \end{bmatrix} \quad ☐$$

Example 7.3.7.

Consider the second order linear ordinary differential equation

$$\Psi'' = \Psi, \quad 0 \le x \le 1 \tag{7.3.27}$$

with *boundary conditions*

$$\Psi(0) = 1, \quad \Psi(1) = 0 \tag{7.3.28}$$

It is known that the unique solution to Eqs. (7.3.27) and (7.3.28) is given by

$$\Psi(x) = \alpha e^x + \beta e^{-x}$$

where

$$\alpha = \frac{1}{1 - e^2}, \quad \beta = \frac{e^2}{e^2 - 1}$$

To approximate the exact solution we *discretize* the differential equation. The interval $[0, 1]$ is replaced by a grid of evenly spaced points

$$x_i = (i - 1)h, \quad 1 \le i \le n$$

such that $(n - 1)h = 1$. The differential equation at each point is approximated by

$$\frac{\Psi_{i-1} + \Psi_{i+1} - 2\Psi_i}{h^2} = \Psi_i, \quad 2 \le i \le n - 1 \tag{7.3.29}$$

where $\Psi_i \equiv \Psi(x_i)$, $1 \le i \le n$. The boundary conditions are

$$\Psi_1 = 1, \quad \Psi_n = 0 \tag{7.3.30}$$

The original problem is thus approximated by Eqs. (7.3.29) and (7.3.30), that is, by the linear system $Ax = b$, where

$$A = \begin{bmatrix} 1 & 0 & 0 & \cdots & & 0 \\ 1 & -(2+h^2) & 1 & \ddots & & \vdots \\ 0 & 1 & -(2+h^2) & 1 & & 0 \\ \vdots & \ddots & & 1 & -(2+h^2) & 1 \\ 0 & \cdots & & 0 & 0 & 1 \end{bmatrix} \tag{7.3.31}$$

and

$$x = \begin{bmatrix} \Psi_1 \\ \Psi_2 \\ \vdots \\ \Psi_n \end{bmatrix}, \quad b = \begin{bmatrix} 1 \\ 0 \\ \vdots \\ 0 \end{bmatrix} \tag{7.3.32}$$

The matrix A is strictly diagonally dominant and can therefore be factorized.   □

A subroutine (TRIDG) for solving tridiagonal systems is found on the attached floppy disk.

PROBLEMS

1. Calculate the LU factorization of the following matrices:

   (a)

   $$\begin{bmatrix} 3 & 1 & -1 \\ 2 & 4 & 0 \\ 1 & 1 & 3 \end{bmatrix}$$

   (b)

   $$\begin{bmatrix} 2 & 0 & 1 \\ 6 & 8 & 1 \\ 1 & 4 & 6 \end{bmatrix}$$

   (c)

   $$\begin{bmatrix} 3 & 1 & 0 & 1 \\ 1 & 4 & 1 & 1 \\ 1 & 0 & 2 & 1/2 \\ 1 & 1 & 1/2 & 3 \end{bmatrix}$$

2. If an LU factorization requires only that L and U be lower and upper triangular matrices, respectively, then the LU factorization is not unique. Let $A = LU$, where L and U are such matrices, and let $D = (d_{ij})$ be a nonsingular diagonal matrix, that is,

   $$d_{ij} = 0, \quad i \neq j$$
   $$d_{ii} \neq 0, \quad 1 \leq i \leq n$$

   Show that $L_1 U_1$, where $L_1 = LD$, $\quad U_1 = D^{-1} U$ is another factorization of A.

3. Verify the following results:

   (a) The product of two lower (upper) triangular matrices is lower (upper) triangular.

   (b) The inverse of a nonsingular lower (upper) triangular matrix is also lower (upper) triangular.

4. Let $L_1 U_1$ and $L_2 U_2$ be two different factorizations of a nonsingular matrix A. Show that $L_2 = L_1 D$ and $U_2 = D^{-1} U_1$ for some nonsingular diagonal matrix D.

5. If A is a symmetric matrix, it is often possible to write it as $A = LL^T$ for some lower triangular matrix L.

   (a) Find a formal expression for L in the case of 3×3 symmetric matrices.

(b) Calculate L for

$$A = \begin{bmatrix} 1 & -1 & 2 \\ -1 & 4 & 3 \\ 2 & 3 & 16 \end{bmatrix}$$

6. Consider the system $Ax = b$, where A is tridiagonal and fulfills the requirements of Theorem 7.3.4. Compute the operations count for solving the system, using the factorization method of Eqs. (7.3.22) through (7.3.24).

7. Find the LU factorization of the tridiagonal matrix

$$A = \begin{bmatrix} 2 & 1 & 0 & 0 \\ 1 & 2 & 1 & 0 \\ 0 & 1 & 2 & 1 \\ 0 & 0 & 1 & 2 \end{bmatrix}$$

and solve the linear system $Ax = f$ in the following cases:

(a) $f = (1, 1, 0, 0)^T$

(b) $f = (1, 1, 1, 1)^T$

(c) $f = (0, 0, 0, 0)^T$

8. Solve the following 4×4 tridiagonal system using LU factorization

$$x_1 + (1/2)x_2 = 1$$
$$x_1 - 2x_2 + x_3 = -1$$
$$x_2 + 3x_3 - x_4 = 2$$
$$(1/3)x_3 + (1/2)x_4 = 0$$

9. Let $A = (a_{ij})$ be an $n \times n$ tridiagonal matrix defined by

$$a_{ii} = 3, \quad 1 \le i \le n$$
$$a_{i,i-1} = 1, \quad 2 \le i \le n$$
$$a_{i,i+1} = 1, \quad 1 \le i \le n-1$$

(a) Factor A, using Eqs. (7.3.22) through (7.3.24), and show that

$$\alpha = \lim_{n \to \infty} \alpha_n, \quad \beta = \lim_{n \to \infty} \beta_n$$

exist and calculate α, β.

(b) Compute the solution to $Ax = f$ in the case $n = 4$, $f = (1, 1, 1, 0)^T$.

10. Obtain a tridiagonal system by discretizing the differential equation

$$\Psi'' = x\Psi, \quad 1 \le x \le 2$$

with boundary conditions $\Psi(1) = 2$, $\Psi(2) = 3$. Use a total of n evenly spaced grid points and solve the system in the particular case $n = 11$.

11. Consider the linear system associated with cubic spline interpolation, given by Eqs. (5.5.8) and (5.5.9), and assume evenly spaced grid points. Show that all the requirements of Theorem 7.3.4 are satisfied.

7.4   Iterative Methods

The methods considered so far for solving systems of linear equations are all direct. In this section we will present some *iterative* techniques. An iterative method for solving $Ax = b$ starts with an initial approximation $x^{(0)}$ to the exact solution x. It then generates a sequence of vectors $\{x^{(m)}\}_{m=1}^{\infty}$ that converges to x.

Because convergence is measured by the "distance" between vectors, we will first introduce the notion of *norm* for vectors and matrices in R^n.

7.4.1   Norms of Vectors and Matrices

Definition 7.4.1.   (Vector norm). *A vector norm (length) is a real function $\|.\|$ defined for all n–dimensional real vectors $x \in R^n$ with the following properties:*

1. $\|x\| \ge 0$, $x \in R^n$; $\|x\| = 0 \Leftrightarrow x = 0$ $[= (0,0,\ldots,0)^T]$

2. $\|\alpha x\| = |\alpha|\,\|x\|$, $x \in R^n$, $-\infty < \alpha < \infty$

3. $\|x + y\| \le \|x\| + \|y\|$, $x, y \in R^n$

The distance between x and y is defined as $\|x - y\|$. The most commonly used are the l_2, l_∞ norms defined as

$$l_2 \text{ norm:} \quad \|x\|_2 = \left(\sum_{i=1}^{n} x_i^2\right)^{1/2} \tag{7.4.1}$$

$$l_\infty \text{ norm:} \quad \|x\|_\infty = \max_{1 \le i \le n} |x_i| \tag{7.4.2}$$

for any given $x = (x_1, \ldots, x_n)^T$.

Example 7.4.1.
For $x = (2, 0, -1, 3, 2)^T \in R^5$, we have

$$\|x\|_2 = \sqrt{18}, \quad \|x\|_\infty = 3 \qquad \square$$

For all $x \in R^n$ we have

$$\|x\|_\infty \leq \|x\|_2 \leq \sqrt{n}\|x\|_\infty \tag{7.4.3}$$

The proof that $\|x\|_2$ and $\|x\|_\infty$ are indeed norms is left as an exercise for the reader.

Definition 7.4.2.   (Convergence). *A sequence of vectors* $\{x^{(m)}\}_{m=0}^\infty$ *in* R^n *is said to converge to* $x \in R^n$ *with respect to a given norm* $\|.\|$, *if*

$$\lim_{m \to \infty} \|x^{(m)} - x\| = 0 \tag{7.4.4}$$

Theorem 7.1.4.

A sequence $\{x^{(m)}\}$ *converges to* x *in* R^n *with respect to* $\|.\|_\infty$, *if and only if*

$$\lim_{m \to \infty} x_k^{(m)} = x_k, \quad 1 \leq k \leq n$$

The proof is straightforward and is left as an exercise for the reader.

Theorem 7.4.2.

(Equivalence of norms). *Any two norms* $\|.\|$, $\|.\|'$ *in* R^n *are equivalent with respect to convergence, that is, for any given sequence of vectors* $\{x^{(m)}\}_0^\infty$

$$\lim_{m \to \infty} \|x^{(m)} - x\| = 0 \iff \lim_{m \to \infty} \|x^{(m)} - x\|' = 0$$

The proof of this theorem is omitted.

Because it will be necessary to measure the distance between matrices as well, we also define a matrix norm.

Definition 7.4.3.   (Matrix norm). *A matrix norm defined for all* $n \times n$ *real matrices is a real function* $\|.\|$ *that, for any given* $n \times n$ *real matrices A and B and arbitrary real number* α, *satisfies*

1. $\|A\| \geq 0$, $\|A\| = 0 \iff A = 0$ (= *a matrix with only zero entries*)
2. $\|\alpha A\| = |\alpha| \|A\|$
3. $\|A + B\| \leq \|A\| + \|B\|$
4. $\|AB\| \leq \|A\| \|B\|$

The distance between A and B is defined as $\|A - B\|$.

Theorem 7.4.3.

For any given vector norm $\|.\|$ *the relation*

$$\|A\| = \max_{\|x\|=1} \|Ax\| \tag{7.4.5}$$

defines a matrix norm.

Proof.

We shall assume that the right-hand side of Eq. (7.4.5) exists and thus $\|A\|$ is well defined.

1. By definition $\|A\| \geq 0$. If $A = 0$ then $Ax = 0$ for all x. Consequently $\|A\| = 0$. On the other hand, if $\|A\| = 0$ then $\|Ax\| = 0$ for all x that satisfy $\|x\| = 1$. Therefore $Ax = 0$ for all such x. Define

$$x^{(i)} = (\underbrace{0, \ldots, 0}_{i-1}, 1, 0, \ldots, 0)^T, \quad y^{(i)} = \frac{x^{(i)}}{\|x^{(i)}\|}$$

Clearly $\|y^{(i)}\| = 1$. Hence $Ay^{(i)} = 0$, leading to $a_{ji} = 0$, $1 \leq j \leq n$. This holds for $1 \leq i \leq n$ and therefore $A = 0$.

2. The norm of αA is given by

$$\|\alpha A\| = \max_{\|x\|=1} \|\alpha Ax\| = |\alpha| \left[\max_{\|x\|=1} \|Ax\| \right] = |\alpha| \|A\|$$

3. For any given $n \times n$ matrices A and B

$$\|A + B\| = \max_{\|x\|=1} \|(A+B)x\| = \max_{\|x\|=1} \|Ax + Bx\| \leq \max_{\|x\|=1} (\|Ax\| + \|Bx\|)$$
$$\leq \max_{\|x\|=1} \|Ax\| + \max_{\|x\|=1} \|Bx\| = \|A\| + \|B\|$$

4. For any $n \times n$ matrix C and nonzero $y \in R^n$ we have

$$Cy = C \left(\|y\| \frac{y}{\|y\|} \right) = \|y\| C \left(\frac{y}{\|y\|} \right)$$

By virtue of Eq. (7.4.5)

$$\|Cy\| = \|y\| \left\| C \left(\frac{y}{\|y\|} \right) \right\| \leq \|y\| \, \|C\|$$

and this clearly holds for $y = 0$ as well. Thus, if A and B are any $n \times n$ matrices and $\|x\| = 1$, then

$$\|(AB)x\| = \|A(Bx)\| \leq \|A\| \, \|Bx\| \leq \|A\| \, \|B\| \, \|x\| = \|A\| \, \|B\|$$

Hence

$$\|AB\| = \max_{\|x\|=1} \|(AB)x\| \leq \|A\| \, \|B\| \qquad\qquad \square$$

Equation (7.4.5) defines a matrix norm that is called a *natural norm*.

Definition 7.4.4. (Convergence). *A sequence of $n \times n$ matrices $\{A_m\}_{m=1}^{\infty}$ is said to converge to a matrix A, if*

$$\lim_{m \to \infty} \|A_m - A\| = 0$$

for some natural matrix norm.

It can be shown that convergence with respect to one natural matrix norm implies convergence with respect to *any* natural matrix norm.

The l_2 and l_{∞} matrix norms are

$$\|A\|_2 = \max_{\|x\|_2=1} \|Ax\|_2 \tag{7.4.6}$$

and

$$\|A\|_{\infty} = \max_{\|x\|_{\infty}=1} \|Ax\|_{\infty} \tag{7.4.7}$$

respectively. A simple way to calculate $\|A\|_{\infty}$ is given next.

Theorem 7.4.4.
(Calculating $\|A\|_{\infty}$). *Let $A = (a_{ij})$ be an $n \times n$ real matrix. Then*

$$\|A\|_{\infty} = \max_{1 \le i \le n} \sum_{j=1}^{n} |a_{ij}| \tag{7.4.8}$$

The proof is left as an exercise for the reader.

Example 7.4.2.
For the matrix

$$A = \begin{bmatrix} 3 & 6 & 0 \\ -1 & 0 & 9 \\ 2 & 1 & 5 \end{bmatrix}$$

we obtain

$$\sum_{j=1}^{3} |a_{1j}| = 9, \quad \sum_{j=1}^{3} |a_{2j}| = 10, \quad \sum_{j=1}^{3} |a_{3j}| = 8$$

leading to $\|A\|_{\infty} = 10$. $\qquad\qquad\square$

Prior to introducing iterative methods we must also define eigenvalues, eigenvectors, the characteristic polynomial, and spectral radius for a given $n \times n$ matrix A.

Definition 7.4.5. (Characteristic polynomial). *Given an $n \times n$ matrix A, the nth order polynomial*

$$p_A(\lambda) = \det(A - \lambda I) = |A - \lambda I|$$

is called the characteristic polynomial of A.

Example 7.4.3.
The characteristic polynomial of

$$A = \begin{bmatrix} 1 & 1 \\ -1 & 5 \end{bmatrix}$$

is

$$p_A(\lambda) = \begin{vmatrix} 1-\lambda & 1 \\ -1 & 5-\lambda \end{vmatrix} = (1-\lambda)(5-\lambda) + 1 = \lambda^2 - 6\lambda + 6$$

For

$$B = \begin{bmatrix} 0 & -1 & 2 \\ 1 & 1 & 0 \\ 2 & 3 & 1 \end{bmatrix}$$

we obtain

$$p_B(\lambda) = \begin{vmatrix} -\lambda & -1 & 2 \\ 1 & 1-\lambda & 0 \\ 2 & 3 & 1-\lambda \end{vmatrix} = -\lambda^3 + 2\lambda^2 + 2\lambda + 3 \qquad \square$$

Definition 7.4.6. (Eigenvalues and eigenvectors). *For a given $n \times n$ matrix A, the zeroes of $p_A(\lambda)$ are called the eigenvalues of A. For each eigenvalue λ, a nonzero vector x that satisfies*

$$(A - \lambda I)x = 0$$

is called an eigenvector *of A.*

Example 7.4.4.
In the previous example, the eigenvalues of A satisfy $\lambda^2 - 6\lambda + 6 = 0$, that is, $\lambda_1 = 3 + \sqrt{3}$, $\lambda_2 = 3 - \sqrt{3}$. Let $x^{(1)} = (x_1^{(1)}, x_2^{(1)})^T$ be an eigenvector of λ_1. Then

$$\begin{bmatrix} -2-\sqrt{3} & 1 \\ -1 & 2-\sqrt{3} \end{bmatrix} \begin{bmatrix} x_1^{(1)} \\ x_2^{(1)} \end{bmatrix} = \begin{bmatrix} 0 \\ 0 \end{bmatrix}$$

leading to

$$x^{(1)} = c \begin{bmatrix} 1 \\ 2+\sqrt{3} \end{bmatrix}$$

where c is an arbitrary nonzero constant. $\qquad \square$

Definition 7.4.7. (Spectral radius). *Let λ_i, $1 \le i \le n$ be the eigenvalues of A. The number*

$$\rho(A) = \max_{1 \le i \le n} |\lambda_i| \qquad (7.4.9)$$

is called the spectral radius of A.

Example 7.4.5.

Consider the matrix

$$A = \begin{bmatrix} 1 & 2 & 0 \\ 2 & 1 & 2 \\ 0 & 2 & 1 \end{bmatrix}$$

The characteristic polynomial is $p_A(\lambda) = (1 - \lambda)^3 - 8(1 - \lambda)$. Thus $\lambda_1 = 1$, $\lambda_2 = 1 + 2\sqrt{2}$, $\lambda_3 = 1 - 2\sqrt{2}$, and $\rho(A) = 1 + 2\sqrt{2}$.   □

The relation between a matrix norm and its spectral radius is our next result.

Theorem 7.4.5.

For any real $n \times n$ matrix A and any natural matrix norm $\|.\|$

$$\rho(A) \leq \|A\|$$

Proof.

Let λ be an eigenvalue of A with an associated eigenvector x such that $\|x\| = 1$. Because $(A - \lambda I)x = 0$, we find $Ax = \lambda x$. Thus

$$|\lambda| = |\lambda|\,\|x\| = \|\lambda x\| = \|Ax\| \leq \|A\|$$

leading to $\rho(A) \leq \|A\|$.   ☐

A relation between $\|A\|_2$ and $\rho(A)$ is given by Theorem 7.4.6.

Theorem 7.4.6.

For any real $n \times n$ matrix A

$$[\rho(A^T A)]^{1/2} = \|A\|_2$$

The proof of this result is omitted.

Example 7.4.6.

For the matrix

$$A = \begin{bmatrix} 1 & 3 \\ -1 & 2 \end{bmatrix}$$

we have

$$A^T A = \begin{bmatrix} 1 & -1 \\ 3 & 2 \end{bmatrix} \begin{bmatrix} 1 & 3 \\ -1 & 2 \end{bmatrix} = \begin{bmatrix} 2 & 1 \\ 1 & 13 \end{bmatrix}$$

The eigenvalues of $A^T A$ solve the equation

$$\begin{vmatrix} 2 - \lambda & 1 \\ 1 & 13 - \lambda \end{vmatrix} = (2 - \lambda)(13 - \lambda) - 1 = \lambda^2 - 15\lambda + 25 = 0$$

that is, $\lambda_1 = (15 + \sqrt{125})/2$, $\lambda_2 = (15 - \sqrt{125})/2$. Thus

$$\|A\|_2^2 = \rho(A^TA) = \frac{15 + \sqrt{125}}{2} \approx 13.09, \quad \|A\|_2 \approx 3.62 \qquad \square$$

An additional notation is still needed.

Definition 7.4.8. (Convergence). *An $n \times n$ matrix A is said to be convergent if*

$$\lim_{m \to \infty} (A^m)_{i,j} = 0, \quad 1 \le i,j \le n$$

Example 7.4.7.
The matrix

$$A = \begin{bmatrix} 1/2 & 1/2 \\ 1/2 & 1/2 \end{bmatrix}$$

is not convergent because $A^m = A$ for all m. On the other hand, for

$$B = \begin{bmatrix} \epsilon & \epsilon \\ \epsilon & \epsilon \end{bmatrix}, \quad \epsilon > 0$$

we have

$$B^{m+1} = 2^m \epsilon^{m+1} \begin{bmatrix} 1 & 1 \\ 1 & 1 \end{bmatrix}$$

and if $\epsilon < 1/2$, B is convergent. $\qquad \square$

A useful result that connects between the spectral radius, a natural norm, and the convergence of a given matrix is the following.

Theorem 7.4.7.
The following statements are equivalent:

1. $\rho(A) < 1$.

2. *A is convergent.*

3. $\lim_{m \to \infty} \|A^m\| = 0$ *for any natural norm* $\|.\|$.

4. $\lim_{m \to \infty} \|A^m x\| = 0$ *for all x.*

The proof of this theorem is omitted.

Example 7.4.8.
Consider the matrix

$$B = \begin{bmatrix} \epsilon & \epsilon \\ \epsilon & \epsilon \end{bmatrix}, \quad \epsilon > 0$$

The eigenvalues of B solve $(\epsilon - \lambda)^2 - \epsilon^2 = 0$, that is, $\lambda_1 = 0$, $\lambda_2 = 2\epsilon$, and $\rho(B) = 2\epsilon$. For $\epsilon < 1/2$ we have $\rho(B) < 1$ and hence $\|B^m\| \to 0$ as $m \to \infty$ for some natural matrix norm. This is in agreement with the previous example, because

$$\|B^m\|_\infty = (2\epsilon)^{m+1} \to 0, \text{ as } m \to \infty \qquad \square$$

We will now present a few popular iterative methods for solving linear equations. The common pattern of these techniques is to replace $Ax = b$ with an equivalent system $x = Mx + c$ for some $n \times n$ matrix M and vector c, such that for any initial vector $x^{(0)}$ the sequence of vectors

$$x^{(m)} = Mx^{(m-1)} + c, \quad m \geq 1 \qquad (7.4.10)$$

converges. Iterative techniques are used mainly for large systems whose entries are mostly zeroes. Otherwise, direct methods are generally more efficient.

7.4.2 Jacobi Iterative Method

A general $n \times n$ set of linear equations given by

$$\sum_{j=1}^{n} a_{ij}x_j = b_i, \quad 1 \leq i \leq n \qquad (7.4.11)$$

can be rewritten as

$$x_i = \sum_{\substack{j=1 \\ j \neq i}}^{n} \left(-\frac{a_{ij}}{a_{ii}}x_j \right) + \frac{b_i}{a_{ii}}, \quad 1 \leq i \leq n \qquad (7.4.12)$$

provided that $a_{ii} \neq 0$, $1 \leq i \leq n$. Thus, the original ith equation is used to express x_i in terms of the remaining variables. To obtain the matrix form of Eq. (7.4.12) we first write

$$Ax = (D + L + U)x = b$$

where L, U, and D are lower triangular, upper triangular, and diagonal matrices, respectively, given by

$$L = \begin{bmatrix} 0 & \cdots & & 0 \\ a_{21} & \ddots & & \vdots \\ \vdots & \ddots & \ddots & \\ a_{n1} & \cdots & a_{n,n-1} & 0 \end{bmatrix}, \quad U = \begin{bmatrix} 0 & a_{12} & \cdots & a_{1n} \\ \vdots & \ddots & \ddots & \vdots \\ & & \ddots & a_{n-1,n} \\ 0 & \cdots & & 0 \end{bmatrix}$$

and

$$D = \begin{bmatrix} a_{11} & 0 & \cdots & 0 \\ 0 & a_{22} & \ddots & \vdots \\ \vdots & \ddots & \ddots & 0 \\ 0 & \cdots & 0 & a_{nn} \end{bmatrix}$$

Thus $Dx = -(L + U)x + b$ and because $a_{ii} \neq 0$, $1 \leq i \leq n$ we obtain

$$x = -D^{-1}(L + U)x + D^{-1}b \qquad (7.4.13)$$

The iterative scheme based on Eq. (7.4.13), called the *Jacobi iterative method*, is given by

$$x^{(m)} = -D^{-1}(L + U)x^{(m-1)} + D^{-1}b \qquad (7.4.14)$$

and is clearly identical to Eq. (7.4.12).

Example 7.4.9.

Consider the linear system

$$\begin{bmatrix} 2 & 0 & 1 \\ 0 & 3 & 1 \\ -1 & 0 & 4 \end{bmatrix} \begin{bmatrix} x_1 \\ x_2 \\ x_3 \end{bmatrix} = \begin{bmatrix} 3 \\ -2 \\ 3 \end{bmatrix}$$

where

$$L = \begin{bmatrix} 0 & 0 & 0 \\ 0 & 0 & 0 \\ -1 & 0 & 0 \end{bmatrix}, \quad U = \begin{bmatrix} 0 & 0 & 1 \\ 0 & 0 & 1 \\ 0 & 0 & 0 \end{bmatrix}, \quad D = \begin{bmatrix} 2 & 0 & 0 \\ 0 & 3 & 0 \\ 0 & 0 & 4 \end{bmatrix}$$

The Jacobi iterative scheme is given by $x^{(m)} = Mx^{(m-1)} + c$, where

$$M = -D^{-1}(L + U) = \begin{bmatrix} 0 & 0 & -1/2 \\ 0 & 0 & -1/3 \\ 1/4 & 0 & 0 \end{bmatrix}, \quad c = D^{-1}b = \begin{bmatrix} 3/2 \\ -2/3 \\ 3/4 \end{bmatrix}$$

Its implementation for $x^{(0)} = (2, 0, 0)^T$ produces Table 7.4.1, which clearly indicates convergence to the exact solution $x = (1, -1, 1)^T$. □

Table 7.4.1. Jacobi Iterative Method Applied to a 3×3 Matrix

m	$x_1^{(m)}$	$x_2^{(m)}$	$x_3^{(m)}$
1	1.500	−0.667	1.250
2	0.875	−1.083	1.125
3	0.938	−1.042	0.969
4	1.016	−0.990	0.984
5	1.008	−0.995	1.004

The next iterative scheme is generally an improvement over the Jacobi method.

7.4.3   Gauss–Seidel Iterative Method

Throughout the iteration process, the components $x_i^{(m)}$, $1 \leq i \leq n$ are supposedly closer than $x_i^{(m-1)}$, $1 \leq i \leq n$ to the exact values x_i, $1 \leq i \leq n$. Therefore, to calculate $x_k^{(m)}$ it is quite reasonable to use the already computed $x_i^{(m)}$, $1 \le i \le k-1$ rather than $x_i^{(m-1)}$, $1 \le i \le k-1$. This approach leads to the scheme

$$x_k^{(m)} = \frac{-\sum_{i=1}^{k-1} a_{ki} x_i^{(m)} - \sum_{i=k+1}^{n} a_{ki} x_i^{(m-1)} + b_k}{a_{kk}} \tag{7.4.15}$$

known as the *Gauss–Seidel iterative method*. The matrix form of Eq. (7.4.15) is

$$(D+L)x^{(m)} = -Ux^{(m-1)} + b \tag{7.4.16}$$

or

$$x^{(m)} = -(D+L)^{-1}Ux^{(m-1)} + (D+L)^{-1}b \tag{7.4.17}$$

Because we still assume $a_{ii} \ne 0$, $1 \le i \le n$, $D+L$ is nonsingular and $x^{(m)}$ is well defined for all m.

Example 7.4.10.

Consider the system in the previous example. For the initial approximation $x^{(0)} = (2,0,0)$, the Gauss–Seidel method provides faster convergence than does the Jacobi method, as demonstrated in Table 7.4.2. □

Table 7.4.2. Gauss–Seidel Iterative Method Applied to a 3×3 Matrix

m	$x_1^{(m)}$	$x_2^{(m)}$	$x_3^{(m)}$
1	1.500	−0.667	1.125
2	0.938	−1.042	0.984
3	1.008	−0.995	1.002
4	0.999	−1.001	1.000
5	1.000	−1.000	1.000

An algorithm based on the Gauss–Seidel iterative method is given next.

Algorithm 7.4.1
(Gauss–Seidel method for solving $Ax = b$)

Input: The matrix $A = (a_{ij})$, $1 \le i,j \le n$ such that $a_{ii} \ne 0$ for all i; the vector $b = (b_1, \ldots, b_n)^T$; an initial approximation $x^{(0)}$ to the exact solution y; a tolerance $\epsilon > 0$; maximum number of iterations N.

Output: An approximation x to y or a message that the number of iterations was exceeded.
Step 1. Set $m = 1$.
Step 2. For $k = 1, \ldots, n$ set

$$x_k = \frac{-\sum_{i=1}^{k-1} a_{ki} x_i - \sum_{i=k+1}^{n} a_{ki} x_i^{(0)} + b_k}{a_{kk}}$$

Step 3. If $\|x - x^{(0)}\| < \epsilon$ then output x and stop. Otherwise $m \leftarrow m+1$, $x^{(0)} \leftarrow x$.
Step 4. If $m \le N$ go to step 2; otherwise output "Number of iterations exceeded" and stop.

A subroutine GS, based on Algorithm 7.4.1, is found on the attached floppy disk.

Both Jacobi and Gauss–Seidel iterative methods do not always converge. Sufficient conditions for convergence are available. However, prior to specifying them we will briefly discuss the convergence problem in the general case $x^{(m)} = Mx^{(m-1)} + c$.

Theorem 7.4.8.

If the spectral radius $\rho(M)$ satisfies $\rho(M) < 1$, then $(I - M)^{-1}$ exists and

$$(I - M)^{-1} = I + M + M^2 + \cdots \tag{7.4.18}$$

that is,

$$\lim_{k \to \infty} \|(I - M)^{-1} - S_k\| = 0 \tag{7.4.19}$$

for any natural matrix norm, where $S_k = I + M + M^2 + \cdots + M^k$.

Proof.

A number λ is an eigenvalue of M, if and only if $1 - \lambda$ is an eigenvalue of $I - M$. Because $\rho(M) < 1$ we have $|1 - \lambda| > 0$ and therefore $I - M$ is nonsingular and has an inverse. We clearly have

$$S_k(I - M) = (I + M + \cdots + M^k)(I - M) = I - M^{k+1}$$

and by multiplying both sides by $(I - M)^{-1}$ one obtains

$$S_k = (I - M)^{-1} - M^{k+1}(I - M)^{-1}$$

Therefore, by virtue of Theorem 7.4.7,

$$\|S_k - (I - M)^{-1}\| \le \|(I - M)^{-1}\| \, \|M^{k+1}\| \to 0, \quad k \to \infty \qquad \Box$$

We are now able to present a necessary and sufficient condition for the convergence of a general iterative method given in the form of Eq. (7.4.10) for solving $Ax = b$.

Theorem 7.4.9.

[A convergence theorem for $x^{(m)} = Mx^{(m-1)} + c$ for any initial approximation $x^{(0)} \in R^n$]. The iterative scheme

$$x^{(m)} = Mx^{(m-1)} + c$$

converges to the unique solution of $x = Mx + c$ (equivalent to $Ax = b$) if and only if $\rho(M) < 1$.

Proof.

1. Let $\rho(M) < 1$. Using Eq. (7.4.10) and induction we find

$$x^{(m)} = M^m x^{(0)} + (M^{m-1} + \cdots + M + I)c$$

Thus

$$\lim_{m \to \infty} [x^{(m)} - (M^{m-1} + \cdots + M + I)c] = \lim_{m \to \infty} (M^m x^0) = 0 \quad \text{(why?)}$$

In addition, in view of Theorem 7.4.8 we have

$$\lim_{m \to \infty} [(I + M + \cdots + M^{m-1})c - (I - M)^{-1}c] = 0$$

Therefore

$$\lim_{m \to \infty} x^{(m)} = (I - M)^{-1}c$$

and by setting $x = \lim_{m \to \infty} x^{(m)}$ we obtain $x = Mx + c$.

2. Let $\{x^{(m)}\}_0^\infty$ converge to x for any initial guess $x^{(0)}$. Consequently, $x = Mx + c$ and

$$x - x^{(m)} = M[x - x^{(m-1)}] = \cdots = M^m[x - x^{(0)}]$$

Therefore,

$$\lim_{m \to \infty} M^m[x - x^{(0)}] = \lim_{m \to \infty} [x - x^{(m)}] = 0$$

Because $x^{(0)}$ is arbitrary we must have $\lim_{m \to \infty} M^m y = 0$ for all $y \in R^n$ and by virtue of Theorem 7.4.7 $\rho(M) < 1$.   □

Corollary 7.4.1.

If $\|M\| < 1$ for some given natural matrix norm $\|.\|$, then the sequence $\{x^{(m)}\}_{m=1}^\infty$ defined by Eq. (7.4.10) converges.

The proof, based on Theorems 7.4.5 and 7.4.9, is left as as exercise for the reader.

A sufficient condition for the convergence of the Jacobi and Gauss–Seidel iterative methods is presented next.

Theorem 7.4.10.

If A is strictly diagonally dominant, then the Jacobi and Gauss–Seidel schemes converge to the unique solution of $Ax = b$.

Proof.

We will restrict ourselves to the Jacobi iterative method and show (using Corollary 7.4.1)

$$\|M\|_\infty = \| - D^{-1}(L + U)\|_\infty < 1$$

Let $M = (m_{ij})$. Then, by Theorem 7.4.4,

$$\|M\|_\infty = \max_{1 \leq i \leq n} \sum_{j=1}^{n} |m_{ij}|$$

For this particular M

$$m_{ij} = \begin{cases} -(a_{ij}/a_{ii}) \ , & i \neq j \\ 0 & , \quad i = j \end{cases}$$

Because A is strictly diagonally dominant we have

$$\sum_{j=1}^{n} |m_{ij}| = \frac{1}{|a_{ii}|} \sum_{j \neq i} |a_{ij}| < 1, \quad 1 \leq i \leq n$$

and therefore by Theorem 7.4.4 $\|M\|_\infty < 1$. □

7.4.4 Speed of Convergence

The speed with which each iterative algorithm converges to the exact solution of $Ax = b$ is a major factor in determining which method to use. Consider a general iterative scheme

$$x^{(m)} = Mx^{(m-1)} + c$$

and let $\|M\| < 1$ for some natural matrix norm $\|.\|$. Then $x^{(m)}$ converges to x and it can be easily shown that

$$\|x - x^{(m)}\| \leq \|M\|^m \|x - x^{(0)}\| \tag{7.4.20}$$

$$\|x - x^{(m)}\| \leq \frac{\|M\|^m}{1 - \|M\|} \|x^{(1)} - x^{(0)}\| \tag{7.4.21}$$

A better error estimate is given by

$$\|x - x^{(m)}\| \approx [\rho(M)]^m \|x - x^{(0)}\| \tag{7.4.22}$$

and suggests that one should select the iterative scheme with the smallest $\rho(M) < 1$.

In most cases it is difficult to determine which of the two iterative schemes, Jacobi or Gauss–Seidel, is better. An exceptional case is given next without a proof.

Theorem 7.4.11.

Consider the linear system $Ax = b$ and let M_J, M_G denote the matrices associated with the Jacobi *and Gauss–Seidel methods, respectively, that is,*

$$M_J = -D^{-1}(L+U), \quad M_G = -(D+L)^{-1}U$$

If $a_{ii} > 0$, $1 \le i \le n$ and $a_{ij} \le 0$, $i \ne j$, then only one of the following statements holds:

1. $0 < \rho(M_G) < \rho(M_J) < 1$
2. $1 < \rho(M_J) < \rho(M_G)$
3. $\rho(M_J) = \rho(M_G) = 0$
4. $\rho(M_J) = \rho(M_G) = 1$

Thus, if one scheme converges so does the other, but the Gauss–Seidel technique is faster.

In view of Eq. (7.4.22) it is desirable to choose an iterative method whose associated matrix M has a minimal spectral radius. The next iterative method is designed in agreement with this principle.

7.4.5   Relaxation Methods

Consider the linear system $Ax = b$ and let y denote an approximate solution to the exact solution x. We define the associated *residual vector* by

$$r = b - Ay \tag{7.4.23}$$

Let $x^{(m-1)}$ and $x^{(m)}$ be two consecutive approximate solutions associated with the Gauss–Seidel method and let

$$r_k^{(m)} = \left[r_{k1}^{(m)}, r_{k2}^{(m)}, \dots, r_{kn}^{(m)} \right]^T$$

denote the residual vector corresponding to a third approximate solution

$$y_k^{(m)} = \left[x_1^{(m)}, x_2^{(m)}, \dots, x_{k-1}^{(m)}, x_k^{(m-1)}, \dots, x_n^{(m-1)} \right]^T$$

The jth component of $r_k^{(m)}$ is

$$r_{kj}^{(m)} = b_j - \sum_{i=1}^{k-1} a_{ji} x_i^{(m)} - \sum_{i=k}^{n} a_{ji} x_i^{(m-1)}$$

$$= b_j - \sum_{i=1}^{k-1} a_{ji} x_i^{(m)} - \sum_{i=k+1}^{n} a_{ji} x_i^{(m-1)} - a_{jk} x_k^{(m-1)}, \quad 1 \le j \le n$$

In the particular case $j = k$ we have

$$r_{kk}^{(m)} = b_k - \sum_{i=1}^{k-1} a_{ki} x_i^{(m)} - \sum_{i=k+1}^{n} a_{ki} x_i^{(m-1)} - a_{kk} x_k^{(m-1)}$$

and from the definition of the Gauss–Seidel scheme given by Eq. (7.4.15) one determines

$$a_{kk}x_k^{(m-1)} + r_{kk}^{(m)} = a_{kk}x_k^{(m)}$$

or

$$x_k^{(m)} = x_k^{(m-1)} + \frac{r_{kk}^{(m)}}{a_{kk}} \tag{7.4.24}$$

Equation (7.4.24) is an alternative form of the Gauss–Seidel scheme, which we now modify and replace by

$$x_k^{(m)} = x_k^{(m-1)} + w[r_{kk}^{(m)}/a_{kk}] \tag{7.4.25}$$

where w is some *relaxation factor*. Any iterative method given by Eq. (7.4.25) is called a *relaxation method*. Any choice of $0 < w < 1$ defines an *underrelaxation method*, whereas a choice of $w > 1$ defines an *overrelaxation method*. An underrelaxation method can be used whenever the Gauss–Seidel scheme fails to converge. If it does converge, an overrelaxation technique may be used to accelerate the convergence. The overrelaxation method is quite useful in solving linear systems associated with partial differential equation and is often called a *successive overrelaxation* (SOR) method.

To obtain the matrix form of the SOR method we rewrite Eq. (7.4.25) as

$$x_k^{(m)} = (1 - w)x_k^{(m-1)} + \frac{w}{a_{kk}}\left[b_k - \sum_{i=1}^{k-1}a_{ki}x_i^{(m)} - \sum_{i=k+1}^{n}a_{ki}x_i^{(m-1)}\right]$$

or

$$a_{kk}x_k^{(m)} + w\sum_{i=1}^{k-1}a_{ki}x_i^{(m)} = (1 - w)a_{kk}x_k^{(m-1)} - w\sum_{i=k+1}^{n}a_{ki}x_i^{(m-1)} + wb_k$$

leading to

$$(D + wL)x^{(m)} = [(1 - w)D - wU]x^{(m-1)} + wb$$

and finally to

$$x^{(m)} = (D + wL)^{-1}[(1 - w)D - wU]x^{(m-1)} + w(D + wL)^{-1}b \tag{7.4.26}$$

The particular choice $w = 1$ yields the Gauss–Seidel method. A comparison between the various iterative methods is given next.

Example 7.4.11.

Consider the 3×3 linear system

$$5x_1 - x_2 + 2x_3 = 5$$
$$-x_1 + 4x_2 + x_3 = 9$$

$$2x_1 + x_2 + 8x_3 = -1$$

whose exact solution is $x = (2, 3, -1)^T$. Assume an initial approximation $x^{(0)} = (1, 1, 1)^T$. The equations associated with the Jacobi method are

$$x_1^{(k)} = 0.2x_2^{(k-1)} - 0.4x_3^{(k-1)} + 1$$
$$x_2^{(k)} = 0.25x_1^{(k-1)} - 0.25x_3^{(k-1)} + 2.25$$
$$x_3^{(k)} = -0.25x_1^{(k-1)} - 0.125x_2^{(k-1)} - 0.125$$

and the first five iterations computed by this scheme are given in Table 7.4.3.

Table 7.4.3. Jacobi Method: Five Iterations

k	0	1	2	3	4	5
$x_1^{(k)}$	1	0.8000	1.6500	1.7575	1.9066	1.9473
$x_2^{(k)}$	1	2.2500	2.5750	2.8141	2.9042	2.9557
$x_3^{(k)}$	1	-0.5000	-0.6063	-0.8594	-0.9161	-0.9647

The equations for the Gauss–Seidel method are

$$x_1^{(k)} = 0.2x_2^{(k-1)} - 0.4x_3^{(k-1)} + 1$$
$$x_2^{(k)} = 0.25x_1^{(k)} - 0.25x_3^{(k-1)} + 2.25$$
$$x_3^{(k)} = -0.25x_1^{(k)} - 0.125x_2^{(k)} - 0.125$$

and its first five iterations are given in Table 7.4.4.

Table 7.4.4. Gauss–Seidel Method: Five Iterations

k	0	1	2	3	4	5
$x_1^{(k)}$	1	0.8000	1.6800	1.9230	1.9811	1.9954
$x_2^{(k)}$	1	2.2000	2.8200	2.9551	2.9891	2.9973
$x_3^{(k)}$	1	-0.6000	-0.8975	-0.9751	-0.9939	-0.9985

Finally, let us apply the SOR method with relaxation factor $w = 1.2$. The iterated equations are

$$x_1^{(k)} = -0.2x_1^{(k-1)} + 0.24x_2^{(k-1)} - 0.48x_3^{(k-1)} + 1.2$$
$$x_2^{(k)} = -0.2x_2^{(k-1)} + 0.3x_1^{(k)} - 0.3x_3^{(k-1)} + 2.7$$
$$x_3^{(k)} = -0.2x_3^{(k-1)} - 0.3x_1^{(k)} - 0.15x_2^{(k)} - 0.15$$

and the first five iterations are given in Table 7.4.5.

Table 7.4.5. Successive Overrelaxation Method with $w = 1.2$: Five Iterations

k	0	1	2	3	4	5
$x_1^{(k)}$	1	0.7600	2.0830	2.0390	1.9937	1.9995
$x_2^{(k)}$	1	2.4280	3.1220	3.0037	2.9978	2.9996
$x_3^{(k)}$	1	-0.9422	-1.0547	-1.0013	-0.9975	-1.0003

Thus, for this particular example, the SOR method with $w = 1.2$ converges significantly faster than both the Jacobi and Gauss–Seidel methods. $\square$

The fundamental question related to the SOR method is, how do we choose w? Partial answers do exist and are given at the end of this section. Prior to this we introduce a new class of matrices.

Definition 7.4.9. (Positive definiteness). *A symmetric matrix A of order n is called positive definite if*

$$(x)^T A(x) > 0 \tag{7.4.27}$$

for all n-dimensional vectors $x \in R^n$, $x \neq 0$.

Example 7.4.12.

The matrix

$$A = \begin{bmatrix} 1 & 0 \\ 0 & 0 \end{bmatrix}$$

is not positive definite, because

$$\begin{bmatrix} 0 & 1 \end{bmatrix} \begin{bmatrix} 1 & 0 \\ 0 & 0 \end{bmatrix} \begin{bmatrix} 0 \\ 1 \end{bmatrix} = 0$$

On the other hand,

$$B = \begin{bmatrix} 2 & 1 \\ 1 & 3 \end{bmatrix}$$

is positive definite because

$$(x_1 x_2) \begin{bmatrix} 2 & 1 \\ 1 & 3 \end{bmatrix} \begin{bmatrix} x_1 \\ x_2 \end{bmatrix} = 2x_1^2 + 2x_1 x_2 + 3x_2^2 = (x_1 + x_2)^2 + x_1^2 + 2x_2^2$$

and the right-hand side vanishes if and only if $x_1 = x_2 = 0$. $\square$

Positive definite matrices are generated in many applications. They possess many useful properties, two of which are listed below.

Theorem 7.4.12.

A positive definite matrix is nonsingular.

Proof.

If A is singular then $Ax = 0$ for some nonzero x. Therefore

$$(x)^T Ax = 0$$

which contradicts the positive definiteness of A.   ▯

Theorem 7.4.13.

Let $A = (a_{ij})$, $1 \leq i,j \leq n$ denote a symmetric matrix. Define the corner matrices

$$A_k = (a_{ij}),   1 \leq i,j \leq k,   1 \leq k \leq n \tag{7.4.28}$$

Then A is positive definite if and only if

$$\det(A_k) > 0,   1 \leq k \leq n \tag{7.4.29}$$

The proof of this theorem is omitted.

Example 7.4.13.

The symmetric matrix

$$A = \begin{bmatrix} 3 & -1 & 1 \\ -1 & 2 & 0 \\ 1 & 0 & 4 \end{bmatrix}$$

is positive definite, because $\det(A_1) = 3 > 0$, $\det(A_2) = 5 > 0$, and $\det(A_3) = \det(A) = 18 > 0$.   ☐

The next results relate to the appropriate choice of the relaxation factor w, and are given without proof.

Theorem 7.4.14.

If A is a positive definite matrix and $0 < w < 2$, then the SOR method converges to the exact solution of $Ax = b$ for any choice of an initial approximate solution $x^{(0)}$.

Theorem 7.4.15.

If A is positive definite and tridiagonal, then $\rho(M_G) = [\rho(M_J)]^2 < 1$, the optimal choice of w for the SOR method is

$$w = \frac{2}{1 + \sqrt{1 - [\rho(M_J)]^2}} \tag{7.4.30}$$

and for this choice we have

$$\rho(M_w) = w - 1 \tag{7.4.31}$$

where

$$M_w = (D + wL)^{-1}[(1 - w)D - wU] \tag{7.4.32}$$

Example 7.4.14.

The matrix in Example 7.4.11 is

$$A = \begin{bmatrix} 5 & -1 & 2 \\ -1 & 4 & 1 \\ 2 & 1 & 8 \end{bmatrix}$$

and can be shown (Theorem 7.4.13) to be positive definite. Therefore the SOR method converges for all w between 0 and 2. □

Example 7.4.15.

Consider the matrix

$$A = \begin{bmatrix} 3 & 1 & 0 \\ 1 & 2 & -1 \\ 0 & -1 & 4 \end{bmatrix}$$

which is both tridiagonal and positive definite. The Jacobi matrix is

$$M_J = -D^{-1}(L+U) = \begin{bmatrix} 1/3 & 0 & 0 \\ 0 & 1/2 & 0 \\ 0 & 0 & 1/4 \end{bmatrix} \begin{bmatrix} 0 & -1 & 0 \\ -1 & 0 & 1 \\ 0 & 1 & 0 \end{bmatrix} = \begin{bmatrix} 0 & -1/3 & 0 \\ -1/2 & 0 & 1/2 \\ 0 & 1/4 & 0 \end{bmatrix}$$

and

$$M_J - \lambda I = \begin{bmatrix} -\lambda & -1/3 & 0 \\ -1/2 & -\lambda & 1/2 \\ 0 & 1/4 & -\lambda \end{bmatrix}$$

The characteristic polynomial of M_J is $-\lambda^3 + (7/24)\lambda$ and the eigenvalues are

$$\lambda_1 = 0, \quad \lambda_2 = \sqrt{7/24}, \quad \lambda_3 = -\sqrt{7/24}$$

Thus, $\rho(M_J) \approx 0.54$ and the optimal relaxation factor for the SOR method is

$$w \approx \frac{2}{1 + \sqrt{1 - 0.54^2}} \approx 1.09$$

Thus $\rho(M_w) \approx 0.09$. □

PROBLEMS

1. Verify Theorem 7.4.5 with respect to the l_∞ norm for the following matrices:
   (a)

$$\begin{bmatrix} 3 & 1 \\ 2 & 4 \end{bmatrix}$$

(b)

$$\begin{bmatrix} 0 & 2 & 0 \\ -2 & 1 & 2 \\ 0 & 3 & 0 \end{bmatrix}$$

(c)

$$\begin{bmatrix} 1 & a \\ a & 0 \end{bmatrix}$$

2. Show that

$$\|A\| = \sum_{i=1}^{n} \sum_{j=1}^{n} |a_{ij}|$$

defines a matrix norm for the set of all $n \times n$ matrices.

3. For any $n \times n$ matrix A show

$$\det(A) = \prod_{i=1}^{n} \lambda_i$$

where $\lambda_1, \ldots, \lambda_n$ are the eigenvalues of A.

4. Let λ be an eigenvalue of an $n \times n$ matrix A, with an associated eigenvector x.

   (a) Show that if A^{-1} exists, then $1/\lambda$ is an eigenvalue of A^{-1} with the same eigenvector x.

   (b) Show that for any integer $k \geq 2$, λ^k is an eigenvalue of A^k with the same eigenvector x.

5. Let A be a positive definite matrix. Show that by defining

$$\|x\| = (x^T A x)^{\frac{1}{2}}, \quad x \in R^n$$

we obtain a norm in R^n.

6. Apply the Jacobi method to approximate the solutions to the systems

   (a)

$$\begin{array}{rrrcl} 2x_1 & -4x_2 & +x_3 & = & 1 \\ & 7x_2 & -x_3 & = & 3 \\ x_1 & +4x_2 & -2x_3 & = & 0 \end{array}$$

   (b)

$$\begin{array}{rrrrcl} 3x_1 & -x_2 & +x_3 & -x_4 & = & 0 \\ x_1 & +2x_2 & +x_3 & -x_4 & = & 1 \\ x_1 & & -x_3 & & = & 2 \\ & x_2 & +2x_3 & +5x_4 & = & 2 \end{array}$$

Use $x^0 = (1, 1, 1)^T$ and $x^{(0)} = (1, 1, 1, 1)^T$, respectively, with $\epsilon = 10^{-4}$.

7. Solve problem 6 by using the Gauss–Seidel method.

8. Use Jacobi, Gauss–Seidel, and SOR methods to solve the linear system

$$
\begin{array}{rrrrr}
3x_1 & -x_2 & +x_3 & & = & 3 \\
-x_1 & +4x_2 & & +x_4 & = & 1 \\
x_1 & & +5x_3 & -x_4 & = & 2 \\
& x_2 & -x_3 & +3x_4 & = & 0
\end{array}
$$

For the SOR method use $w = 1.3$. Calculate five iterations starting with $x^{(0)} = 0$.

9. Find the optimal relaxation factor to the following matrices:
   (a)

$$
\begin{bmatrix}
3 & 1 \\
1 & 2
\end{bmatrix}
$$

   (b)

$$
\begin{bmatrix}
4 & 1 & 0 \\
1 & 4 & -1 \\
0 & -1 & 4
\end{bmatrix}
$$

   (c)

$$
\begin{bmatrix}
4 & 1 & 0 & 0 \\
1 & 4 & -1 & 0 \\
0 & -1 & 4 & 1 \\
0 & 0 & 1 & 4
\end{bmatrix}
$$

10. Use Jacobi, Gauss–Seidel, and SOR methods to approximate the solution of the linear system

$$
\begin{array}{rrrr}
5x_1 & +x_2 & & = & 1 \\
x_1 & +5x_2 & -2x_3 & = & 0 \\
& -2x_2 & +5x_3 & = & 2
\end{array}
$$

   starting with an initial approximation $x^{(0)} = 0$ and calculating five iterations. For the SOR method use the optimal w given by Eq. (7.4.30).

7.5 Error and Stability

Numerical procedures, such as Gaussian elimination, for solving the linear system $Ax = b$ involve rounding errors through each arithmetic operation. We open this section by introducing a method for treating the accumulated round-off error in the approximate solution.

Let $\tilde{x}$ denote the approximate solution. The vector

$$r = b - A\tilde{x} \tag{7.5.1}$$

is called the *residual* associated with $\tilde{x}$. It simply measures the approximation of b by $A\tilde{x}$. If $\tilde{x}$ is the exact solution, then $r = 0$. The relation between r and $x - \tilde{x}$ is given by

$$r = Ax - A\tilde{x} = A(x - \tilde{x}) \tag{7.5.2}$$

If $e = x - \tilde{x}$ denotes the error associated with $\tilde{x}$, then

$$Ae = r \tag{7.5.3}$$

Thus the error vector e solves a linear system with the same matrix A as x. This is helpful: in solving $Ax = b$, the matrices L and U are saved along with all the information related to row interchanges. This information can be used for solving Eq. (7.5.3) with minimal cost. A total of $O(n^2)$ operations is needed to obtain e, a significant reduction from the original $O(n^3)$ operations needed for solving $Ax = b$.

This procedure for solving $Ae = r$ has two disadvantages. First, each element of r is computed as a difference of two almost identical numbers

$$r_i = b_i - \sum_{j=1}^{n} a_{ij}\tilde{x}_j \tag{7.5.4}$$

Consequently, some significant digits may be lost, causing an inaccuracy in r_i. This is treated by using a higher-precision arithmetic throughout the computation. For example, if $Ax = b$ was originally solved using single-precision arithmetic, double-precision arithmetic should be used to obtain r.

The second source of trouble in solving Eq. (7.5.3) is that the same round-off errors that occurred throughout the computation of $\tilde{x}$ still exist! Thus, rather than calculating e we obtain $\tilde{e}$, an approximation to e. How close $\tilde{e}$ is to e depends on the coefficient matrix A. Generally, by calculating $\tilde{e}$ one improves the computed solution by at least one digit of accuracy. The cases for which the matrix is extremely ill behaved are exceptional.

Example 7.5.1.

Consider the 3×3 system

$$1.24x_1 - 0.5x_2 - 2.96x_3 = 0.1035$$
$$x_1 - 2x_2 + 1.55x_3 = 1.1$$
$$0.27x_1 + x_2 - 4.382x_3 = 0.515$$

whose exact solution rounded to four significant digits is given by

$$x_1 = -4.212, \quad x_2 = -3.581, \quad x_3 = -1.194$$

Solving the system by Gaussian elimination without pivoting and assuming a four-digit arithmetic with rounding, we obtain the approximate solution

$$\tilde{x}_1 = -4.210, \quad \tilde{x}_2 = -3.580, \quad \tilde{x}_3 = -1.194$$

We now use double precision, that is, an eight-digit arithmetic, to obtain the associated residual

$$r_1 = -0.00034, \quad r_2 = 0.0007, \quad r_3 = -0.000408$$

Now, by solving Eq. (7.5.3), one finds the approximate error

$$\tilde{e}_1 = -0.001674, \quad \tilde{e}_2 = -0.001451, \quad \tilde{e}_3 = -0.0003411$$

The exact error is

$$e = x - \tilde{x} = (-0.002, -0.001, 0)^T \qquad \qquad \square$$

The procedure for solving Eq. (7.5.3) may be repeated and provides the following correction method.

7.5.1 The Residual Correction Method

Let $x^{(0)}$ denote the solution to $Ax = b$, computed using Gaussian elimination. We define

$$r^{(0)} = b - Ax^{(0)} \tag{7.5.5}$$

and solve $Ae^{(0)} = r^{(0)}$, where $e^{(0)} = x - x^{(0)}$. We actually obtain $\tilde{e}^{(0)}$ and define the next approximate solution to $Ax = b$ as

$$x^{(1)} = x^{(0)} + \tilde{e}^{(0)} \tag{7.5.6}$$

The process is repeated, yielding

$$r^{(1)} = b - Ax^{(1)} \tag{7.5.7}$$
$$x^{(2)} = x^{(1)} + \tilde{e}^{(1)} \tag{7.5.8}$$

where $\tilde{e}^{(1)}$ is the approximate solution to $Ae^{(1)} = r^{(1)}$, $e^{(1)} = x - x^{(1)}$. The iteration stops when $\tilde{e}^{(k)}$ is sufficiently small for some $k \geq 0$. The procedure is called the *residual correction method* and is given by

$$Ae^{(i)} = r^{(i)}, \quad 0 \leq i \leq k \tag{7.5.9}$$

where

$$r^{(i)} = b - Ax^{(i)}, \quad e^{(i)} = x - x^{(i)}, \quad 0 \leq i \leq k \tag{7.5.10}$$

Example 7.5.2.

The 3×3 linear system

$$0.3x_1 + 0.5x_2 + 0.2x_3 = 1$$
$$0.7x_1 - 0.3x_2 - 0.4x_3 = 0$$
$$0.5x_1 + 0.1x_2 + 0.4x_3 = 1$$

has a unique solution $x = (1, 1, 1)^T$. Assuming a four-digit decimal machine with rounding, we obtain

$$x^{(0)} = (1.001, \ 0.9993, \ 1.000)^T$$
$$r^{(0)} = (0.00005, \ -0.00091, \ -0.00043)^T$$
$$\tilde{e}^{(0)} = (-0.001000, \ 0.0007001, \ 0.000002)^T$$
$$x^{(1)} = (1.000, \ 1.000, \ 1.000)^T$$

Thus one iteration is needed to produce the exact solution.   □

We will now define and discuss the concept of stability for linear systems.

7.5.2 Stability

The system $Ax = b$ is called *stable* if small changes in the coefficients of the system do not cause a sizeable change in the solution to the system. Let us examine the effect of a small change in the right-hand side column b on the solution x.

Let $\tilde{b}$ denote an approximation to b and let

$$A\tilde{x} = \tilde{b} \tag{7.5.11}$$

Thus

$$b - \tilde{b} = Ax - A\tilde{x} = A(x - \tilde{x}) \tag{7.5.12}$$

and we are interested in estimating $x - \tilde{x}$, once $b - \tilde{b}$ is known. The following example will demonstrate how a small change in b may lead to a large change in x.

Example 7.5.3.

Consider the 2×2 system

$$2x_1 + 5x_2 = 7$$
$$4x_1 + 10.001x_2 = 14.001$$

whose unique solution is $x = (1, 1)^T$. If $b = (7, 14.001)^T$ is replaced by $\tilde{b} = (7.001, 14)^T$ we obtain a perturbed system

$$2x_1 + 5x_2 = 7.001$$

$$4x_2 + 10.001x_2 = 14$$

whose unique solution $\tilde{x} = (8.5005, -2)^T$ presents a large change in x.   □

To estimate $x - \tilde{x}$ in terms of $b - \tilde{b}$, vector and matrix norms must be used. We will use any vector norm and its associated natural matrix norm.

Theorem 7.5.1.

(Error estimate). *Let A be a nonsingular matrix and let $x, b, \tilde{x},$ and $\tilde{b}$ satisfy*

$$Ax = b, \quad A\tilde{x} = \tilde{b} \tag{7.5.13}$$

Then

$$\frac{\|x - \tilde{x}\|}{\|x\|} \le \|A\|\|A^{-1}\| \frac{\|b - \tilde{b}\|}{\|b\|} \tag{7.5.14}$$

Proof.

From Eq. (7.5.13) we obtain

$$A(x - \tilde{x}) = b - \tilde{b}$$

Consequently

$$x - \tilde{x} = A^{-1}(b - \tilde{b})$$

and therefore

$$\|x - \tilde{x}\| = \|A^{-1}(b - \tilde{b})\| \le \|A^{-1}\|\|b - \tilde{b}\|$$

We now divide by $\|x\|$ to obtain

$$\frac{\|x - \tilde{x}\|}{\|x\|} \le \frac{\|A^{-1}\|\|b - \tilde{b}\|}{\|x\|} = \|A\|\|A^{-1}\| \frac{\|b - \tilde{b}\|}{\|A\|\|x\|} \tag{7.5.15}$$

By virtue of Eq. (7.5.13) we have

$$\|b\| = \|Ax\| \le \|A\|\|x\|$$

and by substituting in Eq. (7.5.15) we conclude Eq. (7.5.14).   ◻

Theorem 7.5.1 provides a bound to the relative error in $\tilde{x}$ in terms of the relative error in $\tilde{b}$ and the *condition number* defined as

$$\text{cond}(A) = \|A\|\|A^{-1}\| \tag{7.5.16}$$

If the condition number is small, the linear system is not sensitive to small changes in b. If $\text{cond}(A)$ is large, the relative error in $\tilde{x}$ may be significantly larger than the relative error in $\tilde{b}$. A matrix with a small condition number defines a well-conditioned system, whereas a matrix with a large condition number defines an *ill-conditioned* system.

Example 7.5.4.

The system given in the previous example has the matrix

$$A = \begin{bmatrix} 2 & 5 \\ 4 & 10.001 \end{bmatrix}$$

whose inverse is

$$A^{-1} = \begin{bmatrix} 5000.5 & -2500 \\ -2000 & 1000 \end{bmatrix}$$

If one uses the norms

$$\|x\| = \max_{1 \le i \le n} |x_i|$$

$$\|A\| = \max_{1 \le i \le n} \sum_{j=1}^{n} |a_{ij}|$$

for error estimate, one finds

$$\|A\| = 14.001, \quad \|A^{-1}\| = 7500.5$$

Thus, $\text{cond}(A) \approx 10^5$ and the system is obviously ill-conditioned. □

Our next aim is to determine the sensitivity of a linear system in response to small changes in the system matrix. We first prove the following lemma.

Lemma 7.5.1.

If $\|A\| < 1$ then $I + A$ is nonsingular and

$$\|(I + A)^{-1}\| \le \frac{\|I\|}{1 - \|A\|} \tag{7.5.17}$$

Proof.

If $I + A$ is singular, then $(I + A)x = 0$ for some $x \neq 0$. Thus, $x = -Ax$ and

$$\|x\| = \| - Ax\| = \|Ax\| \le \|A\|\|x\| < \|x\|$$

which is impossible. Thus, $I + A$ is nonsingular.

Now, by definition

$$I = (I + A)^{-1}(I + A) = (I + A)^{-1} + (I + A)^{-1}A$$

Therefore

$$\|I\| \ge \|(I + A)^{-1}\| - \|(I + A)^{-1}A\|$$

$$\geq \|(I+A)^{-1}\| - \|(I+A)^{-1}\|\|A\| = \|(I+A)^{-1}\|[1 - \|A\|]$$

which concludes the proof.   ⬜

The next theorem provides a bound to $\|x - \tilde{x}\|$ in terms of $\|A - \tilde{A}\|$, where $\tilde{A}$ represents a small perburbation of A.

Theorem 7.5.2.

Let x and $\tilde{x}$ solve the linear systems

$$Ax = b, \quad \tilde{A}\tilde{x} = b \tag{7.5.18}$$

where A and $\tilde{A}$ are nonsingular matrices and let

$$\|A - \tilde{A}\| < \frac{1}{\|A^{-1}\|} \tag{7.5.19}$$

Then, if $x \neq 0$,

$$\frac{\|x - \tilde{x}\|}{\|x\|} \leq \frac{\|I\| \text{cond}(A)}{1 - \text{cond}(A)(\|A - \tilde{A}\|/\|A\|)} \frac{\|A - \tilde{A}\|}{\|A\|} \tag{7.5.20}$$

[Equation (7.5.19) presents a bound on the perturbation in A.]

Proof.

By Eq. (7.5.18) we have

$$0 = Ax - \tilde{A}\tilde{x} = Ax - A\tilde{x} + A\tilde{x} - \tilde{A}\tilde{x}$$

and therefore

$$A(x - \tilde{x}) = (\tilde{A} - A)\tilde{x}$$

or

$$x - \tilde{x} = A^{-1}(\tilde{A} - A)\tilde{x}$$

Consequently

$$\|x - \tilde{x}\| = \|A^{-1}(\tilde{A} - A)\tilde{x}\| \leq \|A^{-1}\|\|A - \tilde{A}\|\|\tilde{x}\|$$

and by dividing by $\|x\|$ we obtain

$$\frac{\|x - \tilde{x}\|}{\|x\|} \leq \frac{\|A^{-1}\|\|A - \tilde{A}\|\|\tilde{x}\|}{\|x\|} = \frac{\|A\|\|A^{-1}\|\|A - \tilde{A}\|\|\tilde{x}\|}{\|A\|\|x\|} \tag{7.5.21}$$

Now, because $Ax = \tilde{A}\tilde{x}$ we also obtain

$$\tilde{A}^{-1}Ax = \tilde{x}$$

which implies $\|\tilde{x}\| \leq \|x\|\|\tilde{A}^{-1}A\|$, or

$$\frac{\|\tilde{x}\|}{\|x\|} \leq \|\tilde{A}^{-1}A\| \tag{7.5.22}$$

By virtue of Eq. (7.5.19) we find

$$\|A^{-1}\tilde{A} - I\| = \|A^{-1}(\tilde{A} - A)\| \leq \|A^{-1}\|\|\tilde{A} - A\| < 1$$

Therefore, by implementing Lemma 7.5.1,

$$\begin{aligned}
\|\tilde{A}^{-1}A\| &= \|[I + (A^{-1}\tilde{A} - I)]^{-1}\| \leq \frac{\|I\|}{1 - \|A^{-1}\tilde{A} - I\|} \\
&= \frac{\|I\|}{1 - \|A^{-1}\tilde{A} - A^{-1}A\|} \leq \frac{\|I\|}{1 - \|A^{-1}\|\|\tilde{A} - A\|} \\
&= \frac{\|I\|}{1 - \text{cond}(A)(\|A - \tilde{A}\|/\|A\|)}
\end{aligned} \tag{7.5.23}$$

By substituting Eqs. (7.5.22) and (7.5.23) into Eq. (7.5.21) we finally obtain Eq. (7.5.20).   □

For a matrix norm defined as $\|A\| = \max_{1 \leq i \leq n} \sum_{j=1}^{n} |a_{ij}|$, we have $\|I\| = 1$ and Eq. (7.5.20) is somewhat simplified.

Example 7.5.5.

Consider the 3×3 matrix

$$A = \begin{bmatrix} 2 & 1 & 0 \\ 1 & 0 & 1 \\ 0 & 1 & 3 \end{bmatrix}$$

whose inverse is

$$A^{-1} = \begin{bmatrix} 0.2 & 0.6 & -0.2 \\ 0.6 & -1.2 & 0.4 \\ -0.2 & 0.4 & 0.2 \end{bmatrix}$$

The condition number of A is

$$\text{cond}(A) = \|A\|\|A^{-1}\| = 4 \times 2.2 = 8.8$$

indicating that the system is stable. Define a perturbed matrix

$$\tilde{A} = \begin{bmatrix} 2.1 & 1 & -0.1 \\ 1 & 0.1 & 0.9 \\ 0 & 1.1 & 3.2 \end{bmatrix}$$

for which

$$A - \tilde{A} = \begin{bmatrix} -0.1 & 0 & 0.1 \\ 0 & -0.1 & 0.1 \\ 0 & -0.1 & -0.2 \end{bmatrix}, \quad \|A - \tilde{A}\| = 0.3$$

Equation (7.5.19) holds because $\|A^{-1}\|^{-1} \approx 0.45 > 0.3$. Therefore

$$\frac{\|x - \tilde{x}\|}{\|x\|} \leq \frac{8.8}{1 - 8.8(0.3/4)} \left(\frac{0.3}{4}\right) \approx 1.94 \qquad \square$$

Example 7.5.6.

A typical example of an ill-conditioned matrix is the Hilbert matrix defined as

$$H_n = \begin{bmatrix} 1 & 1/2 & 1/3 & \cdots & 1/n \\ 1/2 & 1/3 & 1/4 & \cdots & 1/(n+1) \\ \vdots & & & & \vdots \\ 1/n & 1/(n+1) & 1/(n+2) & \cdots & 1/(2n-1) \end{bmatrix} \qquad (7.5.24)$$

In the particular case $n = 3$ we have

$$H_3 = \begin{bmatrix} 1 & 1/2 & 1/3 \\ 1/2 & 1/3 & 1/4 \\ 1/3 & 1/4 & 1/5 \end{bmatrix}, \quad \|H_3\| = 1 + 1/2 + 1/3 = 11/6$$

$$H_3^{-1} = \begin{bmatrix} 9 & -36 & 30 \\ -36 & 192 & -180 \\ 30 & -180 & 180 \end{bmatrix}, \quad \|H_3^{-1}\| = 36 + 192 + 180 = 408$$

and therefore

$$\text{cond}(H_3) = (11/6) \times 408 = 748$$

$\square$

indicating instability for the system $H_3 x = b$.

In the next example we will examine an ordinary application in which ill-conditioning occurs. Generally, we are unable to predict ill-conditioning *a priori*. We must therefore use a linear equation solver that detects it or alternatively implements the residual correction method to estimate and improve the accuracy of the computed solution.

Example 7.5.7.

Let $\{(x_i, y_i)\}_{i=1}^n$ be an experimental data set. We search for a least-squares polynomial approximation

$$p_m(x) = a_m x^m + a_{m-1} x^{m-1} + \cdots + a_1 x + a_0 \qquad (7.5.25)$$

for the given data. The coefficients $a_0, a_1, \ldots, a_m$ satisfy [Eq. (5.9.11)] the linear system

$$a_0 \sum_{i=1}^{n} x_i^0 + a_1 \sum_{i=1}^{n} x_i^1 + a_2 \sum_{i=1}^{n} x_i^2 + \cdots + a_m \sum_{i=1}^{n} x_i^m = \sum_{i=1}^{n} y_i x_i^0$$

$$a_0 \sum_{i=1}^{n} x_i^1 + a_1 \sum_{i=1}^{n} x_i^2 + a_2 \sum_{i=1}^{n} x_i^3 + \cdots + a_m \sum_{i=1}^{n} x_i^{m+1} = \sum_{i=1}^{n} y_i x_i^1$$

$$\vdots$$

$$a_0 \sum_{i=1}^{n} x_i^m + a_1 \sum_{i=1}^{n} x_i^{m+1} + a_2 \sum_{i=1}^{n} x_i^{m+2} + \cdots + a_m \sum_{i=1}^{n} x_i^{2m} = \sum_{i=1}^{n} y_i x_i^m$$

and the matrix of the system can be shown to be nonsingular provided that $\{x_i\}_{i=1}^{n}$ are all distinct points. However, for large m, the matrix is ill-conditioned. This will be demonstrated for the particular data set

$$x_i = ih, \quad 1 \le i \le n; \quad nh = 1$$

where n is sufficiently large. We then have

$$\sum_{i=1}^{n} x_i^k = \frac{1}{h} \sum_{i=1}^{n} (h x_i^k) \approx \frac{1}{h} \int_0^1 x^k \, dx = \frac{1}{h} \frac{1}{k+1} \tag{7.5.26}$$

and the matrix of the system can be approximated by a Hilbert matrix of order $m + 1$. Consequently, the use of high-order least-squares polynomial approximations is not recommended. □

PROBLEMS

1. Use Gaussian elimination to solve the linear system

$$2.2x_1 + 1.4x_2 = 3.6$$
$$1.9x_1 + 1.2x_2 = 3.1$$

   assuming

   (a) two-digit arithmetic with rounding

   (b) three-digit arithmetic with rounding

   Compare the computed solutions with the exact solution $x = (1, 1)^T$.

2. (a) Use Gaussian elimination without pivoting to solve the 3×3 linear system

$$2.13x_1 - 1.5x_2 + 0.0315x_3 = 0.6615$$
$$2x_1 - x_2 + 1.555x_3 = 2.555$$
$$2.5x_1 - 1.333x_2 + 2x_3 = 3.167$$

   assuming four-digit arithmetic with rounding.

(b) Compare the computed solution with the exact solution $x = (1, 1, 1)^T$, calculate the residual r (using double-precision arithmetic), and improve your results by solving the original system with r at the right-hand side.

3. Use the residual correction method and two-digit arithmetic to solve the 2×2 system

$$7x_1 + 4x_2 = 3$$
$$1.5x_1 - 2x_2 = 2$$

whose exact solution is $x = (0.7, -0.475)^T$.

4. Calculate the condition number of

$$A = \begin{bmatrix} a & 1 \\ 1 & a \end{bmatrix}$$

and determine for what values of a the matrix is ill-conditioned.

5. Find the condition number for the following matrices

(a)

$$\begin{bmatrix} 1 & -1 & 3 \\ -1 & 2 & 0 \\ 3 & 0 & 4 \end{bmatrix}$$

(b)

$$\begin{bmatrix} -1 & 1 & 2 \\ 2 & 2 & 3 \\ 0 & 0 & -1.1 \end{bmatrix}$$

6. Show that $\operatorname{cond}(A) \geq 1$ for all A. (*Hint*: First show $\|I\| \geq 1$.)

7. Let $Ax = b$, $A\tilde{x} = \tilde{b}$, where $\tilde{b}$ represents a perturbed b and

$$A = \begin{bmatrix} 1 & 2 & 1 \\ -1 & 3 & 7 \\ 2 & 1 & 0 \end{bmatrix}, \quad \|b\| = 4, \quad \|x\| = 1$$

Find a bound on $\|b - \tilde{b}\|$, if $\|x - \tilde{x}\|$ must not exceed 10^{-3}. (Use l_∞ vector and matrix norms.)

8. Let $Ax = b$, $\tilde{A}\tilde{x} = b$, where

$$A = \begin{bmatrix} 1 & 1/2 & 1/3 & 1/4 \\ 1/2 & 1/3 & 1/4 & 1/5 \\ 1/3 & 1/4 & 1/5 & 1/6 \\ 1/4 & 1/5 & 1/6 & 1/7 \end{bmatrix}, \quad \|x\| = 1$$

and $\tilde{A}$ represents a perturbed A. Assume l_∞ vector and matrix norms and determine a bound on $\|A - \tilde{A}\|$, if $\|x - \tilde{x}\|$ must not exceed 10^{-1}.

7.6   The Eigenvalue Problem

The concept of eigenvalues and eigenvectors, introduced previously, arises naturally in most sciences, particularly in physics. A typical example is finding the natural frequences of vibration of some musical instrument. These frequencies are the eigenvalues of a specific matrix associated with the problem.

We start with basic notation. Let A denote a square $n \times n$ matrix.

Definition 7.6.1.   *Let a number λ and a vector $x \neq 0$ satisfy*

$$Ax = \lambda x \tag{7.6.1}$$

Then λ is called an eigenvalue of A and x is called an eigenvector of A.

The set of pairs $\{\lambda, x\}$ of eigenvalues and their corresponding eigenvectors will be the main subject of this section.

One can replace Eq. (7.6.1) by

$$(A - \lambda I)x = 0 \tag{7.6.2}$$

and find all the numbers λ for which the matrix $A - \lambda I$ is singular.

Example 7.6.1.

Consider the 2×2 matrix

$$A = \begin{bmatrix} 2 & 2 \\ 2 & 5 \end{bmatrix}$$

It has exactly two eigenvalues: $\lambda_1 = 1$ and $\lambda_2 = 6$. Two eigenvectors associated with these eigenvalues are

$$x_1 = \begin{bmatrix} -2 \\ 1 \end{bmatrix}, \quad x_2 = \begin{bmatrix} 1 \\ 2 \end{bmatrix}$$

respectively.   □

Clearly, if x is an eigenvector that corresponds to λ, so is αx, $\alpha \neq 0$. Thus, for an arbitrary eigenvalue, the corresponding eigenvector is determined, at the most, up to a constant.

7.6.1   The Characteristic Polynomial

By virtue of Eq. (7.6.2) a necessary and sufficient condition for a number λ to be an eigenvalue of a matrix A is

$$|A - \lambda I| = 0 \tag{7.6.3}$$

The expression $f(\lambda) = |A - \lambda I|$, which is a polynomial of degree n, is called the *characteristic polynomial* of A. It may be seen that

$$f(\lambda) = (-1)^n \lambda^n + \alpha_{n-1} \lambda^{n-1} + \cdots + \alpha_1 \lambda + \alpha_0 \tag{7.6.4}$$

Example 7.6.2.

The characteristic polynomial of the matrix in the previous example is given by

$$f(\lambda) = \begin{vmatrix} 2 - \lambda & 2 \\ 2 & 5 - \lambda \end{vmatrix} = (2 - \lambda)(5 - \lambda) - 4 = \lambda^2 - 7\lambda + 6$$

It has two zeroes: $\lambda_1 = 1$ and $\lambda_2 = 6$.   □

Example 7.6.3.

For a general 2×2 matrix $A = (a_{ij})$ we have

$$f(\lambda) = \begin{vmatrix} a_{11} - \lambda & a_{12} \\ a_{21} & a_{22} - \lambda \end{vmatrix} = (a_{11} - \lambda)(a_{22} - \lambda) - a_{12}a_{21}$$

$$= \lambda^2 - (a_{11} + a_{22})\lambda + (a_{11}a_{22} - a_{12}a_{21})$$

The two eigenvalues can be distinct or equal, real or complex. For example, if $a_{11} = a_{22} = 0$, $a_{12} = 1$, $a_{21} = -1$ then $\lambda_1 = i$, $\lambda_2 = -i$.   □

Example 7.6.4.

Consider the 3×3 matrix

$$A = \begin{bmatrix} 2 & 1 & 0 \\ 0 & 1 & 0 \\ 1 & -1 & 2 \end{bmatrix}$$

for which

$$f(\lambda) = \begin{vmatrix} 2 - \lambda & 1 & 0 \\ 0 & 1 - \lambda & 0 \\ 1 & -1 & 2 - \lambda \end{vmatrix} = (2 - \lambda)^2(1 - \lambda)$$

The eigenvalues are $\lambda_1 = \lambda_2 = 2$, $\lambda_3 = 1$. Let $x^{(1)} = [x_1^{(1)}, x_2^{(1)}, x_3^{(1)}]^T$ be an eigenvector associated with λ_1. Then

$$\begin{bmatrix} 0 & 1 & 0 \\ 0 & -1 & 0 \\ 1 & -1 & 0 \end{bmatrix} \begin{bmatrix} x_1^{(1)} \\ x_2^{(1)} \\ x_3^{(1)} \end{bmatrix} = \begin{bmatrix} 0 \\ 0 \\ 0 \end{bmatrix}$$

leading to $x_2^{(1)} = 0$, $x_1^{(1)} - x_2^{(1)} = 0$.

Thus $x_1^{(1)} = x_2^{(1)} = 0$ and we must choose $x_3^{(1)} \neq 0$. If we normalize the eigenvectors so that the maximum component is 1, we find $x^{(1)} = (0, 0, 1)$. Let $x^{(3)} = [x_1^{(3)}, x_2^{(3)}, x_3^{(3)}]^T$ be the eigenvector associated with λ_3. Then

$$\begin{bmatrix} 1 & 1 & 0 \\ 0 & 0 & 0 \\ 1 & -1 & 1 \end{bmatrix} \begin{bmatrix} x_1^{(3)} \\ x_2^{(3)} \\ x_3^{(3)} \end{bmatrix} = \begin{bmatrix} 0 \\ 0 \\ 0 \end{bmatrix}$$

that is,

$$x_1^{(3)} + x_2^{(3)} = 0, \quad x_1^{(3)} - x_2^{(3)} + x_3^{(3)} = 0$$

Clearly, if $x_3^{(3)} = 0$ then $x_1^{(3)} = x_2^{(3)} = 0$ and $x^{(3)}$ is not an eigenvector. Let us choose $x_3^{(3)} = a \neq 0$. Then $x_1^{(3)} = -a/2$, $x_2^{(3)} = a/2$, and $x^{(3)} = a(-0.5, 0.5, 1)^T$. The normalized eigenvector corresponds to $a = 1$. $\qquad\square$

7.6.2   Eigenvalues of Symmetric Matrices

Real symmetric matrices, that is, matrices whose entries are real and satisfy

$$a_{ij} = a_{ji}, \quad 1 \leq i, j \leq n$$

occur in many problems and computing their eigenvalues and eigenvectors is relatively simpler than in the general case. The basic features of these eigenvalues and eigenvectors are summarized by the next result.

Theorem 7.6.1.

Every real $n \times n$ symmetric matrix A has the following properties:

1. *The eigenvalues $\lambda_1, \ldots, \lambda_n$ of A are real.*

2. *There is a set of corresponding real eigenvectors $x^{(1)}, \ldots, x^{(n)}$ that are linearly independent, of length 1, and mutually orthogonal, that is*

$$x_1^{(i)} x_1^{(j)} + \cdots + x_n^{(i)} x_n^{(j)} = \delta_{ij}, \quad 1 \leq i, j \leq n \tag{7.6.5}$$

3. *The matrix X with columns $x^{(1)}, \ldots, x^{(n)}$ yields*

$$X^T A X = D = \begin{bmatrix} \lambda_1 & 0 & \cdots & 0 \\ 0 & \lambda_2 & \ddots & \vdots \\ \vdots & \ddots & \lambda_{n-1} & 0 \\ 0 & \cdots & 0 & \lambda_n \end{bmatrix} \tag{7.6.6}$$

and $X^{-1} = X^T$, that is,

$$X^T X = I \tag{7.6.7}$$

4. *Any arbitrary $x = (x_1, \ldots, x_n)^T$ can be expressed as*

$$x = \sum_{i=1}^{n} c_i x^{(i)} \tag{7.6.8}$$

where

$$c_i = \sum_{j=1}^{n} x_j x_j^{(i)} \tag{7.6.9}$$

The proof of this theorem is omitted.

Any matrix X that satisfies Eq. (7.6.7) is called *orthogonal*.

Example 7.6.5.

Consider the 3×3 real symmetric matrix

$$A = \begin{bmatrix} 2 & \sqrt{2} & 1 \\ \sqrt{2} & 0 & 0 \\ 1 & 0 & 0 \end{bmatrix}$$

whose characteristic polynomial is

$$f(\lambda) = \begin{vmatrix} 2-\lambda & \sqrt{2} & 1 \\ \sqrt{2} & -\lambda & 0 \\ 1 & 0 & -\lambda \end{vmatrix} = -\lambda^3 + 2\lambda^2 + 3\lambda$$

The three eigenvalues are $\lambda_1 = 0$, $\lambda_2 = 3$, $\lambda_3 = -1$. An eigenvector $x^{(1)} = [x_1^{(1)}, x_2^{(1)}, x_3^{(1)}]^T$ of λ_1 satisfies

$$\begin{bmatrix} 2 & \sqrt{2} & 1 \\ \sqrt{2} & 0 & 0 \\ 1 & 0 & 0 \end{bmatrix} \begin{bmatrix} x_1^{(1)} \\ x_2^{(1)} \\ x_3^{(1)} \end{bmatrix} = \begin{bmatrix} 0 \\ 0 \\ 0 \end{bmatrix}$$

or

$$\begin{aligned} 2x_1^{(1)} + \sqrt{2}x_2^{(1)} + x_3^{(1)} &= 0 \\ \sqrt{2}x_1^{(1)} &= 0 \\ x_1^{(1)} &= 0 \end{aligned}$$

and, after normalizing it to length 1 (in l_∞ norm), we have $x^{(1)} = (0, -1/\sqrt{3}, \sqrt{2}/\sqrt{3})^T$. Similarly we obtain

$$x^{(2)} = (\sqrt{3}/2, 1/\sqrt{6}, \sqrt{3}/6)^T, x^{(3)} = (-1/2, 1/\sqrt{2}, 1/2)^T$$

and the reader may verify that the eigenvectors are mutually orthogonal and that $X = [x^{(1)}, x^{(2)}, x^{(3)}]$ satisfies Eq. (7.6.6). $\quad\square$

If A is nonsymmetric, its characteristic polynomial may have complex roots.

Example 7.6.6.

Consider the 2×2 matrix

$$A = \begin{bmatrix} 0 & 1 \\ -1 & 1 \end{bmatrix}$$

Here

$$f(\lambda) = \begin{vmatrix} -\lambda & 1 \\ -1 & 1 - \lambda \end{vmatrix} = \lambda^2 - \lambda + 1$$

and the eigenvalues are $\lambda_1 = (1/2)(1 + i\sqrt{3})$, $\lambda_2 = (1/2)(1 - i\sqrt{3})$. □

Let a real symmetric matrix A possess an eigenvalue λ with multiplicity m. Then by virtue of Theorem 7.6.1 one can find a set of m linearly independent corresponding eigenvectors $x^{(1)}, \ldots, x^{(m)}$, of length 1 and mutually orthogonal, associated with λ. This may not be the case for nonsymmetric matrices, as demonstrated by the next example.

Example 7.6.7.

Consider the nonsymmetric matrices

$$A = \begin{bmatrix} 1 & 1 \\ 0 & 1 \end{bmatrix}, \quad B = \begin{bmatrix} 1 & 0 & 1 \\ 0 & 1 & 0 \\ 0 & 0 & 2 \end{bmatrix}$$

Matrix A possesses a single eigenvalue $\lambda = 1$ with multiplicity 2. Let $x = (x_1, x_2)^T$ be an eigenvector of A. Then $(A - I)x = 0$, that is,

$$\begin{bmatrix} 0 & 1 \\ 0 & 0 \end{bmatrix} \begin{bmatrix} x_1 \\ x_2 \end{bmatrix} = \begin{bmatrix} 0 \\ 0 \end{bmatrix}$$

Thus $x_2 = 0$ and only *one* linearly independent eigenvector exists, for example $x = (1, 0)^T$.

The second matrix, B, has an eigenvalue $\lambda_1 = 1$ with multiplicity 2 and a simple eigenvalue $\lambda_2 = 2$. Let $x = (x_1, x_2, x_3)^T$ be an eigenvector associated with λ_1. Then

$$(B - I)x = \begin{bmatrix} 0 & 0 & 1 \\ 0 & 0 & 0 \\ 0 & 0 & 1 \end{bmatrix} \begin{bmatrix} x_1 \\ x_2 \\ x_3 \end{bmatrix} = \begin{bmatrix} 0 \\ 0 \\ 0 \end{bmatrix}$$

that is, $x_3 = 0$ and we obtain *two* linearly independent eigenvectors, for example $x^{(1)} = (1, 0, 0)^T$, $x^{(2)} = (0, 1, 0)^T$. □

We shall now present an iterative algorithm for calculating the *largest* eigenvalue of a matrix, if it has a single dominant eigenvalue.

7.6.3 The Power Method

Let A be an $n \times n$ matrix and let its eigenvalues $\{\lambda_i, \ 1 \leq i \leq n\}$ satisfy

$$|\lambda_1| > |\lambda_2| \geq \cdots \geq |\lambda_n| \tag{7.6.10}$$

We also assume that the associated eigenvectors $x^{(i)}$, $1 \leq i \leq n$ are linearly independent. For arbitrary $x = (x_1, x_2, \ldots, x_n)^T$ we define

$$\|x\| = \max_{1 \leq i \leq n} |x_i| \tag{7.6.11}$$

Because $x^{(i)}$, $1 \leq i \leq n$ are linearly independent, x can be uniquely written as

$$x = c_1 x^{(1)} + \cdots + c_n x^{(n)}$$

Thus

$$Ax = A \sum_{i=1}^{n} c_i x^{(i)} = \sum_{i=1}^{n} c_i A x^{(i)} = \sum_{i=1}^{n} c_i \lambda_i x^{(i)} \tag{7.6.12}$$

and by using induction

$$A^k x = \sum_{i=1}^{n} c_i \lambda_i^k x^{(i)}, \quad k \geq 1 \tag{7.6.13}$$

We can now rewrite Eq. (7.6.13) as

$$A^k x = \lambda_1^k \left[c_1 x^{(1)} + \sum_{i=2}^{n} c_i \left(\frac{\lambda_i}{\lambda_1} \right)^k x^{(i)} \right] \tag{7.6.14}$$

and by virtue of Eq. (7.6.10) conclude

$$A^k x \approx \lambda_1^k c_1 x^{(1)}, \quad k \to \infty \tag{7.6.15}$$

provided that $c_1 \neq 0$. If the powers $A^k x$ are properly scaled, Eq. (7.6.15) can be used to approximate λ_1 and its associated eigenvector $x^{(1)}$. We start by choosing an initial guess $x = y^{(0)}$ to $x^{(1)}$, where

1. $\|y^{(0)}\| = 1$
2. $y_{m_0}^{(0)} = 1$ for some $1 \leq m_0 \leq n$

Let $z^{(1)} = Ay^{(0)}$ and define

$$\lambda^{(1)} = z_{m_0}^{(1)} = \frac{z_{m_0}^{(1)}}{y_{m_0}^{(0)}} = \frac{c_1 \lambda_1 x_{m_0}^{(1)} + c_2 \lambda_2 x_{m_0}^{(2)} + \cdots + c_n \lambda_n x_{m_0}^{(n)}}{c_1 x_{m_0}^{(1)} + c_2 x_{m_0}^{(2)} + \cdots + c_n x_{m_0}^{(n)}}$$

$$= \lambda_1 \left[\frac{c_1 x_{m_0}^{(1)} + \sum_{i=2}^{n} c_i \left(\frac{\lambda_i}{\lambda_1} \right) x_{m_0}^{(i)}}{c_1 x_{m_0}^{(1)} + \sum_{i=2}^{n} c_i x_{m_0}^{(i)}} \right]$$

Let m_1 be the first integer, $1 \leq m_1 \leq n$, for which

$$|z_{m_1}^{(1)}| = \|z^{(1)}\|$$

and define

$$y^{(1)} = \frac{z^{(1)}}{z_{m_1}^{(1)}} = \frac{1}{z_{m_1}^{(1)}} A y^{(0)}$$

which clearly implies $y_{m_1}^{(1)} = 1$. Also define

$$z^{(2)} = A y^{(1)} = \frac{1}{z_{m_1}^{(1)}} A^2 y^{(0)}$$

and

$$\lambda^{(2)} = z_{m_1}^{(2)} = \frac{z_{m_1}^{(2)}}{y_{m_1}^{(1)}} = \frac{c_1 \lambda_1^2 x_{m_1}^{(1)} + c_2 \lambda_2^2 x_{m_1}^{(2)} + \cdots + c_n \lambda_n^2 x_{m_1}^{(n)}}{c_1 \lambda_1 x_{m_1}^{(1)} + c_2 \lambda_2 x_{m_1}^{(2)} + \cdots + c_n \lambda_n x_{m_1}^{(n)}}$$

$$= \lambda_1 \left[\frac{c_1 x_{m_1}^{(1)} + \sum_{i=2}^n c_i (\lambda_i/\lambda_1)^2 x_{m_1}^{(i)}}{c_1 x_{m_1}^{(1)} + \sum_{i=2}^n c_i (\lambda_i/\lambda_1) x_{m_1}^{(i)}} \right]$$

Now let m_2 be the first integer, $1 \le m_2 \le n$, for which

$$|z_{m_2}^{(2)}| = \|z^{(2)}\|$$

and define

$$y^{(2)} = \frac{z^{(2)}}{z_{m_2}^{(2)}} = \frac{1}{z_{m_2}^{(2)} z_{m_1}^{(1)}} A^2 y^{(0)}$$

which implies $y_{m_2}^{(2)} = 1$.

Using induction we generate simultaneously two sequences of vectors $\{y^{(0)}, y^{(1)}, \ldots\}$, $\{z^{(1)}, z^{(2)}, \ldots\}$ and a sequence of numbers $\{\lambda^{(1)}, \lambda^{(2)}, \ldots\}$ that, for $k \ge 1$, satisfy

$$z^{(k)} = A y^{(k-1)} \tag{7.6.16}$$

$$\lambda^{(k)} = \lambda_1 \left[\frac{c_1 x_{m_{k-1}}^{(1)} + \sum_{i=2}^n c_i (\lambda_i/\lambda_1)^k x_{m_{k-1}}^{(i)}}{c_1 x_{m_{k-1}}^{(1)} + \sum_{i=2}^n c_i (\lambda_i/\lambda_1)^{k-1} x_{m_{k-1}}^{(i)}} \right] \tag{7.6.17}$$

$$y^{(k)} = \frac{z^{(k)}}{z_{m_k}^{(k)}} = \frac{A^k y^{(0)}}{\prod_{j=1}^k z_{m_j}^{(j)}} \tag{7.6.18}$$

where m_k is the first integer, $1 \le m_k \le n$, for which

$$|z_{m_k}^{(k)}| = \|z^{(k)}\| \tag{7.6.19}$$

In view of Eq. (7.6.10) we find

$$\lim_{k \to \infty} \lambda^{(k)} = \lambda_1 \tag{7.6.20}$$

provided that $c_1 \neq 0$. In addition, m_k becomes invariant for large k and $y^{(k)}$ converges to an eigenvector of λ_1 with norm 1.

The procedure defined by Eqs. (7.6.16) through (7.6.19) is called the *power method* and leads to convergence, if λ_1 is a single dominant eigenvalue. The restriction $c_1 \neq 0$ is not crucial, because after several iterations we should get some contribution from $x^{(1)}$ due to round-off errors, and from then on $c_1 \neq 0$.

7.6.4   Error Analysis

From Eq. (7.6.17) one can obtain

$$\lambda^{(k)} - \lambda_1 = O\left[\left(\frac{\lambda_2}{\lambda_1}\right)^k\right] \tag{7.6.21}$$

In the particular case

$$|\lambda_1| > |\lambda_2| > |\lambda_3| \geq \ldots \geq |\lambda_n| \tag{7.6.22}$$

the error in the kth iteration is

$$\lambda^{(k)} - \lambda_1 \approx K\left(\frac{\lambda_2}{\lambda_1}\right)^k \tag{7.6.23}$$

where

$$K = \frac{\lambda_1 c_2 x_m^{(2)}(\lambda_2 - \lambda_1)}{\lambda_2 c_1 x_m^{(1)}} \tag{7.6.24}$$

and $m = \lim_{k \to \infty} m_k$. Similar conclusions can be derived with respect to the rate of convergence of $y^{(k)}$ as $k \to \infty$.

Example 7.6.8.

Consider the 3×3 symmetric matrix

$$A = \begin{bmatrix} 3 & 0 & 2 \\ 0 & 0 & 0 \\ 2 & 0 & 6 \end{bmatrix}$$

The characteristic polynomial is

$$f(\lambda) = \begin{vmatrix} 3 - \lambda & 0 & 2 \\ 0 & -\lambda & 0 \\ 2 & 0 & 6 - \lambda \end{vmatrix} = -\lambda^3 + 9\lambda^2 - 14\lambda$$

and the three eigenvalues are

$$\lambda_1 = 7, \quad \lambda_2 = 2, \quad \lambda_3 = 0$$

Let us generate $z^{(k)}$, $y^{(k)}$, $\lambda^{(k)}$ from $y^{(0)} = (0.2, 0.2, 1)^T$. The results are given in Table 7.6.1 and clearly indicate $\lambda^{(k)} \to 7$ and the convergence of $y^{(k)}$ to $x^{(1)} = (0.5, 0, 1)^T$, which is an associated eigenvector with norm 1. $\qquad\qquad\qquad\qquad\qquad\qquad\qquad\qquad\qquad\qquad\qquad\qquad\qquad\qquad\qquad\qquad\square$

Table 7.6.1. The Power Method for a 3×3 Symmetric Matrix

k	$y_1^{(k)}$	$y_2^{(k)}$	$y_3^{(k)}$	$\lambda^{(k)}$	$\|x^{(1)} - y^{(k)}\|$	$\|\lambda_1 - \lambda^{(k)}\|$
0	0.20000	0.2	1		0.30000	
1	0.40625	0	1	6.40000	0.09375	0.60000
2	0.47248	0	1	6.81250	0.02752	0.19750
3	0.49207	0	1	6.94495	0.00793	0.05505
4	0.49773	0	1	6.98415	0.00227	0.01585
5	0.49935	0	1	6.99546	0.00065	0.00454

The power method is, of course, also applicable to nonsymmetric matrices, as demonstrated by the next example.

Example 7.6.9.

Consider the 4×4 nonsymmetric matrix

$$A = \begin{bmatrix} 4 & 3 & 0 & 0 \\ 1 & 1 & 0 & 1 \\ 0 & -1 & 0 & 0 \\ 0 & 10 & 0 & 1 \end{bmatrix}$$

whose eigenvalues, rounded to six digits, are

$$\lambda_1 = 5.40512, \quad \lambda_2 = 3, \quad \lambda_3 = -2.40512, \quad \lambda_4 = 0$$

The results in Table 7.6.2 follow from the initial guess $y^{(0)} = (0.5, 1, 0.5, 1)^T$.

Table 7.6.2. The Power Method Applied to a Nonsymmetric Matrix

k	$y_1^{(k)}$	$y_2^{(k)}$	$y_3^{(k)}$	$y_4^{(k)}$	$\lambda_1^{(k)}$	$\|\lambda_1 - \lambda_1^{(k)}\|$
0	0.50000	1.00000	0.50000	1		
1	0.45455	0.22727	−0.09091	1	11.00000	5.59488
2	0.76389	0.51389	−0.06944	1	3.27273	2.13239
3	0.74887	0.37104	−0.08371	1	6.13889	0.73377
4	0.87224	0.45005	−0.07877	1	4.71041	0.69471
5	0.87976	0.42220	−0.08182	1	5.50048	0.09536
6	0.91644	0.44082	−0.08085	1	5.22197	0.18315
7	0.92234	0.43587	−0.08151	1	5.40821	0.00309
8	0.93250	0.44007	−0.08134	1	5.35867	0.04645
9	0.93510	0.43931	−0.08148	1	5.40074	0.00438
10	0.93793	0.44027	−0.08146	1	5.39306	0.01206

To obtain 5-digit accuracy, 22 iterations are needed and the final iteration yields

$$y^{(22)} = (0.94051, 0.44051, -0.08150, 1)^T, \quad \lambda^{(22)} = 5.40512$$ □

Assuming Eq. (7.6.22) holds, we can use Eq. (7.6.23) to obtain

$$\lambda^{(k+1)} - \lambda_1 \approx \frac{\lambda_2}{\lambda_1}(\lambda^{(k)} - \lambda_1), \quad k \to \infty \tag{7.6.25}$$

and apply the Aitken method to form a sequence $\mu^{(k)}$ that converges to λ_1 faster than $\lambda^{(k)}$. We define

$$\mu^{(k)} = \lambda^{(k)} - \frac{[\Delta\lambda^{(k)}]^2}{\Delta^2\lambda^{(k)}} = \lambda^{(k)} - \frac{[\lambda^{(k+1)} - \lambda^{(k)}]^2}{\lambda^{(k+2)} - 2\lambda^{(k+1)} + \lambda^{(k)}}$$

and, due to Theorem 3.3.2, obtain

$$\mu^{(k+1)} - \lambda_1 \approx \left(\frac{\lambda_2}{\lambda_1}\right)^2(\mu^{(k)} - \lambda_1), \quad k \to \infty \tag{7.6.26}$$

Because $|\lambda_2/\lambda_1| < 1$, $\mu^{(k)}$ converges faster than $\lambda^{(k)}$ to λ_1, although the convergence is still linear.

Example 7.6.10.
For the matrix of Example 7.6.8 we have

$$\lambda_1 = 7, \quad \lambda_2 = 2, \quad \lambda_2/\lambda_1 = 0.28571, \quad (\lambda_2/\lambda_1)^2 = 0.08163$$

Denote

$$\alpha_k = \frac{\lambda^{(k+1)} - \lambda_1}{\lambda^{(k)} - \lambda_1}, \quad \beta_k = \frac{\mu^{(k+1)} - \lambda_1}{\mu^{(k)} - \lambda_1}$$

Then $\lim_{k\to\infty} \alpha_k = 0.28571$ and $\lim_{k\to\infty} \beta_k = 0.08163$ as indicated by Table 7.6.3. □

Table 7.6.3. Accelerating the Power Method by Using the Aitken Technique

k	$\lambda^{(k)}$	$\mu^{(k)}$	α_k	β_k
1	6.40000	7.00760	0.313	0.0815
2	6.81250	7.00062	0.294	0.0816
3	6.94495	7.00005	0.288	
4	6.98415		0.286	
5	6.99546			

The subroutine POWER (power method), which evaluates a single dominant eigenvalue and its associated eigenvector, is found on the attached floppy disk.

For real symmetric matrices, the power method can be modified in a way that will increase its rate of convergence from $O[(\lambda_2/\lambda_1)^k]$ to $O[(\lambda_2/\lambda_1)^{2k}]$. We generate the sequences $\{z^{(k)}\}$ and $\{y^{(k)}\}$

as before, but define $\lambda^{(k)}$ by

$$\lambda^{(k)} = \frac{[y^{(k)}]^T Ay^{(k)}}{[y^{(k)}]^T y^{(k)}}, \quad k \geq 0 \tag{7.6.27}$$

rather than by Eq. (7.6.17). In view of Eq. (7.6.18) and the symmetry of A (i.e., $A^T = A$) we obtain

$$\lambda^{(k)} = \frac{[A^k y^{(0)}]^T A[A^k y^{(0)}]}{[A^k y^{(0)}]^T [A^k y^{(0)}]} = \frac{[y^{(0)}]^T A^{2k+1} y^{(0)}}{[y^{(0)}]^T A^{2k} y^{(0)}} \tag{7.6.28}$$

and by using Eq. (7.6.14)

$$\lambda^{(k)} = \frac{[y^{(0)}]^T \lambda_1^{2k+1} [c_1 x^{(1)} + \sum_{i=2}^{n} c_i(\lambda_i/\lambda_1)^{2k+1} x^{(i)}]}{[y^{(0)}]^T \lambda_1^{2k} [c_1 x^{(1)} + \sum_{i=2}^{n} c_i(\lambda_i/\lambda_1)^{2k} x^{(i)}]}$$

We therefore have

$$\lambda^{(k)} - \lambda_1 = O[(\lambda_2/\lambda_1)^{2k}] \tag{7.6.29}$$

and, if Eq. (7.6.22) holds,

$$\lambda^{(k)} - \lambda_1 \approx K(\lambda_2/\lambda_1)^{2k} \tag{7.6.30}$$

and the error decreases by $(\lambda_2/\lambda_1)^2$ rather than by (λ_2/λ_1) per iteration. The additional computations needed to obtain $\lambda^{(k)}$ of Eq. (7.6.27) are negligible, because the vector $Ay^{(k)}$ must be computed by the original power method in order to obtain $z^{(k+1)}$.

It should be noted that the power method, modified to symmetric matrices by Eq. (7.6.27), does not increase the rate of convergence for the eigenvector associated with λ_1. The subroutine MPOWER (modified power method), is found on the attached floppy disk.

Example 7.6.11.

Let A denote the symmetric matrix previously considered in Example 7.6.8. The choice $y^{(0)} = (0.2, 0.2, 1)^T$ and the use of Eq. (7.6.27) for defining $\lambda^{(k)}$ provide the results in Table 7.6.4, which validate Eq. (7.6.30). □

Table 7.6.4. The Modified Power Method Applied to a Symmetric Matrix

k	$y_1^{(k)}$	$y_2^{(k)}$	$y_3^{(k)}$	$\lambda^{(k)}$	$\|x^{(1)} - y^{(k)}\|$	$|\lambda_1 - \lambda^{(k)}|$
0	0.20000	0.2	1	6.40741	0.30000	0.59259
1	0.40625	0	1	6.96982	0.09375	0.03018
2	0.47248	0	1	6.99752	0.02752	0.00248
3	0.49207	0	1	6.99980	0.00793	0.00020
4	0.49773	0	1	6.99998	0.00227	0.00002
5	0.49935	0	1	7.00000	0.00065	0.00000

We will next discuss the problem of calculating the remaining eigenvalues of a matrix, once the single dominant one is known.

7.6.5 The Deflation Method

Let A denote a matrix whose eigenvalues $\{\lambda_i\}_{i=1}^n$ are distinct, nonzero, and monotonically decreasing, that is,

$$|\lambda_1| > |\lambda_2| > \cdots > |\lambda_n| > 0 \qquad (7.6.31)$$

and let $\{x^{(i)}\}_{i=1}^n$ denote the associated eigenvectors. Once λ_1 is computed by the power method, A can be replaced by a matrix B whose eigenvalues are $\lambda_2, \lambda_3, \ldots, \lambda_n, 0$. The process of generating B is called *deflation*.

Theorem 7.6.2.

(A deflation technique). *Let $\{\lambda_i\}_{i=1}^n$ satisfy Eq. (7.6.31) and let $\{x^{(i)}\}_{i=1}^n$ be their associated eigenvectors. If $z^T x^{(1)} = 1$ for some given vector z, then the matrix*

$$B = A - \lambda_1 x^{(1)} z^T \qquad (7.6.32)$$

has the eigenvalues $\lambda_2, \lambda_3, \ldots, \lambda_n, 0$.

Proof.
Because

$$Bx^{(1)} = [A - \lambda_1 x^{(1)} z^T] x^{(1)} = Ax^{(1)} - \lambda_1 x^{(1)}[z^T x^{(1)}] = Ax^{(1)} - \lambda_1 x^{(1)} = 0$$

the matrix B has the eigenvalue zero, with the associated eigenvector $x^{(1)}$. For $\lambda_i, i \neq 1$ to be an eigenvalue of B, an eigenvector $y^{(i)} \neq 0$ must be found such that $By^{(i)} = \lambda_i y^{(i)}$, $2 \leq i \leq n$. The vector

$$y^{(i)} = \alpha_i x^{(i)} + \beta_i x^{(1)} \qquad (7.6.33)$$

satisfies
$$(B - \lambda_i I)y^{(i)} = [A - \lambda_1 x^{(1)} z^T - \lambda_i I][\alpha_i x^{(i)} + \beta_i x^{(1)}] = -\lambda_1 \alpha_i x^{(1)} z^T x^{(i)} - \lambda_i \beta_i x^{(1)}$$

and the particular choice

$$\alpha_i = \frac{1}{\lambda_i - \lambda_1}, \quad \beta_i = -\frac{\lambda_1}{\lambda_i(\lambda_i - \lambda_1)} z^T x^{(i)} \qquad (7.6.34)$$

guarantees $(B - \lambda_i I)y^{(i)} = 0$. Because $x^{(i)}$ and $x^{(1)}$ are linearly independent and $\alpha_i \neq 0$, we have $y^{(i)} \neq 0$, $2 \leq i \leq n$. Thus λ_i is an eigenvalue of B and $y^{(i)}$ is its associated eigenvector (with respect to B). $\qquad \square$

The relation that exists between the eigenvectors of A and B is our next result.

Theorem 7.6.3.
For each i, $2 \leq i \leq n$ we have

$$x^{(i)} = (\lambda_i - \lambda_1)y^{(i)} + \lambda_1 \left[z^T y^{(i)} \right] x^{(1)} \qquad (7.6.35)$$

Proof.

Denote the right-hand side of Eq. (7.6.35) by $w^{(i)}$. We use Eqs. (7.6.33) and (7.6.34) and rewrite $w^{(i)}$ as

$$
\begin{aligned}
w^{(i)} &= (\lambda_i - \lambda_1)\left\{ \frac{1}{\lambda_i - \lambda_1} x^{(i)} - \frac{\lambda_1}{\lambda_i(\lambda_i - \lambda_1)}\left[z^T x^{(i)}\right] x^{(1)} \right\} \\
&\quad + \lambda_1\left(z^T \left\{ \frac{1}{\lambda_i - \lambda_1} x^{(i)} - \frac{\lambda_1}{\lambda_i(\lambda_i - \lambda_1)}\left[z^T x^{(i)}\right] x^{(1)} \right\} \right) x^{(1)} \\
&= x^{(i)} - \left(\frac{\lambda_1}{\lambda_i}\right)\left[z^T x^{(i)}\right] x^{(1)} + \left(\frac{\lambda_1}{\lambda_i - \lambda_1}\right)\left[z^T x^{(i)}\right] x^{(1)} - \frac{\lambda_1^2}{\lambda_i(\lambda_i - \lambda_1)}\left[z^T x^{(i)}\right] x^{(1)} \\
&= x^{(i)} - \left[\frac{\lambda_1}{\lambda_i} - \frac{\lambda_1}{\lambda_i - \lambda_1} + \frac{\lambda_1^2}{\lambda_i(\lambda_i - \lambda_1)}\right]\left[z^T x^{(i)}\right] x^{(1)} \\
&= x^{(i)}
\end{aligned}
$$

$\square$

The results of Theorems 7.6.2 and 7.6.3 enable us to reapply the power method several times, computing each time a smaller eigenvalue of A (which is the single dominant eigenvalue of a new matrix), along with its associated eigenvector with respect to the *original* matrix A.

We will now present a popular deflation procedure.

7.6.6   Wielandt's Deflation Technique

Because $x^{(1)}$ is a nonzero vector, at least one of its components, say $x_k^{(1)}$, is nonzero. Define

$$
z = \frac{1}{\lambda_1 x_k^{(1)}}
\begin{bmatrix}
a_{k1} \\
a_{k2} \\
\vdots \\
a_{kn}
\end{bmatrix}
\tag{7.6.36}
$$

where a_{kj}, $1 \le j \le n$ are the entries of the kth row of A. Consequently

$$
z^T x^{(1)} = \frac{1}{\lambda_1 x_k^{(1)}}(a_{k1}, \ldots, a_{kn})\left[x_1^{(1)}, \ldots, x_n^{(1)}\right]^T = \frac{1}{\lambda_1 x_k^{(1)}}\sum_{j=1}^{n} a_{kj} x_j^{(1)}
$$

Because $Ax^{(1)} = \lambda_1 x^{(1)}$, we have

$$
\sum_{j=1}^{n} a_{ij} x_j^{(1)} = \lambda_1 x_i^{(1)}, \quad 1 \le i \le n
$$

and in particular

$$
\sum_{j=1}^{n} a_{kj} x_j^{(1)} = \lambda_1 x_k^{(1)}
$$

that is, $z^T x^{(1)} = 1$. Therefore the matrix $B = (b_{ij})$, defined by Eq. (7.6.32), has the eigenvalues $\lambda_2, \lambda_3, \ldots, \lambda_n, 0$ and the associated eigenvectors $y^{(2)}, y^{(3)}, \ldots, y^{(n)}, x^{(1)}$, respectively, and $y^{(i)}, 2 \leq i \leq n$ are given by Eqs. (7.6.33) and (7.6.34). Also, for this particular choice of z,

$$b_{kj} = a_{kj} - \lambda_1 x_k^{(1)} z_j = a_{kj} - \lambda_1 x_k^{(1)} \left[\frac{1}{\lambda_1 x_k^{(1)}} \right] a_{kj} = 0$$

that is, the entire kth row of B vanishes.

Consider now the relation $B y^{(i)} = \lambda_i y^{(i)}$, $2 \leq i \leq n$. Because $\lambda_i \neq 0$ and $b_{kj} = 0, 1 \leq j \leq n$, we obtain $y_k^{(i)} = 0$. Thus, for all computational purposes B can be replaced by an $(n-1) \times (n-1)$ matrix B', by deleting the kth row and column of B. The new matrix B' will have the eigenvalues $\lambda_2, \ldots, \lambda_n$ and the associated eigenvectors $y'^{(2)}, \ldots, y'^{(n)}$, which are obtained from $y^{(i)}$, $2 \leq i \leq n$ by deleting their kth vanishing components.

We can now reapply the power method to find λ_2 and an associated eigenvector $y'^{(2)}$. To find an associated eigenvector with respect to A, we first construct $y^{(2)}$ by inserting 0 between the $(k-1)$th and kth components of $y'^{(2)}$, and then use Eq. (7.6.35) to obtain $x^{(2)}$.

The deflation procedure based on Eq. (7.6.36) is called *Wielandt's deflation technique*.

Example 7.6.12.

Consider the matrix

$$A = \begin{bmatrix} 2 & 1 & 0 \\ 1 & 3 & 1 \\ 0 & 1 & 2 \end{bmatrix}$$

whose eigenvalues $\lambda_1 = 4$, $\lambda_2 = 2$, $\lambda_3 = 1$ satisfy Eq. (7.6.31). Assume that the dominant eigenvalue $\lambda_1 = 4$ and its associated eigenvector $x^{(1)} = (1, 2, 1)^T$ have already been calculated. Because $x_1^{(1)} \neq 0$ we can choose

$$z = \frac{1}{4}(2, 1, 0)^T$$

and obtain

$$B = A - \lambda_1 x^{(1)} z^T = \begin{bmatrix} 2 & 1 & 0 \\ 1 & 3 & 1 \\ 0 & 1 & 2 \end{bmatrix} - \begin{bmatrix} 1 \\ 2 \\ 1 \end{bmatrix} (2\ 1\ 0) = \begin{bmatrix} 0 & 0 & 0 \\ -3 & 1 & 1 \\ -2 & 0 & 2 \end{bmatrix}$$

We now delete the first row and column of B to obtain

$$B' = \begin{bmatrix} 1 & 1 \\ 0 & 2 \end{bmatrix}$$

whose eigenvalues are $\lambda_2 = 2$, $\lambda_3 = 1$. The associated eigenvectors with respect to B' are

$$y'^{(2)} = (1, 1)^T, \quad y'^{(3)} = (1, 0)^T$$

and thus

$$y^{(2)} = (0, 1, 1)^T, \quad y^{(3)} = (0, 1, 0)^T$$

By applying Eq. (7.6.35) we determine

$$
\begin{aligned}
x^{(2)} &= -2y^{(2)} + 4[z^T y^{(2)}]x^{(1)} \\
&= -2(0, 1, 1)^T + [(2, 1, 0)(0, 1, 1)^T](1, 2, 1)^T = (1, 0, -1)^T \\
x^{(3)} &= -3y^{(3)} + 4[z^T y^{(3)}]x^{(1)} \\
&= -3(0, 1, 0)^T + [(2, 1, 0)(0, 1, 0)^T](1, 2, 1)^T = (1, -1, 1)^T \qquad \square
\end{aligned}
$$

PROBLEMS

1. Find the eigenvalues of the following matrices:
   (a)

$$A = \begin{bmatrix} 1 & 2 \\ 2 & -1 \end{bmatrix}$$

   (b)

$$B = \begin{bmatrix} 1 & 0 \\ 0 & 0 \end{bmatrix}$$

   (c)

$$C = \begin{bmatrix} 1 & 0 & 1 \\ 3 & -1 & 0 \\ 2 & 0 & 5 \end{bmatrix}$$

2. Find the eigenvalues of the general 2×2 matrix

$$A = \begin{bmatrix} a & b \\ c & d \end{bmatrix}$$

3. Two $n \times n$ matrices A and B are called *similar*, if

$$B = TAT^{-1}$$

   for some nonsingular matrix T. Show that similar matrices must have the same eigenvalues.

4. Let A be a real symmetric matrix. Show that each eigenvalue of A is real and that distinct eigenvalues possess orthogonal eigenvectors.

5. Show that a set of mutually orthogonal vectors is also linearly independent.

6. A useful result, related to the location of the eigenvalues of a general matrix, is Gerschgorin's theorem: let $A = (a_{ij})$ be an $n \times n$ matrix and let R_i denote a circle in the complex plane, centered at a_{ii}, with radius

$$r_i = \sum_{\substack{j=1 \\ j \neq i}}^{n} |a_{ij}|$$

that is,

$$R_i = \{z : |z - a_{ii}| \leq r_i\}$$

Then all the eigenvalues of A are located within $R = \bigcup_{i=1}^{n} R_i$. Furthermore, *any* union of k circles that does not intersect the union of the remaining $(n - k)$ circles contains exactly k eigenvalues. Use Gerschgorin's theorem to determine bounds and location for the eigenvalues of the following matrices:

(a)

$$\begin{bmatrix} 1 & 1 & -1 \\ -1 & 0 & 0 \\ 1 & 1 & 3 \end{bmatrix}$$

(b)

$$\begin{bmatrix} 7 & 1 & -1 \\ 1 & 2 & 0.5 \\ 0 & 1 & 1 \end{bmatrix}$$

(c)

$$\begin{bmatrix} 1 & 0 & 0.4 & 0 \\ 0.5 & i & 0 & 0 \\ 1 & 0 & 3 & 0 \\ 0.5 & 0 & 0.5 & -i \end{bmatrix}$$

7. Use the power method to determine the single dominant eigenvalue and its associated eigenvector of the following matrices:

(a)

$$\begin{bmatrix} 4 & 2 & 0 \\ -1 & 8 & 1 \\ 0 & -1 & 1 \end{bmatrix}$$

(b)

$$\begin{bmatrix} 1 & 0 & 0 \\ 1 & 5 & 2 \\ 0 & 2 & 6 \end{bmatrix}$$

(c)

$$\begin{bmatrix} 8 & 2 & 1 \\ 2 & 0 & 1 \\ 1 & 1 & 3 \end{bmatrix}$$

8. Apply the modified power method to calculate the dominant eigenvalue in problem 7 [case (c)].

9. Apply Aitken's method to accelerate the convergence of the power method in problem 7 [cases (a) and (b)].

10. Apply the power method and the Wielandt deflation technique to evaluate the two largest eigenvalues and the associated eigenvectors of the following matrices:

   (a)

$$\begin{bmatrix} 18 & -1 & 0 \\ -1 & 5 & 1 \\ 0 & 1 & 1 \end{bmatrix}$$

   (b)

$$\begin{bmatrix} 1 & -1 & 0 \\ -2 & 4 & -2 \\ 0 & -1 & 2 \end{bmatrix}$$

   (c)

$$\begin{bmatrix} -7 & 13 & -16 \\ 13 & -10 & 13 \\ -16 & 13 & -7 \end{bmatrix}$$

8

Numerical Solutions of Differential Equations

Mathematical models that are associated with problems in science and engineering often lead to some relation between an unknown function and several of its derivatives. Such a relation is called a *differential equation*. In this chapter we discuss the problem of finding numerical solutions to differential equations.

8.1 Preliminaries

Let $f(x, y)$ be a real-valued function defined over the domain

$$R = \{(x, y) \mid a \leq x \leq b, \quad -\infty < y < \infty\}$$

The equation

$$y' = f(x, y) \tag{8.1.1}$$

is called an *ordinary differential equation* of the first order. This equation represents the following problem: find a function $y(x) \in C^1[a, b]$ such that

$$y'(x) = f[x, y(x)], \quad a \leq x \leq b$$

If $y(x)$ is such a function, it is called a *solution* to the differential equation [given by Eq. (8.1.1)].

Example 8.1.1.
The solutions of the differential equation $y' = y$ are given by $y(x) = Ce^x$, where C is arbitrary constant. ☐

Example 8.1.2.
Let $y' = y^{2/3}$. All the solutions of this differential equation are given by

$$\begin{cases} y = \left(\dfrac{x+C}{3}\right)^3, & -\infty < C < \infty \\ y = 0 \end{cases}$$
☐

Generally we are interested only in a specific solution that cuts through a given point, say (x_0, y_0), that is, $y(x_0) = y_0$.

Definition 8.1.1.   (An initial-value problem). *The problem of finding a solution to*

$$y' = f(x, y)$$
$$y(x_0) = y_0 \qquad (8.1.2)$$

is called an initial-value problem, *with the initial condition* $y(x_0) = y_0$.

Example 8.1.3.

The initial value problem $y' = y, y(0) = 2$ has the unique solution $y = 2e^x$.   □

If certain requirements are fulfilled, the initial value problem defined by Eq. (8.1.2) possesses a unique solution. However, exceptions exist. The next example is an initial value problem with two different solutions.

Example 8.1.4.

Consider the problem $y' = y^{2/3}, y(0) = 0$. The functions $y_1(x) = x^3/27, y_2(x) = 0$ both satisfy the differential equation and the initial condition.   □

In some cases, solutions to differential equations may be expressed in terms of elementary functions or indefinite integrals of such functions. But generally this is not possible and we must count on numerical methods to *approximate* the solution.

The next theorem provides the requirements to existence and uniqueness of a solution to the initial-value problem.

Theorem 8.1.1.

Let $f(x, y)$ be continuous in the domain

$$R = \{(x, y) \mid a \leq x \leq b, \quad -\infty < y < \infty\}$$

and let there exist a constant L such that

$$|f(x, y_1) - f(x, y_2)| \leq L|y_1 - y_2| \qquad (8.1.3)$$

for all $a \leq x \leq b$ and $-\infty < y_1, y_2 < \infty$. Then the initial-value problem defined by Eq. (8.1.2) possesses a unique solution.

The *Lipschitz condition* given by Eq. (8.1.3) guarantees the uniqueness of the solution. Its existence follows from the continuity of $f(x, y)$.

The proof of Theorem 8.1.1 is found in most textbooks on differential equations. The method frequently used to prove this theorem is the method of successive approximations. It is a constructive method that consumes an enormous amount of computing time and is therefore numerically impractical.

A particularly important case of Eq. (8.1.1) is

$$y' = p(x)y(x) + q(x) \tag{8.1.4}$$

This is a *linear equation* of the first order. It is generally required that $p(x)$ and $q(x)$ be continuous over the interval $[a, b]$. Thus the Lipschitz condition exists (why?) and the associated initial-value problem must yield a unique solution.

Theorem 8.1.2.

(Solution to a linear initial-value problem). *The unique solution to*

$$y' = p(x)y(x) + q(x), \quad y(x_0) = y_0 \tag{8.1.5}$$

where $p(x)$ and $q(x)$ are continuous over $[x_0, b]$, is given by

$$y(x) = \frac{\int_{x_0}^x r(t)q(t)\, dt + y_0}{r(x)} \tag{8.1.6}$$

where

$$r(x) = e^{-\int_{x_0}^x p(t)\, dt} \tag{8.1.7}$$

Proof.

Substitute $y(x)$ of Eqs. (8.1.6) and (8.1.7) into Eq. (8.1.5). ▯

Example 8.1.5.

Consider the linear initial-value problem

$$y' = \frac{y}{x} + 1, \quad 1 \le x \le 2; \quad y(1) = 0$$

Substitute $p(t) = 1/t$ into Eq. (8.1.7) to obtain $r(x) = 1/x$. Then, by Eq. (8.1.6), $y = x \ln x$. ▯

Example 8.1.6.

Consider the case in which $p(x)$ is a constant α. Here $r(x) = e^{-\int_{x_0}^x \alpha\, dt} = e^{\alpha(x_0 - x)}$ and therefore $y(x) = e^{\alpha(x - x_0)}\left[\int_{x_0}^x e^{\alpha(x_0 - t)}q(t)\, dt + y_0\right]$. If, for example, $\alpha = 1$, $q(x) = 1, x_0 = 0$, $y_0 = 0$ we obtain $y' = y + 1$ and the unique solution is $y = e^x - 1$. ▯

Another interesting and important class of differential equations of the first order is

$$y' = p(x)y^2 + q(x)y + r(x) \tag{8.1.8}$$

This is clearly a class of *nonlinear equations* and the various choices of $p(x)$, $q(x)$, and $r(x)$ yield classes of solutions with different behaviors. If a particular solution $y_1(x)$ of Eq. (8.1.8) is known, the general solution $y(x)$ may be written as

$$y(x) = y_1(x) + 1/v(x) \tag{8.1.9}$$

and by substituting in Eq. (8.1.8) we obtain a linear equation of the first order for $v(x)$. The proof is left as an exercise for the reader.

Example 8.1.7.

Consider the nonlinear equation $y' = y^2 + 3 - x^2$. It is easily seen that $y_1(x) = x - (1/x)$ is a solution. Define the general solution as $y(x) = y_1(x) + [1/v(x)]$ and obtain $v' = -2[x - (1/x)]v - 1$.

$\square$

8.1.1   Stability

The concept of *stability* in differential equations is defined in relation to

1. The initial-value problem [Eq. (8.1.2)]

2. The numerical procedures for solving Eq. (8.1.2)

In this section we will discuss stability of the first type.

Consider the initial-value problems

$$y' = f(x, y), \quad y(x_0) = y_0; \quad x_0 \le x \le b$$
$$y' = f(x, y), \quad y(x_0) = y_0 + \epsilon; \quad x_0 \le x \le b \tag{8.1.10}$$

which we assume to possess the unique solutions $y(x)$ and $y_\epsilon(x)$, respectively.

Definition 8.1.2.   (Stability of the initial-value problem). *The initial-value problem*

$$y' = f(x, y), \quad y(x_0) = y_0; \quad x_0 \le x \le b$$

is said to be stable *if*

$$\max_{x_0 \le x \le b} |y(x) - y_\epsilon(x)| \to 0 \text{ as } \epsilon \to 0 \tag{8.1.11}$$

It can be shown that under the requirements of Theorem 8.1.1

$$\max_{x_0 \le x \le b} |y(x) - y_\epsilon(x)| \le c|\epsilon|, \quad |\epsilon| \le \epsilon_0 \tag{8.1.12}$$

for some $\epsilon_0 > 0$ and $c > 0$. This automatically implies that the general initial-value problem is stable, that is, a "small" change in the initial-value y_0 causes a "small" change in the solution.

Example 8.1.8.

The initial-value problem

$$y' = -y + e^{-x}, \quad 0 \le x \le 1, \quad y(0) = 0$$

yields the unique solution $y(x) = xe^{-x}$. The *perturbed* problem is by definition

$$y'_\epsilon = -y_\epsilon + e^{-x}, \quad 0 \le x \le 1, \quad y_\epsilon(0) = \epsilon$$

and is solved uniquely by $y_\epsilon(x) = xe^{-x} + \epsilon e^{-x}$. Thus,

$$|y(x) - y_\epsilon(x)| = |\epsilon e^{-x}| \le |\epsilon| \to 0 \text{ as } \epsilon \to 0$$

and the problem is classified as *stable*.   □

By virtue of Eq. (8.1.12) we are motivated to believe that every initial-value problem is stable. This is quite deceiving, as shown by the next example.

Example 8.1.9.

Consider the initial-value problem

$$y' = \alpha y + e^{\alpha x}, \quad 0 \le x \le b, \quad y(0) = 0$$

whose unique solution is $y = xe^{\alpha x}$. The perturbed problem is

$$y'_\epsilon = \alpha y_\epsilon + e^{\alpha x}, \quad 0 \le x \le b, \quad y_\epsilon(0) = \epsilon$$

and has the solution $y_\epsilon = xe^{\alpha x} + \epsilon e^{\alpha x}$. The error is $y - y_\epsilon = -\epsilon e^{\alpha x}$ and satisfies

$$\max |y - y_\epsilon| = |\epsilon| \max(e^{\alpha x}) = \begin{cases} |\epsilon| , & \alpha \le 0 \\ |\epsilon| e^{\alpha b}, & \alpha > 0 \end{cases}$$

Because Eq. (8.1.12) holds with $c = \max\{1, e^{\alpha b}\}$, the initial-value problem is stable. However, let us distinguish between two cases:

1. $\alpha \le 0$: The error decreases as x increases and is *uniformly* bounded by the "small" number $|\epsilon|$. The problem is not only stable but also *well conditioned*.

2. $\alpha > 0$: The error increases with x. If αb is sufficiently large, so is the error at $x = b$. The problem, although formally stable, is classified as *ill conditioned* for large αb and well conditioned for small αb.   □

Let us now consider the general initial-value problems with initial values y_0 and $y_0 + \epsilon$ [Eq.(8.1.10)]. It can be shown that

$$y(x) - y_\epsilon(x) \approx -\epsilon e^{\int_{x_0}^x g(t)\, dt} \tag{8.1.13}$$

where $g(t) = \partial f[t, y(t)]/\partial y$ provided that x is sufficiently close to x_0.

Example 8.1.10.

The initial-value problem

$$y' = -y^2, \quad y(0) = 1, \quad x \geq 0$$

possesses a unique solution $y = 1/(x+1)$. Define a perturbed problem by

$$y'_\epsilon = -y^2_\epsilon, \quad y_\epsilon(0) = 1 + \epsilon$$

Because $g(t) = -2y(t) = -2/(1+t)$ we have

$$y(x) - y_\epsilon(x) \approx -\epsilon \exp\left\{-\int_0^x [2/(1+t)]\, dt\right\} = -\frac{\epsilon}{(1+x)^2}, \quad x \geq 0$$

and the problem is definitely well conditioned, because $|y(x) - y_\epsilon(x)| \leq \epsilon$. Alternatively, we may find the perturbed solution $y_\epsilon(x) = 1/\{x + [1/(1+\epsilon)]\}$ and obtain

$$|y(x) - y_\epsilon(x)| = \left| \frac{1}{x+1} - \frac{1}{x + [1/(1+\epsilon)]} \right| = -\frac{\epsilon}{(1+\epsilon)(x+1)(x+\frac{1}{1+\epsilon})} \approx -\frac{\epsilon}{(1+x)^2}$$

in agreement with the previous result. □

In general, if

$$\frac{\partial f[x, y(x)]}{\partial y} \leq 0, \quad x_0 \leq x \leq b \tag{8.1.14}$$

it can be shown that $|y(x) - y_\epsilon(x)|$ decreases when x increases and the initial-value problem is said to be well conditioned. For example, the problem

$$y' = -(1+x^2)y + \sin^2 x, \quad y(0) = 1$$

is well conditioned, because $\partial f/\partial y = -(1+x^2) < 0$.

8.1.2   Differential Equations of Higher Order

Ordinary differential equations (ODEs) of the first order are just one class among many. In general, the equation

$$y^{(n)} = f[x, y', \ldots, y^{(n-1)}] \tag{8.1.15}$$

is called an ordinary differential equation of order n. For example,

$$y''' = (y'')^2 - xy' + xy^2$$

is a third order ODE. In particular, the equation

$$y^{(n)} = \sum_{i=0}^{n-1} p_i(x) y^{(i)}(x) + q(x) \tag{8.1.16}$$

is called a linear ODE of order n. An initial-value problem of order n is defined by an ODE of order n [Eq. (8.1.15)] associated with n initial values of $y, y', \ldots, y^{(n-1)}$ given at some initial point x_0:

$$y^{(n)} = f[x, y', \ldots, y^{(n-1)}]$$
$$y(x_0) = y_0, \quad y'(x_0) = y'_0, \ldots, y^{(n-1)}(x_0) = y_0^{(n-1)} \tag{8.1.17}$$

Example 8.1.11.

Consider the second order initial-value problem

$$y'' = -y, \quad y(0) = 1, \quad y'(0) = 2$$

The general solution is given by $y = A\cos x + B\sin x$, whereas the unique solution that also satisfies the initial conditions is $y(x) = \cos x + 2\sin x$. □

We next define a system of simultaneous differential equations.

Definition 8.1.3. (System of ODEs). *A system*

$$y'_i = f_i(x, y_1, \ldots, y_n), 1 \le i \le n \tag{8.1.18}$$

with n unknown functions $y_1(x), \ldots, y_n(x)$ is called a system of n simultaneous ODEs of the first order. The associated initial-value problem is Eq. (8.1.18), with given initial conditions at some initial point x_0,

$$y_1(x_0) = y_{10}, \quad y_2(x_0) = y_{20}, \ldots, y_n(x_0) = y_{n0} \tag{8.1.19}$$

Example 8.1.12.

Consider

$$y'_1 = 2y_1 - y_2 + x, \quad y'_2 = y_1 y_2 + x^2$$

This is a system of two simultaneous nonlinear ODEs of the first order. The nonlinearity follows only from the second equation, but reflects on the nature of the whole system. To define an initial-value problem we must impose initial conditions at some given point, for example,

$$y_1(1) = -2, \quad y_2(1) = 3.5$$ □

It should be noted that an ODE of order n, given by Eq. (8.1.15), may be alternatively defined by a system of n simultaneous ODEs of the first order. Denote

$$y_1 = y, \quad y_2 = y', \ldots, y_n = y^{(n-1)}$$

Then Eq. (8.1.15) can be replaced by

$$y_i' = y_{i+1}, \quad 1 \le i \le n-1$$
$$y_n' = f(x, y_1, \ldots, y_n) \tag{8.1.20}$$

which is a special case of Eq. (8.1.18). The initial-value problem associated with Eq. (8.1.20) is defined by the requirements

$$y_1(x_0) = y_0, \quad y_2(x_0) = y_0', \ldots, y_n(x_0) = y_0^{(n-1)} \tag{8.1.21}$$

Example 8.1.13.
Consider the initial-value problem

$$y''' = -\frac{1}{2}xy'' - \frac{1}{2}xy - 3\sin x$$
$$y(0) = 0, \quad y'(0) = 0, \quad y''(0) = 2$$

whose unique solution is $y = x\sin x$. It can be rewritten as

$$y_1' = y_2, \quad y_2' = y_3, \quad y_3' = -\frac{1}{2}xy_3 - \frac{1}{2}xy_1 - 3\sin x$$

with initial conditions $y_1(0) = 0$, $y_2(0) = 0$, $y_3(0) = 2$. The unique solution of the system is

$$y_1 = x\sin x, \quad y_2 = \sin x + x\cos x, \quad y_3 = 2\cos x - x\sin x \qquad \square$$

Because we are actually interested only in $y_1(x)$, we may consider the calculation of $y_2(x)$ and $y_3(x)$ unnecessary. Yet y_2 and y_3 are the first and second derivatives of the solution and might often be of use. Also, solving an ODE of the first order rather than of order n is a significant reduction in problem complexity and is therefore an advantage.

8.1.3 Boundary Value Problems

Quite often the information associated with a solution $y(x)$ of an ODE is related to several points rather that to one initial point. Such a problem is called a *boundary value problem*. It generally consists of an ODE associated with two points x_0 and x_1 where *boundary values*, that is, the numerical values of the solution and/or its derivatives, are specified. The solution is then calculated at the interval $[x_0, x_1]$.

Example 8.1.14.
Consider the problem $y'' = y$, $0 \le x \le 1$, $y(0) = 1$, $y(1) = 2$. The general solution is $y = Ae^x + Be^{-x}$, and once the boundary conditions are imposed we obtain

$$y = \frac{2e-1}{e^2-1}e^x + \frac{e(e-2)}{e^2-1}e^{-x}$$

Another set of boundary conditions is, for example, $y(0) = 1$, $y'(1) = 0$, leading to the solution

$$y = \frac{1}{1 + e^2} e^x + \frac{e^2}{1 + e^2} e^{-x} \qquad \square$$

In contrast to an initial-value problem, which under some "general" requirements has a unique solution, a boundary value problem may often possess more than one solution. For example, all the functions $y(x) = A \sin x$, $-\infty < A < \infty$ solve the boundary value problem

$$y'' = -y, \quad y(0) = y(\pi) = 0$$

We will next discuss various algorithms for solving a single ODE, then extend them to systems.

PROBLEMS

1. Find the general solution to the following differential equations of the first order.
   (a) $y' = y - 1$
   (b) $y' = xy^\alpha, \alpha > 0$
   (c) $y' = f(x)g(y)$

2. Solve the following initial-value problems:
   (a) $y' = y - 1$, $y(0) = 3$
   (b) $y' = xy$, $y(0) = 2$
   (c) $y' = x^2 y^3$, $y(1) = 1$

3. Check for the existence of a Lipschitz condition in the following cases:
   (a) $f(x, y) = x \sin y$, $0 \le x \le 1$
   (b) $f(x, y) = (1 + x^2 + y^2)^{1/2}$, $0 \le x \le 1$
   (c) $f(x, y) = g(x)y^\alpha$, $a \le x \le b$

4. Give explicit solutions to the following initial-value problems, using Eqs. (8.1.6) and (8.1.7).
   (a) $y' = 2y + 5$, $y(0) = 0$
   (b) $y' = y \ln x + x^x$, $y(1) = 0$
   (c) $y' = (1/x)y + 1/x^2$, $y(1) = -1/2$

5. Find a particular solution $y_1(x) = x^\alpha$ to

$$y' = y^2 + xy - \frac{2}{x^2} - 1$$

   and use $y_1(x)$ in determining the general solution.

6. Check the following initial-value problems for stability and well conditioning
   (a) $y' = y + \alpha x$, $y(0) = 1$, $x \ge 0$
   (b) $y' = y^2$, $y(0) = 1$, $0 \le x \le b < 1$

(c) $y' = -y - y^2$, $y(0) = 1$, $x \geq 0$

7. Write the initial-value problem

$$y''' = xy''y' + x^2y$$
$$y(0) = 0, \quad y'(0) = 1, \quad y''(0) = 0$$

in the form of simultaneous ODEs of the first order.

8. Solve directly the initial-value problem

$$y'' = -y'$$
$$y(0) = 1, \quad y'(0) = 1$$

by first substituting $y' = u$. How does the solution change if the imposed initial conditions are $y(0) = 1$, $y'(0) = 0$?

9. The general solution of $y'' = -y$ is given by

$$y(x) = A \sin x + B \cos x, \quad -\infty < x < \infty$$

(a) Find a necessary and sufficient condition for the boundary value problem

$$y'' = -y, \quad y(a) = \alpha, \quad y(b) = \beta$$

to possess a unique solution.

(b) Impose boundary conditions for which there can be no solution.

8.2   Euler's Method

The simplest numerical procedure for solving initial-value problems is Euler's method. Although numerically inefficient, it involves many concepts and ideas that are related to numerical solutions of ODE in general and is worth a detailed discussion.

We consider the initial-value problem given by Eq. (8.1.2),

$$y' = f(x, y), \quad x_0 \leq x \leq b$$
$$y(x_0) = y_0 \tag{8.2.1}$$

where $f(x, y)$ fulfills the requirements of Theorem 8.1.1, that is, continuity and the Lipschitz condition. We denote the exact solution of Eq. (8.2.1) by $Y(x)$, that is,

$$Y'(x) = f[x, Y(x)], \quad x_0 \leq x \leq b$$
$$Y(x_0) = y_0$$

In most cases, finding $Y(x)$ is impossible. Instead, we use numerical methods that replace the interval $[x_0, b]$ by a set of discrete *grid points* (or *mesh points*)

$$x_0 < x_1 < \cdots < x_N = b$$

at which $Y(x)$ is approximated by some $y(x)$. The exact and approximate solutions at x_n, $0 \le n \le N$ are denoted by Y_n, y_n, $0 \le n \le N$, respectively, that is,

$$Y(x_n) = Y_n, \quad 0 \le n \le N$$
$$y(x_n) = y_n, \quad 0 \le n \le N \tag{8.2.2}$$

Clearly, $Y_0 = y_0$ but $Y_n \ne y_n$, $1 \le n \le N$ in general. Unless otherwise specified the discrete points $\{x_n, 0 \le n \le N\}$ are assumed to be evenly spaced, that is,

$$x_n = x_0 + nh, \quad 0 \le n \le N \tag{8.2.3}$$

where $h = (b - x_0)/N$ is the *step size* (or *mesh size*). The approximate solution $y(x)$ will be denoted by $y_h(x)$. The *Euler method* is based on the first order approximation for $Y'(x)$, given by

$$Y'(x) \approx \frac{Y(x+h) - Y(x)}{h} \tag{8.2.4}$$

which leads to

$$Y(x+h) \approx Y(x) + hY'(x) \tag{8.2.5}$$

or

$$Y(x+h) \approx Y(x) + hf[x, Y(x)] \tag{8.2.6}$$

By substituting $x = x_n$ we obtain

$$Y(x_{n+1}) \approx Y(x_n) + hf[x_n, Y(x_n)] = Y_{n+1} + \Delta_{n+1} \tag{8.2.7}$$

The geometric interpretation of Eq. (8.2.7) is given in Fig. 8.2.1.

The solution between x_n and x_{n+1} is approximated by the tangent line to $Y(x)$ at $x = x_n$. It intersects the straight line $x = x_{n+1}$ at $(x_{n+1}, Y_{n+1} + \Delta_{n+1})$ and the quality of this approximation at x_{n+1} is measured by Δ_{n+1}.

However, because the exact solution $Y(x)$ is known only at the initial point, we may use Eq. (8.2.7) only once to approximate Y_1. For $n \ge 1$ it is replaced by

$$Y_{n+1} \approx y(x_n) + hf[x_n, y(x_n)] \tag{8.2.8}$$

where $y(x_n)$ is the available approximate value to $Y(x_n)$ at $x = x_n$. We now denote the right-hand side of Eq. (8.2.8) by y_{n+1} and, by Euler's method,

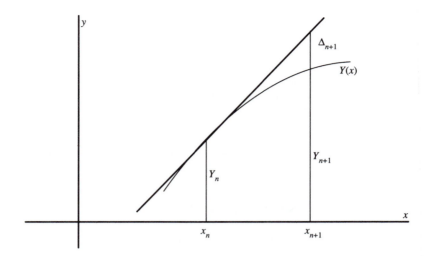

Figure 8.2.1. Derivation of Euler's method.

$$y_{n+1} = y_n + hf(x_n, y_n), \quad n \geq 0$$
$$y_0 = Y_0 \qquad\qquad\qquad\qquad (8.2.9)$$

Note that Eq. (8.2.9) determines the approximate values only at the grid points. The approximate solution $y(x)$ over each subinterval $[x_n, x_{n+1}]$ is defined by

$$y(x) = y_n + (x - x_n)f(x_n, y_n), \quad x_n \leq x \leq x_{n+1} \qquad (8.2.10)$$

Thus, over the entire interval $[x_0, b]$, $Y(x)$ is approximated by a polygon $y_h(x)$ (Fig. 8.2.2). The numerical procedure based on Eq. (8.2.9) is given by Algorithm 8.2.1.

Algorithm 8.2.1.
[Euler's method for solving $y' = f(x, y)$, $y(x_0) = Y_0$].

Step 1. Insert the initial and end points x_0 and b, the initial value Y_0, the desired number of grid points $N + 1$, and set $h = (b - x_0)/N$.
Step 2. Set $y_0 = Y_0$ and for $0 \leq n \leq N - 1 : x_{n+1} \leftarrow x_n + h, \ y_{n+1} \leftarrow y_n + hf(x_n, y_n)$.

Example 8.2.1.
Consider the initial-value problem

$$y' = y, \quad y(0) = 2, \quad 0 \leq x \leq 1$$

whose exact solution is $Y(x) = 2e^x$. By setting $h = 0.2$ we obtain an approximate solution whose comparison with $Y(x)$ is given in Table 8.2.1.

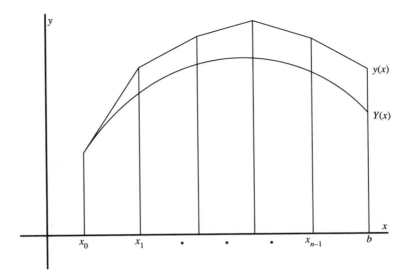

Figure 8.2.2. A global approximation (Euler's method).

Table 8.2.1. Using Euler's Method to Solve $y' = y$,
$y_0 = 2$, $0 \le x \le 1$, $h = 0.2$

x_n	Y_n	y_n	$Y_n - y_n$	$(Y_n - y_n)/Y_n$
0	2.000	2.000	0	0
0.2	2.443	2.400	0.043	0.018
0.4	2.984	2.880	0.104	0.035
0.6	3.644	3.456	0.188	0.052
0.8	4.451	4.147	0.304	0.068
1.0	5.437	4.977	0.460	0.085

Table 8.2.2. Behavior of $y_h(x)$ as $h \to 0$

h	$y_h(1)$	$Y(1) - y_h(1)$	$[Y(1) - y_h(1)]/Y(1)$
0.2	4.977	0.460	0.085
0.1	5.187	0.250	0.046
0.05	5.307	0.130	0.024
0.025	5.370	0.067	0.012

It demonstrates an increase in the relative error with x up to 8.5% at $x = 1$. The behavior of $y_h(x)$ as $h \to 0$ for this particular example is given in Table 8.2.2, where $y(1)$ is calculated for four grids: $h = 0.2$, 0.1, 0.05, and 0.025.

It is quite clear that $\lim_{h \to 0} y_h(1) = Y(1)$ and that the convergence is linear, that is,

$$Y(1) - y_h(1) \approx Ch, \quad h \to 0$$

for some constant C. The convergence claim is easily verified by writing

$$y_{n+1} = y_n + hy_n = (1 + h)y_n$$

This yields $y_n = (1 + h)^n Y_0 = 2(1 + h)^n$ and, in particular,

$$y_N = y_h(1) = 2[1 + (1/N)]^N$$

Because $N \to \infty$ as $h \to 0$, we have $\lim_{h \to \infty} y_h(1) = Y(1) = 2e$. The linear convergence can be shown by a clever application of the L'Hospital rule. $\square$

8.2.1   Error Analysis

We will now show that in general

$$\lim_{h \to 0} y_h(x) = Y(x), \quad x_0 \le x \le b \tag{8.2.11}$$

that is, Euler's method converges to the exact solution of Eq. (8.2.1). We start by using Taylor's theorem to write

$$Y(x_{n+1}) = Y(x_n) + hY'(x_n) + \frac{h^2}{2}Y''(\xi_n) \tag{8.2.12}$$

where ξ_n is some interim point between x_n and x_{n+1}. Because $Y' = f(x, Y)$, Eq. (8.2.12) can be written as

$$Y(x_{n+1}) = Y(x_n) + hf[x_n, Y(x_n)] + T_{n+1} \tag{8.2.13}$$

where

$$T_{n+1} = \frac{h^2}{2}Y''(\xi_n) \tag{8.2.14}$$

is the *truncation error* for Euler's method at the $(n + 1)$th step. By subtracting Eq. (8.2.9) from Eq. (8.2.13) one obtains

$$Y_{n+1} - y_{n+1} = Y_n - y_n + h[f(x_n, Y_n) - f(x_n, y_n)] + \frac{h^2}{2}Y''(\xi_n) \tag{8.2.15}$$

Thus, the total error at $x = x_{n+1}$ consists of

1. The truncation error T_{n+1}

2. The propagated error $Y_n - y_n + h[f(x_n, Y_n) - f(x_n, y_n)]$

At $x = x_0$ the propagated error vanishes, but it exists for all $1 \le n \le N$. It is composed of two parts: the first due to the previous total error, the second due to using $f(x_n, y_n)$ as an approximate value to

$Y'_n = f(x_n, Y_n)$. By using the mean-value theorem we find

$$f(x_n, Y_n) - f(x_n, y_n) = (Y_n - y_n)\frac{\partial f(x_n, \zeta_n)}{\partial y}$$

where ζ_n is an interim point between y_n and Y_n. Therefore the total error at $x = x_{n+1}$ is given by

$$z_{n+1} = \left[1 + h\frac{\partial f(x_n, \zeta_n)}{\partial y}\right] z_n + \frac{h^2}{2}Y''(\xi_n) \tag{8.2.16}$$

where $z_n = Y_n - y_n$, $n \geq 0$.

The next result is useful in many proofs that are related to error bound calculations.

Lemma 8.2.1.

Let a sequence of numbers $\{W_n\}_0^N$ satisfy

$$|W_{n+1}| \leq A|W_n| + B, \quad 0 \leq n \leq N - 1 \tag{8.2.17}$$

for some given positive constants A and B. Then

$$|W_n| \leq A^n|W_0| + B\frac{A^n - 1}{A - 1}, \quad 1 \leq n \leq N \tag{8.2.18}$$

The proof, based on mathematical induction, is left as an exercise for the reader.

We can now calculate an error bound for the Euler method.

Theorem 8.2.1.

(Error bound for Euler's method). *Consider the initial-value problem given by Eq. (8.2.1) and let $f(x, y)$ satisfy the requirement*

$$M = \sup |f_y(x, y)| < \infty, \quad -\infty < y < \infty, \quad x_0 \leq x \leq b \tag{8.2.19}$$

Then the approximate solution $y_h(x)$, calculated by Euler's method, satisfies

$$|Y_n - y_n| \leq e^{M(x_n - x_0)}|Y_0 - y_0| + h\left[\frac{e^{M(x_n - x_0)} - 1}{2M}\right]M_2, \quad 0 \leq n \leq N \tag{8.2.20}$$

where

$$M_2 = \max_{x_0 \leq x \leq b} |Y''(x)| \tag{8.2.21}$$

and $Y_0 - y_0$ is the empirical error in Y_0 (which often occurs).

Proof.

By using Eqs. (8.2.16), (8.2.19), and (8.2.21) we obtain

$$|z_{n+1}| \leq A|z_n| + B, \quad 0 \leq n \leq N - 1$$

where $A = 1 + hM$, $\quad B = (h^2/2)M_2$.

Thus, by virtue of Lemma 8.2.1 and because $A^n \leq e^{M(x_n - x_0)}$ (why?),

$$|z_n| \leq e^{M(x_n - x_0)}|z_0| + \frac{h^2}{2}M_2 \frac{e^{M(x_n - x_0)} - 1}{hM}$$

$$= e^{M(x_n - x_0)}|z_0| + h\left[\frac{e^{M(x_n - x_0)} - 1}{2M}\right]M_2$$

in agreement with Eq. (8.2.20).   □

Note that if the initial value is exact, that is, $y_0 = Y_0$, we have

$$|Y_n - y_n| \leq Ch, \quad 0 \leq n \leq N \tag{8.2.22}$$

where

$$C = \frac{e^{M(b - x_0)} - 1}{2M}M_2 \tag{8.2.23}$$

Definition 8.2.1.   (Order of convergence). *If a numerical procedure for solving differential equations satisfies*

$$|Y(x_n) - y_h(x_n)| \leq Ch^p, \quad 0 \leq n \leq N \tag{8.2.24}$$

the method is said to be convergent with (at least) order p.

Corollary 8.2.1.

Euler's method is convergent with order 1, *provided that* $y_0 = Y_0$.

By virtue of Eq. (8.2.22) the proof is straightforward.

It should be noted that Theorem 8.2.1 comprises only the discrete points $x_n, 0 \leq n \leq N$. Let x be any point between x_n and x_{n+1}. Then, by slightly modifying the proof to the previous theorem, we find that Eq. (8.2.11) holds as well.

Example 8.2.2.

The initial-value problem

$$y' = -y + e^{-x}, \quad y(0) = 0, \quad 0 \leq x \leq 1$$

has the exact solution $Y(x) = xe^{-x}$. We have

$$M = \sup\left|\frac{\partial f(x, y)}{\partial y}\right| = 1, \quad M_2 = \max|Y''(x)| = 2$$

Therefore,

$$|Y_n - y_n| \le hM_2 \frac{e^{M(b-x_0)} - 1}{2M} = 2h\frac{e-1}{2} \approx 1.718h$$

Table 8.2.3 presents the Euler and exact solutions for $h = 1/20$. The absolute error should not exceed 0.086, and this is perfectly confirmed by the results. □

Table 8.2.3. Using Euler's Method to Solve $y' = -y + e^{-x}$, $y_0 = 0$, $h = 1/20$

n	x_n	y_n	Y_n	$|Y_n - y_n|$
2	0.1	0.095	0.090	0.005
4	0.2	0.172	0.164	0.008
6	0.3	0.233	0.222	0.011
8	0.4	0.281	0.268	0.012
10	0.5	0.317	0.303	0.014
12	0.6	0.344	0.329	0.014
14	0.7	0.362	0.348	0.015
16	0.8	0.374	0.359	0.015
18	0.9	0.380	0.366	0.015
20	1.0	0.382	0.368	0.014

The bound for $|z_n|$, given by Eq. (8.2.20), involves $M = \sup |f_y|$ and $M_2 = \max |Y''|$, which may be difficult to derive. Also, the right-hand side of Eq. (8.2.20) is an upper bound for $|z_n|$ and does not necessarily reflect on the size of $|z_n|$. In the previous example this bound is 0.086, while $\max |z_n| = 0.015$. The next result provides an *asymptotic* error estimate for Euler's method.

Theorem 8.2.2.
(Asymptotic error estimate). *If $f(x,y)$ is sufficiently smooth, then the discrete error $z_n = Y_n - y_n$ associated with Euler's method satisfies*

$$z_n = hZ(x_n) + O(h^2) \tag{8.2.25}$$

where $Z(x)$ is the solution to the initial-value problem

$$Z'(x) = f_y[x, Y(x)]Z(x) + \frac{1}{2}Y''(x), \quad Z(x_0) = 0 \tag{8.2.26}$$

Proof.
First, rewrite Eq. (8.2.15) as

$$z_{n+1} = z_n + h[f_y(x_n, Y_n)z_n + O(z_n^2)] + \frac{h^2}{2}[Y''(x_n) + O(h)] \tag{8.2.27}$$

By virtue of Eq. (8.2.22) we have $z_n = O(h)$ and therefore

$$z_{n+1} = z_n + hz_n f_y(x_n, Y_n) + \frac{h^2}{2}Y''(x_n) + O(h^3) \tag{8.2.28}$$

Let $Z(x)$ be the unique solution to Eq. (8.2.26). By using Taylor's theorem

$$Z(x_{n+1}) = Z(x_n) + hZ'(x_n) + O(h^2) \tag{8.2.29}$$

Now, multiply Eq. (8.2.29) by h and subtract from Eq. (8.2.28) to obtain

$$z_{n+1} - hZ(x_{n+1}) = z_n - hZ(x_n) + hz_n f_y(x_n, Y_n) + \frac{h^2}{2} Y''(x_n) - h^2 Z'(x_n) + O(h^3) \tag{8.2.30}$$

If we denote $w_n = z_n - hZ(x_n)$ and use Eq. (8.2.26) to replace $Z'(x_n)$ by $f_y(x_n, Y_n)Z(x_n) + (1/2)Y''(x_n)$, we finally have

$$w_{n+1} = [1 + hf_y(x_n, Y_n)]w_n + O(h^3) \tag{8.2.31}$$

Thus $|w_{n+1}| \leq (1 + hM)|w_n| + B$, $n \geq 0$, where $w_0 = 0$ and $B = O(h^3)$. By implementing Lemma 8.2.1 we conclude

$$|w_n| = |z_n - hZ(x_n)| \leq \frac{e^{M(b-x_0)} - 1}{hM} B = O(h^2) \qquad \square$$

Let y_h and $y_{h/2}$ denote the approximate values at some point to the exact solution Y, using Euler's method with step sizes h and $h/2$, respectively. Then, by using Theorem 8.2.2, one obtains

$$Y - y_h = hZ + O(h^2), \quad Y - y_{h/2} = \frac{h}{2} Z + O(h^2) \tag{8.2.32}$$

which leads to

$$Y = 2y_{h/2} - y_h + O(h^2) \tag{8.2.33}$$

and to Richardson's extrapolation formula for Euler's method,

$$Y \approx 2y_{h/2} - y_h \tag{8.2.34}$$

Equation (8.2.34) provides a better approximation to Y than do either y_h or $y_{h/2}$.

Example 8.2.3.

Consider the initial-value problem $y' = y + 1$, $y(0) = 0$, $0 \leq x \leq 1$ with the exact solution $Y(x) = e^x - 1$. The associated problem for $Z(x)$ is

$$Z' = Z + \frac{1}{2} e^x, \quad Z(0) = 0, \quad 0 \leq x \leq 1$$

and has the solution $Z(x) = (1/2)xe^x$. Thus, using Euler's method, one should have

$$z_n - hZ(x_n) = Y_n - y_n - \frac{h}{2} x_n e^{x_n} = O(h^2) \tag{8.2.35}$$

Table 8.2.4 contains the results for $h = 1/10$ and confirms the validity of Eq. (8.2.35). $\square$

Table 8.2.4. Using Euler's Method to Evaluate the Asymptotic Error in $y' = y + 1$,
$y_0 = 0$, $h = 0.1$

x_n	y_n	Y_n	z_n	$hZ(x_n)$
0.1	0.1000	0.1052	0.0052	0.0055
0.2	0.2100	0.2214	0.0114	0.0122
0.3	0.3310	0.3499	0.0189	0.0202
0.4	0.4641	0.4918	0.0277	0.0298
0.5	0.6105	0.6487	0.0382	0.0412
0.6	0.7716	0.8221	0.0506	0.0547
0.7	0.9487	1.0138	0.0650	0.0705
0.8	1.1436	1.2255	0.0820	0.0890
0.9	1.3579	1.4596	0.1017	0.1107
1.0	1.5937	1.7183	0.1245	0.1359

In the next example we consider an initial-value problem in which $f(x, y)$ is nonlinear and does not satisfy a Lipschitz condition. Still, the asymptotic error estimate theorem is effective.

Example 8.2.4.

The initial-value problem $y' = y^2$, $y(0) = 1$, $0 \le x \le 0.5$ is uniquely solved by $Y(x) = 1/(1 - x)$. Thus $Y''(x) = 2/(1 - x)^3$ and the associated problem for $Z(x)$,

$$Z'(x) = 2Y(x)Z(x) + \frac{1}{2}Y''(x) = \frac{2}{1 - x}Z(x) + \frac{1}{(1 - x)^3}, \quad Z(0) = 0$$

has the solution $Z(x) = -\{[\ln(1 - x)]/(1 - x)^2\}$. The computed and estimated errors for Euler's method, using $h = 0.05$, are given in Table 8.2.5. $\square$

Table 8.2.5. Using Euler's Method to Evaluate the Asymptotic Error in $y' = y^2$,
$y_0 = 1$, $h = 0.05$

x_n	y_n	Y_n	z_n	$hZ(x_n)$
0.05	1.0500	1.0526	0.0026	0.0028
0.10	1.1051	1.1111	0.0060	0.0065
0.15	1.1662	1.1765	0.0103	0.0112
0.20	1.2342	1.2500	0.0158	0.0174
0.25	1.3104	1.3333	0.0230	0.0256
0.30	1.3962	1.4286	0.0324	0.0364
0.35	1.4937	1.5385	0.0448	0.0510
0.40	1.6052	1.6667	0.0614	0.0709
0.45	1.7341	1.8182	0.0841	0.0988
0.50	1.8844	2.0000	0.1156	0.1386

Next, we apply Richardson's extrapolation method to the initial-value problem previously discussed in Example 8.2.3.

Example 8.2.5.

The approximate solutions to $y' = y + 1$, $y(0) = 0$, $0 \leq x \leq 1$, using Euler's method and $h = 0.1, 0.05$ satisfy $y_{0.1}(1) = 1.5937$, $Y_{0.05}(1) = 1.6533$.

By using Eq. (8.2.34) we obtain a better approximation, $Y(1) \approx 2y_{0.05}(1) - y_{0.1}(1) = 1.7129$. The exact value is 1.7183. $\qquad\square$

PROBLEMS

1. Solve the following initial-value problems by using Euler's method with $h = 0.1$. Find $y(1)$.

    (a) $y' = xy$, $y(0) = 1$, $0 \leq x \leq 1$

    (b) $y' = x^2 \sin y + y$, $y(0) = -1$, $0 \leq x \leq 1$

    (c) $y' = y - x^2$, $y(0) = 0$, $0 \leq x \leq 1$

2. Solve the initial-value problem $y' = e^y$, $y(0) = 0$, $0 \leq x \leq 1$, using Euler's method with $h = 1/5$, $1/10$, $1/20$. Explain the behavior of $y_h(1)$ as $h \to 0$.

3. Find $y_h(1)$ for case (a) of problem 1, using $h = 1/10$, $1/20$, $1/40$, and discuss the behavior of $Y(1) - y_h(1)$ as $h \to 0$.

4. Derive Eq. (8.2.18) from Eq. (8.2.17).

5. Find an upper bound for $|z_n| = |Y_n - y_n|$ in the following cases:

    (a) $y' = x^2 y$, $y(0) = 1$, $0 \leq x \leq 1$, $h = 1/10$

    (b) $y' = \cos y$, $y(0) = 0$, $0 \leq x \leq 2$, $h = 1/20$

    (c) $y' = 2y - x^2$, $y(0) = 0$, $0 \leq x \leq 1$, $h = 1/10$

6. Find the *asymptotic error equation* given by Eq. (8.2.26) in the following cases:

    (a) $y' = x \sin^2 y + x$

    (b) $y' = y + x^4$

    (c) $y' = x \cos y + 2y \sin x$

7. For a given initial-value problem $y' = f(x, y)$, $y(x_0) = y_0$, $x_0 \leq x \leq b$ derive an algorithm, based on Euler's method, to calculate y_n, $1 \leq n \leq N$ and approximate the asymptotic error function $Z(x)$ at the discrete points x_n, $1 \leq n \leq N$.

8. Use the result of problem 7 and approximate $Z(x_n)$ in the following cases:

    (a) $y' = xy + \sin x$, $y(0) = 1$, $0 \leq x \leq 1$, $h = 1/10$

    (b) $y' = (1 + \sin y)x^2$, $y(0) = 0$, $0 \leq x \leq 1$, $h = 1/10$

    (c) $y' = y^{3/2}$, $y(0) = 1$, $0 \leq x \leq 1$, $h = 1/10$

9. (a) Use Euler's method to solve the initial-value problem $y' = -y^2 x - (y/x)$, $y(1) = 1$, $1 \leq x \leq 2$, choosing $h = 1/10$ and $h = 1/20$.

    (b) Apply Richardson's extrapolation method and use $y_{0.1}(2)$ and $y_{0.05}(2)$ to obtain a better approximation to $Y(2)$. Compare with the exact solution $Y(x) = 1/x^2$.

10. (a) Use Euler's method to solve

$$y' = \frac{2y^2}{1+y^2}x + \frac{2x}{1+x^4}, \quad y(0) = 0, \quad 0 \le x \le 2$$

for $h = 1/10$ and $h = 1/20$.

(b) Apply Richardson's extrapolation method to obtain an improved approximate value to $Y(2)$. Compare with $Y(x) = x^2$.

(c) What is the main difference between problems 9 and 10?

8.3 Taylor's Method

The previously discussed Euler's method primarily serves as an introduction to numerical integration of differential equations. It involves important ideas and features that are related to error anaylsis. It is not, however, an efficient numerical tool. The convergence, of $O(h)$ type, is definitely slow for all practical purposes. For example, to solve

$$y' = y, \quad y(0) = 1, \quad 0 \le x \le 1$$

and obtain $y(1)$ with five-digit accuracy, we must choose $h = 10^{-5}$, which means $100,000$ function evaluations! We therefore search for other procedures that are less costly than Euler's method. In this section we introduce Taylor's method, a generalization of Euler's approach.

We consider the initial-value problem

$$y' = f(x, y), \quad y(x_0) = y_0, \quad x_0 \le x \le b \tag{8.3.1}$$

where $f(x, y)$ is sufficiently differentiable. Because $f(x, y)$ is a given function, the exact derivatives of the solution can be expressed in terms of f and its partial derivatives, that is, in terms of x and y. For example,

$$Y'' = f_x + f_y Y' = f_x + ff_y \tag{8.3.2}$$

$$
\begin{aligned}
Y''' &= f_{xx} + f_{xy}Y' + (f_{yx} + f_{yy}Y')f + f_y(f_x + f_yY') \\
&= f_{xx} + 2ff_{xy} + f^2 f_{yy} + f_x f_y + ff_y^2
\end{aligned} \tag{8.3.3}
$$

In general $Y^{(n+1)} = f^{(n)}(x, y)$, where

$$
\begin{aligned}
f^{(0)}(x, y) &= f(x, y) \\
f^{(n)}(x, y) &= f_x^{(n-1)} + f_y^{(n-1)}f
\end{aligned} \tag{8.3.4}
$$

Example 8.3.1.

Let $f(x, y) = x^2 + y^2$. Then

$$f_x = 2x, \quad f_y = 2y, \quad f_{xx} = 2, \quad f_{xy} = 0, \quad f_{yy} = 2$$

Thus

$$Y'' = 2x + 2Y(x^2 + Y^2), \quad Y''' = 2 + 2(x^2 + Y^2)^2 + 4xY + 4Y^2(x^2 + Y^2)$$

If the initial value is $Y(0) = 0$, then $Y'(0) = 0$, $Y''(0) = 0$, $Y'''(0) = 2$ and by replacing $Y(x)$ by its third order Taylor's expansion we obtain $Y(x) \approx x^3/3$ near $x = 0$. □

Example 8.3.2.

The unique solution to $y' = y^2 e^x$, $y(0) = -1$ is $Y(x) = -e^{-x}$. The Taylor series of $Y(x)$ about $x = 0$ is therefore

$$Y(x) = -\left(1 - x + \frac{x^2}{2} - \frac{x^3}{6} + \cdots\right) = -1 + x - \frac{x^2}{2} + \frac{x^3}{6} - \cdots$$

Alternatively we may find this expansion by calculating the partial derivatives of $f(x, y) = y^2 e^x$ and substituting in Eqs. (8.3.1) through (8.3.3) to obtain $Y^{(n)}(0)$, $1 \le n \le 3$. We have

$$f_x = y^2 e^x, \quad f_y = 2y e^x, \quad f_{xx} = y^2 e^x, \quad f_{xy} = 2y e^x, \quad f_{yy} = 2e^x$$

and at the initial point $(0, -1)$

$$f(0, -1) = 1, \quad f_x(0, -1) = 1, \quad f_y(0, -1) = -2$$
$$f_{xx}(0, -1) = 1, \quad f_{xy}(0, -1) = -2, \quad f_{yy}(0, -1) = 2$$

Thus $Y'(0) = 1$, $Y''(0) = -1$, $Y'''(0) = 1$, and

$$Y(x) = \sum_{n=0}^{\infty} \frac{Y^{(n)}(0)}{n!} x^n = -1 + x - \frac{x^2}{2} + \frac{x^3}{6} - \cdots$$ □

We now replace $[x_0, b]$ by the finite grid $x_0 < x_1 < \cdots < x_N = b$, where

$$x_n = x_0 + nh, \quad 0 \le n \le N$$

are evenly spaced points. Instead of approximating Y_{n+1} by $y_n + hf(x_n, y_n)$, that is, by the first two terms of a Taylor series about x_n, we use $(p + 1)$ terms and represent y_{n+1} as

$$y_{n+1} = y_n + hT_p(x_n, y_n; h) \tag{8.3.5}$$

where

$$T_p(x,y;h) = f(x,y) + \frac{f^{(1)}(x,y)}{2!}h + \cdots + \frac{f^{(p-1)}(x,y)}{p!}h^{p-1} \tag{8.3.6}$$

We define the *local truncation error* of the method as

$$R_{n+1} = Y_{n+1} - [Y_n + hT_p(x_n, Y_n; h)] \tag{8.3.7}$$

which is the error at x_{n+1}, provided that $y_n = Y_n$. Clearly,

$$R_{n+1} = \frac{h^{p+1}}{(p+1)!}Y^{(p+1)}(\xi_n) = \frac{h^{p+1}}{(p+1)!}f^{(p)}[\xi_n, Y(\xi_n)] \tag{8.3.8}$$

where ξ_n is an interim point between x_n and x_{n+1}. The procedure given by Eqs. (8.3.5) and (8.3.6) is known as Taylor's method of order p. The particular case $p = 1$ is the previously discussed Euler's method. The general case is described as follows.

Algorithm 8.3.1.

[Taylor's method of order p: solving $y' = f(x,y)$, $y(x_0) = Y_0$, $x_0 \le x \le b$].

Step 1. Insert the initial and end points x_0 and b, the initial value Y_0, the desired number of grid points $N + 1$, and set $h = (b - x_0)/N$.

Step 2. Set $y_0 = Y_0$ and for $0 \le n \le N - 1$: $x_{n+1} \leftarrow x_n + h$, $y_{n+1} \leftarrow y_n + hT_p(x_n, y_n; h)$, where $T_p(x_n, y_n; h)$ is computed from Eq. (8.3.6), using $f(x,y)$ and its partial derivatives to calculate $f^{(m)}(x,y)$, $1 \le m \le p - 1$.

A subroutine TAYLOR, based on Taylor's algorithm of order $p \le 2$, is found on the attached floppy disk.

We shall now find a bound for the error $z_n = Y_n - y_n$. The following assumptions are necessary:

1. In addition to $f(x,y)$, the function $T_p(x,y;h)$, defined by Eq. (8.3.6), satisfies a Lipschitz condition

$$|T_p(x,y_1;h) - T_p(x,y_2;h)| \le L|y_1 - y_2| \tag{8.3.9}$$

for some constant L independent of x, y_1, y_2, and h, provided that x and $x + h$ are in $[x_0, b]$.

2. $Y^{(p+1)}(x) \in C[x_0, b]$.

If these requirements are satisfied we obtain the following result.

Theorem 8.3.1.

(Convergence of Taylor's algorithm). *Under the previous assumptions (1) and (2), the error $z_n = Y_n - y_n$ associated with Taylor's algorithm of order p satisfies*

$$|z_n| \le h^p \left[\frac{e^{L(x_n - x_0)} - 1}{L(p+1)!} \right] M_{p+1} \tag{8.3.10}$$

where

$$M_{p+1} = \max_{x_0 \le x \le b} |Y^{(p+1)}(x)| \tag{8.3.11}$$

Proof.

The exact solution $Y(x)$ satisfies

$$Y_{n+1} = Y_n + hT_p(x_n, Y_n; h) + \frac{h^{p+1}}{(p+1)!} Y^{(p+1)}(\xi_n) \tag{8.3.12}$$

where ξ_n is between x_n and x_{n+1}. By subtracting Eq. (8.3.5) from Eq. (8.3.12) we find

$$z_{n+1} = z_n + h[T_p(x_n, Y_n; h) - T_p(x_n, y_n; h)] + \frac{h^{p+1}}{(p+1)!} Y^{(p+1)}(\xi_n) \tag{8.3.13}$$

Therefore

$$|z_{n+1}| \le |z_n| + h|T_p(x_n, Y_n; h) - T_p(x_n, y_n; h)| + \frac{h^{p+1}}{(p+1)!} |Y^{(p+1)}(\xi_n)|$$

and by virtue of Eqs. (8.3.9) and (8.3.11) we have

$$|z_{n+1}| \le |z_n| + hL|z_n| + \frac{h^{p+1}}{(p+1)!} M_{p+1}$$

or

$$|z_{n+1}| \le (1 + hL)|z_n| + \frac{h^{p+1}}{(p+1)!} M_{p+1} \tag{8.3.14}$$

We now implement Lemma 8.2.1 and obtain

$$|z_n| \le (1 + hL)^n |z_0| + \frac{h^{p+1}}{(p+1)!} M_{p+1} \frac{(1 + hL)^n - 1}{hL}$$

which, because $z_0 = 0$, leads to Eq. (8.3.10). $\qquad\qquad\qquad\square$

Corollary 8.3.1.

The order of convergence of Taylor's algorithm of order p is at least p.

Example 8.3.3.

Let $y' = y$, $y(0) = 2$, $0 \le x \le 1$. Here $Y^{(n)} = Y$, $n > 0$. By applying Taylor's algorithm of order p we obtain

$$y_{n+1} = y_n + h\left(y_n + \frac{y_n}{2!}h + \cdots + \frac{y_n}{p!}h^{p-1}\right)$$

$$= y_n \left(1 + h + \frac{h^2}{2!} + \cdots + \frac{h^p}{p!} \right), \quad 0 \le n \le N - 1$$

and thus $y_N = 2[1 + h + (h^2/2!) + \cdots + (h^p/p!)]^N$. By choosing $p = 2$ and $h = 1/10$ we obtain

$$y_{10} = y(1) = 2 \left(1 + 0.1 + \frac{0.01}{2} \right)^{10} = 5.428$$

The exact solution, $2e^x$, rounded to four digits, is 5.437 and the relative error is 0.17% compared with that produced by Euler's method (8.5%). By taking $p = 3$ and $h = 1/10$ we obtain

$$y_{10} = 2 \left(1 + 0.1 + \frac{0.01}{2} + \frac{0.001}{6} \right)^{10} = 5.436$$

and the relative error is 0.02%. □

Example 8.3.4.

Consider the problem $y' = e^y$, $y(0) = 0$, $0 \le x \le 0.5$. Here

$$f = e^y, \quad f_x = f_{xx} = f_{xy} = 0, \quad f_y = e^y, \quad f_{yy} = e^y$$

Therefore $f^{(1)}(x, y) = e^{2y}$, $f^{(2)}(x, y) = f_x^{(1)} + f f_y^{(1)} = 2e^{3y}$ and Taylor's algorithm of order 3 is given by

$$y_0 = 0, \quad y_{n+1} = y_n + h \left(e^{y_n} + \frac{e^{2y_n}}{2} h + \frac{e^{3y_n}}{3} h^2 \right)$$

Table 8.3.1 presents the approximate solution for $h = 1/10$, $1/20$ compared with the exact solution $Y = -\ln(1-x)$. All the numbers are rounded to six significant digits. The accuracy of both solutions is shown in Table 8.3.2.

Table 8.3.2 clearly indicates

$$\frac{Y - y_{0.1}}{Y - y_{0.05}} \approx \text{constant}$$

and, as we will see below, this is no coincidence [Eq. (8.3.17)].

Table 8.3.1. Taylor's Algorithm of Order 3 for $y' = e^y$,
$y_0 = 0$, $h = 0.1, 0.05$: Selected Points

x	$y_{0.1}(x)$	$y_{0.05}(x)$	$Y(x)$
0.1	0.105333	0.105357	0.105361
0.2	0.223071	0.223134	0.223144
0.3	0.356524	0.356654	0.356675
0.4	0.510532	0.510784	0.510826
0.5	0.692573	0.693065	0.693147

Table 8.3.2. Error Behavior for $y_{0.1}$ and $y_{0.05}$

x	$Y - y_{0.1}$	$Y - y_{0.05}$	$(Y - y_{0.1})/(Y - y_{0.05})$
0.1	$0.28 \cdot 10^{-4}$	$0.40 \cdot 10^{-5}$	7.0
0.2	$0.73 \cdot 10^{-4}$	$0.10 \cdot 10^{-4}$	7.3
0.3	$0.15 \cdot 10^{-3}$	$0.21 \cdot 10^{-4}$	7.1
0.4	$0.29 \cdot 10^{-3}$	$0.42 \cdot 10^{-4}$	6.9
0.5	$0.57 \cdot 10^{-3}$	$0.82 \cdot 10^{-4}$	7.0

□

The following result is a generalization to Theorem 8.2.2 and represents an asymptotic error estimate for Taylor's algorithm of order p.

Theorem 8.3.2.

(Taylor's method: asymptotic error estimate). *Let $f(x, y)$ be sufficiently smooth and let Y_n and y_n be the exact and approximate values, respectively, of the unique solution of Eq. (8.3.1), as calculated using Algorithm 8.3.1. Then the error $z_n = Y_n - y_n$, at each discrete point x_n, satisfies*

$$z_n = h^p Z(x_n) + O(h^{p+1}) \tag{8.3.15}$$

where $Z(x)$ solves the initial-value problem

$$Z'(x) = f_y[x, Y(x)]Z(x) + \frac{1}{(p+1)!} Y^{(p+1)}(x)$$
$$Z(x_0) = 0 \tag{8.3.16}$$

The proof is similar to that of Theorem 8.2.2 and is left as an exercise for the interested reader.

Example 8.3.5.

The unique solution to the initial-value problem

$$y' = -y/x, \quad y(1) = 1, \quad 1 \le x \le 2$$

is $Y(x) = 1/x$. To apply Taylor's algorithm of order 2 we substitute

$$f = -y/x, \quad f_x = y/x^2, \quad f_y = -1/x$$

in Eq. (8.3.6) and obtain

$$y_{n+1} = y_n + h \left[f(x_n, y_n) + \frac{h}{2}(f_x + ff_y)(x_n, y_n) \right]$$

$$= y_n + h \left[-\frac{y_n}{x_n} + \frac{h}{2} \left(\frac{y_n}{x_n^2} + \frac{y_n}{x_n^2} \right) \right] = y_n \left(1 - \frac{h}{x_n} + \frac{h^2}{x_n^2} \right)$$

The equation for $Z(x)$ is

$$Z' = f_y Z + \frac{1}{6} Y''' = -\frac{1}{x} Z - \frac{1}{x^4}, \quad Z(1) = 0$$

and its solution $Z(x) = (1 - x^2)/2x^3$ must satisfy $z_n = Y_n - y_n = h^2 Z(x_n) + O(h^3)$.

Table 8.3.3 contains the exact and approximate discrete solutions for $h = 1/10$, and in addition a comparison between the error z_n and $h^2 Z(x_n)$. The results clearly validate Eq. (8.3.15). ☐

Table 8.3.3. Taylor's Algorithm of Order 2 for $y' = -y/x$, $y(1) = 1$: z_n vs. $Z(x_n)$

x_n	Y_n	y_n	z_n	$h^2 Z(x_n)$
1.1	0.9091	0.9100	$-0.9 \cdot 10^{-3}$	$-0.8 \cdot 10^{-3}$
1.2	0.8333	0.8348	$-1.5 \cdot 10^{-3}$	$-1.3 \cdot 10^{-3}$
1.3	0.7692	0.7710	$-1.8 \cdot 10^{-3}$	$-1.6 \cdot 10^{-3}$
1.4	0.7143	0.7163	$-2.0 \cdot 10^{-3}$	$-1.7 \cdot 10^{-3}$
1.5	0.6667	0.6688	$-2.1 \cdot 10^{-3}$	$-1.9 \cdot 10^{-3}$
1.6	0.6250	0.6272	$-2.2 \cdot 10^{-3}$	$-1.9 \cdot 10^{-3}$
1.7	0.5882	0.5904	$-2.2 \cdot 10^{-3}$	$-1.9 \cdot 10^{-3}$
1.8	0.5556	0.5577	$-2.2 \cdot 10^{-3}$	$-1.9 \cdot 10^{-3}$
1.9	0.5263	0.5285	$-2.1 \cdot 10^{-3}$	$-1.9 \cdot 10^{-3}$
2.0	0.5000	0.5021	$-2.1 \cdot 10^{-3}$	$-1.9 \cdot 10^{-3}$

As in the case of Euler's method, it is in general inconvenient to evaluate $Z(x)$. Instead we again apply Richardson's method, and given two solutions, y_h and $y_{h/2}$, derive a third approximate solution significantly closer than both y_h and $y_{h/2}$ to Y. By virtue of Eq. (8.3.15),

$$Y - y_h = h^p Z + O(h^{p+1})$$
$$Y - y_{h/2} = (h/2)^p Z + O(h^{p+1}) \tag{8.3.17}$$

and thus

$$Y \approx \frac{2^p y_{h/2} - y_h}{2^p - 1} \tag{8.3.18}$$

at any point common to both grids. This is Richardson's extrapolation formula applied to Taylor's algorithm of order p.

Example 8.3.6.

Consider the problem $y' = \sin y$, $y(0) = \pi/2$, $0 \leq x \leq 1$, whose exact solution can be found and equals $Y(x) = 2 \tan^{-1}(e^x)$. To apply Taylor's algorithm of order 2, we calculate

$$f^{(1)}(x, y) = (\cos y)(\sin y) = \sin(2y)/2$$

and obtain the relation

$$y_{n+1} = y_n + h \left[\sin(y_n) + \frac{h}{2} \frac{\sin(2y_n)}{2} \right]$$

$$= y_n + h \left[\sin(y_n) + \frac{h}{4} \sin(2y_n) \right], \quad n \geq 0$$

where $y_0 = \pi/2$. For $h = 1/5$ we have $y_{0.2}(1) = 2.43919$ and for $h = 1/10$, $y_{0.1}(1) = 2.43705$. The new approximate value for $Y(1)$ is $Y^*(1) = (4 \cdot 2.43705 - 2.43919)/3 = 2.43634$. The exact value rounded to six digits is 2.43657.   □

PROBLEMS

1. Calculate $f^{(1)}(x, y)$ and $f^{(2)}(x, y)$ in the following cases:

    (a) $f(x, y) = xe^y$

    (b) $f(x, y) = x^3 + y^3$

    (c) $f(x, y) = \sin(x + y)$

    (d) $f(x, y) = \sin(xy)$

2. Let $Y(x)$ solve the initial-value problem

$$y' = 2x^2 y + 1, \quad y(0) = 1$$

    Derive the third order Taylor polynomial approximation to $Y(x)$ about $x = 0$.

3. Present the Taylor's algorithm of order 2 in the following cases:

    (a) $y' = x \cos y$

    (b) $y' = xy^2 + y$

    (c) $y' = 1/y$

4. Use Taylor's algorithm of order 2 to solve the following initial-value problems. In each case calculate the approximate solution at each discrete point.

    (a) $y' = (y/x) + x, \quad y(1) = 0, \quad 1 \leq x \leq 2, \quad h = 0.1$

    (b) $y' = x^2 + y^2, \quad y(0) = 1, \quad 0 \leq x \leq 1, \quad h = 0.1$

    (c) $y' = x^2 e^{-y}, \quad y(0) = 0, \quad 0 \leq x \leq 2, \quad h = 0.1$

5. Apply Taylor's algorithm of order p, $1 \leq p \leq 3$ to solve

$$y' = \left(\frac{3}{x} - 1 \right) y + x - 2, \quad y(1) = 1 + \frac{1}{e}, \quad 1 \leq x \leq 2, \quad h = 0.1$$

    Calculate the approximate solutions at $x = 2$ and compare with the exact solution

$$Y(x) = x^3 e^{-x} + x$$

6. Consider the initial-value problem

$$y' = y + 1, \quad y(0) = 0, \quad 0 \leq x \leq 1$$

whose unique solution is $Y(x) = e^x - 1$.

(a) Apply Taylor's algorithm of order 2 to calculate $y_{0.1}(1)$.

(b) Apply Euler's method to approximate $Z(x)$ of Eq. (8.3.16), using $h = 0.1$. Denote this approximation by $Z_{0.1}(x)$.

(c) Compute the exact value of $Z(x)$ and compare $Y(1) - y_{0.1}(1)$ with $(0.1)^2 Z_{0.1}(1)$ and $(0.1)^2 Z(1)$.

7. Apply Taylor's algorithm of order 2 to solve

$$y' = 2y - x^2, \quad y(0) = 0, \quad 0 \leq x \leq 2$$

Calculate $y_{0.1}(x)$ and $y_{0.05}(x)$ and use Richardson's extrapolation formula to obtain a better approximation.

8. (a) Apply Taylor's algorithm of order 3 to solve the initial-value problem given in problem 7, using $h = 0.1, 0.05$.

(b) Compare the results of (a) with the Richardson extrapolation solution of problem 7.

8.4 Runge–Kutta Methods

In most cases the expressions for $f^{(n)}(x, y)$ are complicated and their calculation is time consuming. Thus the Taylor method, although conceptually easy to manipulate, is generally not recommended.

The *Runge–Kutta methods* replace the evaluation of higher order derivatives by calculating $f(x, y)$ at more points. The number and locations of these points depend on the desired degree of accuracy. The general form of a Runge–Kutta method is given by

$$y_{n+1} = y_n + hK_p(x_n, y_n; h), \quad n \geq 0$$
$$y_0 = Y_0 \tag{8.4.1}$$

where $K_p(x_n, y_n; h)$ is some *average slope* of the solution between x_n and x_{n+1}. It is calculated as a weighted average of $f(x, y)$ at p points.

8.4.1 A Second Order Method

The most general second order Runge–Kutta (RK) method is defined by

$$K_2(x, y; h) \equiv a_1 f(x, y) + a_2 f[x + \alpha h, y + \beta h f(x, y)] \tag{8.4.2}$$

where $a_1, a_2, \alpha,$ and β are chosen in a way that minimizes the local truncation error

$$R_{n+1} = Y_{n+1} - [Y_n + hK_2(x_n, Y_n; h)] \tag{8.4.3}$$

We first apply Taylor's theorem and rewrite

$$K_2(x, y; h) = a_1 f(x, y) + a_2 \left[f(x, y) + (\alpha h f_x + \beta h f f_y)(x, y) + O(h^2) \right]$$
$$= (a_1 + a_2)f(x, y) + (a_2 \alpha f_x + a_2 \beta f f_y)h + O(h^2)$$

Then, using the notation of Eq. (8.3.6), we compare $T_2(x, y; h)$ given by

$$T_2(x, y; h) = f(x, y) + \frac{(f_x + f f_y)(x, y)}{2!} h$$

with $K_2(x, y; h)$ and determine a_1, a_2, α, and β so that

$$T_2(x, y; h) - K_2(x, y; h) = O(h^2) \tag{8.4.4}$$

We are thus led to choose

$$a_1 + a_2 = 1, \quad \alpha a_2 = 1/2, \quad \beta a_2 = 1/2 \tag{8.4.5}$$

that is,

$$\alpha = 1/2a_2, \quad \beta = 1/2a_2, \quad a_1 = 1 - a_2 \tag{8.4.6}$$

where a_2 is a free parameter. Now, because $Y_{n+1} = Y_n + hT_2(x_n, y_n; h) + O(h^3)$ we may substitute in Eq. (8.4.3), obtain $R_{n+1} = h[T_2(x_n, Y_n; h) - K_2(x_n, Y_n; h)] + O(h^3)$, and, in view of Eq. (8.4.4),

$$R_{n+1} = O(h^3) \tag{8.4.7}$$

Popular choices for a_2 are 1/2, 3/4, and 1. The choice $a_2 = 1/2$ provides the *modified Euler method*

$$y_{n+1} = y_n + \frac{h}{2} \left\{ f(x_n, y_n) + f\left[x_n + h, y_n + hf(x_n, y_n) \right] \right\}, \quad n \geq 0 \tag{8.4.8}$$

This is just one of many second order RK methods. It presents $K_2(x_n, y_n; h)$ as the average of two slopes, at (x_n, y_n) and at (x_{n+1}, y_{n+1}^E), where y_{n+1}^E is the approximate value of Y_{n+1} determined using Euler's method. Starting with $y_n = Y_n$, the Eq. (8.4.8) interpretation is given by Fig. 8.4.1.

Another popular choice is $a_2 = 1$. It yields the *midpoint method* (Fig. 8.4.2)

$$y_{n+1} = y_n + hf\left[x_n + \frac{h}{2}, y_n + \frac{h}{2}f(x_n, y_n) \right], \quad n \geq 0 \tag{8.4.9}$$

Example 8.4.1.

Consider the initial-value problem

$$y' = xy, \quad y(0) = 1, \quad 0 \leq x \leq 1, \quad h = 0.1$$

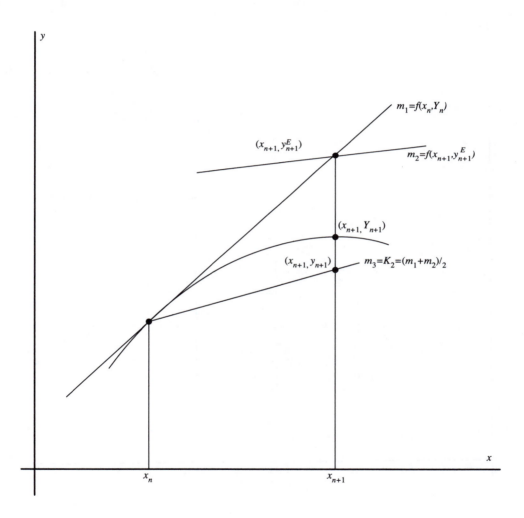

Figure 8.4.1.   The modified Euler method: geometric interpretation.

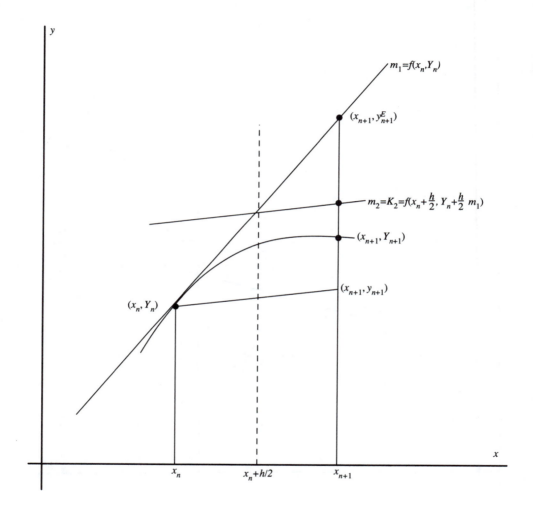

Figure 8.4.2.   Another second order Runge–Kutta method given by Eq. (8.4.9): the midpoint method.

The modified Euler method yields

$$y_{n+1} = y_n + \frac{h}{2}[x_n y_n + (x_n + h)(y_n + hx_n y_n)]$$

$$= y_n \left\{ 1 + \frac{h}{2} \left[x_n + (x_n + h)(1 + hx_n) \right] \right\}$$

The results compared with the exact solution $Y(x) = e^{x^2/2}$ are presented along with their errors $z_n = Y_n - y_n$ in Table 8.4.1. ☐

Table 8.4.1. A Second Order Runge–Kutta Method (Modified Euler) Applied to $y' = xy$, $y_0 = 1$, $h = 0.1$

n	x_n	y_n	Y_n	z_n
2	0.2	1.0202	1.0202	$< 10^{-4}$
4	0.4	1.0832	1.0833	$1 \cdot 10^{-4}$
6	0.6	1.1971	1.1972	$1 \cdot 10^{-4}$
8	0.8	1.3768	1.3771	$4 \cdot 10^{-4}$
10	1.0	1.6479	1.6487	$8 \cdot 10^{-4}$

Example 8.4.2.

Applying the midpoint method to the previous example, we obtain

$$y_{n+1} = y_n + h\left(x_n + \frac{h}{2}\right)\left(y_n + \frac{h}{2}x_n y_n\right)$$

$$= y_n \left[1 + h\left(x_n + \frac{h}{2}\right)\left(1 + \frac{h}{2}x_n\right) \right], \quad n \geq 0$$

The results are given in Table 8.4.2. ☐

Table 8.4.2. The Midpoint Method Applied to $y' = xy$, $y_0 = 1$, $h = 0.1$

n	x_n	y_n	Y_n	z_n
2	0.2	1.0202	1.0202	$< 10^{-4}$
4	0.4	1.0831	1.0833	$2 \cdot 10^{-4}$
6	0.6	1.1966	1.1972	$6 \cdot 10^{-4}$
8	0.8	1.3759	1.3771	$1.3 \cdot 10^{-3}$
10	1.0	1.6462	1.6487	$2.6 \cdot 10^{-3}$

We can now apply Eq. (8.4.4) to show that the error associated with a second order RK method satisfies Eq. (8.2.24) and Eq. (8.3.15), both with $p = 2$. The first result is as follows.

Theorem 8.4.1.

The error $z_n = Y_n - y_n$ associated with a second order RK method satisfies

$$|z_n| \leq Ch^2 \tag{8.4.10}$$

for some $C > 0$, provided that a Lipschitz condition holds for $T_2(x, y; h)$ [Eq. (8.3.9)].

Proof.

In view of Eq. (8.3.12) we have

$$Y_{n+1} = Y_n + hT_2(x_n, Y_n; h) + O(h^3) \tag{8.4.11}$$

Also, by definition,

$$y_{n+1} = y_n + hK_2(x_n, y_n; h) \tag{8.4.12}$$

Thus, by subtracting Eq. (8.4.12) from Eq. (8.4.11), we obtain

$$\begin{aligned} z_{n+1} &= z_n + h[T_2(x_n, Y_n; h) - K_2(x_n, y_n; h)] + O(h^3) \\ &= z_n + h[T_2(x_n, Y_n; h) - T_2(x_n, y_n; h) + T_2(x_n, y_n; h) - K_2(x_n, y_n; h)] + O(h^3) \end{aligned}$$

leading to

$$\begin{aligned} |z_{n+1}| &\leq |z_n| + h|T_2(x_n, Y_n; h) - T_2(x_n, y_n; h)| \\ &\quad + h|T_2(x_n, y_n; h) - K_2(x_n, y_n; h)| + C_1 h^3 \end{aligned}$$

for some $C_1 > 0$. Because Eq. (8.3.9) holds for $p = 2$, we have

$$|T_2(x_n, Y_n; h) - T_2(x_n, y_n; h)| \leq L|z_n|$$

and from Eq. (8.4.4) $|T_2(x_n, y_n; h) - K_2(x_n, y_n; h)| \leq C_2 h^2$ for some $C_2 > 0$. Therefore

$$|z_{n+1}| \leq (1 + hL)|z_n| + C_3 h^3 \tag{8.4.13}$$

where $z_0 = 0$ and $C_3 = C_1 + C_2$. By virtue of Lemma 8.2.1, Eq. (8.4.10) follows immediately. ⬜

Example 8.4.3.

Consider the initial-value problem $y' = -y$, $y(0) = 1$, $0 \leq x \leq 1 [Y(x) = e^{-x}]$ and let $h = 0.2$. By applying the midpoint method, we obtain

$$y_{n+1} = y_n + h\left\{-\left[y_n + \frac{h}{2}(-y_n)\right]\right\} = y_n\left(1 - h + \frac{h^2}{2}\right)$$

Substituting $h = 0.2, 0.1$ yields $y_{0.2}(1) = 0.3707$, $y_{0.1}(1) = 0.3685$, respectively, and an implementation of Richardson's extrapolation formula [Eq. (8.3.18) with $p = 2$] leads to a new value, $[4y_{0.1}(1) - y_{0.2}(1)]/3 = 0.3678$, which approximates the exact solution $Y(1) = 0.3679$ better than both $y_{0.2}(1)$ and $y_{0.1}(1)$. ☐

8.4.2 The Classical Runge–Kutta Method

Higher order RK methods are also available. A popular classic method of the fourth order is given by

$$y_{n+1} = y_n + hK_4(x_n, y_n; h), \quad n \geq 0$$
$$K_4(x, y; h) = \frac{1}{6}(k_1 + 2k_2 + 2k_3 + k_4) \qquad (8.4.14)$$

where

$$k_1 = f(x, y), \quad k_2 = f\left(x + \frac{h}{2}, y + \frac{h}{2}k_1\right),$$
$$k_3 = f\left(x + \frac{h}{2}, y + \frac{h}{2}k_2\right), \quad k_4 = f(x + h, y + hk_3) \qquad (8.4.15)$$

It can be shown that

$$T_4(x, y; h) - K_4(x, y; h) = O(h^4) \qquad (8.4.16)$$

and that consequently Eqs. (8.2.24) and (8.3.15) are satisfied with $p = 4$. If $f(x, y)$ is independent of y, the classical RK method reduces to Simpson's integration rule.

In general, a Runge–Kutta method of order p yields a local truncation error $O(h^{p+1})$ provided that $f(x, y)$ is sufficiently differentiable, and consequently Eqs. (8.2.24) and (8.3.15) hold as well. Thus, the Richardson extrapolation method can be applied and the error estimated. Using Eqs. (8.3.17) and (8.3.18) we have

$$Y - y_{h/2} \approx \frac{y_{h/2} - y_h}{2^p - 1} \qquad (8.4.17)$$

Example 8.4.4.
The classical RK method for $y' = y$, $y(0) = 1$, $0 \leq x \leq 1$ yields

$$k_1 = y, \quad k_2 = y + \frac{h}{2}k_1 = y\left(1 + \frac{h}{2}\right)$$
$$k_3 = y + \frac{h}{2}k_2 = y\left[1 + \frac{h}{2}\left(1 + \frac{h}{2}\right)\right]$$
$$k_4 = y + hk_3 = y\left\{1 + h\left[1 + \frac{h}{2}\left(1 + \frac{h}{2}\right)\right]\right\}$$

The results for $h = 0.1$ and for $h = 0.05$ compared with the exact solution $Y(x) = e^x$ are given in Table 8.4.3. The approximate solution, computed by the Richardson extrapolation method and denoted by $Y_R(x)$, is given as well. By virtue of Eq. (8.4.16) we have

$$Y(x) \approx Y_R(x) = \frac{16y_{0.05}(x) - y_{0.1}(x)}{15}$$

Table 8.4.3. Classical Runge–Kutta Method Applied to $y' = y$, $y_0 = 1$

x	$y_{0.1}(x)$	$y_{0.05}(x)$	$Y_R(x)$	$Y(x)$
0.2	1.221402571	1.221402746	1.221402758	1.221402758
0.4	1.491824240	1.491824668	1.491824696	1.491824698
0.6	1.822117962	1.822118746	1.822118798	1.822118800
0.8	2.225539563	2.225540840	2.225540925	2.225540928
1.0	2.718279744	2.718281693	2.718281823	2.718281828

□

It should be noted that Richardson's extrapolation method is effective only if h is sufficiently small. If in the previous example we replace $h = 0.1, 0.05$ by $h = 0.2, 0.1$, Richardson's method does not produce a better solution. In fact, $y_{0.1}(x)$ is closer than $Y_R(x)$ to e^x.

Example 8.4.5.
Consider the initial-value problem $y' = -y + x$, $y(0) = 0$, $0 \le x \le 1$, which possesses the solution $Y(x) = e^{-x} + x - 1$. Let $y_E(x)$, $y_T(x)$, and $y_{RK}(x)$ denote the approximate solutions computed by Euler's method, Taylor's method of order 2, and the classical RK procedure, respectively, with $h = 0.1$. Their discrete values at selected points are given in Table 8.4.4, compared with $Y(x)$.   □

Table 8.4.4. First, Second, and Fourth Order Methods Applied to $y' = -y + x$, $y_0 = 0$, $h = 1/10$

x	y_E	y_T	y_{RK}	Y	$Y - y_E$	$Y - y_T$	$Y - y_{RK}$
0.2	0.0100	0.01903	0.01873090	0.01873075	$0.87 \cdot 10^{-2}$	$-0.29 \cdot 10^{-3}$	$-0.15 \cdot 10^{-6}$
0.4	0.0561	0.07080	0.07032029	0.07032005	$0.14 \cdot 10^{-1}$	$-0.48 \cdot 10^{-3}$	$-0.24 \cdot 10^{-6}$
0.6	0.1314	0.14940	0.14881193	0.14881164	$0.17 \cdot 10^{-1}$	$-0.59 \cdot 10^{-3}$	$-0.30 \cdot 10^{-6}$
0.8	0.2305	0.24998	0.24932929	0.24932896	$0.19 \cdot 10^{-1}$	$-0.65 \cdot 10^{-3}$	$-0.33 \cdot 10^{-6}$
1.0	0.3487	0.36854	0.36787977	0.36787944	$0.19 \cdot 10^{-1}$	$-0.66 \cdot 10^{-3}$	$-0.33 \cdot 10^{-6}$

The subroutine RK24, for solving initial-value problems using Runge–Kutta methods of the second and fourth order, is found on the attached floppy disk.

PROBLEMS

1. Apply the modified Euler method to the following initial-value problems. Use $h = 0.1$, 0.05 and evaluate $y(b)$.
   (a) $y' = -2y + x$, $y(0) = 1$, $b = 1$
   (b) $y' = \sin y - x^2$, $y(0) = -1$, $b = 2$
   (c) $y' = xy + y$, $y(0) = 2$, $b = 3$

2. Apply the midpoint method in problem 1.

3. Solve the initial-value problem $y' = 2y - 2x^2 + 1$, $y(0) = 0$, $0 \le x \le 1$ using the modified Euler, midpoint, and second order Taylor methods with $h = 0.1$. Compare the approximate and the exact solutions at $x_i = ih$, $1 \le i \le 10$ [$Y(x) = x^2 + x$].

4. Substitute $a_2 = 3/4$ in Eq. (8.4.6) and determine the associated second order RK method. Apply it to solve $y' = x - y^2$, $y(0) = 1$, $0 \le x \le 1$ and calculate $y(1)$ for $h = 0.1, 0.05$.

5. Apply the classical RK method to solve $y' = y - 2e^{-x}$, $y(0) = 2$, $0 \leq x \leq 2$ using $h = 0.1, 0.05$. Determine $y_{0.1}(x)$ and $y_{0.05}(x)$ at $x_n = 0.2n$, $1 \leq n \leq 10$ and compare with the exact solution $Y(x) = e^x + e^{-x}$.

6. In the previous problem, apply the Richardson extrapolation method and calculate an improved approximate solution to $Y(x)$.

7. Apply the second order Taylor method, the modified Euler method, and the classical RK method to solve $y' = -y + (x+1)^2$, $y(0) = 1$, $0 \leq x \leq 1$ using $h = 0.05$. Compare the results with the exact solution $Y(x) = x^2 + 1$ at $x_n = 0.1n$, $1 \leq n \leq 10$.

8. Consider a numerical procedure for which $Y(x_n) - y_h(x_n) = h^p Z(x_n) + O(h^{p+1})$. Use $y_h(x_n)$ and $y_{2h}(x_n)$ to derive the following asymptotic formula for the error at x_n:

$$Y(x_n) - y_h(x_n) \approx \frac{2^p y_h(x_n) - y_{2h}(x_n)}{2^p - 1}$$

9. Apply the classical RK method to solve $y' = y - 2\sin x$, $y(0) = 1$, $0 \leq x \leq \pi$ $[Y(x) = \sin x + \cos x]$, using $h = \pi/10, \pi/20$, and approximate $Y(x) - y_{(\pi/20)}(x)$ at $x_n = (\pi/10)n$, $1 \leq n \leq 10$.

10. (a) Find the truncation error of the midpoint method and compare with the error calculated using a second order Taylor method.

    (b) Compare the two errors in the case $f(x, y) = xy$.

11. A particle of mass m falls vertically in the gravitational field of the earth. Its downward motion is opposed by a frictional force $f(v)$, where v denotes the velocity of the particle. The equation of motion is given by $mv'(t) = -mg + f(v)$, $t \geq 0$. Let $v(0) = 0$, $m = 1\text{kg}$, $g = 9.81\text{m/sec}^2$. Solve the equation for $0 \leq t \leq 5$, using the classical RK method in the following cases:

    (a) $f(v) = -0.1v$

    (b) $f(v) = 0.1v^2$

    In both cases $f(v)$ is positive (v is negative at all times). Use time steps $h = \Delta t = 0.2, 0.1$ and apply the Richardson extrapolation method to produce a better approximation.

12. Use the modified Euler method to solve $y' = -y + \sqrt{x}(x + 1.5)$, $y(0) = 0$, $0 \leq x \leq 1$.

    (a) Use $h = 0.2, 0.1, 0.05$ and compare the results with the exact solution $Y(x) = x\sqrt{x}$.

    (b) Is it possible to apply a second order Taylor method?

8.5 Error Control: Runge–Kutta–Fehlberg Method

Consider a general initial-value problem

$$y' = f(x, y), \quad y(x_0) = y_0, \quad x_0 \leq x \leq b \tag{8.5.1}$$

A numerical method for solving Eq. (8.5.1), given by

$$y_{n+1} = y_n + h_n K(x_n, y_n; h_n), \quad 0 \leq n \leq N - 1 \tag{8.5.2}$$

would be ideal if, given a tolerance $\epsilon > 0$, a mesh $\{x_n,\ 0 \leq n \leq N\}$ with a minimal number of points can be designed such that the *global error*, $Y_n - y_n$, satisfies

$$|Y_n - y_n| < \epsilon, \quad 0 \leq n \leq N \tag{8.5.3}$$

Controlling the global error while using a minimal number of mesh points is impossible, unless a nonuniform mesh is considered. Clearly, if the solution and its derivatives vary rapidly at some neighborhood, a finer mesh is necessary in that particular region to ensure a small global error. Thus h_n rather than h is introduced in Eq. (8.5.2).

However, in general we cannot determine the global error of a given method. Instead we work with the local truncation error. In several cases we have already established the fact that if the local truncation error is uniformly small, so is the global error. The general theorem will be given later on, when we discuss stability. To *estimate* the local error for a given method M_1 of order p, we work simultaneously with M_1 and M_2, another method of order $(p + 1)$.

Let M_1 denote the scheme

$$M_1 \begin{cases} u_{n+1} = u_n + h_n K_1(x_n, u_n; h_n), & 0 \leq n \leq N - 1 \\ u_0 = y_0 \end{cases} \tag{8.5.4}$$

with a local truncation error $t_{n+1} = O(h_n^{p+1})$, whereas M_2 is given by

$$M_2 \begin{cases} \tilde{u}_{n+1} = \tilde{u}_n + h_n K_2(x_n, \tilde{u}_n; h_n), & 0 \leq n \leq N - 1 \\ \tilde{u}_0 = y_0 \end{cases} \tag{8.5.5}$$

and has a local truncation error $\tilde{t}_{n+1} = O(h_n^{p+2})$. In essence, we integrate the differential equation, using M_1, and apply M_2 *only* to estimate t_{n+1} and determine a proper h_n at any given mesh point x_n. Let y_n denote the already computed approximate value of Y_n at some x_n, using M_1. Substituting it for u_n and $\tilde{u}_n$ in Eqs. (8.5.4) and (8.5.5) and using a step size h, we obtain

$$y_{n+1} = y_n + h K_1(x_n, y_n; h) \tag{8.5.6}$$

and

$$\tilde{y}_{n+1} = y_n + h K_2(x_n, y_n; h) \tag{8.5.7}$$

respectively. The local errors at x_{n+1} are given by

$$t_{n+1} = Y_{n+1} - Y_n - h K_1(x_n, Y_n; h) \tag{8.5.8}$$

and

$$\tilde{t}_{n+1} = Y_{n+1} - Y_n - h K_2(x_n, Y_n; h) \tag{8.5.9}$$

respectively. Therefore $t_{n+1} - \tilde{t}_{n+1} = h[K_2(x_n, Y_n; h) - K_1(x_n, Y_n; h)]$ and because the schemes M_1

and M_2 are of orders p and $(p+1)$, respectively, we have

$$t_{n+1} \approx h[K_2(x_n, Y_n; h) - K_1(x_n, Y_n; h)] \tag{8.5.10}$$

It can be shown that Y_n in Eq. (8.5.10) may be replaced by y_n, provided that the maximum step size during the integration process approaches zero. Thus, in view of Eqs. (8.5.6) and (8.5.7), the relation

$$t_{n+1} \approx h[K_2(x_n, y_n; h) - K_1(x_n, y_n; h)] = \tilde{y}_{n+1} - y_{n+1} \tag{8.5.11}$$

holds asymptotically, giving an estimate for the local error. We want to find h_n such that

$$|t_{n+1}| \leq \epsilon h_n \tag{8.5.12}$$

Suppose that for some initial step size h, Eq. (8.5.12) is not satisfied. The local truncation error given by Eq. (8.5.11) can also be expressed by

$$t_{n+1} \approx A h^{p+1} \tag{8.5.13}$$

for some given constant A. Let us decrease h and replace it by αh, $0 < \alpha < 1$, denoting the corresponding local error by $t_{n+1}(\alpha h)$. Then

$$t_{n+1}(\alpha h) \approx A(\alpha h)^{p+1} = \alpha^{p+1}(A h^{p+1}) \approx \alpha^{p+1} t_{n+1} \approx \alpha^{p+1}(\tilde{y}_{n+1} - y_{n+1})$$

To satisfy Eq. (8.5.12), we must choose α such that $|\alpha^{p+1}(\tilde{y}_{n+1} - y_{n+1})| \leq \epsilon(\alpha h)$ or

$$\alpha \leq \left(\frac{\epsilon h}{|\tilde{y}_{n+1} - y_{n+1}|} \right)^{1/p} \tag{8.5.14}$$

Thus a proper choice for h_n is

$$h_n = \left(\frac{\epsilon h}{|\tilde{y}_{n+1} - y_{n+1}|} \right)^{1/p} h \tag{8.5.15}$$

where h is the initial step size at x_n, y_{n+1} and $\tilde{y}_{n+1}$ are the approximate values to Y_{n+1} determined using schemes M_1 and M_2, respectively, ϵ is an *a priori* given tolerance, and p is the order of M_1.

Example 8.5.1.

Let M_1 be the standard Euler method ($p = 1$) and let M_2 be the modified Euler method ($p = 2$), applied to $y' = y$, $y(0) = 1$, $0 \leq x \leq 1$ $[Y(x) = e^x]$. In this case $y_{n+1} = (1 + h)y_n$, $\tilde{y}_{n+1} = [1 + h + (h^2/2)]y_n$, and $\tilde{y}_{n+1} - y_{n+1} = (h^2/2)y_n$. The exact local truncation error produced by M_1 at $x_{n+1} = (n+1)h$ is

$$t_{n+1} = Y_{n+1} - (1 + h)Y_n = e^{(n+1)h} - (1 + h)e^{nh}$$

$$= e^{nh}(e^h - 1 - h) = e^{nh} \cdot \frac{h^2}{2} e^{\theta h}$$

where θ is between 0 and 1. Because $\tilde{y}_{n+1} - y_{n+1} = (h^2/2)y_n$ and $e^{nh}/y_n \approx 1$, $e^{\theta h} \approx 1, h \rightarrow 0$ we obtain $\lim_{h \rightarrow 0} \left[t_{n+1}/(\tilde{y}_{n+1} - y_{n+1}) \right] = 1$, in agreement with Eq. (8.5.11). ☐

In the next example, given (x_n, y_n) and an initial step size h, a *proper* step size h_n is determined.

Example 8.5.2.

Consider the previous problem and let y_n denote the numerical solution at x_n, using the standard Euler method of order 1. For a given initial step size h at x_n, and a prefixed tolerance $\epsilon > 0$, we obtain

$$\alpha \leq \left[\frac{\epsilon h}{(h^2/2)y_n} \right] = \frac{2\epsilon}{hy_n}$$

In the particular case $x_n = 0.5$ $(y_n \approx e^{0.5})$, $\epsilon = 10^{-5}$, $h = 0.1$ we have $\alpha \leq 0.00012$ and the optimal choice is $h_n = 0.000012$. ☐

8.5.1 The Runge–Kutta–Fehlberg Method

A popular scheme that uses Eq. (8.5.14) for its error control is the *Runge–Kutta–Fehlberg method*. It uses a Runge–Kutta method of order 5 to estimate the local truncation error of a Runge–Kutta method of order 4. The fourth order method, denoted by M_1, is given by

$$y_{n+1} = y_n + h \sum_{i=1}^{5} a_i k_i(x_n, y_n; h) \tag{8.5.16}$$

and the fifth order method, M_2, is given by

$$\tilde{y}_{n+1} = y_n + h \sum_{i=1}^{6} b_i k_i(x_n, y_n; h) \tag{8.5.17}$$

where

$$k_1(x, y; h) = f(x, y)$$

$$k_i(x, y; h) = f(x + c_i h, y + h \sum_{j=1}^{i-1} d_{ij} k_j), \quad 2 \leq i \leq 6 \tag{8.5.18}$$

The coefficients associated with this procedure are given in Tables 8.5.1 and 8.5.2. They provide schemes M_1 and M_2 with local truncation errors of $O(h^5)$ and $O(h^6)$, respectively.

Table 8.5.1. Fehlberg Coefficients a_i, b_i

i	a_i	b_i
1	25/216	16/135
2	0	0
3	1408/2565	6656/12825
4	2197/4104	28561/56430
5	−1/5	−9/50
6		2/55

Table 8.5.2. Fehlberg Coefficients c_i, d_{ij}

i	c_i	d_{i1}	d_{i2}	d_{i3}	d_{i4}	d_{i5}
2	1/4	1/4				
3	3/8	3/32	9/32			
4	12/13	1932/2197	−7200/2197	7296/2197		
5	1	439/216	−8	3680/513	−845/4104	
6	1/2	−8/27	2	−3544/2565	1859/4104	−11/40

An important advantage of the Runge–Kutta–Fehlberg method is that only *six* function evaluations per step are needed, whereas an arbitrary combination of Runge–Kutta methods of orders 4 and 5 requires a total of 10 evaluations: 4 for the fourth order method and 6 for the fifth order method.

If the right-hand side of Eq. (8.5.14) exceeds 1, Eq. (8.5.12) already holds with the initial h, and the next step size is initially taken as $\alpha h > h$. If $\alpha < 1$, h is replaced by αh and another six function evaluations are performed to determine the final approximate value y_{n+1}. This increase in the computing costs suggests that we choose α somewhat differently. Rather than taking the right-hand side of Eq. (8.5.14) we choose a smaller value, for example,

$$\alpha = \left(\frac{\epsilon h}{2|\tilde{y}_{n+1} - y_{n+1}|} \right)^{1/4} = 0.84 \left(\frac{\epsilon h}{|\tilde{y}_{n+1} - y_{n+1}|} \right)^{1/4}$$

that *may* provide an appropriate initial value for h at the next step.

In the next example, we apply the Runge–Kutta–Fehlberg method without error control, that is, using a fixed step size h.

Example 8.5.3.

Consider the initial-value problem $y' = -y + x + 1$, $y(0) = 1$, $0 \le x \le 2$. Table 8.5.3 provides the approximate solution $y_h(x)$, using the fourth order Runge–Kutta–Fehlberg method with $h = 0.2$, compared with the exact solution $Y(x) = x + e^{-x}$. The right-hand column is the approximate local truncation error $\tilde{y}_h - y_h$, where $\tilde{y}_h$ is computed from Eq. (8.5.17). ☐

Table 8.5.3. Fourth Order Runge–Kutta–Fehlberg Method Applied to
$y' = -y + x + 1$, $y_0 = 1$, $h = 0.2$

x	$y_h(x)$	$Y(x) - y_h(x)$	$\tilde{y}_h(x) - y_h(x)$
0.4	1.070319233	$0.813 \cdot 10^{-6}$	$0.361 \cdot 10^{-6}$
0.8	1.249327874	$0.109 \cdot 10^{-5}$	$0.242 \cdot 10^{-6}$
1.2	1.501193116	$0.110 \cdot 10^{-5}$	$0.162 \cdot 10^{-6}$
1.6	1.801895538	$0.980 \cdot 10^{-6}$	$0.109 \cdot 10^{-6}$
2.0	2.135334462	$0.821 \cdot 10^{-6}$	$0.729 \cdot 10^{-7}$

If we decide to keep $\tilde{y}_{n+1}$ rather than y_{n+1} as a final approximation to Y_{n+1}, we simply replace y_n with $\tilde{y}_n$ in Eqs. (8.5.6) and (8.5.7). We still estimate (and control) the local truncation error associated with M_1, but produce a better numerical solution at the mesh points, using the method M_2.

Example 8.5.4.

Table 8.5.4 contains the numerical solution to the previous problem, when the fifth order Runge–Kutta–Fehlberg method is applied.

Table 8.5.4. Fifth Order Runge–Kutta–Fehlberg Method Applied to $y' = -y + x + 1$

x	$\tilde{y}_h(x)$	$Y(x) - \tilde{y}_h(x)$	$\tilde{y}_h(x) - y_h(x)$
0.4	1.070319594	$0.452 \cdot 10^{-6}$	$0.361 \cdot 10^{-6}$
0.8	1.249328600	$0.364 \cdot 10^{-6}$	$0.242 \cdot 10^{-6}$
1.2	1.501193927	$0.285 \cdot 10^{-6}$	$0.162 \cdot 10^{-6}$
1.6	1.801896299	$0.219 \cdot 10^{-6}$	$0.109 \cdot 10^{-6}$
2.0	2.135335118	$0.165 \cdot 10^{-6}$	$0.729 \cdot 10^{-7}$

The last columns in Tables 8.5.3 and 8.5.4 are not really identical but can be shown to coincide asymptotically as $h \to 0$. $\qquad \square$

We will now formulate an algorithm based on the Runge–Kutta–Fehlberg method.

Algorithm 8.5.1.

[Runge–Kutta–Fehlberg method for approximating the solution to $y' = f(x, y)$, $y(x_0) = y_0$, $x_0 \leq x \leq b$].

Input: End points x_0, b; initial condition y_0; maximum step size h_{max}; minimum step size h_{min}; tolerance ϵ; maximum increase factor of step size α_1; minimum decrease factor of step size α_2 ($\alpha_1 > 1$, $\alpha_2 < 1$).

Output:

1. Computation completed successfully: A set of pairs $\{(x_n, y_n)\}_{n=0}^{N}$, where $x_N = b$ and y_n approximates the exact solution Y_n at x_n

2. Computation completed unsuccessfully: A set of pairs $\{(x_n, y_n)\}_{n=0}^{M}$, where $x_M < b$ and y_n approximates the exact solution Y_n at x_n. In addition, there is a message that the minimum step size exceeded at x_M

Step 1. Set $x = x_0$, $y = y_0$, $n = 0$, $h = \min\{h_{\max}, b - x\}$.

Step 2. Compute $k_i(x, y; h)$, $1 \leq i \leq 6$ by using Eq. (8.5.18) and Table 8.5.2. Compute the right-hand sides of Eqs. (8.5.16) and (8.5.17), u and $\tilde{u}$ respectively, using Table 8.5.1.

Step 3. Set $t = |\tilde{u} - u|/h = |(1/360)k_1 - (128/4275)k_3 - (2197/75, 240)k_4 + (1/50)k_5 + (2/55)k_6|$.

Step 4. Set $\alpha = 0.84(\epsilon/t)^{1/4}$.

Step 5. If $t \leq \epsilon$ then set $n \leftarrow n + 1$, $x_n \leftarrow x + h$, $y_n \leftarrow u$, $x \leftarrow x_n$, $y \leftarrow y_n$.

Step 6. If $\alpha \leq \alpha_2$ then set $h \leftarrow \alpha_2 h$; else if $\alpha \geq \alpha_1$ then set $h \leftarrow \alpha_1 h$; else set $h \leftarrow \alpha h$.

Step 7. If $h > h_{\max}$ then set $h = h_{\max}$.

Step 8. If $h < h_{\min}$ then write "minimum step size exceeded" and stop.

Step 9. If $b - x \geq h_{\min}$ set $h \leftarrow \min\{h, b - x\}$ and go to Step 2; else stop.

The reason for step 6 is to exclude large modifications in h. Thus we avoid high computing costs by working with small step sizes in regions where the solution and its derivatives vary rapidly. We also eliminate large step sizes that may cause us to skip some sensitive neighborhoods.

A subroutine RKF45, based on Algorithm 8.5.1, is found on the attached floppy disk.

Example 8.5.5.

We apply Algorithm 8.5.1 with $\alpha_1 = 5$, $\alpha_2 = 0.1$, $h_{\max} = 1$, $h_{\min} = 0.01$, and $\epsilon = 10^{-5}$ to solve $y' = -y + e^{-x}$, $y(0) = 0$, $0 \leq x \leq 2$. The results compared with the exact solution $Y(x) = xe^{-x}$ are given in Table 8.5.5. We assumed 12 digits of accuracy, but the final numbers were rounded to 6 digits. $\qquad\square$

Table 8.5.5. Applying Algorithm 8.5.1 to Solve $y' = -y + e^{-x}$, $y_0 = 0$, $0 \leq x \leq 2$

n	h_n	x_n	y_n	Y_n	$\tilde{y}_n - y_n$	$Y_n - y_n$
0	0.192405	0	0	0		0
1	0.203592	0.192405	0.158730	0.158729	$-0.764 \cdot 10^{-6}$	$-0.866 \cdot 10^{-6}$
2	0.218590	0.395997	0.266511	0.266509	$-0.763 \cdot 10^{-6}$	$-0.158 \cdot 10^{-5}$
3	0.236260	0.614587	0.332410	0.332408	$-0.797 \cdot 10^{-6}$	$-0.218 \cdot 10^{-5}$
4	0.257827	0.850847	0.363360	0.363357	$-0.829 \cdot 10^{-6}$	$-0.269 \cdot 10^{-5}$
5	0.285061	1.108674	0.365861	0.365858	$-0.859 \cdot 10^{-6}$	$-0.309 \cdot 10^{-5}$
6	0.321342	1.393735	0.345854	0.345851	$-0.879 \cdot 10^{-6}$	$-0.338 \cdot 10^{-5}$
7	0.284924	1.715076	0.308632	0.308628	$-0.868 \cdot 10^{-6}$	$-0.352 \cdot 10^{-5}$
8		2	0.270673	0.270671	$-0.187 \cdot 10^{-6}$	$-0.288 \cdot 10^{-5}$

The choice of a large $h_{\max}$ ($h_{\max} = 1$ in the previous example) is practical, if one is mainly interested in approximating $Y(b)$. If, however, we want to produce a numerical solution over a uniform mesh, we must choose a smaller $h_{\max}$ to ensure $h > h_{\max}$ in step 7 of the algorithm. In the previous example, $h_{\max} = 0.2$ is already a proper choice.

Example 8.5.6.

Consider the nonlinear problem $y' = -y^2 + x^2 + 3$, $y(1) = 2$, $1 \leq x \leq 10$. Applying Algorithm 8.5.1 with the parameters of Example 8.5.5 we obtain $y(10) = 10.09999938$. The exact solution is $Y(x) = x + (1/x)$ and satisfies $Y(10) - y(10) = 0.62 \cdot 10^{-6}$. $\qquad\square$

PROBLEMS

1. Apply Algorithm 8.5.1 to approximate the solutions to the following initial-value problems:

   (a) $y' = y$, $y(0) = 1$, $0 \le x \le 5$ $[Y(x) = e^x]$

   (b) $y' = (y/x) + 1$, $y(1) = 0$, $1 \le x \le 3$ $[Y(x) = x \ln(x)]$

   (c) $y' = y^2$, $y(1) = -1$, $1 \le x \le 3$ $[Y(x) = -1/x]$

   using $h_{max} = 1$, $h_{min} = 0.02$, $\epsilon = 10^{-5}$, $\alpha_1 = 5$, $\alpha_2 = 0.1$. Compare the results with the exact solutions.

2. Apply Algorithm 8.5.1 to approximate the solution to

$$y' = y\left(1 - \frac{1}{x}\right), \quad y(1) = e, \quad 1 \le x \le 3 \quad \left[Y(x) = \frac{e^x}{x}\right]$$

   using $h_{max} = 0.2$, $h_{min} = 0.02$, $\alpha_1 = 5$, $\alpha_2 = 0.1$. Consider the tolerances

   (a) $\epsilon = 10^{-4}$

   (b) $\epsilon = 10^{-5}$

   (c) $\epsilon = 10^{-6}$

   and compare the results with the exact solution.

3. Modify Algorithm 8.5.1 so that $\tilde{u}$ rather than u will be considered as the final approximate value at x_n. Use the modified algorithm to solve case (a) of problem 1.

4. (a) Apply Algorithm 8.5.1 to approximate the solution to the initial-value problem

$$y' = x^2 + y^2, \quad y(0) = 0, \quad 0 \le x \le 1.5$$

   using $h_{max} = 1$, $h_{min} = 0.01$, $\epsilon = 10^{-5}$, $\alpha_1 = 5$, $\alpha_2 = 0.1$.

   (b) Decrease h_{max}, 0.1 at a time, until you receive evenly spaced mesh points.

5. (a) Construct an algorithm similar to Algorithm 8.5.1, based on the Euler method and the modified Euler method.

   (b) Use the algorithm to solve the initial-value problems of problem 1.

6. A simplified model that describes the spread of a contagious disease is given by the *Bernoulli equation*

$$dy/dt = ky(m - y)$$

   where $y(t)$ is the number of infectives at time t, m is the total population, and k is a constant determined by the specific disease.

   (a) Assume $m = 10^6$, $y(0) = 10^3$, $k = 10^{-7}$ and that t is measured in days. Approximate the number of infectives after 10 days, using Algorithm 8.5.1 with a tolerance $\epsilon = 1$.

   (b) Repeat the calculation, assuming $k = 10^{-6}$.

   (c) By substituting $z = 1/y$ in the Bernoulli equation, one obtains a linear equation for which an exact solution can be found. Compare the true value of $y(10)$ with the approximate values computed in cases (a) and (b).

8.6 Multistep Methods

The previous methods for solving initial-value problems are all *one-step methods*. To approximate the exact solution at x_{n+1}, they all use information related only to x_n. Even the Runge–Kutta methods, which involve function evaluations at several interim points between x_n and x_{n+1} to obtain y_{n+1}, use them only once and not in future calculations.

Because approximations prior to y_n are likely to possess higher accuracy than y_n, it seems reasonable to develop methods that use these previous data to obtain y_{n+1}. Methods that use data related to more than one previous mesh point to approximate the solution at x_{n+1} are called *multistep methods*. The precise definition is given below.

> **Definition 8.6.1.** (Multistep methods). *A multistep method for solving the initial-value problem*
>
> $$y' = f(x, y), \quad y(x_0) = y_0, \quad x_0 \leq x \leq b \tag{8.6.1}$$
>
> *is an integration procedure given by*
>
> $$\begin{aligned} y_{n+1} &= a_0 y_n + a_1 y_{n-1} + \cdots + a_{m-1} y_{n-m+1} \\ &\quad + h[b_{-1} f(x_{n+1}, y_{n+1}) + b_0 f(x_n, y_n) + \cdots + b_{m-1} f(x_{n-m+1}, y_{n-m+1})] \end{aligned} \tag{8.6.2}$$
>
> *where $m > 1$ and $n = m - 1, m, \ldots, N - 1$. It is an m-step method provided that $a_{m-1}^2 + b_{m-1}^2 > 0$. If $b_{-1} = 0$, the method is* explicit *or an* open *method, that is, y_{n+1} is determined explicitly by previously computed data. In the case of $b_{-1} \neq 0$, the method is* implicit *or a* closed *method and y_{n+1}, which appears on both sides of Eq. (8.6.2), is determined implicitly. The step size is $h = (b - a)/N$ and the values $y_0, y_1, \ldots, y_{m-1}$, which are essential for implementing the procedure, must be specified a priori. The right-hand side of Eq. (8.6.2) defines the* difference equation *of the multistep method.*

Example 8.6.1.
The multistep method given by

$$\begin{aligned} y_{n+1} &= y_{n-3} + \frac{4h}{3}[2f(x_n, y_n) - f(x_{n-1}, y_{n-1}) + 2f(x_{n-2}, y_{n-2})] \\ &= y_{n-3} + \frac{4h}{3}(2y'_n - y'_{n-1} + 2y'_{n-2}) \end{aligned} \tag{8.6.3}$$

is a four-step method, of order 4, called the *Milne method*. To implement this procedure we must first specify y_0, y_1, y_2, and y_3. This can be done by using a starting method, for example a Runge–Kutta method of order 4. Note that Eq. (8.6.3) defines an open method.

The equation

$$\begin{aligned} y_{n+1} &= y_n + \frac{h}{24}[9f(x_{n+1}, y_{n+1}) + 19f(x_n, y_n) - 5f(x_{n-1}, y_{n-1}) \\ &\quad + f(x_{n-2}, y_{n-2})] = y_n + \frac{h}{24}(9y'_{n+1} + 19y'_n - 5y'_{n-1} + y'_{n-2}) \end{aligned} \tag{8.6.4}$$

defines a three-step closed multistep method: the *fourth order Adams–Moulton method*. ☐

To formulate a popular general derivation of multistep methods, we first integrate the differential equation over $[x_n, x_{n+1}]$ to obtain

$$Y_{n+1} = Y_n + \int_{x_n}^{x_{n+1}} f[x, Y(x)] \, dx \tag{8.6.5}$$

where $Y(x)$ denotes the exact solution to Eq. (8.6.1). The idea is to approximate and replace the integrand $Y'(x) = f[x, Y(x)]$ by the interpolating polynomial determined by (x_n, Y'_n), (x_{n-1}, Y'_{n-1}), ..., (x_{n-m+1}, Y'_{n-m+1}). However, prior to this, we will use the presentation of the Lagrange interpolator by divided differences (Section 5.4) to obtain *Newton's forward and backward difference formulas*.

Let $p_n(x)$ denote the Lagrange interpolator of $F(x)$, determined by $(x_0, F_0), \ldots, (x_n, F_n)$ where $F_i = F(x_i)$, $0 \le i \le n$. Then

$$p_n(x) = F[x_0] + F[x_0, x_1](x - x_0) + F[x_0, x_1, x_2](x - x_0)(x - x_1) + \cdots$$
$$+ F[x_0, x_1, \ldots, x_n](x - x_0)(x - x_1) \cdots (x - x_{n-1})$$

If x_i, $0 \le i \le n$ are evenly spaced and $x_{i+1} - x_i = h$, then

$$F[x_0, x_1] = \frac{F_1 - F_0}{x_1 - x_0} = \frac{1}{h}(\triangle F_0)$$

$$F[x_0, x_1, x_2] = \frac{F[x_1, x_2] - F[x_0, x_1]}{x_2 - x_0} = \frac{(1/h)(\triangle F_1) - (1/h)(\triangle F_0)}{2h} = \frac{1}{2h^2}(\triangle^2 F_0)$$

and by using induction we conclude

$$F[x_0, x_1, \ldots, x_k] = \frac{1}{(k!)h^k}(\triangle^k F_0) \tag{8.6.6}$$

We now define $s = (x - x_0)/h$ and obtain *Newton's forward difference formula*:

$$p_n(x) = \sum_{k=0}^{n} \binom{s}{k} (\triangle^k F_0) \tag{8.6.7}$$

where $\binom{s}{k}$ is the extended binomial coefficient defined as $\binom{s}{k} = [s(s-1) \cdots (s-k+1)]/k!$. We deduce *Newton's backward difference formula* similarly

$$p_n(x) = \sum_{k=0}^{n} (-1)^k \binom{-s}{k} (\nabla^k F_n), \quad s = \frac{x - x_n}{h} \tag{8.6.8}$$

8.6.1   The Adams–Bashforth Method

Returning now to Eq. (8.6.5) we express the integrand as

$$f[x, Y(x)] = p_{m-1}(x) + \frac{f^{(m)}[\xi_x, Y(\xi_x)]}{m!}(x - x_n)(x - x_{n-1}) \cdots (x - x_{n-m+1})$$

where $p_{m-1}(x)$ is the Lagrange interpolator of order $(m-1)$ to $f[x, Y(x)]$ determined by the data set $\{(x_i, Y_i'), n - m + 1 \leq i \leq n\}$. Using Eq. (8.6.8) we find

$$
\int_{x_n}^{x_{n+1}} f[x, Y(x)]\, dx = \int_{x_n}^{x_{n+1}} \sum_{k=0}^{m-1} (-1)^k \binom{-s}{k} (\nabla^k Y_n')\, dx
$$

$$
+ \int_{x_n}^{x_{n+1}} \frac{f^{(m)}\left[\xi_x, Y(\xi_x)\right]}{m!} (x - x_n)(x - x_{n-1}) \cdots (x - x_{n-m+1})\, dx
$$

$$
= h \sum_{k=0}^{m-1} (-1)^k (\nabla^k Y_n') \int_0^1 \binom{-s}{k}\, ds
$$

$$
+ \frac{h^{m+1}}{m!} \int_0^1 s(s+1) \cdots (s + m - 1) f^{(m)}\left[\xi_s, Y(\xi_s)\right]\, ds
$$

The numbers $(-1)^k \int_0^1 \binom{-s}{k}\, ds$ are positive and easily computed. They are given in Table 8.6.1.

Table 8.6.1. The Integrals
$(-1)^k \int_0^1 \binom{-s}{k}\, ds,\ 0 \leq k \leq 5$

k	$(-1)^k \int_0^1 \binom{-s}{k}\, ds$
0	1
1	1/2
2	5/12
3	3/8
4	251/720
5	95/288

Thus,

$$
\int_{x_n}^{x_{n+1}} f[x, Y(x)]\, dx = h \left[Y_n' + \frac{1}{2}(\nabla Y_n') + \frac{5}{12}(\nabla^2 Y_n') + \cdots \right]
$$

$$
+ \frac{h^{m+1}}{m!} \int_0^1 s(s+1) \cdots (s + m - 1) f^{(m)}[\xi_s, Y(\xi_s)]\, ds \qquad (8.6.9)
$$

Equation (8.6.9) produces a multistep method of order m given by

$$
y_{n+1} = y_n + h \left[y_n' + \frac{1}{2}(\nabla y_n') + \frac{5}{12}(\nabla^2 y_n') + \cdots + c_{m-1}(\nabla^{m-1} y_n') \right] \qquad (8.6.10)
$$

where $c_{m-1} = (-1)^{m-1} \int_0^1 \binom{-s}{m-1}\, ds$. The local truncation error at x_n is

$$
t_{n+1} = \frac{h^{m+1}}{m!} \int_0^1 s(s+1) \cdots (s + m - 1) f^{(m)}[\xi_s, Y(\xi_s)]\, ds
$$

Because $s(s+1) \cdots (s+m-1)$ does not change sign within $[0,1]$ we can apply a mean-value theorem for integrals and rewrite

$$t_{n+1} = \frac{h^{m+1}}{m!} f^{(m)}[\xi, Y(\xi)] \int_0^1 s(s+1) \cdots (s+m-1) \, ds$$

$$= h^{m+1} f^{(m)}[\xi, Y(\xi)](-1)^m \int_0^1 \binom{-s}{m} \, ds \qquad (8.6.11)$$

for some interim point ξ. The multistep method defined by Eq. (8.6.10) is open and called the *Adams–Bashforth method* of order m (AB-m).

Example 8.6.2.

The AB-2 method is derived from Eq. (8.6.10) by substituting $m = 2$, that is,

$$y_{n+1} = y_n + h \left[y'_n + \frac{1}{2}(y'_n - y'_{n-1}) \right] = y_n + h \left[\frac{3}{2} y'_n - \frac{1}{2} y'_{n-1} \right]$$

$$= y_n + h \left[\frac{3}{2} f(x_n, y_n) - \frac{1}{2} f(x_{n-1}, y_{n-1}) \right]$$

The local truncation error is $t_{n+1} = (5/12) h^3 f^{(2)}[\xi, Y(\xi)]$. Consider the problem

$$y' = 2y, \quad y(0) = 1, \quad 0 \le x \le 1$$

The approximate value at x_{n+1}, using the AB-2 method, is $y_{n+1} = y_n + h[3y_n - y_{n-1}]$, $n \ge 0$.

Table 8.6.2 provides a comparison between the numerical solution $y_{0.1}(x)$ and the exact solution $Y(x) = e^{2x}$. The midpoint method is applied for calculating y_1. □

Table 8.6.2. Adams–Bashforth Method of Order 2
Applied to $y' = 2y$, $y_0 = 2$, $h = 0.1$

n	y_n	Y_n	$Y_n - y_n$
2	1.4860	1.4918	0.0058
4	2.2041	2.2255	0.0214
6	3.2693	3.3201	0.0508
8	4.8492	4.9530	0.1038
10	7.1927	7.3891	0.1964

The AB-m methods for $3 \le m \le 5$ are derived from Eqs. (8.6.10) and (8.6.11) by substituting $m = 3$, 4, and 5 and are given below.

1. AB-3 Method:

$$\begin{aligned} y_{n+1} &= y_n + h\left[y_n' + \frac{1}{2}(y_n' - y_{n-1}') + \frac{5}{12}(y_n' - 2y_{n-1}' + y_{n-2}')\right] \\ &= y_n + \frac{h}{12}(23y_n' - 16y_{n-1}' + 5y_{n-2}') \\ &= y_n + \frac{h}{12}[23f(x_n, y_n) - 16f(x_{n-1}, y_{n-1}) + 5f(x_{n-2}, y_{n-2})] \end{aligned} \tag{8.6.12}$$

where $2 \le n \le N - 1$ and y_0, y_1, and y_2 are prefixed. The local truncation error is

$$t_{n+1} = \frac{3}{8}h^4 f^{(3)}\left[\xi, Y(\xi)\right], \quad x_{n-2} < \xi < x_{n+1} \tag{8.6.13}$$

2. AB-4 Method:

$$y_{n+1} = y_n + \frac{h}{24}[55f(x_n, y_n) - 59f(x_{n-1}, y_{n-1}) + 37f(x_{n-2}, y_{n-2}) - 9f(x_{n-3}, y_{n-3})] \tag{8.6.14}$$

where $3 \le n \le N - 1$ and y_0, y_1, y_2, and y_3 are prefixed. The local truncation error is

$$t_{n+1} = \frac{251}{720}h^5 f^{(4)}\left[\xi, Y(\xi)\right], \quad x_{n-3} < \xi < x_{n+1} \tag{8.6.15}$$

3. AB-5 Method:

$$\begin{aligned} y_{n+1} &= y_n + \frac{h}{720}[1901f(x_n, y_n) - 2774f(x_{n-1}, y_{n-1}) \\ &\quad + 2616f(x_{n-2}, y_{n-2}) - 1274f(x_{n-3}, y_{n-3}) + 251f(x_{n-4}, y_{n-4})] \end{aligned} \tag{8.6.16}$$

where $4 \le n \le N - 1$ and $y_i, \ 0 \le i \le 4$ are prefixed. The local truncation error is

$$t_{n+1} = \frac{95}{288}h^6 f^{(5)}\left[\xi, Y(\xi)\right], \quad x_{n-4} < \xi < x_{n+1} \tag{8.6.17}$$

8.6.2 The Adams–Moulton Method

The previously discussed Adams–Bashforth methods are generated by using open multistep formulas. To derive similar implicit schemes we use the point (x_{n+1}, Y_{n+1}') as well, to interpolate $f[x, Y(x)]$ in Eq. (8.6.5) $[Y_{n+1}' = f(x_{n+1}, Y_{n+1})]$. It leads to the Adams–Moulton methods (AM-m), some of which are given below.

1. AM-2 Method:

$$y_{n+1} = y_n + \frac{h}{12}[5f(x_{n+1}, y_{n+1}) + 8f(x_n, y_n) - f(x_{n-1}, y_{n-1})] \tag{8.6.18}$$

where $1 \leq n \leq N - 1$ and y_0, y_1 are prefixed. The local truncation error is

$$t_{n+1} = -\frac{1}{24}h^4 f^{(3)}\left[\xi, Y(\xi)\right], \quad x_{n-1} < \xi < x_{n+1} \tag{8.6.19}$$

2. AM-3 Method:

$$
\begin{aligned}
y_{n+1} = y_n + \frac{h}{24}[9f(x_{n+1}, y_{n+1}) + 19f(x_n, y_n) \\
-5f(x_{n-1}, y_{n-1}) + f(x_{n-2}, y_{n-2})]
\end{aligned} \tag{8.6.20}
$$

where $2 \leq n \leq N - 1$ and y_0, y_1, y_2 are prefixed. The local truncation error is

$$t_{n+1} = -\frac{19}{720}h^5 f^{(4)}\left[\xi, Y(\xi)\right], \quad x_{n-2} < \xi < x_{n+1} \tag{8.6.21}$$

3. AM-4 Method:

$$
\begin{aligned}
y_{n+1} = y_n + \frac{h}{720}[251f(x_{n+1}, y_{n+1}) + 646f(x_n, y_n) \\
-264f(x_{n-1}, y_{n-1}) + 106f(x_{n-2}, y_{n-2}) - 19f(x_{n-3}, y_{n-3})]
\end{aligned} \tag{8.6.22}
$$

where $3 \leq n \leq N - 1$ and y_0, y_1, y_2, y_3 are prefixed. The local truncation error is

$$t_{n+1} = -\frac{3}{160}h^6 f^{(5)}\left[\xi, Y(\xi)\right], \quad x_{n-3} < \xi < x_{n+1} \tag{8.6.23}$$

A comparison between the AB-3 method and its corresponding (errorwise) AM-2 method is given next.

Example 8.6.3.

Consider the initial-value problem $y' = xy$, $y(0) = 1$, $0 \leq x \leq 1$, whose exact solution is $Y(x) = e^{x^2/2}$. The AB-3 procedure is given by

$$y_{n+1} = y_n + \frac{h}{12}(23x_n y_n - 16x_{n-1}y_{n-1} + 5x_{n-2}y_{n-2}), \quad 2 \leq n \leq N - 1$$

The corresponding AM-2 scheme generates the numbers

$$y_{n+1} = y_n + \frac{h}{12}(5x_{n+1}y_{n+1} + 8x_n y_n - x_{n-1}y_{n-1}), \quad 1 \leq n \leq N - 1$$

or

$$y_{n+1} = \frac{y_n + (h/12)(8x_n y_n - x_{n-1}y_{n-1})}{1 - (5h/12)x_{n+1}}$$

Let us choose $h = 0.1$ and prefix the initial values, using the exact solution. The results are presented in Table 8.6.3. The approximate solution calculated by the AM-2 method provides smaller error due to the following:

1. Smaller coefficient in the expression for t_{n+1}
2. Smaller coefficients in Eq. (8.6.18) than in Eq. (8.6.12), which reduces the accumulated round-off error

Table 8.6.3. Adams–Bashforth Method 3 vs. Adams–Moulton Method 2 in the Case $y' = xy$, $0 \le x \le 1$, $y_0 = 1$

x_n	AB-3	|Error(AB)|	AM-2	|Error(AM)|
0.2	Exact		1.020214	$1.3 \cdot 10^{-5}$
0.3	1.045909	$1.2 \cdot 10^{-4}$	1.046055	$2.8 \cdot 10^{-5}$
0.4	1.083031	$2.6 \cdot 10^{-4}$	1.083332	$4.5 \cdot 10^{-5}$
0.5	1.132729	$4.2 \cdot 10^{-4}$	1.133214	$6.6 \cdot 10^{-5}$
0.6	1.196594	$6.2 \cdot 10^{-4}$	1.197309	$9.2 \cdot 10^{-5}$
0.7	1.276737	$8.8 \cdot 10^{-4}$	1.277748	$1.3 \cdot 10^{-4}$
0.8	1.375904	$1.2 \cdot 10^{-3}$	1.377299	$1.7 \cdot 10^{-4}$
0.9	1.497629	$1.7 \cdot 10^{-3}$	1.499534	$2.3 \cdot 10^{-4}$
1.0	1.646445	$2.3 \cdot 10^{-3}$	1.649034	$3.1 \cdot 10^{-4}$

$\square$

8.6.3 A Predictor–Corrector Method

The previous example demonstrates the fact that an AM method is generally preferable to an AB method of the same order.

A useful technique for solving initial-value problems, which combines explicit and implicit multistep schemes, is the *predictor–corrector method*. It first incorporates an explicit scheme, the *predictor*, to generate an initial approximation $y_{n+1}^{(0)}$ to y_{n+1}. Then an implicit scheme, the *corrector*, is applied to improve $y_{n+1}^{(0)}$ iteratively until a final y_{n+1} is determined. The way the full procedure works is demonstrated by the next example.

Example 8.6.4.
Define the AB-3 method as the predictor of the procedure. Then

$$y_{n+1}^{(0)} = y_n + \frac{h}{12}[23f(x_n, y_n) - 16f(x_{n-1}, y_{n-1}) + 5f(x_{n-2}, y_{n-2})] \tag{8.6.24}$$

is the initial approximation to y_{n+1}. If the AM-2 method is our choice for the corrector, then the final y_{n+1} must satisfy Eq. (8.6.18). We generate the iterations $\{y_{n+1}^{(j)}\}$, $j \ge 1$ defined by

$$y_{n+1}^{(j)} = y_n + \frac{h}{12}\left\{5f\left[x_{n+1}, y_{n+1}^{(j-1)}\right] + 8f(x_n, y_n) - f(x_{n-1}, y_{n-1})\right\} \tag{8.6.25}$$

until $|y_{n+1}^{(j)} - y_{n+1}^{(j-1)}|$ is sufficiently small and accept the last $y_{n+1}^{(j)}$ as our final y_{n+1}. The criteria for the convergence of $y_{n+1}^{(j)}$ can be found as follows. In view of Eq. (8.6.25)

$$y_{n+1}^{(j+1)} - y_{n+1}^{(j)} = \frac{5h}{12}\left\{f\left[x_{n+1}, y_{n+1}^{(j)}\right] - f\left[x_{n+1}, y_{n+1}^{(j-1)}\right]\right\}$$

$$= \frac{5h}{12} \left[y_{n+1}^{(j)} - y_{n+1}^{(j-1)} \right] \frac{\partial f}{\partial y}(x_{n+1}, \eta)$$

for some η between $y_{n+1}^{(j-1)}$ and $y_{n+1}^{(j)}$. We require $|(5h/12)(\partial f/\partial y)(x_{n+1}, \eta)| < 1$ and, because $\eta \approx y_{n+1}$,

$$\left| \frac{5h}{12} \frac{\partial f}{\partial y}(x_{n+1}, y_{n+1}) \right| < 1 \tag{8.6.26}$$

Generally, it is sufficient to evaluate $y_{n+1}^{(1)}$ and accept it as y_{n+1}. $\qquad\square$

The predictor–corrector (PC) scheme defined by Eqs. (8.6.24) and (8.6.25) is our next example.

Example 8.6.5.

Consider the initial-value problem $y' = y$, $y(0) = 1$, $0 \leq x \leq 1$ with the exact solution $Y = e^x$. The PC scheme of Example 8.6.4 yields

$$y_{n+1}^{(0)} = y_n + \frac{h}{12}\left(23y_n - 16y_{n-1} + 5y_{n-2}\right)$$

$$y_{n+1}^{(1)} = y_n + \frac{h}{12}\left[5y_{n+1}^{(0)} + 8y_n - y_{n-1}\right]$$

and $y_{n+1}^{(1)}$ is substituted for y_{n+1}. Let $h = 0.1$ and prefix $y_0 = 1$, $y_1 = e^{0.1}$, $y_2 = e^{0.2}$.

Table 8.6.4 provides a comparison between the exact solution Y_n, the PC solution y_n, the AB-3 solution u_n, given by $u_{n+1} = u_n + (h/12)(23u_n - 16u_{n-1} + 5u_{n-2})$, and the AM-2 solution v_n

$$v_{n+1} = \frac{v_n + (h/12)(8v_n - v_{n-1})}{1 - (5h/12)}$$

As expected, y_n and v_n are much closer to each other (and to Y_n) than is u_n. $\qquad\square$

Table 8.6.4. Third Order Predictor-Corrector Method Applied to $y' = y$, $y_0 = 1$, $0 \leq x \leq 1$

x_n	Y_n	y_n	v_n	u_n
0.3	1.349859	1.349862	1.349864	1.349815
0.4	1.491825	1.491832	1.491837	1.491725
0.5	1.648721	1.648734	1.648741	1.648556
0.6	1.822119	1.822137	1.822148	1.821874
0.7	2.013753	2.013778	2.013793	2.013415
0.8	2.225541	2.225574	2.225595	2.225093
0.9	2.459603	2.459646	2.459672	2.459025
1.0	2.718282	2.718336	2.718369	2.717551

A fourth order PC procedure based on the AB-4 and the AM-3 methods is given next. The initial points needed to start the computation are determined by the classic Runge–Kutta method.

Algorithm 8.6.1.

(A fourth order PC method). This algorithm approximates the solution to

$$y' = f(x, y), \quad y(x_0) = y_0, \quad x_0 \le x \le b$$

Input: End points x_0, b; initial condition y_0; number of evenly spaced grid points $(N + 1)$.
Output: A set of pairs $\{(x_n, y_n)\}_{n=0}^{N}$, where $x_N = b$ and y_n approximates the exact Y_n at x_n.

Step 1. Set $h = (b - x_0)/N$; $n = 0$.
Step 2. Output (x_n, y_n).
Step 3. For $0 \le n \le 2$ do steps 4 through 6 (using RK to start the procedure).
Step 4. Set

$$k_1 = f(x_n, y_n), \qquad\qquad k_2 = f\left(x_n + \frac{h}{2}, y_n + \frac{hk_1}{2}\right)$$

$$k_3 = f\left(x_n + \frac{h}{2}, y_n + \frac{hk_2}{2}\right), \quad k_4 = f(x_n + h, y_n + hk_3)$$

Step 5. Set $y_{n+1} = y_n + (h/6)(k_1 + 2k_2 + 2k_3 + k_4), x_{n+1} = x_n + h$.
Step 6. Output (x_{n+1}, y_{n+1}); $n \leftarrow n + 1$.
Step 7. For $3 \le n \le N - 1$ do steps 8 through 10.
Step 8. Set (computing the predictor using the AB-4 method)

$$y_{n+1}^{(0)} = y_n + \frac{h}{24}[55f(x_n, y_n) - 59f(x_{n-1}, y_{n-1})$$
$$+37f(x_{n-2}, y_{n-2}) - 9f(x_{n-3}, y_{n-3})]$$

Step 9. Set $x_{n+1} = x_n + h$ and (computing the corrector using the AM-3 method)

$$y_{n+1} = y_n + \frac{h}{24}\left\{9f\left[x_{n+1}, y_{n+1}^{(0)}\right] + 19f(x_n, y_n)\right.$$
$$\left. -5f(x_{n-1}, y_{n-1}) + f(x_{n-2}, y_{n-2})\right\}$$

Step 10. Output (x_{n+1}, y_{n+1}); $n \leftarrow n + 1$ and stop.

The subroutine PC4, based on Algorithm 8.6.1, is found on the attached floppy disk.

Other multistep methods can be constructed by integrating the interpolating polynomial to $f[x, Y(x)]$ over the interval $[x_k, x_{n+1}]$, $k < n$ instead of $[x_n, x_{n+1}]$. One popular PC method of the fourth order is obtained for $k = n - 3$. It is given by Milne's predictor [Eq. (8.6.3)]

$$y_{n+1}^{(0)} = y_{n-3} + \frac{4h}{3}[2f(x_n, y_n) - f(x_{n-1}, y_{n-1}) + 2f(x_{n-2}, y_{n-2})] \tag{8.6.27}$$

and by *Simpson's corrector*

$$y_{n+1}^{(j+1)} = y_{n-1} + \frac{h}{3}\left\{f\left[x_{n+1}, y_{n+1}^{(j)}\right] + 4f(x_n, y_n) + f(x_{n-1}, y_{n-1})\right\} \tag{8.6.28}$$

The local truncation errors are $(14/45)h^5 f^{(4)}[\xi_1, Y(\xi_1)]$ and $-(h^5/90)f^{(4)}[\xi_2, Y(\xi_2)]$ for the predictor and the corrector, respectively.

Example 8.6.6.

The Milne–Simpson method for $y' = y$, $0 \le x \le 1$, $y(0) = 1$ is

$$y_{n+1}^{(0)} = y_{n-3} + \frac{4h}{3}(2y_n - y_{n-1} + 2y_{n-2})$$

$$y_{n+1}^{(j+1)} = y_{n-1} + \frac{h}{3}\left[y_{n+1}^{(j)} + 4y_n + y_{n-1}\right], \quad j \ge 0$$

Let $h = 0.1$ and prefix $y_0 = 1$, $y_1 = e^{0.1}$, $y_2 = e^{0.2}$, $y_3 = e^{0.3}$. Table 8.6.5 presents the Milne–Simpson (MS) approximate solution compared with Y. The impressive accuracy is related to the small truncation error associated with the Simpson rule. □

Table 8.6.5. Milne–Simpson Procedure Applied to $y' = y$, $y_0 = 1$, $0 \le x \le 1$

x	Y	MS	$|Y-\text{MS}|$
0.4	1.49182470	1.49182472	$2.3 \cdot 10^{-8}$
0.5	1.64872127	1.64872130	$2.9 \cdot 10^{-8}$
0.6	1.82211880	1.82211886	$5.6 \cdot 10^{-8}$
0.7	2.01375271	2.01375278	$6.9 \cdot 10^{-8}$
0.8	2.22554093	2.22554103	$1.0 \cdot 10^{-7}$
0.9	2.45960311	2.45960324	$1.3 \cdot 10^{-7}$
1.0	2.71828183	2.71828200	$1.7 \cdot 10^{-7}$

PROBLEMS

1. Use Milne's method [Eq. (8.6.3)] to solve the following initial-value problems. Derive the needed initial data from the exact solutions.

   (a) $y' = -y$, $0 \le x \le 1$, $y(0) = 1$, $h = 0.1$ $[Y(x) = e^{-x}]$

   (b) $y' = y - x + 1$, $0 \le x \le 1$, $y(0) = 1$, $h = 0.1$ $[Y(x) = x + e^x]$

   (c) $y' = -y^2/2$, $1 \le x \le 3$, $y(1) = 2$, $h = 0.2$ $[Y(x) = 2/x]$

2. Verify Eq. (8.6.6).

3. (a) Apply the AB-2 method to solve $y' = \sin y$, $0 \le x \le 1$, $y(0) = \pi/2$, using $h = 0.1$, and compare the approximate and exact solutions $[Y(x) = 2 \tan^{-1}(e^x)]$.

   (b) Obtain the maximum step size for which the local truncation error is less than 10^{-6}.

4. Repeat and solve problem 3 using the AB-3 method.

5. Repeat and solve problem 3 using the AM-2 method.

6. (a) Apply the PC procedure given by Eqs. (8.6.24) and (8.6.25) to solve the following initial-value problems. Derive the needed initial data from the exact solution and at each x_n use the corrector only twice.

   i. $y' = (y/x) + 1$, $1 \le x \le 2$, $y(1) = 0$, $h = 0.1$ $[Y(x) = x \ln x]$

ii. $y' = (y/x) - (2/x^2)$, $1 \le x \le 2$, $y(1) = 2$, $h = 0.1$ $[Y(x) = x + (1/x)]$

iii. $y' = (1/y)$, $1 \le x \le 3$, $y(1) = \sqrt{2}$, $h = 0.2$ $[Y(x) = \sqrt{2x}]$

(b) For each problem compare the approximate and exact solutions at the right end point of the interval.

7. Apply the AB-2 method to solve

$$y' = y + 1, \quad 0 \le x \le 1, \quad y(0) = 0, \quad h = 0.1$$

and compare the approximate and exact solutions $[Y(x) = e^x - 1]$. Use the modified Euler method to start the computation.

8. Apply Algorithm 8.6.1 to solve the initial-value problem

$$y' = y(1 - \tan x), \quad 0 \le x \le \pi/4, \quad y(0) = 1, \quad h = \pi/40$$

and compare the results with the exact solution $Y(x) = e^x \cos x$.

9. (a) Apply Algorithm 8.6.1 to solve the initial-value problem

$$y' = -y, \quad y(0) = 1, \quad 0 \le x \le 1$$

using $h = 0.2, 0.1$.

(b) Use the two approximations and Richardson's extrapolation method to obtain a more accurate approximate solution and compare with $Y(x) = e^{-x}$.

10. Apply a PC procedure that is a combination of Milne's predictor and Simpson's corrector and obtain an approximate solution to problem 8. Use the exact solution $Y(x) = e^x \cos x$ to determine y_i, $0 \le i \le 3$.

11. Repeat problem 10, using the classic Runge–Kutta method to start the computation.

8.7   Stability of Numerical Methods

As previously stated in Section 8.1, we distinguish between stability of the initial-value problem itself and stability of the numerical procedures for solving this problem. The main result with respect to the first type of stability is given by Eq. (8.1.12).

In this section we will discuss stability of numerical methods. Consider, for example, Euler's method

$$y_{n+1} = y_n + hf(x_n, y_n), \quad 0 \le n \le N - 1 \tag{8.7.1}$$

applied to

$$y' = y + 1, \quad y(0) = 1, \quad 0 \le x \le 1 \tag{8.7.2}$$

We find that $y_{n+1} = y_n + h(y_n + 1) = (1 + h)y_n + h$, leading to

$$y_{n+1} = (1 + h)^{n+1} y_0 + h[1 + (1 + h) + \cdots + (1 + h)^n]$$

$$= (1+h)^{n+1} y_0 + (1+h)^{n+1} - 1$$

This implies $\lim_{h \to 0} y_N = 2e - 1 [Y(x) = 2e^x - 1]$. Let us assume a perturbed initial value $y_{\epsilon,0} = 1+\epsilon$. Then $\lim_{h \to 0} y_{\epsilon,N} = e(1+\epsilon) + e - 1 = 2e - 1 + e\epsilon$ and consequently

$$\lim_{h \to 0} |y_{\epsilon,N} - y_N| = e\epsilon = O(\epsilon)$$

indicating that Euler's method is *stable*. Now, consider another numerical procedure

$$y_{n+1} = 3y_n - 2y_{n-1} + \frac{h}{2}[f(x_n, y_n) - 3f(x_{n-1}, y_{n-1})] \tag{8.7.3}$$

which is a two-step method with a local truncation error of $O(h^3)$. Applying this method to solve

$$y' = 1, \quad y(0) = 1, \quad 0 \le x \le 1 \tag{8.7.4}$$

with an exact solution $Y(x) = 1+x$, we obtain $y_{n+1} = 3y_n - 2y_{n-1} - h$. If we assume the initial data $y_0 = 1$, $y_1 = 1 + h$ it may be seen that $y_n = Y_n$, $0 \le n \le N - 1$. However, in the case of perturbed initial data, such as $y_{\epsilon,0} = 1 + \epsilon$, $y_{\epsilon,1} = 1 + h + 2\epsilon$ we have $y_{\epsilon,n} = 1 + nh + 2^n \epsilon$. Because ϵ is fixed, $\lim_{h \to 0} y_{\epsilon,N} = \infty$ and the scheme is *unstable*. A method is called *stable* if its approximate solution depends *continuously* on the initial data, uniformly in h.

We shall now present the main result concerning convergence and stability of one-step numerical methods. Prior to this we define *consistency*.

Definition 8.7.1. (Consistency). *A numerical procedure with local truncation error t_n at x_n is consistent with the differential equation if*

$$\lim_{h \to 0} \left[\max_{0 \le n \le N} (t_n/h) \right] = 0 \tag{8.7.5}$$

For a one-step method of solving $y' = f(x, y)$, given by $y_{n+1} = y_n + hK(x_n, y_n; h)$, consistency clearly occurs if and only if (why?)

$$K(x, y; 0) = f(x, y) \tag{8.7.6}$$

Example 8.7.1.
For the classical RK method, $K(x, y; h) = (1/6)(k_1 + 2k_2 + 2k_3 + k_4)$ and $k_i(x, y; 0) = f(x, y)$. Therefore $K(x, y; 0) = f(x, y)$ and the method is consistent.   □

The concept of "convergence," already used in previous discussions, is defined next.

Definition 8.7.2. (Convergence). *A numerical procedure for solving an initial-value problem is said to be convergent if*

$$\lim_{h \to 0} \left(\max_{0 \le n \le N} |Y_n - y_n| \right) = 0 \tag{8.7.7}$$

Example 8.7.2.

By Theorem 8.4.1, the error $z_n = Y_n - y_n$ associated with a second order Runge–Kutta method satisfies $|z_n| \leq Ch^2$ for some constant C. This method is therefore convergent and the *order of convergence* is at least 2.   □

Theorem 8.7.1.

(Stability of a one-step method). *Let the solution $Y(x)$ to the problem*

$$y' = f(x, y), \quad y(x_0) = y_0, \quad x_0 \leq x \leq b \tag{8.7.8}$$

be approximated by a one-step numerical procedure

$$y_{n+1} = y_n + hK(x_n, y_n; h), \quad 0 \leq n \leq N - 1 \tag{8.7.9}$$

which satisfies the following requirements.

1. *For some given $H > 0, K(x, y; h)$ is continuous over*

$$R = \{(x, y; h) \mid x_0 \leq x \leq b, \ -\infty < y < \infty, \ 0 \leq h \leq H\} \tag{8.7.10}$$

2. *$K(x, y; h)$ satisfies a Lipschitz condition over R with respect to y. Then*
   (a) *The procedure is convergent if and only if it is consistent [Eq. (8.7.6)].*
   (b) *The procedure is stable, that is,*

$$\lim_{\epsilon \to 0} \left(\max_{0 \leq n \leq N} |y_{\epsilon,n} - y_n| \right) = 0 \text{ uniformly for } N \tag{8.7.11}$$

   where $y_{\epsilon,n}$ is the approximate solution associated with the initial value $y(0) = y_0 + \epsilon$.
   (c) *If the local truncation error t_n satisfies $|t_n| \leq t(h)$, $0 \leq n \leq N$, $0 \leq h \leq H$ then*

$$|Y_n - y_n| \leq \frac{t(h)}{hM} e^{M(x_n - x_0)}, \quad 0 \leq n \leq N \tag{8.7.12}$$

   where M is the Lipschitz constant.

The global error bound given by Eq. (8.7.12) is in agreement with Eq. (8.2.20), which provides an error bound in the particular case of Euler's method.

Example 8.7.3.

Consider the one-step numerical procedure

$$y_{n+1} = y_n + hf\left[x_n + \frac{h}{2}, \ y_n + \frac{h}{2}f(x_n, y_n)\right], \quad 0 \leq n \leq N - 1 \tag{8.7.13}$$

known as the midpoint method. We assume $f(x, y)$ to be continuous over the region

$$D = \{(x, y) \mid x_0 \leq x \leq b, \ -\infty < y < \infty\}$$

and that

$$\left|\frac{\partial f}{\partial y}(x, y)\right| \le M, \quad (x, y) \in D$$

which immediately implies an existing Lipschitz condition for $f(x, y)$ with the constant M. Here $K(x, y; h) = f[x + (h/2), y + (h/2)f(x, y)]$ and therefore $K(x, y; 0) = f(x, y)$, implying the consistency of the midpoint method. Because $f(x, y)$ is continuous, so is $K(x, y; h)$ and, in addition,

$$
\begin{aligned}
|K(x, y_1; h) - K(x, y_2; h)| &= \left| f\left[x + \frac{h}{2}, \ y_1 + \frac{h}{2}f(x, y_1)\right] \right. \\
&\quad \left. -f\left[x + \frac{h}{2}, \ y_2 + \frac{h}{2}f(x, y_2)\right] \right| \\
&\le M \left| y_1 + \frac{h}{2}f(x, y_1) - y_2 - \frac{h}{2}f(x, y_2) \right| \\
&\le M|y_1 - y_2| + \frac{Mh}{2}|f(x, y_1) - f(x, y_2)| \\
&\le M|y_1 - y_2| + \frac{M^2 h}{2}|y_1 - y_2| = M\left(1 + \frac{Mh}{2}\right)|y_1 - y_2|
\end{aligned}
$$

Thus a Lipschitz condition with the constant $M(1 + Mh/2)$ holds for $K(x, y; h)$ with respect to y. By Theorem 8.7.1 the method is convergent and stable. Also

$$|Y_n - y_n| \le \frac{t(h)}{h\tilde{M}} e^{\tilde{M}(x_n - x_0)}$$

where $\tilde{M} = M(1 + Mh/2)$ and $t(h)$ is the largest local truncation error. If all the second partial derivatives of $f(x, y)$ are bounded we obtain $t(h) \le Ch^3$ for some $C > 0$ and $|Y_n - y_n| = O(h^2)$. $\square$

The next result is related to a particular type of multistep method.

Theorem 8.7.2.
(Stability for the Adams–Bashforth method). *Let the exact solution $Y(x)$ to the problem $y' = f(x, y)$, $y(x_0) = y_0$, $x_0 \le x \le b$ be approximated by the AB method of order m, that is,*

$$y_{n+1} = y_n + h\left[y_n' + \frac{1}{2}(\nabla y_n') + \frac{5}{12}(\nabla^2 y_n') + \cdots + c_{m-1}(\nabla^{m-1} y_n')\right] \tag{8.7.14}$$

where

$$c_{m-1} = (-1)^{m-1} \int_0^1 \binom{-s}{m-1} ds \tag{8.7.15}$$

Assume that $f(x, y), (\partial f/\partial y)(x, y) \in C(D), D = \{(x, y) \mid x_0 \le x \le b, \ -\infty < y < \infty\}$ and that $|(\partial f/\partial y)(x, y)| \le M, (x, y) \in D$.

Let $y_n, m \le n \le N$ denote the approximate solution using the initial data

$$y_i = Y_i, \quad 0 \le i \le m - 1 \tag{8.7.16}$$

and let $y_{\epsilon,n}$, $m \leq n \leq N$ denote the approximate solution generated from the initial perturbed data $y_{\epsilon,i}$, $0 \leq i \leq m-1$ for which

$$|y_{\epsilon,i} - Y_i| \leq \epsilon, \quad 0 \leq i \leq m-1 \tag{8.7.17}$$

Then there exist $\epsilon_0 > 0$ and $\delta > 0$ such that

$$\max_{m \leq n \leq N} |y_{\epsilon,n} - y_n| \leq c\epsilon \tag{8.7.18}$$

for some $c > 0$ independent of h and ϵ, provided that $h \leq \delta$ and $\epsilon \leq \epsilon_0$.

This theorem establishes the stability of the Adams–Bashforth methods. A similar result exists for the Adams–Moulton methods.

8.7.1 Stability Regions

A numerical procedure that is found to be stable may still not be *unconditionally stable*, that is, stable for *any* choice of h. The previous result, for example, guarantees stability only if h is sufficiently small. If two numerical procedures are of identical order, one is expected to prefer the scheme with the larger *region of stability*. Finding these regions for the general initial-value problem is naturally complicated. However, it can be shown that to derive most conclusions that are related to stability, it is enough to examine the particular class of linear problems

$$y' = ky + f(x), \quad y(x_0) = y_0 \tag{8.7.19}$$

Let $Y(x), Y_\epsilon(x)$ be the exact solutions to Eq. (8.7.19) and to the perturbed problem

$$y' = ky + f(x), \quad y(x_0) = y_0 + \epsilon \tag{8.7.20}$$

respectively. Then the difference $Z_\epsilon(x) = Y_\epsilon(x) - Y(x)$ satisfies

$$Z_\epsilon'(x) = kZ_\epsilon(x), \quad Z_\epsilon(x_0) = \epsilon \tag{8.7.21}$$

Consequently $Z_\epsilon(x) = \epsilon e^{k(x-x_0)}$. Thus if $k < 0$ the effect of an initial perturbation in Y_0 vanishes as $x \to \infty$. If a numerical procedure is applied to solve Eqs. (8.7.19) and (8.7.20), it is desirable to carry this feature along. For example, apply Euler's method to obtain

$$y_{n+1} = y_n + h[ky_n + f(x_n)], \quad y_0 = Y_0$$
$$y_{\epsilon,n+1} = y_{\epsilon,n} + h[ky_{\epsilon,n} + f(x_n)], \quad y_0 = Y_0 + \epsilon$$

and denote $z_{\epsilon,n} = y_{\epsilon,n} - y_n$. Then

$$z_{\epsilon,n+1} = z_{\epsilon,n} + khz_{\epsilon,n} = z_{\epsilon,n}(1 + kh) = \cdots = z_{\epsilon,0}(1 + kh)^{n+1} = \epsilon(1 + kh)^{n+1} \tag{8.7.22}$$

Therefore, $\lim_{n\to\infty} z_{\epsilon,n} = 0$ if and only if $|1 + kh| < 1$ or

$$-2 < hk < 0 \tag{8.7.23}$$

We define the *region of absolute stability* as the set of all values of hk for which Eq. (8.7.23) holds. If, for example, $k = -10$, then the region of absolute stability is $0 < h < 0.2$. An Adams–Moulton method of order m has a larger region of absolute stability than an Adams–Bashforth method of the same order. Consequently it is more popular.

PROBLEMS

1. Show that the two-step method given by Eq. (8.7.3) has a local truncation error of $O(h^3)$.

2. (a) Apply the method of Eq. (8.7.3) to solve $y' = y$, $0 \le x \le 0.5$, $y(0) = 1$, $h = 0.1$. Use $Y(x) = e^x$ to prefix y_0 and y_1, that is, $y_0 = 1$, $y_1 = 1.10517092$.

   (b) Repeat part (a), using perturbed initial data $y_0 = 1.02$, $y_1 = 1.12$.

   (c) Compare the approximate solutions of parts (a) and (b) vs. the exact solution.

3. Use Euler's method to solve $y' = ky$, $0 \le x \le 1$, $y(0) = 1$ for $k = -1$, -10, and -20, using $h = 0.2$. Which cases are within the region of absolute stability?

4. Apply the Adams–Moulton method of order 1 given by

$$y_{n+1} = y_n + \frac{h}{2}[f(x_n, y_n) + f(x_{n+1}, y_{n+1})], \quad n \ge 0$$

   to Eq. (8.7.19) and show that the region of absolute stability is $-\infty < hk < 0$.

5. Show that the point $hk = -2/3$ is within the stability region of the Adams–Bashforth method of order 2. What about $hk = -2$?

6. Show that the region of absolute stability for the *backward Euler's method* $y_{n+1} = y_n + hf(x_{n+1}, y_{n+1})$, $n \ge 0$ is $-\infty < hk < 0$.

7. Apply the AB-2 method to solve $y' = -20y$, $0 \le x \le 1$, $y(0) = 1$, $y(h) = e^{-20h}$, using $h = 1/10$, $1/30$ and compare the approximate and exact solutions. Explain.

8.8 Systems of Differential Equations

Many applications in various sciences involve a system of differential equations that must be solved. In this section we introduce such systems and discuss numerical procedures for treating them.

The most general initial-value problem for a system of the first order is defined by

$$y_i' = f_i(x, y_1, \ldots, y_n), \quad y_i(x_0) = y_{i,0}, \quad 1 \le i \le n, \quad x_0 \le x \le b \tag{8.8.1}$$

Example 8.8.1.

The initial-value problem

$$y_1' = y_2, \quad y_1(0) = 0; \quad y_2' = -y_1, \quad y_2(0) = 1$$

is represented by a system of two differential equations of the first order. It has the unique solution

$$Y_1(x) = \sin x, \quad Y_2(x) = \cos x \qquad \square$$

By denoting

$$y = \begin{bmatrix} y_1 \\ y_2 \\ \vdots \\ y_n \end{bmatrix}, \quad f(x,y) = \begin{bmatrix} f_1(x, y_1, \dots, y_n) \\ f_2(x, y_1, \dots, y_n) \\ \vdots \\ f_n(x, y_1, \dots, y_n) \end{bmatrix}, \quad y_0 = \begin{bmatrix} y_{1,0} \\ y_{2,0} \\ \vdots \\ y_{n,0} \end{bmatrix}$$

we can rewrite Eq. (8.8.1) as

$$y' = f(x, y), \quad y(x_0) = y_0 \tag{8.8.2}$$

that is, by using the identical form of the original initial-value problem. If each f_i is a linear combination of $y_1, \dots, y_n$, we obtain $f(x, y) = Ay$ for some square matrix A.

8.8.1   Numerical Methods for Systems

The various numerical methods that were previously generated and applied for solving a single equation can be used without any change to treat systems of equations. Their derivation is the same as for a single equation, and their features related to convergence and stability are preserved. For example, to solve a system of two equations by using Euler's method, we write

$$\begin{aligned} y_{1,n+1} &= y_{1,n} + h f_1(x_n, y_{1,n}, y_{2,n}), \quad n \geq 0 \\ y_{2,n+1} &= y_{2,n} + h f_2(x_n, y_{1,n}, y_{2,n}), \quad n \geq 0 \end{aligned} \tag{8.8.3}$$

that is, $y_{1,n+1}$ and $y_{2,n+1}$ are computed "simultaneously" using *all* the available information at x_n. Under some requirements of smoothness $|Y_{1,n} - y_{1,n}| \leq ch$, $|Y_{2,n} - y_{2,n}| \leq ch$ for $0 \leq n \leq N$ and some constant c. Also, similarly to Eq. (8.2.35), we obtain an asymptotic error formula

$$z_{1,n} - h Z_1(x_n) = O(h^2), \quad z_{2,n} - h Z_2(x_n) = O(h^2) \tag{8.8.4}$$

where $z_{1,n} = Y_{1,n} - y_{1,n}$, $z_{2,n} = Y_{2,n} - y_{2,n}$, and $Z_1(x)$, $Z_2(x)$ solve a particular linear system of two equations.

In view of Eq. (8.8.4), Richardson's extrapolation method can also be applied to obtain an improved approximation to Y, if y_h and $y_{h/2}$ are available.

Example 8.8.2.

Consider the initial-value problem of the previous example. By implementing the standard Euler method we find

$$y_{1,n+1} = y_{1,n} + hy_{2,n}, \quad y_{1,0} = 0$$
$$y_{2,n+1} = y_{2,n} - hy_{1,n}, \quad y_{2,0} = 1$$

The results for $0 \le x \le \pi/2$ and $h = \pi/10, \pi/20$ are given in Tables 8.8.1 and 8.8.2. The columns denoted by R_1 and R_2 consist of the improved Richardson approximations for Y_1 and Y_2, respectively.

Table 8.8.1. Euler's Method for a System: First Component

x/π	$y_{1,\pi/10}$	$y_{1,\pi/20}$	R_1	Y_1
0.1	0.31416	0.31416	0.31416	0.30902
0.2	0.62832	0.61282	0.59731	0.58779
0.3	0.91147	0.86554	0.81960	0.80902
0.4	1.13261	1.04493	0.95725	0.95106
0.5	1.26379	1.12952	0.99524	1.00000

Table 8.8.2. Euler's Method for a System: Second Component

x/π	$y_{2,\pi/10}$	$y_{2,\pi/20}$	R_2	Y_2
0.1	1.00000	0.97533	0.95065	0.95106
0.2	0.90130	0.85256	0.80383	0.80902
0.3	0.70391	0.63901	0.57410	0.58779
0.4	0.41756	0.35132	0.28508	0.30902
0.5	0.06174	0.01438	-0.03298	0.00000

□

Next, we outline the modified Euler method [Eq. (8.4.8)] for a system of two equations.

Example 8.8.3.

Consider the initial-value problem

$$y_1' = f_1(x, y_1, y_2), \quad y_1(x_0) = y_{1,0}$$
$$y_2' = f_2(x, y_1, y_2), \quad y_2(x_0) = y_{2,0}$$

The modified Euler method is given by

$$y_{1,n+1} = y_{1,n} + \frac{h}{2} \{ f_1(x_n, y_{1,n}, y_{2,n})$$
$$+ f_1[x_n + h, y_{1,n} + hf_1(x_n, y_{1,n}, y_{2,n}), y_{2,n} + hf_2(x_n, y_{1,n}, y_{2,n})] \}$$
$$y_{2,n+1} = y_{2,n} + \frac{h}{2} \{ f_2(x_n, y_{1,n}, y_{2,n})$$
$$+ f_2[x_n + h, y_{1,n} + hf_1(x_n, y_{1,n}, y_{2,n}), y_{2,n} + hf_2(x_n, y_{1,n}, y_{2,n})] \} \qquad \square$$

An initial-value problem that involves a single differential equation of order n, with n initial values, can be replaced by a system of n equations of the first order. Its solution provides approximations to the solution of the original equation and to its first $(n-1)$ derivatives.

Example 8.8.4.

The initial-value problem

$$y^{(4)} - \sin x - \cos x = 0, \quad 0 \le x \le \pi/2$$
$$y(0) = 1, \quad y'(0) = 1, \quad y''(0) = -1, \quad y'''(0) = -1$$

is solved by $Y(x) = \sin x + \cos x$. By substituting $y_1 = y$, $y_2 = y'$, $y_3 = y''$, $y_4 = y'''$ we obtain the four-equation system

$$y_1' = y_2, \quad y_{1,0} = 1$$
$$y_2' = y_3, \quad y_{2,0} = 1$$
$$y_3' = y_4, \quad y_{3,0} = -1$$
$$y_4' = \sin x + \cos x, \quad y_{4,0} = -1$$

The modified Euler method yields

$$y_{1,n+1} = y_{1,n} + \frac{h}{2}(y_{2,n} + y_{2,n} + hy_{3,n})$$
$$= y_{1,n} + \frac{h}{2}(2y_{2,n} + hy_{3,n})$$
$$y_{2,n+1} = y_{2,n} + \frac{h}{2}(y_{3,n} + y_{3,n} + hy_{4,n})$$
$$= y_{2,n} + \frac{h}{2}(2y_{3,n} + hy_{4,n})$$
$$y_{3,n+1} = y_{3,n} + \frac{h}{2}\{y_{4,n} + y_{4,n} + h[\sin(x_n) + \cos(x_n)]\}$$
$$= y_{3,n} + \frac{h}{2}\{2y_{4,n} + h[\sin(x_n) + \cos(x_n)]\}$$
$$y_{4,n+1} = y_{4,n} + \frac{h}{2}[\sin(x_n) + \cos(x_n) + \sin(x_n + h) + \cos(x_n + h)]$$

The approximate solution y_1 and the first approximate derivative y_2 are given in Table 8.8.3, compared with the exact $Y(x) = \sin x + \cos x$ and $Y'(x) = \cos x - \sin x$. The step size is $\pi/10$. $\qquad \square$

Table 8.8.3. Modified Euler Method Applied to a Fourth Order Differential Equation

x/π	y_1	Y	$|Y - y_1|$	y_2	Y'	$|Y' - y_2|$
0.1	1.2648	1.2601	$4.7 \cdot 10^{-3}$	0.6365	0.6420	$5.5 \cdot 10^{-3}$
0.2	1.4024	1.3968	$5.6 \cdot 10^{-3}$	0.2073	0.2212	$1.4 \cdot 10^{-2}$
0.3	1.3981	1.3968	$1.3 \cdot 10^{-3}$	-0.2454	-0.2212	$2.4 \cdot 10^{-2}$
0.4	1.2516	1.2601	$8.5 \cdot 10^{-3}$	-0.6773	-0.6420	$3.5 \cdot 10^{-2}$
0.5	0.9760	1	$2.4 \cdot 10^{-2}$	-1.0460	-1	$4.6 \cdot 10^{-2}$

PROBLEMS

1. Let

$$A = \begin{bmatrix} 1 & -1 & 0 \\ -1 & 2 & 3 \\ 0 & 2 & 3 \end{bmatrix}, \quad Y = \begin{bmatrix} y_1 \\ y_2 \\ y_3 \end{bmatrix}, \quad Y' = \begin{bmatrix} y_1' \\ y_2' \\ y_3' \end{bmatrix}, \quad F = \begin{bmatrix} x \\ -x \\ x^2 \end{bmatrix}, \quad Y_0 = \begin{bmatrix} 0 \\ 1 \\ 0 \end{bmatrix}$$

Write three linear differential equations with initial conditions that are equivalent to

$$Y' = AY + F, \quad Y(x_0) = Y_0$$

2. Replace the initial-value problem

$$y^{(5)} - 2y'' + y = x \ln x, \quad 1 \le x \le 2; \quad y^{(i)}(1) = 0, \quad 0 \le i \le 4$$

   by a system of five differential equations of the first order with initial conditions.

3. Apply Euler's method to solve the initial-value problem of problem 2, using $h = 0.2$.

4. Apply the classic Runge–Kutta method to approximate the solution to the initial-value problem

$$y_1' = y_1 y_2, \quad 0 \le x \le 1, \quad y_1(0) = 1$$
$$y_2' = y_1 - y_2, \quad 0 \le x \le 1, \quad y_2(0) = 1$$

   using $h = 0.2$.

5. Apply the modified Euler method to solve problem 4 using $h = 0.1, 0.2$ and use the approximate solutions and Richardson's method to obtain an improved approximation.

6. Apply the AB-2 method to approximate the solution of

$$y_1' = y_1 + y_2, \quad 1 \le x \le 2, \quad y_1(1) = e$$
$$y_2' = y_1/x, \quad 1 \le x \le 2, \quad y_2(1) = e$$

   using $h = 0.2, 0.1$ and compare the approximate solutions and the improved approximation (using Richardson's extrapolation) to the exact solution

$$y_1 = xe^x, \quad y_2 = e^x$$

7. (a) Replace

$$y''' - y = 0, \quad y(0) = 0, \quad y'(0) = 1, \quad y''(0) = -1, \quad 0 \le x \le 1$$

by a system of three differential equations.

(b) Solve the system of part (a) by using Taylor's method of order 2, and $h = 0.1$.

9

Numerical Solutions of
Partial Differential Equations

Partial differential equations generally describe processes that involve more than one variable. Thus they involve *partial derivatives*. For example, let us consider an *isotropic* body, that is, a body whose thermal conductivity at each point (x, y, z) is independent of the direction of heat flow through that point. Let $k(x, y, z)$, $\rho(x, y, z)$, and $c(x, y, z)$ denote the thermal conductivity, density, and specific heat of the body, respectively. Then the temperature $u(x, y, z, t)$ at each point (x, y, z) at time t is governed by the *partial differential equation* known as the *heat equation*:

$$\frac{\partial}{\partial x}\left(k\frac{\partial u}{\partial x}\right) + \frac{\partial}{\partial y}\left(k\frac{\partial u}{\partial y}\right) + \frac{\partial}{\partial z}\left(k\frac{\partial u}{\partial z}\right) = c\rho\frac{\partial u}{\partial t}$$

In the particular case when k, ρ, and c are constants we have

$$\frac{\partial^2 u}{\partial x^2} + \frac{\partial^2 u}{\partial y^2} + \frac{\partial^2 u}{\partial z^2} = \frac{c\rho}{k}\frac{\partial u}{\partial t}$$

In either case this is a *second order* partial differential equation (PDE), because the highest order of the particular derivatives involved is 2. It is also a *three-dimensional* PDE in space and *time dependent*.

If k, ρ, and c are constants and if the boundary of the body is relatively simple, an analytical solution for $u(x, y, z, t)$ can be found by using the Fourier series method. Otherwise, *numerical* schemes that approximate $u(x, y, z, t)$ are implemented.

In general, a PDE is an equation $Lu = 0$, where L is a given expression that involves u, its partial derivatives, and the various coordinates. The *order* of the PDE is the highest order of a partial derivative in L. Throughout this chapter we will concentrate on techniques for solving several types of second order PDEs, which mostly occur in applications. For the sake of simplicity, we confine ourselves to two-dimensional PDEs.

9.1   Various Types of Partial Differential Equations

9.1.1   Poisson and Laplace Equations

We proceed with a popular PDE,

$$\nabla^2 u \equiv \frac{\partial^2 u}{\partial x^2} + \frac{\partial^2 u}{\partial y^2} = p(x, y) \tag{9.1.1}$$

known as the two-dimensional *Poisson equation*, where $p(x, y)$ is a function specified over the region where a solution $u(x, y)$ is sought. The particular homogeneous case $p(x, y) = 0$ is

$$\nabla^2 u \equiv \frac{\partial^2 u}{\partial x^2} + \frac{\partial^2 u}{\partial y^2} = 0 \tag{9.1.2}$$

and is equally important. It is known as the *Laplace equation*. Both equations arise in the study of many *time-independent* (*steady-state*) physical problems. For example, the electrostatic potential function over a two-dimensional region satisfies a Poisson equation, where $p(x, y)$ is proportional to the electric charge density. Another example is the problem of a steady-state heat distribution, where the temperature $u(x, y)$ satisfies the Laplace equation.

In general, Poisson and Laplace equations possess an infinite number of solutions. To obtain a unique solution, some *constraints* must be imposed on the solution. They generally relate to the *behavior* of the solution on the boundary and are therefore called *boundary conditions* (BCs). The two major types of BC are listed below.

Let R denote a two-dimensional region with a boundary B.

Definition 9.1.1.   (Dirichlet BC). *For the PDE*

$$Lu = 0, \quad (x, y) \in R \tag{9.1.3}$$

the constraint

$$u(x, y) = f(x, y), \quad (x, y) \in B \tag{9.1.4}$$

where $f(x, y)$ is a specified *function over B, is called the* Dirichlet boundary condition *(Fig. 9.1.1).*

The system given by Eqs. (9.1.3) and (9.1.4) is called a *Dirichlet boundary value problem*. We are naturally interested in those problems that possess a *unique solution*.

Example 9.1.1.
Let $u(x, y)$ denote a steady-state temperature distribution over the unit circle (Fig. 9.1.2) $R = \{(x, y) \mid x^2 + y^2 < 1\}$ with Dirichlet boundary conditions

$$u(x, y) = u(r, \theta) = \cos(2\theta), \quad r = 1, \quad 0 \le \theta \le 2\pi$$

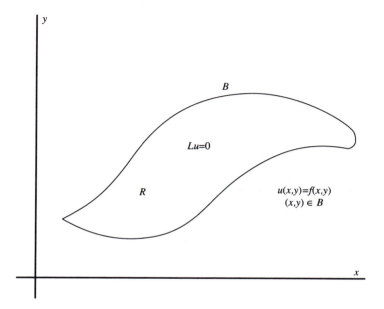

Figure 9.1.1. Dirichlet boundary condition.

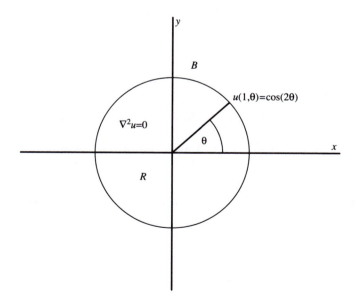

Figure 9.1.2. Laplace equation: a boundary value problem.

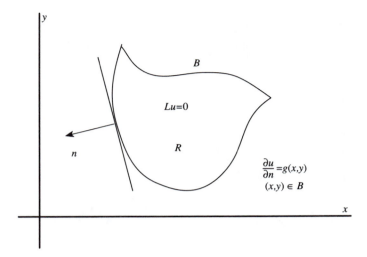

Figure 9.1.3.   Neumann boundary condition.

The unique solution to $\nabla^2 u = 0$ that satisfies the boundary conditions is

$$u(x,y) = x^2 - y^2, \quad (x,y) \in R + B \qquad \square$$

Another type of constraint is obtained by specifying the normal derivative of u (generally in the direction of the outward normal) on the boundary.

Definition 9.1.2.   (Neumann BC). *Given $Lu = 0$, the constraint*

$$\frac{\partial u}{\partial n} = g(x,y), \quad (x,y) \in B \tag{9.1.5}$$

where $g(x,y)$ is a function specified on B, is called the Neumann *boundary condition (Fig. 9.1.3).*

The system given by Eqs. (9.1.3) and (9.1.5) defines a *Neumann boundary value problem.*

Example 9.1.2.

Let $u(x,y)$ solve the Laplace equation over the region of the previous example and consider the Neumann boundary conditions

$$\frac{\partial u}{\partial n}(r,\theta)|_{r=1} = 2\cos(2\theta)$$

These conditions are satisfied by a class of solutions

$$u(x,y) = x^2 - y^2 + c, \quad -\infty < c < \infty$$

If we further impose $u(1,0) = 2$, we obtain $c = 1$ and a unique solution $u(x,y) = x^2 - y^2 + 1$. $\square$

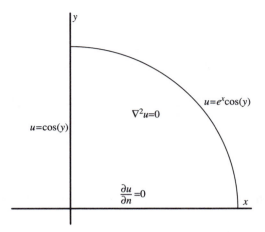

Figure 9.1.4. A boundary value problem of mixed type.

If $f(x,y)$ or $g(x,y)$ vanish, we obtain *homogeneous* Dirichlet or Neumann BCs, respectively. Otherwise, the BCs are *inhomogeneous*.

Example 9.1.3.

Let R be the first quarter of the unit circle (Fig. 9.1.4), let B be its boundary, and consider the Laplace equation $\nabla^2 u = 0$ with the following boundary conditions:

$$u(x,y) = e^x \cos y, \quad x, y \geq 0, \quad x^2 + y^2 = 1$$
$$u(x,y) = \cos y, \quad x = 0, \quad 0 \leq y \leq 1$$
$$\frac{\partial u}{\partial n} = 0, \quad y = 0, \quad 0 \leq x \leq 1$$

This is a boundary value problem of mixed type whose unique solution is $u(x,y) = e^x \cos y$. $\quad\square$

Poisson and Laplace equations are important examples of *elliptic* PDEs for which BCs are essential to obtain a unique solution.

9.1.2 The Heat Equation

A PDE of another type, given by

$$\frac{\partial u}{\partial t} - c\frac{\partial^2 u}{\partial x^2} = 0, \quad c > 0 \tag{9.1.6}$$

is the *heat equation*. It represents a class of PDEs known as *parabolic* partial differential equations.

Let us consider the problem of heat flow along a rod of length l, where the temperature u is assumed to be uniform over each cross-section of the rod (Fig. 9.1.5). Then u, a function of a

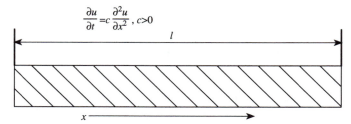

Figure 9.1.5.   Heat flow along a rod.

space coordinate x and time t, must satisfy Eq. (9.1.6), where c is deteremined by heat conduction properties of the particular rod.

We generally assume that the temperature distribution is known at $t = 0$, that is,

$$u(x, 0) = f(x), \quad 0 \leq x \leq l \qquad (9.1.7)$$

where $f(x)$ is a function specified along the rod. This constraint determines *initial conditions* for $u(x, t)$. To obtain a unique solution, additional constraints are needed. Usually we specify boundary conditions at the end points of the rod.

Example 9.1.4.

If the end points are held at constant temperatures, the set of constraints is

$$u(x, 0) = f(x), \quad 0 \leq x \leq l$$
$$u(0, t) = U_1, \quad u(l, t) = U_2, \quad t \geq 0$$

where $U_1 = f(0)$ and $U_2 = f(l)$. Here, the temperature distribution along the rod can be shown to be asymptotically linear as $t \to \infty$. $\square$

Example 9.1.5.

Consider a rod whose initial temperature distribution is known and whose end points are insulated. Then the set of constraints is

$$u(x, 0) = f(x), \quad 0 \leq x \leq l$$
$$\frac{\partial u}{\partial x}(0, t) = \frac{\partial u}{\partial x}(l, t) = 0, \quad t \geq 0$$

and the temperature along the rod is asymptotically constant as $t \to \infty$. $\square$

If we are interested in solving the heat equation for $0 \leq t \leq T$, the complete region is a two-dimensional rectangle (Fig. 9.1.6) and the set of constraints is a combination of "initial conditions" at $\{(x, t) \mid t = 0,\ 0 \leq x \leq l\}$ and "boundary conditions" along the vertical sides $\{(x, t) \mid x = 0,\ l;\ 0 \leq t \leq T\}$. No conditions are specified along the fourth side $\{(x, t) \mid t = T,\ 0 \leq x \leq l\}$.

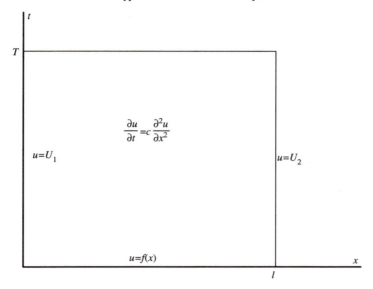

Figure 9.1.6. Heat-flow problem in the (x,t) plane.

9.1.3 The Wave Equation

Consider the problem of a vibrating elastic string of length l, stretched between end points that are held fixed at equal heights. Then under certain conditions the quantity $u(x, t)$, which denotes the vertical displacement of a point x, $0 \leq x \leq l$ at time t (Fig. 9.1.7), satisfies the *wave equation*

$$\frac{\partial^2 u}{\partial t^2} - c^2 \frac{\partial^2 u}{\partial x^2} = 0, \quad 0 < x < l \tag{9.1.8}$$

This is a particular case of a class of PDEs called *hyperbolic* partial differential equations.

To obtain a unique solution we impose a set of constraints. The initial conditions generally consist

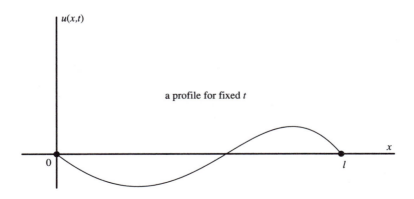

Figure 9.1.7. A vibrating elastic string.

of the position and velocity of the string at time $t = 0$, that is,

$$u(x,0) = f(x), \quad \frac{\partial u}{\partial t}(x,0) = g(x), \quad 0 \le x \le l \tag{9.1.9}$$

Because the end points are kept fixed we have

$$u(0,t) = u(l,t) = 0, \quad 0 \le t \le T \tag{9.1.10}$$

In the next section we present several examples of PDEs with initial and/or boundary conditions for which *analytical* solutions can be obtained.

PROBLEMS

1. Determine the order and dimension of the following PDEs:

   (a) $\dfrac{\partial^2 u}{\partial x^2} + \dfrac{\partial^2 u}{\partial y^2} - \dfrac{\partial^2 u}{\partial z^2} = 0$

   (b) $\dfrac{\partial^4 u}{\partial x^4} + \dfrac{\partial^4 u}{\partial y^4} - u^2 \left(\dfrac{\partial u}{\partial z}\right)^2 - \dfrac{\partial u}{\partial t} = 0$

   (c) $\dfrac{\partial u}{\partial t} - A\dfrac{\partial^2 u}{\partial x^2} - B\dfrac{\partial^2 u}{\partial y^2} = 0$

2. Let $u(x,y)$ solve the Laplace equation over the unit square $R = \{(x,y) \mid 0 \le x,y \le 1\}$ with the boundary conditions

   $$u(x,0) = u(x,1) = 0, \quad 0 \le x \le 1$$
   $$\frac{\partial u}{\partial x}(0,y) = f_1(y), \quad \frac{\partial u}{\partial x}(1,y) = f_2(y), \quad 0 \le y \le 1$$

   Obtain necessary conditions for $f_1(y)$ and $f_2(y)$ if $u \in C^2(R)$ (i.e., if u and its partial derivatives of the first and second order are continuous).

3. The second order PDE,

   $$a\frac{\partial^2 u}{\partial x^2} + 2b\frac{\partial^2 u}{\partial x\partial y} + c\frac{\partial^2 u}{\partial y^2} + d\frac{\partial u}{\partial x} + e\frac{\partial u}{\partial y} + fu + g = 0$$

   whose coefficients $a, b, c, d, e, f,$ and g are functions of x and y, is called elliptic, parabolic, or hyperbolic at (x,y) provided that $ac - b^2 > 0$, $ac - b^2 = 0$, or $ac - b^2 < 0$, respectively, at that particular point. For each of the following PDEs and their associated regions, find the subregions where the particular equation is elliptic, parabolic, and hyperbolic.

   (a) $\dfrac{\partial^2 u}{\partial x^2} + (x^2 + y^2)\dfrac{\partial^2 u}{\partial y^2} - 2u = 0, \quad x^2 + y^2 < 1$

   (b) $(1 + x^2)\dfrac{\partial^2 u}{\partial x^2} - 2xy\dfrac{\partial^2 u}{\partial y^2} + u^2 = 0, \quad 0 \le x \le 1, \quad 0 \le y \le 1$

   (c) $\sin(xy)\dfrac{\partial^2 u}{\partial x^2} + 2\cos(xy)\dfrac{\partial^2 u}{\partial x\partial y} - \sin(xy)\dfrac{\partial^2 u}{\partial y^2} = 0, \quad |x| + |y| \le 1$

4. Solve the heat equation

$$\frac{\partial u}{\partial t} - \frac{\partial^2 u}{\partial x^2} = 0$$

assuming a solution of the form $u(x,t) = f(t) + g(x)$.

5. Show that the wave equation [Eq. (9.1.8)] has the solutions

$$u(x,t) = f(x - ct) + g(x + ct)$$

where $f(z)$ and $g(z)$ are any given functions in $C^2(-\infty, \infty)$.

6. Consider the heat equation when the two end points are kept at constant temperatures U_1 and U_2. Show that if

$$\lim_{t \to \infty} u(x,t) = U(x), \quad 0 \le x \le l$$

then

$$U(x) = U_1 + \frac{U_2 - U_1}{l}x, \quad 0 \le x \le l$$

9.2 Analytical Solutions I: Fourier Series

Solutions to PDEs can sometimes be represented analytically, using Fourier series expansions, thus skipping the need for complex numerical algorithms. Whether or not this can be done depends on the type of the particular PDE, the region over which a solution is sought, and the complexity of the boundary and/or initial conditions.

We open this section with a brief discussion of the basics of Fourier series theory, followed by several representative cases for which this method is applicable.

9.2.1 Fourier Series

If the series

$$\frac{a_0}{2} + \sum_{n=1}^{\infty} \left[a_n \cos\left(\frac{n\pi x}{l}\right) + b_n \sin\left(\frac{n\pi x}{l}\right) \right] \tag{9.2.1}$$

converges to a function $f(x)$ over a set of points A, it is called the *Fourier series* of $f(x)$ over A. A study of this series must consider the special properties of the trigonometric functions $\cos(n\pi x/l)$ and $\sin(\pi x/l)$. We start with their periodicity.

Definition 9.2.1. *A function $f(x)$ is said to be* periodic *with period T if*

$$f(x + T) = f(x) \tag{9.2.2}$$

for all x.

We can always maintain a positive T, because otherwise we reverse the direction of the x axis. If $f(x)$ is periodic with period T, then kT is also a period of $f(x)$, for any given integer k. The smallest positive value T for which Eq. (9.2.2) holds is called the *fundamental period* of $f(x)$.

If $f(x)$ and $g(x)$ are periodic with period T, their sum $F = f + g$ is also periodic with period T. Indeed,

$$F(x + T) = f(x + T) + g(x + T) = f(x) + g(x) = F(x)$$

Using induction, this result is easily extended to a finite number of functions and then to a convergent infinite series.

We will now confine ourselves to the trigonometric functions in Eq. (9.2.1).

Theorem 9.2.1.

The functions $\cos(n\pi x/l)$ and $\sin(n\pi x/l)$ are periodic for all $n \geq 1$, with fundamental period $T = 2l/n$.

Proof.

Let T be a positive number for which

$$\cos\left[\frac{n\pi(x + T)}{l}\right] = \cos\left(\frac{n\pi x}{l}\right)$$

Then

$$\cos\left(\frac{n\pi x}{l}\right)\cos\left(\frac{n\pi T}{l}\right) - \sin\left(\frac{n\pi x}{l}\right)\sin\left(\frac{n\pi T}{l}\right) = \cos\left(\frac{n\pi x}{l}\right)$$

for all x. Therefore

$$\cos\left(\frac{n\pi T}{l}\right) = 1, \quad \sin\left(\frac{n\pi T}{l}\right) = 0$$

which implies $n\pi T/l = 2k\pi$ for some integer k. We thus obtain

$$T = 2lk/n, \quad k = 1, 2, \ldots$$

and the smallest T corresponds to $k = 1$, that is, $T = 2l/k$. The same conclusion holds for $\sin(n\pi x/l)$.

□

As previously confirmed in Chapter 5, the functions $1/2$, $\cos(n\pi x/l)$, and $\sin(n\pi x/l)$ are mutually orthogonal over the interval $[-l, l]$. In particular (the integration is left for the reader),

$$\int_{-l}^{l} \cos\left(\frac{n\pi x}{l}\right)\cos\left(\frac{m\pi x}{l}\right) dx = \begin{cases} 0, & m \neq n \\ l, & m = n \end{cases}$$

$$\int_{-l}^{l} \sin\left(\frac{n\pi x}{l}\right)\sin\left(\frac{m\pi x}{l}\right) dx = \begin{cases} 0, & m \neq n \\ l, & m = n \end{cases}$$

$$\int_{-l}^{l} \sin\left(\frac{n\pi x}{l}\right)\cos\left(\frac{m\pi x}{l}\right) dx = 0$$

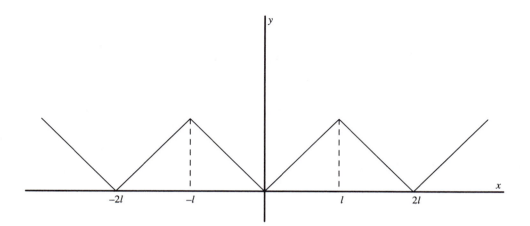

Figure 9.2.1. A triangular wave.

$$\int_{-l}^{l} \cos\left(\frac{n\pi x}{l}\right) dx = \int_{-l}^{l} \sin\left(\frac{n\pi x}{l}\right) dx = 0$$

$$\int_{-l}^{l} \frac{1}{2}\, dx = l \qquad (9.2.3)$$

Consequently, if the Fourier series of Eq. (9.2.1) converges to $f(x)$, then

$$a_n = \frac{1}{l} \int_{-l}^{l} f(x) \cos\left(\frac{n\pi x}{l}\right) dx, \quad n \geq 0 \qquad (9.2.4)$$

$$b_n = \frac{1}{l} \int_{-l}^{l} f(x) \sin\left(\frac{n\pi x}{l}\right) dx, \quad n \geq 1 \qquad (9.2.5)$$

Example 9.2.1.
Define the function

$$f(x) = \begin{cases} -x, & -l \leq x < 0 \\ x, & 0 \leq x < l \end{cases}$$

and extend it periodically to all x by $f(x + 2l) = f(x)$. We obtain the triangular wave shown in Fig. 9.2.1. Let us now assume that a Fourier series in the form of Eq. (9.2.1) converges to $f(x)$. Then Eqs. (9.2.4) and (9.2.5) imply

$$a_0 = l, \quad a_n = \frac{2l}{n^2 \pi^2}[\cos(n\pi) - 1], \quad n \geq 1$$
$$b_n = 0, \quad n \geq 1$$

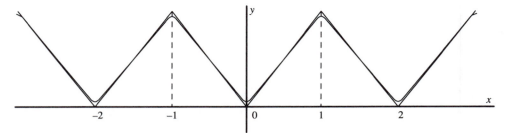

Figure 9.2.2.   A three-term Fourier approximate for a triangular wave, $l = 1$.

and the Fourier series is represented by

$$f(x) = \frac{l}{2} - \frac{4l}{\pi^2} \sum_{n=1}^{\infty} \frac{1}{(2n-1)^2} \cos\left[\frac{(2n-1)\pi x}{l}\right]$$

If we replace the infinite series by its first three terms, we have

$$f(x) \approx \frac{l}{2} - \frac{4l}{\pi^2}\left[\cos\left(\frac{\pi x}{l}\right) + \frac{1}{9}\cos\left(\frac{3\pi x}{l}\right)\right]$$

as demonstrated in Fig. 9.2.2.   □

9.2.2   The Fourier Theorem

We previously obtained a relation [Eqs. (9.2.4) and (9.2.5)] between the *Fourier coefficients* a_i, b_i
and the function $f(x)$, which holds whenever the Fourier series of Eq. (9.2.1) converges to $f(x)$.
A somewhat more natural approach is as follows: given a periodic function $f(x)$, use Eqs. (9.2.4)
and (9.2.5) to define its Fourier coefficients and find a criterion for the associated Fourier series to
converge to $f(x)$.

   We first introduce the concept of *piecewise continuity*.

Definition 9.2.2.   *A function $f(x)$ is said to be piecewise continuous over the interval $a \leq
x \leq b$ if there exists a finite partition*

$$a = x_0 < x_1 < \cdots < x_n = b$$

such that for each i, $0 \leq i \leq n-1$,

1. *$f(x)$ is continuous over the open interval (x_i, x_{i+1}).*

2. *$f(x)$ approaches finite limits $f(x_i+)$ and $f(x_{i+1}-)$ at the end points x_i and x_{i+1}, respectively, that
   is,*

$$\lim_{x \to x_i+} f(x) = f(x_i+), \qquad \lim_{x \to x_{i+1}-} f(x) = f(x_{i+1}-)$$

It should be noted that a piecewise continuous function need not be defined at any of the partition points. It should only converge from either side of these points. Thus, $f(x)$ and $f'(x)$ may be piecewise continuous with respect to the same partition set of points, although $f'(x)$ is not even defined at any point of discontinuity of $f(x)$.

Example 9.2.2.

The function

$$f(x) = \begin{cases} x & , \quad -\infty < x < 0 \\ 1 & , \quad 0 < x < 1 \\ 0.5, & \quad x = 1 \\ x & , \quad 1 < x < \infty \end{cases}$$

is piecewise continuous over $(-\infty, \infty)$. At the partition point $x = 0$ $f(x)$ is not defined. The derivative is given by

$$f'(x) = \begin{cases} 1, & -\infty < x < 0 \\ 0, & 0 < x < 1 \\ 1, & 1 < x < \infty \end{cases}$$

and is also piecewise continuous over $(-\infty, \infty)$. It is not defined at the partition points $x = 0$ and $x = 1$. Both f and f' are shown in Fig. 9.2.3a and b, respectively. $\square$

We now present the main result.

Theorem 9.2.2.

(Fourier theorem). *Let $f(x)$ and $f'(x)$ be piecewise continuous over the interval $-l \leq x < l$ and let $f(x)$ be defined outside the interval so that $f(x)$ is periodic with period 2l. Then the associated Fourier series*

$$\frac{a_0}{2} + \sum_{n=1}^{\infty} \left[a_n \cos\left(\frac{n\pi x}{l}\right) + b_n \sin\left(\frac{n\pi x}{l}\right) \right]$$

where a_i, b_i are derived from $f(x)$ by using Eqs. (9.2.4) and (9.2.5), converges to $f(x)$ at any point where $f(x)$ is continuous. Elsewhere it converges to $[f(x+) + f(x-)]/2$.

The proof of this theorem is beyond the scope of this book.

Example 9.2.3.

Consider the function (a *sawtooth wave*)

$$f(x) = \begin{cases} 1 + x, & -1 \leq x < 0 \\ x & , \quad 0 < x < 1 \end{cases}$$

extended periodically over the whole x axis (Fig. 9.2.4).

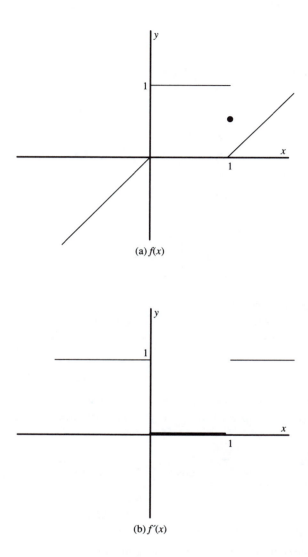

Figure 9.2.3. Piecewise continuous functions f and f'.

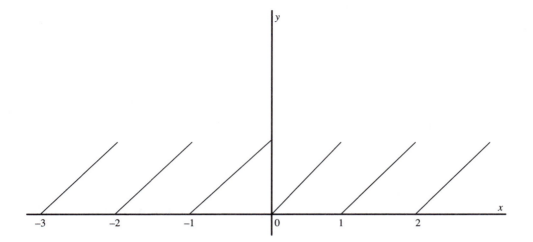

Figure 9.2.4. A sawtooth wave.

Here $l = 1$ and the associated Fourier coefficients are

$$a_n = \int_{-1}^{0} (1+x)\cos(n\pi x)\,dx + \int_{0}^{1} x\cos(n\pi x)\,dx = \begin{cases} 1, & n = 0 \\ 0, & n \geq 1 \end{cases}$$

$$b_n = \int_{-1}^{0} (1+x)\sin(n\pi x)\,dx + \int_{0}^{1} x\sin(n\pi x)\,dx = -\frac{1+\cos(n\pi)}{n\pi}, \quad n \geq 1$$

By using the Fourier theorem we obtain

$$f(x) = \frac{1}{2} - \sum_{n=1}^{\infty} \frac{1+\cos(n\pi)}{n\pi}\sin(n\pi x)$$

$$= \frac{1}{2} - 2\left[\frac{\sin(2\pi x)}{2\pi} + \frac{\sin(4\pi x)}{4\pi} + \cdots\right], \quad -1 < x < 1$$

At $x = 0$, $f(x)$ has a discontinuity and the Fourier series must converge to 0.5, which is indeed the case. At $x = 0.25$, $f(x)$ is continuous and equals 0.25. This implies

$$\frac{1}{2} - 2\left(\frac{1}{2\pi} - \frac{1}{6\pi} + \frac{1}{10\pi} - \cdots\right) = \frac{1}{4}$$

or

$$\frac{\pi}{4} = 1 - \frac{1}{3} + \frac{1}{5} - \cdots \qquad \square$$

9.2.3 Even and Odd Functions

The particular case when a Fourier series of $f(x)$ is either a pure sine or cosine series, is of great importance and frequently used. This occurs whenever the function $f(x)$ is either an even or an odd function. The basic properties of such functions are as follow:

1. The sum, difference, product, and quotient of two even functions are even functions.

2. The sum and difference of two odd functions are odd functions.

3. The product and quotient of two odd functions are even functions.

4. If $f(x)$ and $g(x)$ are even and odd functions, respectively, then

$$\int_{-l}^{l} f(x)\, dx = 2 \int_{0}^{l} f(x)\, dx, \quad \int_{-l}^{l} g(x)\, dx = 0 \qquad (9.2.6)$$

We shall now examine the Fourier coefficients and series of even and odd functions.

9.2.3.1 Even Functions

Let $f(x)$ be an even periodic function with period $2l$. Because

$$f(-x) = f(x) \qquad (9.2.7)$$

the functions $f(x) \cos(n\pi x/l)$ and $f(x) \sin(n\pi x/l)$ are even and odd, respectively. Thus the Fourier coefficients are

$$a_n = \frac{2}{l} \int_{0}^{l} f(x) \cos\left(\frac{n\pi x}{l}\right) dx, \quad n \geq 0$$
$$b_n = 0, \quad n \geq 1 \qquad (9.2.8)$$

and the Fourier series is a *Fourier cosine series*

$$f(x) = \frac{a_0}{2} + \sum_{n=1}^{\infty} a_n \cos\left(\frac{n\pi x}{l}\right) \qquad (9.2.9)$$

Example 9.2.4.

Consider the triangular wave given in Example 9.2.1. It is an even function and consequently $b_n = 0$, $n \geq 1$ and

$$a_0 = \frac{2}{l} \int_{0}^{l} x\, dx = l, \quad a_n = \frac{2}{l} \int_{0}^{l} x \cos\left(\frac{n\pi x}{l}\right) dx = \frac{2l}{n^2 \pi^2}[\cos(n\pi) - 1], \quad n \geq 1$$

For $l = 1$ we have

$$f(x) = \frac{1}{2} - \frac{4}{\pi^2}\left[\cos(\pi x) + \frac{1}{3^2}\cos(3\pi x) + \frac{1}{5^2}\cos(5\pi x) + \cdots\right]$$

and at $x = 0$, $f(x)$ is continuous and the Fourier theorem implies

$$0 = \frac{1}{2} - \frac{4}{\pi^2} \left(1 + \frac{1}{3^2} + \frac{1}{5^2} + \cdots \right)$$

or

$$\frac{\pi^2}{8} = 1 + \frac{1}{3^2} + \frac{1}{5^2} + \cdots \qquad \square$$

9.2.3.2 Odd Functions

Let $f(x)$ be an odd periodic function with period $2l$. Because

$$f(-x) = -f(x) \qquad (9.2.10)$$

the functions $f(x) \cos(n\pi x/l)$ and $f(x) \sin(n\pi x/l)$ are odd and even, respectively. Therefore,

$$a_n = 0, \quad n \geq 0$$
$$b_n = \frac{2}{l} \int_0^l f(x) \sin\left(\frac{n\pi x}{l}\right) dx \qquad (9.2.11)$$

and we have an associated *Fourier sine series*

$$f(x) = \sum_{n=1}^{\infty} b_n \sin\left(\frac{n\pi x}{l}\right) \qquad (9.2.12)$$

Example 9.2.5.

Consider the function

$$f(x) = x, \quad -1 \leq x < 1$$

extended periodically to a sawtooth wave over the x axis. Because $f(x)$ is odd its Fourier coefficients are

$$a_n = 0, \quad n \geq 0; \quad b_n = 2 \int_0^1 x \sin(n\pi x) \, dx = -\frac{2\cos(n\pi)}{n\pi}, \quad n \geq 1$$

The associated Fourier sine series is

$$f(x) = \frac{2}{\pi} \left[\sin(\pi x) - \frac{1}{2}\sin(2\pi x) + \frac{1}{3}\sin(3\pi x) - \cdots \right]$$

$$\square$$

If we combine Examples 9.2.4 and 9.2.5 we observe that the function

$$f(x) = x, \quad 0 \leq x < 1$$

can be expanded to a Fourier cosine series as well as to a Fourier sine series. Specifically,

$$
\begin{aligned}
x &= \frac{1}{2} - \frac{4}{\pi^2} \left[\cos(\pi x) + \frac{1}{3^2} \cos(3\pi x) + \frac{1}{5^2} \cos(5\pi x) + \cdots \right] \\
&= \frac{2}{\pi} \left[\sin(\pi x) - \frac{1}{2} \sin(2\pi x) + \frac{1}{3} \sin(3\pi x) - \cdots \right]
\end{aligned}
\tag{9.2.13}
$$

This is not at all surprising. In Example 9.2.4, $f(x)$, originally defined as x, $0 \le x \le 1$, is defined over $-1 < x \le 0$ as $|x|$ and *then* extended periodically for $-\infty < x < \infty$. Thus $f(x)$ is *even* and generates a Fourier cosine series. In Example 9.2.5, the same $f(x)$ is defined as x over $-1 < x \le 0$ before being extended. It is therefore an odd function and has an associated Fourier sine series. In either case the Fourier coefficients depend on the values of $f(x)$ over the *whole* interval $(-1, 1)$ and thus provide two different Fourier series.

In general, given $f(x)$, $0 \le x < l$ we may extend it for $-l < x \le 0$ in any desirable manner and, after defining it periodically over the whole x axis, obtain the associated Fourier series that converges (in the sense of the Fourier theorem) to $f(x)$ for *all* x. The *even* and *odd* extensions are popular. However, other extensions of $f(x)$ may also be useful, for example,

$$
f(x) = 0, \quad -l < x \le 0
$$

Generally, one is motivated to choose an extension that would generate a fast-converging Fourier series.

Throughout the next section we will present several examples of PDEs for which analytical solutions can be found by using Fourier series.

PROBLEMS

1. Determine which of the following functions is periodic and find the associated fundamental periods. All the functions are defined for $-\infty < x < \infty$.
   (a) x
   (b) $\sin(3x)$
   (c) e^{-x}
   (d) $\sin(\pi x) - \cos(2\pi x)$
   (e) $\sin x + \cos(\pi x)$
   (f) $x^3 - x^2 - 1$
2. Let $f(x)$ be an integrable function with period T.
   (a) Show that

$$
\int_0^a f(x)\,dx = \int_T^{a+T} f(x)\,dx
$$

   (b) Use part (a) to show that

$$
\int_0^T f(x)\,dx = \int_a^{a+T} f(x)\,dx
$$

(c) For all a and b show that

$$\int_a^{a+T} f(x)\,dx = \int_b^{b+T} f(x)\,dx$$

3. Find the Fourier series associated with each of the following functions.

(a)

$$f(x) = 1 - x, \quad -l \le x < l; \quad f(x + 2l) = f(x)$$

(b)

$$f(x) = \begin{cases} 0, & -l \le x < 0 \\ x, & 0 \le x < l \end{cases} \quad ; \quad f(x + 2l) = f(x)$$

(c)

$$f(x) = \begin{cases} -x - 1, & -1 \le x < 0 \\ 0, & 0 \le x < 1 \end{cases} \quad ; \quad f(x + 2) = f(x)$$

(d)

$$f(x) = \begin{cases} x + 1, & -1 \le x < 0 \\ (1/2)x - (1/2), & 0 \le x < 1 \end{cases} \quad ; \quad f(x + 2) = f(x)$$

(e)

$$f(x) = x^2, \quad -l \le x < l; \quad f(x + 2l) = f(x)$$

4. Assuming that

$$f(x) = \frac{a_0}{2} + \sum_{n=1}^{\infty} \left[a_n \cos\left(\frac{n\pi x}{l}\right) + b_n \sin\left(\frac{n\pi x}{l}\right) \right]$$

perform a "formal calculation" to obtain Parseval's equation

$$\frac{1}{l} \int_{-l}^{l} [f(x)]^2\,dx = \frac{a_0^2}{2} + \sum_{n=1}^{\infty} (a_n^2 + b_n^2)$$

5. Determine which of the following functions are even, odd, or neither.

(a) x^3

(b) $(x - 1)^2$

(c) $e^{-x^2} \cos x$

(d) $x^4 - x^3$

(e) $|x + 1| + |x - 1|$

6. Use the properties of even and odd functions to evaluate the following integrals.

   (a) $\int_{-\pi}^{\pi} \sin(2x)\,dx$

   (b) $\int_{-1}^{1}(x^2 + x^4)\,dx$

   (c) $\int_{-\pi}^{\pi} x \sin x\,dx$

   (d) $\int_{-\pi}^{\pi} x \sin(x^2)\,dx$

7. Show that if $f(x)$ and $g(x)$ are odd and even functions, respectively, then $f(0) = 0$, $g'(0) = 0$.

8. Show that the derivatives of even and odd functions are odd and even functions, respectively.

9. Show that any function $f(x)$, $-\infty < x < \infty$ can be represented as a sum of even and odd functions. [*Hint*: Assume $f(x) = g(x) + h(x)$, where $g(x)$ and $h(x)$ are even and odd functions, respectively, and find $g(x)$ and $h(x)$ using $f(x)$ and $f(-x)$].

10. For each of the following functions find the associated Fourier cosine series.
    (a)

$$f(x) = \begin{cases} 0, & 0 < x < 1 \\ 1, & 1 < x < 2 \end{cases}, \quad \text{period 4}$$

(b)

$$f(x) = \begin{cases} x, & 0 < x < 1 \\ 1, & 1 < x < 2 \end{cases}, \quad \text{period 4}$$

(c)

$$f(x) = \begin{cases} 1 - x, & 0 < x < 1 \\ x - 1, & 1 < x < 3 \end{cases}, \quad \text{period 6}$$

11. Replace each of the following functions by its associated Fourier sine series.
    (a)

$$f(x) = \begin{cases} 1, & 0 < x < 1 \\ 2, & 1 < x < 2 \end{cases}, \quad \text{period 4}$$

(b)

$$f(x) = \begin{cases} x, & 0 < x < 1 \\ 2 - x, & 1 < x < 2 \end{cases}, \quad \text{period 4}$$

(c)

$$f(x) = \begin{cases} 0, & 0 < x < \pi \\ 1, & \pi < x < 2\pi \\ -1, & 2\pi < x < 3\pi \end{cases}, \quad \text{period } 6\pi$$

12. Let $f(x)$ be represented by a Fourier sine series, that is,

$$f(x) = \sum_{n=1}^{\infty} b_n \sin\left(\frac{n\pi x}{l}\right), \quad 0 \le x \le l$$

Show formally that

$$\frac{2}{l} \int_0^l [f(x)]^2 \, dx = \sum_{n=1}^{\infty} b_n^2$$

13. Apply the result of problem 12 to $f(x) = x$, $0 \le x \le 1$ and show that

$$\frac{\pi^2}{6} = \sum_{n=1}^{\infty} \frac{1}{n^2}$$

14. Let the function

$$f(x) = \begin{cases} 0, & -1 \le x < 0 \\ x, & 0 \le x < 1 \end{cases}$$

be extended periodically for all x. Find its Fourier series first directly, then by using the result of problem 9.

15. Let the function $f(x)$, $0 \le x \le l$ be extended in the following order (see Fig. 9.2.5):

(a) Symmetrically about $x = l$, that is,

$$f(2l - x) = f(x), \quad 0 \le x \le l$$

(b) As an odd function for $-2l \le x \le 0$

(c) Periodically for all x

Show that the Fourier series of the resulting function is

$$f(x) = \sum_{n=1}^{\infty} b_n \sin\left[\frac{(2n-1)\pi x}{2l}\right]$$

where

$$b_n = \frac{2}{l} \int_0^l f(x) \sin\left[\frac{(2n-1)\pi x}{2l}\right] dx$$

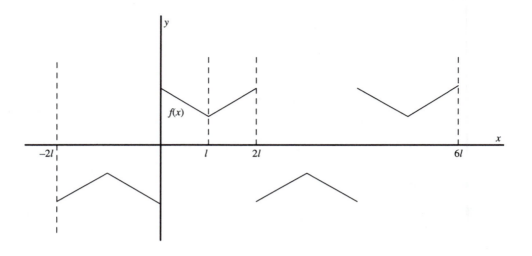

Figure 9.2.5. A symmetric–odd–periodic extension of $f(x)$, $0 \le x \le l$.

16. Use the result of problem 15 in the particular case $f(x) = x$, $0 \le x \le 1$.

17. Find the Fourier series associated with a symmetric–even–periodic extension of $f(x)$, $0 \le x \le l$ and apply the result to $f(x) = x$, $0 \le x \le 1$.

9.3   Analytical Solutions II: Separation of Variables

We shall now apply the results of the previous section to present the solutions of several problems in partial differential equations in the form of a Fourier series, using the method of separation of variables.

9.3.1   Heat Conduction

Consider the heat equation

$$\frac{\partial u}{\partial t} = c\frac{\partial^2 u}{\partial x^2}, \quad 0 < x < l, \quad t > 0, \quad c > 0 \tag{9.3.1}$$

with homogeneous boundary conditions

$$u(0,t) = 0, \quad u(l,t) = 0, \quad t > 0 \tag{9.3.2}$$

and initial condition

$$u(x,0) = f(x), \quad 0 \le x \le l \tag{9.3.3}$$

We use the technique of *separation of variables* and search for a *nontrivial* (i.e., not identically zero) solution to Eq. (9.3.1) of the form

$$u(x, t) = X(x)T(t), \quad 0 < x < l, \quad t > 0 \tag{9.3.4}$$

By substituting in Eq. (9.3.1) we have

$$\frac{T'}{T} = c\frac{X''}{X} \tag{9.3.5}$$

Each side of Eq. (9.3.5) is a function of a different coordinate and therefore

$$\frac{T'}{T} = K, \quad c\frac{X''}{X} = K \tag{9.3.6}$$

for some *constant K*. We shall now seek K for which $u = XT$ satisfies the boundary conditions [Eq. (9.3.2)].

1. $K = 0$: By Eq. (9.3.6) we have $T' = 0$ and $X'' = 0$, that is,

$$T = \alpha, \quad X = \beta x + \gamma$$

   for some constants α, β, and γ. To satisfy the boundary conditions, either α or β and γ must vanish. In either case we obtain the *trivial* solution $u(x, t) = 0$. Therefore $K \neq 0$.

2. $K > 0$: In this case

$$T = \alpha e^{Kt}, \quad X = \beta e^{\lambda x} + \gamma e^{-\lambda x}$$

   where $\lambda = \sqrt{K/c}$ and α, β, and γ are constants. To satisfy Eq. (9.3.2) we must have

$$\alpha e^{Kt}(\beta + \gamma) = 0, \quad \alpha e^{Kt}(\beta e^{\lambda l} + \gamma e^{-\lambda l}) = 0, \quad t > 0$$

   and therefore obtain $\alpha = 0$ or $\beta = \gamma = 0$ (why?), that is, the trivial solution.

3. $K < 0$: This is the only choice left, and by Eq. (9.3.6) we have

$$T = \alpha e^{Kt}, \quad X = \beta \cos(\lambda x) + \gamma \sin(\lambda x)$$

   where α, β, and γ are constants and $\lambda = \sqrt{-K/c}$. To satisfy the boundary conditions without obtaining a trivial solution we must have

$$\beta = 0, \quad \lambda l = n\pi$$

   where n is an integer. This determines K as

$$K = -\frac{cn^2\pi^2}{l^2}, \quad n = 1, 2, \ldots \tag{9.3.7}$$

with an associated solution

$$u_n(x, t) = \alpha_n \exp\left(-\frac{cn^2\pi^2 t}{l^2}\right) \sin\left(\frac{n\pi x}{l}\right), \quad \alpha_n \neq 0 \tag{9.3.8}$$

The only condition left to satisfy is the initial condition of Eq. (9.3.3). Because any linear combination of $u_n(x, t)$ satisfies Eqs. (9.3.1) and (9.3.2), it is sufficient to find a function

$$u(x, t) = \sum_{n=1}^{\infty} \alpha_n \exp\left(-\frac{cn^2\pi^2 t}{l^2}\right) \sin\left(\frac{n\pi x}{l}\right) \tag{9.3.9}$$

for which $u(x, 0) = f(x)$.

Assume $f(x)$ to have the Fourier sine series

$$f(x) = \sum_{n=1}^{\infty} b_n \sin\left(\frac{n\pi x}{l}\right) \tag{9.3.10}$$

To ensure $u(x, 0) = f(x)$, $0 \leq x \leq l$ we simply choose $\alpha_n = b_n$ for $n \geq 1$. Thus

$$u(x, t) = \sum_{n=1}^{\infty} b_n \exp\left(-\frac{cn^2\pi^2 t}{l^2}\right) \sin\left(\frac{n\pi x}{l}\right) \tag{9.3.11}$$

where

$$b_n = \frac{2}{l} \int_0^l f(x) \sin\left(\frac{n\pi x}{l}\right) dx \tag{9.3.12}$$

It should be noted that the solution given by Eqs. (9.3.11) and (9.3.12) must be regarded as a *formal* solution to Eqs. (9.3.1) through (9.3.3). A rigorous proof that the series on the right-hand side of Eq. (9.3.11) converges to a continuous function $u(x, t)$, that u_t and u_{xx} exist and satisfy the heat equation, and that $u(x, t)$ satisfies the boundary and initial conditions is beyond the scope of this book.

Example 9.3.1.

Let $u(x, t)$ be the solution to Eqs. (9.3.1) through (9.3.3) for $c = 1$, $l = 1$ and

$$f(x) = 2\sin(\pi x) + \sin(2\pi x)$$

By substituting in Eq. (9.3.11) we have

$$u(x, t) = 2e^{-\pi^2 t}\sin(\pi x) + e^{-4\pi^2 t}\sin(2\pi x) \qquad \square$$

Although the initial temperature distribution function, $f(x)$, may possess a finite number of jump discontinuities, the solution $u(x, t)$ is continuous for *all* $t > 0$. This is in agreement with the physical reality and emphasizes the fact that heat conduction is a *diffusive process*. Also, for a piecewise continuous $f(x)$ (which is always assumed to be the case), $f(x)$ and therefore b_n are bounded. Consequently

$$\lim_{t \to \infty} u(x,t) = 0, \quad 0 \le x \le l \tag{9.3.13}$$

as expected.

We will now consider another heat conduction problem along a given rod, where there is no heat flow at either end point. Instead of Eq. (9.3.2) we have

$$\frac{\partial u}{\partial x}(0,t) = 0, \quad \frac{\partial u}{\partial x}(l,t) = 0, \quad t > 0 \tag{9.3.14}$$

As before we assume a solution of the form $u(x,t) = X(x)T(t)$ and obtain

$$\frac{T'}{T} = K, \quad c\frac{X''}{X} = K$$

for some *separation constant* K. If $K = 0$ we have

$$T = \alpha, \quad X = \beta x + \gamma$$

and by virtue of Eq. (9.3.14) $\beta = 0$, that is, $u(x,t) = \alpha\gamma$. Thus we obtain a nontrivial solution $u(x,t) = \alpha_0$, where α_0 is a nonzero constant.

If $K > 0$ we have $T = \alpha e^{Kt}$, $X = \beta e^{\lambda x} + \gamma e^{-\lambda x}$, where $\lambda = \sqrt{K/c}$. To satisfy the boundary conditions we must have

$$\alpha e^{Kt}(\lambda\beta - \lambda\gamma) = 0, \quad \alpha e^{Kt}(\lambda\beta e^{\lambda l} - \lambda\gamma e^{-\lambda l}) = 0$$

and because $\lambda \ne 0$ we obtain either $\alpha = 0$ or $\beta = \gamma = 0$. In both cases we end up with a trivial solution. Finally, if $K < 0$ we have $T = \alpha e^{Kt}$, $X = \beta \cos(\lambda x) + \gamma \sin(\lambda x)$ where α, β, and γ are constants and $\lambda = \sqrt{-K/c}$. In view of Eq. (9.3.14) we obtain $\gamma = 0$, $\lambda l = n\pi$ and the associated solution

$$u_n(x,t) = \alpha_n \exp\left(-\frac{cn^2\pi^2 t}{l^2}\right)\cos\left(\frac{n\pi x}{l}\right), \quad \alpha_n \ne 0 \tag{9.3.15}$$

We now seek to present the solution to the complete problem as

$$u(x,t) = \sum_{n=0}^{\infty} \alpha_n \exp\left(-\frac{cn^2\pi^2 t}{l^2}\right)\cos\left(\frac{n\pi x}{l}\right) \tag{9.3.16}$$

Assume $f(x)$ to have the Fourier cosine series

$$f(x) = \frac{a_0}{2} + \sum_{n=1}^{\infty} a_n \cos\left(\frac{n\pi x}{l}\right) \tag{9.3.17}$$

Then, by choosing $\alpha_0 = a_0/2$, $\alpha_n = a_n$, $n \ge 1$, we enforce the initial condition as well as the boundary conditions. Thus

$$u(x,t) = \frac{a_0}{2} + \sum_{n=1}^{\infty} a_n \exp\left(-\frac{cn^2\pi^2 t}{l^2}\right) \cos\left(\frac{n\pi x}{l}\right) \tag{9.3.18}$$

where

$$a_n = \frac{2}{l} \int_0^l f(x) \cos\left(\frac{n\pi x}{l}\right) dx, \quad n \geq 0 \tag{9.3.19}$$

The asymptotic behavior of $u(x,t)$ is given by

$$\lim_{t \to \infty} u(x,t) = \frac{a_0}{2} = \frac{1}{l} \int_0^l f(x)\, dx \tag{9.3.20}$$

Example 9.3.2.

For $c = 1$, $l = 1$, and an initial temperature distribution given by

$$f(x) = 1 + 3\cos(2\pi x) - \cos(3\pi x)$$

the solution to the heat equation with boundary conditions given by Eq. (9.3.14) is

$$u(x,t) = 1 + 3e^{-4\pi^2 t}\cos(2\pi x) - e^{-9\pi^2 t}\cos(3\pi x) \qquad \square$$

Example 9.3.3.

In the previous example consider an initial temperature distribution along the rod given by

$$f(x) = \begin{cases} 1, & 0 \leq x < 1/2 \\ 2, & 1/2 \leq x \leq 1 \end{cases}$$

To determine the solution $u(x,t)$ we first find the Fourier cosine series of $f(x)$. Using Eq. (9.3.19) we obtain

$$a_0 = 2\left(\int_0^{1/2} dx + \int_{1/2}^1 2\, dx\right) = 3$$

$$a_n = 2\left[\int_0^{1/2} \cos(n\pi x)\, dx + \int_{1/2}^1 2\cos(n\pi x)\, dx\right]$$

$$= 2\left[\frac{\sin(n\pi/2)}{n\pi} - 2\frac{\sin(n\pi/2)}{n\pi}\right] = -\frac{2\sin(n\pi/2)}{n\pi}$$

and finally

$$u(x,t) = \frac{3}{2} - \frac{2}{\pi}e^{-\pi^2 t}\cos(\pi x) + \frac{2}{3\pi}e^{-9\pi^2 t}\cos(3\pi x) - \cdots$$

$$= \frac{3}{2} + \frac{2}{\pi} \sum_{n=1}^{\infty} \frac{(-1)^n}{(2n-1)} e^{-(2n-1)^2\pi^2 t} \cos[(2n-1)\pi x] \qquad \square$$

The next example of a Fourier series solution to a PDE involves a hyperbolic PDE. It provides a solution to the problem of a vibrating elastic string. However, the technique given below can be applied to the general analysis of propagation of waves.

9.3.2   The Wave Equation

Consider a vibrating elastic string stretched between the end points $x = 0$ and $x = l$, which are held fixed at identical heights. If $u(x, t)$ denotes the vertical displacement of a point x, $0 \le x \le l$ at time t, then

$$\frac{\partial^2 u}{\partial t^2} = c^2 \frac{\partial^2 u}{\partial x^2}, \quad 0 < x < l, \quad t > 0 \tag{9.3.21}$$

Here $c^2 = F/\rho$, where F is the tension along the string and ρ denotes the mass per unit length. The boundary conditions are

$$u(0, t) = 0, \quad u(l, t) = 0, \quad t \ge 0 \tag{9.3.22}$$

Let us assume a string that is given a nonzero initial displacement from equilibrium, followed by a release at zero velocity, that is,

$$u(x, 0) = f(x), \quad \frac{\partial u}{\partial t}(x, 0) = 0, \quad 0 < x < l \tag{9.3.23}$$

By using separation of variables, we first seek a nontrivial solution to Eqs. (9.3.21) and (9.3.22) of the form

$$u(x, t) = X(x)T(t)$$

Substituting in Eq. (9.3.21) we have

$$T''X = c^2 X''T$$

and therefore

$$\frac{T''}{T} = c^2 \frac{X''}{X} = K \tag{9.3.24}$$

for some constant K.

1. $K = 0$: In view of Eq. (9.3.24) we have

$$T'' = 0, \ X'' = 0$$

that is,

$$T = \alpha t + \beta, \quad X = \gamma x + \delta$$

for some constants α, β, γ, and δ, and to satisfy the boundary conditions we must have

$$\delta(\alpha t + \beta) = 0, \quad (\gamma l + \delta)(\alpha t + \beta) = 0, \quad t > 0$$

that is, $\delta = 0$ and $\gamma = 0$ or $\alpha = \beta = 0$, leading to the trivial solution.

2. $K > 0$: In this case

$$T = \alpha e^{\sqrt{K}t} + \beta e^{-\sqrt{K}t}, \quad X = \gamma e^{\sqrt{\lambda}x} + \delta e^{-\sqrt{\lambda}x}$$

where $\lambda = K/c^2$ and α, β, γ, and δ are constants. To satisfy the boundary conditions we impose

$$\left(\alpha e^{\sqrt{K}t} + \beta e^{-\sqrt{K}t} \right) (\gamma + \delta) = 0$$

$$\left(\alpha e^{\sqrt{K}t} + \beta e^{-\sqrt{k}t} \right) \left(\gamma e^{\sqrt{\lambda}l} + \delta e^{-\sqrt{\lambda}l} \right) = 0$$

and obtain $\alpha = \beta = 0$ or $\gamma = \delta = 0$ (why?). In either case one ends with the trivial solution.

3. $K < 0$: The solution of Eq. (9.3.24) is

$$T = \alpha \cos\left(\sqrt{-K}t \right) + \beta \sin\left(\sqrt{-K}t \right), \quad X = \gamma \cos\left(\sqrt{\lambda}x \right) + \delta \sin\left(\sqrt{\lambda}x \right)$$

where $\lambda = \sqrt{-K/c^2}$. The boundary conditions enforce

$$\gamma = 0, \quad \sqrt{\lambda}l = n\pi$$

where n is an integer, that is,

$$K = -c^2 n^2 \pi^2 / l^2, \quad n \geq 1 \tag{9.3.25}$$

Thus the functions

$$u_n(x,t) = \sin\left(\frac{n\pi x}{l} \right) \cos\left(\frac{n\pi ct}{l} \right), \quad n \geq 1$$

$$v_n(x,t) = \sin\left(\frac{n\pi x}{l} \right) \sin\left(\frac{n\pi ct}{l} \right), \quad n \geq 1 \tag{9.3.26}$$

satisfy both the wave equation and the boundary conditions.

We now seek constants a_n, b_n such that

$$u(x,t) = \sum_{n=1}^{\infty} (a_n u_n + b_n v_n)$$

$$= \sum_{n=1}^{\infty} \sin\left(\frac{n\pi x}{l} \right) \left[a_n \cos\left(\frac{n\pi ct}{l} \right) + b_n \sin\left(\frac{n\pi ct}{l} \right) \right]$$

satisfies the initial conditions of Eq. (9.3.23). The condition $u(x,0) = f(x)$ is replaced by

$$u(x,0) = f(x) = \sum_{n=1}^{\infty} a_n \sin\left(\frac{n\pi x}{l}\right) \tag{9.3.27}$$

The right-hand side of Eq. (9.3.27) is the Fourier sine series of $f(x)$ and we thus choose

$$a_n = \frac{2}{l} \int_0^l f(x) \sin\left(\frac{n\pi x}{l}\right) dx \tag{9.3.28}$$

The second initial condition is

$$\frac{\partial u}{\partial t}(x,0) = \frac{c\pi}{l} \sum_{n=1}^{\infty} n b_n \sin\left(\frac{n\pi x}{l}\right) = 0$$

that is, $b_n = 0$, $n \geq 1$. Thus the final solution to Eqs. (9.3.21) through (9.3.23) is given by

$$u(x,t) = \sum_{n=1}^{\infty} a_n \sin\left(\frac{n\pi x}{l}\right) \cos\left(\frac{n\pi c t}{l}\right) \tag{9.3.29}$$

where a_n are given by Eq. (9.3.28).

Example 9.3.4.
Consider the wave equation with $c = 1$, $l = 1$, homogeneous boundary conditions, and initial conditions given by

$$u(x,0) = \sin(\pi x) + \sin(3\pi x), \quad \frac{\partial u}{\partial t}(x,0) = 0$$

The solution given by Eq. (9.3.29) is

$$u(x,t) = \sin(\pi x)\cos(\pi t) + \sin(3\pi x)\cos(3\pi t) \qquad \square$$

The quantities $f_n = n\pi c/l$, $n \geq 1$ in Eq. (9.3.29) are the *natural frequencies* of the string. These are the frequencies at which the string can vibrate periodically. Each initial displacement of the form

$$u(x,0) = \sin\left(\frac{n\pi x}{l}\right)$$

determines a *natural mode* of vibration with a frequency f_n and a *wavelength* (i.e., the spatial period) $2l/n$. For example, the third natural mode of an elastic string has an initial displacement $\sin(3\pi x/l)$, a frequency $3\pi c/l$, and a wavelength $2l/3$.

The method of separation of variables that so far has been implemented in the case of the heat and wave equations is also applicable in solving elliptic partial differential equations. In our next example we solve a Laplace equation with Dirichlet boundary conditions.

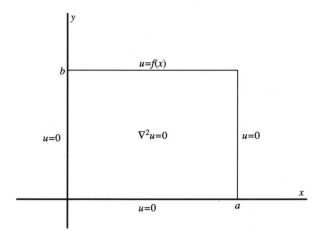

Figure 9.3.1. A Dirichlet problem for a rectangle.

9.3.3 Dirichlet Problem for a Rectangle

Let $u(x, y)$ be a solution to the Laplace equation

$$\frac{\partial^2 u}{\partial x^2} + \frac{\partial^2 u}{\partial y^2} = 0 \tag{9.3.30}$$

for a rectangle (Fig. 9.3.1)

$$R = \{(x, y) \mid 0 < x < a,\ 0 < y < b\}$$

satisfying the following Dirichlet boundary conditions

$$u(0, y) = u(a, y) = 0, \quad 0 \le y \le b \tag{9.3.31}$$

$$u(x, 0) = 0, \quad 0 \le x \le a \tag{9.3.32}$$

$$u(x, b) = f(x), \quad 0 < x < a \tag{9.3.33}$$

where $f(x)$ is some given piecewise continuous function.

We first seek a nontrivial solution $u(x, y) = X(x)Y(y)$ to the Laplace equation that satisfies the homogeneous boundary conditions [Eqs. (9.3.31) and (9.3.32)]. By substituting in Eq. (9.3.30) we obtain

$$\frac{X''}{X} + \frac{Y''}{Y} = 0$$

which implies

$$\frac{X''}{X} = K, \quad \frac{Y''}{Y} = -K \tag{9.3.34}$$

for some constant K.

1. $K = 0$: Equation (9.3.34) yields

$$X = \alpha x + \beta, \quad Y = \gamma y + \delta$$

for some constants α, β, γ, and δ and the only way to satisfy Eq. (9.3.31) is to choose $\alpha = \beta = 0$ or $\gamma = \delta = 0$, that is, the trivial solution (why?).

2. $K > 0$: In this case

$$X = \alpha e^{\sqrt{K}x} + \beta e^{-\sqrt{K}x}, \quad Y = \gamma \cos\left(\sqrt{K}y\right) + \delta \sin\left(\sqrt{K}y\right)$$

for some constants α, β, γ, and δ and once again, unless $\alpha = \beta = 0$ or $\gamma = \delta = 0$, Eq. (9.3.31) cannot hold.

3. $K < 0$: We now have

$$X = \alpha \cos(\sqrt{-K}x) + \beta \sin\left(\sqrt{-K}x\right), \quad Y = \gamma e^{\sqrt{-K}y} + \delta e^{-\sqrt{-K}y}$$

and to satisfy Eq. (9.3.31) with a nontrivial solution we must have

$$\alpha = 0, \quad \sqrt{-K}a = n\pi$$

for some integer n. The additional boundary condition of Eq. (9.3.32) holds, provided that we choose $\gamma = -\delta$. Thus for each n we obtain an associated solution

$$u_n(x, y) = \sin\left(\frac{n\pi x}{a}\right) \sinh\left(\frac{n\pi y}{a}\right) \tag{9.3.35}$$

which satisfies the Laplace equation and the homogeneous boundary conditions as well. Assuming the complete solution to be a linear combination of $u_n(x, y)$, $n \geq 1$ we must find constants b_n, such that

$$\sum_{n=1}^{\infty} b_n \sin\left(\frac{n\pi x}{a}\right) \sinh\left(\frac{n\pi b}{a}\right) = f(x)$$

We replace $f(x)$ by its Fourier sine series with respect to the interval $[0, a]$ and obtain

$$b_n \sinh\left(\frac{n\pi b}{a}\right) = \frac{2}{a} \int_0^a f(x) \sin\left(\frac{n\pi x}{a}\right) dx$$

that is,

$$b_n = \frac{\dfrac{2}{a} \displaystyle\int_0^a f(x) \sin\left(\dfrac{n\pi x}{a}\right) dx}{\sinh\left(\dfrac{n\pi b}{a}\right)} \tag{9.3.36}$$

The final solution is $u(x, y) = \sum_{n=1}^{\infty} b_n u_n(x, y)$.

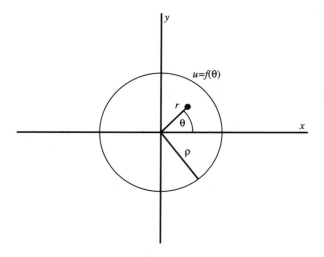

Figure 9.3.2.   A Dirichlet problem for a circle.

9.3.4   Dirichlet Problem for a Circle

Another example for which the method of separation of variables is applicable is the Laplace equation over a circular region, given Dirichlet boundary conditions.

The differential equation, given in polar coordinates, is (Fig. 9.3.2):

$$\frac{\partial^2 u}{\partial r^2} + \frac{1}{r}\frac{\partial u}{\partial r} + \frac{1}{r^2}\frac{\partial^2 u}{\partial \theta^2} = 0 \tag{9.3.37}$$

Consider a circular region of radius ρ centered at the origin and Dirichlet boundary conditions given by

$$u(\rho, \theta) = f(\theta), \quad 0 \le \theta < 2\pi \tag{9.3.38}$$

where $f(\theta)$ is a piecewise continuous function.

We apply the method of separation of variables with respect to the polar coordinates and assume a solution $u(r, \theta) = R(r)\Theta(\theta)$. Thus

$$R''\Theta + \frac{1}{r}R'\Theta + \frac{1}{r^2}R\Theta'' = 0$$

implying

$$\frac{r^2 R'' + rR'}{R} = K, \quad \frac{\Theta''}{\Theta} = -K \tag{9.3.39}$$

for some constant K.

1. $K = 0$: In view of Eq. (9.3.39)

$$\frac{R''}{R'} = -\frac{1}{r}, \quad \Theta'' = 0$$

leading to

$$R = \alpha \ln r + \beta, \quad \Theta = \gamma\theta + \delta$$

where α, β, γ, and δ are constants. For $u(r,\theta)$ to remain bounded as $r \to 0$, we must have $\alpha = 0$. Also, $\gamma = 0$ because $\Theta(\theta)$ is periodic. We thus obtain a fundamental solution $u(r,\theta) = 1$.

2. $K < 0$. Here

$$\Theta(\theta) = \alpha e^{\sqrt{-K}\theta} + \beta e^{-\sqrt{-K}\theta}$$

and unless $\alpha = \beta = 0$ (providing the trivial solution), $\Theta(\theta)$ cannot be periodic.

3. $K > 0$. In this case

$$R = \alpha r^{\sqrt{K}} + \beta r^{-\sqrt{K}}, \quad \Theta = \gamma \cos\left(\sqrt{-K}\theta\right) + \delta \sin\left(\sqrt{-K}\theta\right)$$

for some constants α, β, γ, and δ. The choice $\beta = 0$ guarantees a bounded solution at the origin and, for Θ to have a period 2π, we must have $\sqrt{-K} = n$, where n is an integer. Thus we obtain a set of fundamental solutions (which includes $u_0 = 1$ of case 1):

$$u_n = r^n \cos(n\theta), \quad n \geq 0$$
$$v_n = r^n \sin(n\theta), \quad n \geq 1 \tag{9.3.40}$$

We now assume the solution to the boundary value problem to be a linear combination of the fundamental solutions, that is,

$$u(r,\theta) = \frac{a_0}{2} + \sum_{n=1}^{\infty} r^n[a_n \cos(n\theta) + b_n \sin(n\theta)]$$

The coefficients must be chosen so that

$$f(\theta) = \frac{a_0}{2} + \sum_{n=1}^{\infty} \rho^n[a_n \cos(n\theta) + b_n \sin(n\theta)]$$

We now extend $f(\theta)$ periodically, with a period 2π, and find

$$a_n = \frac{1}{\pi\rho^n} \int_0^{2\pi} f(\theta) \cos(n\theta)\, d\theta, \quad n \geq 0$$
$$b_n = \frac{1}{\pi\rho^n} \int_0^{2\pi} f(\theta) \sin(n\theta)\, d\theta, \quad n \geq 1 \tag{9.3.41}$$

Unlike in the previous examples, the boundary conditions that are given on an interval $[0, l)$, are extended periodically with a period l rather than $2l$, and the generated expansion is a *full* Fourier series.

The method of separation of variables, together with the Fourier series technique, provides a powerful tool for solving PDEs. Still, for many problems analytical solutions are not available because of the complexity of either the equation or the boundary and the boundary conditions. To solve such problems one must switch to numerical techniques.

PROBLEMS

1. Solve the heat equation $u_t = cu_{xx}$ for a rod of length l, assuming homogeneous boundary conditions $u(0,t) = u(l,t) = 0$ and initial temperature distribution $f(x)$, in the following cases.

   (a) $c = 1$, $\quad l = 1$,

   $$f(x) = \begin{cases} 0, & 0 \le x < 1/2 \\ 1, & 1/2 \le x \le 1 \end{cases}$$

   (b) $c = 1$, $\quad l = 1$, $\quad f(x) = x$

   (c) $c = 3$, $\quad l = 2$,

   $$f(x) = \begin{cases} x & , 0 \le x < 1 \\ 2 - x, & 1 \le x \le 2 \end{cases}$$

2. Solve problem 1, assuming no heat flow at the end points of the rod instead of zero temperature.

3. Use the method of separation of variables to solve the heat equation, assuming initial temperature distribution $f(x)$ and boundary conditions

   $$u(0,t) = U_1, \quad u(l,t) = U_2, \quad t > 0$$

   Hint: Replace $u(x,t)$ by

   $$v(x,t) = u(x,t) - U_1 - \frac{U_2 - U_1}{l}x$$

4. Solve the heat equation with $c = 1$, for a rod of length 4, assuming boundary conditions

   $$u(0,t) = 1, \quad u(4,t) = 3, \quad t > 0$$

   and initial temperature distribution

   $$f(x) = \begin{cases} x & , 0 \le x < 1 \\ 1 & , 1 \le x < 3 \\ 4 - x, & 3 \le x \le 4 \end{cases}$$

5. Consider the heat equation

$$\frac{\partial u}{\partial t} = c\frac{\partial^2 u}{\partial x^2}, \quad 0 < x < l, \quad t > 0$$

given the boundary conditions

$$u(0,t) = 0, \quad \frac{\partial u}{\partial x}(l,t) + \alpha u(l,t) = 0, \quad t > 0$$

and determine all its solutions of the form $u(x,t) = X(x)T(t)$.

6. Solve the problem of a vibrating elastic string whose end points are held fixed and whose initial displacement from equilibrium is given by

$$f(x) = \begin{cases} x & , 0 \le x < l/2 \\ l - x, & l/2 \le x \le l \end{cases}$$

Assume that the string is set in motion with no velocity.

7. Solve the previous problem with initial displacement

$$f(x) = \begin{cases} x & , 0 \le x < l/4 \\ (l/2) - x, & l/4 \le x < 3l/4 \\ x - l & , 3l/4 \le x \le l \end{cases}$$

8. Consider the problem

$$\frac{\partial^2 u}{\partial t^2} = 9\frac{\partial^2 u}{\partial x^2}, \quad 0 < x < 2, \quad t > 0$$

with boundary conditions

$$u(0,t) = 0, \quad u(2,t) = 0, \quad t > 0$$

and initial conditions

$$u(x,0) = x(2 - x), \quad \frac{\partial u}{\partial t}(x,0) = 0, \quad 0 < x < 2$$

(a) What are the natural frequencies and their wavelengths?

(b) Find $u(x,t)$ for all x and t.

9. Use the method of separation of variables and solve $u_{tt} = c^2 u_{xx}$, $0 < x < l$, $t > 0$ with boundary conditions

$$u(0,t) = 0, \quad u(l,t) = 0, \quad t > 0$$

and initial conditions

$$u(x,0) = f(x), \quad \frac{\partial u}{\partial t}(x,0) = g(x), \quad 0 < x < l$$

where $f(x)$ and $g(x)$ are given piecewise continuous functions.

10. Solve the problem of a vibrating elastic string with fixed end points, given $c = 1$, $l = 1$, and the initial conditions

$$u(x,0) = f(x) = \begin{cases} x & , 0 \le x < 1/2 \\ 1 - x, & 1/2 \le x \le 1 \end{cases}$$

$$\frac{\partial u}{\partial t}(x,0) = g(x) = \begin{cases} x & , 0 \le x < 1/4 \\ 1/4 & , 1/4 \le x < 3/4 \\ 1 - x, & 3/4 \le x \le 1 \end{cases}$$

11. Solve the Laplace equation $u_{xx} + u_{yy} = 0$ over the rectangle

$$R = \{(x,y) \mid 0 < x < 1,\ 0 < y < 1\}$$

given the Dirichlet boundary conditions

$$u(0,y) = u(1,y) = 0, \quad 0 < y < 1$$
$$u(x,0) = 0, \quad u(x,1) = x, \quad 0 < x < 1$$

12. Solve the Laplace equation over the rectangle

$$R = \{(x,y) \mid 0 < x < a,\ 0 < y < b\}$$

given the Dirichlet boundary conditions

$$u(0,y) = u(a,y) = 0, \quad 0 < y < b$$
$$u(x,0) = f(x), \quad u(x,b) = g(x), \quad 0 < x < a$$

[*Hint*: Solve first for $g(x) = 0$, then for $f(x) = 0$, and add the two solutions.]

13. Solve the Laplace equation over R in problem 12, given the Dirichlet boundary conditions

$$u(0,y) = p(y), \quad u(a,y) = q(y), \quad 0 < y < b$$
$$u(x,0) = f(x), \quad u(x,b) = g(x), \quad 0 < x < a$$

14. Solve the Laplace equation over $\{(x,y) \mid 0 < x < 1,\ 0 < y < 2\}$ given the boundary conditions

$$u(0,y) = 1, \quad u(1,y) = 2, \quad 0 < y < 2$$
$$u(x,0) = 1 + x, \quad 0 < x < 1$$
$$u(x,2) = 1 + x^2, \quad 0 < x < 1$$

15. Solve the Laplace equation over the strip (Fig. 9.3.3)

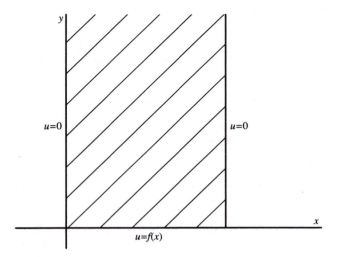

Figure 9.3.3.   A Dirichlet problem for a semiinfinite strip.

$$R = \{(x, y) \mid 0 < x < a,\ 0 < y < \infty\}$$

given the boundary conditions

$$u(0, y) = u(a, y) = 0, \quad y > 0$$
$$u(x, 0) = f(x), \quad 0 < x < a$$

16. Solve the Laplace equation in polar coordinates

$$u_{rr} + \frac{1}{r}u_r + \frac{1}{r^2}u_{\theta\theta} = 0$$

over a circle centered at the origin with radius ρ, given Neumann boundary conditions

$$\frac{\partial u}{\partial r}(\rho, \theta) = f(\theta), \quad 0 \le \theta < 2\pi$$

(The solution is determined up to an additive constant.) Which restriction must be imposed on $f(\theta)$?

9.4   Finite Differences: Parabolic Partial Differential Equations

The method of *finite difference* for solving partial differential equations consists of the following stages:

1. Replacing the given region by a discrete *grid* of points at which an approximate solution is calculated

2. Replacing each derivative in the differential equation by an appropriate finite difference expression that *approximates* the derivative

3. Solving a set of linear or nonlinear algebraic equations whose unknowns are the values of the approximate solution at the grid points

At each grid point in stage 2, the differential equation is replaced by a finite difference equivalent expression that involves some of the neighbors of the point at the grid. This expression forms an *algebraic* equation. The set of these equations is then solved directly or iteratively.

In this section we apply the finite difference approach to parabolic partial differential equations and choose the heat equation as a case study.

9.4.1   Forward-Difference Method

Consider the heat equation

$$\frac{\partial u}{\partial t} = c \frac{\partial^2 u}{\partial x^2}, \quad 0 < x < l, \, t > 0, \quad c > 0 \tag{9.4.1}$$

with boundary conditions

$$u(0, t) = 0, \quad u(l, t) = 0, \quad t > 0 \tag{9.4.2}$$

and initial conditions

$$u(x, 0) = f(x), \quad 0 \le x \le l \tag{9.4.3}$$

We first select two mesh constants, each called *mesh size h and k*, and replace the semiinfinite strip $\{(x, t) \mid 0 \le x \le l, \, t \ge 0\}$ by a grid

$$G = \{(x_i, t_j) \mid x_i = ih, \, t_j = jk; \, 0 \le i \le n, \, 0 \le j\} \tag{9.4.4}$$

where $nh = l$ (Fig. 9.4.1). Each grid point (x_i, t_j) is denoted by the pair of integers (i, j). Also, for any given function $f(x, t)$, the number $f(x_i, t_j)$ is denoted by $f_{i,j}$.

The heat equation involves two derivatives that satisfy

$$c \frac{\partial u}{\partial t}(x_i, t_j) = \frac{u_{i,j+1} - u_{i,j}}{k} - \frac{k}{2} \frac{\partial^2 u}{\partial t^2}(x_i, \eta_j) \tag{9.4.5}$$

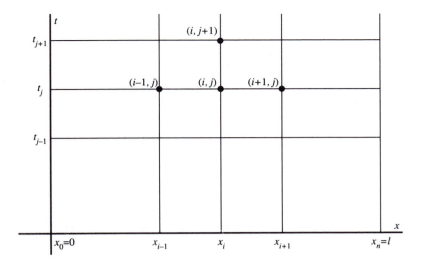

Figure 9.4.1. Replacing a semiinfinite strip by a grid of points.

$$\frac{\partial^2 u}{\partial x^2}(x_i, t_j) = \frac{u_{i-1,j} + u_{i+1,j} - 2u_{i,j}}{h^2} - \frac{h^2}{12}\frac{\partial^4 u}{\partial x^4}(\xi_i, t_j) \qquad (9.4.6)$$

for some $\eta_j \in (t_j, t_{j+1})$ and $\xi_i \in (x_{i-1}, x_{i+1})$. We may therefore replace the differential equation [Eq. (9.4.1)] at each grid point (i,j) with the finite difference expression

$$\frac{u_{i,j+1} - u_{i,j}}{k} = c\frac{u_{i-1,j} + u_{i+1,j} - 2u_{i,j}}{h^2} \qquad (9.4.7)$$

neglecting the local truncation error

$$\tau_{i,j} = \frac{k}{2}\frac{\partial^2 u}{\partial t^2}(x_i, \eta_j) - \frac{ch^2}{12}\frac{\partial^4 u}{\partial x^4}(\xi_i, t_j) \qquad (9.4.8)$$

Because Eq. (9.4.7) only *approximates* the differential equation, we denote the exact solution at (i,j) by $U_{i,j}$ and use $u_{i,j}$ for the approximate one.

The relation given by Eq. (9.4.7) between $u_{i,j}$ and some of its neighbors implies

$$u_{i,j+1} = \left(1 - \frac{2ck}{h^2}\right)u_{i,j} + \frac{ck}{h^2}(u_{i-1,j} + u_{i+1,j}) \qquad (9.4.9)$$

The initial conditions given by Eq. (9.4.3) determine $u_{i,0}$, $0 \le i \le n$ and by virtue of Eq. (9.4.9) we can calculate $u_{i,1}$, $1 \le i \le n - 1$ as well. Because $u_{0,1}$ and $u_{n,1}$ are *specified* boundary conditions, we can reapply Eq. (9.4.9) to obtain $u_{i,2}$, $1 \le i \le n - 1$, and so on.

Let $a = ck/h^2$ and denote the vector $[f(x_1), f(x_2), \ldots, f(x_{n-1})]^T$ by $u^{(0)}$. Then the numerical scheme for evaluating u_{ij}, $1 \le i \le n - 1$, $0 < j$ can also be represented in the matrix form

$$u^{(j)} = Au^{(j-1)}, \quad j \ge 1 \qquad (9.4.10)$$

where A is the tridiagonal matrix

$$
A = \begin{bmatrix}
1-2a & a & 0 & \cdots & & 0 \\
a & 1-2a & a & \ddots & & \vdots \\
0 & \ddots & \ddots & \ddots & & 0 \\
\vdots & & & & 1-2a & a \\
0 & \cdots & & 0 & a & 1-2a
\end{bmatrix} \tag{9.4.11}
$$

and

$$
u^{(j)} = (u_{1,j}, \ldots, u_{n-1,j})^T, \quad j \geq 1 \tag{9.4.12}
$$

The scheme given by Eq. (9.4.9) is known as the *forward-difference* method.

Example 9.4.1.

Consider the heat equation with $c = 1$, subject to homogeneous boundary conditions

$$
u(0,t) = u(1,t) = 0, \quad t > 0
$$

and initial conditions

$$
u(x,0) = \sin(\pi x), \quad 0 \leq x \leq 1
$$

The exact solution as given in the previous section is

$$
u(x,t) = e^{-\pi^2 t} \sin(\pi x)
$$

Because the local truncation error $\tau_{i,j}$, is of the order $O(k + h^2)$, one may expect the approximate solution to converge at that rate to the exact solution. Table 9.4.1 presents of the results for $h = 0.1$, $k = 0.001$ at $t = 0.5$ (i.e., $j = 500$). $\qquad\square$

Table 9.4.1. The Forward-Difference Method, Using "Proper" Mesh Sizes h and k

x_i	$U_{i,500}$	$u_{i,500}$	$\lvert U_{i,500} - u_{i,500} \rvert$
0.1	0.0022224	0.0022590	$0.366 \cdot 10^{-4}$
0.3	0.0058184	0.0059142	$0.959 \cdot 10^{-4}$
0.5	0.0071919	0.0073104	$0.118 \cdot 10^{-3}$
0.7	0.0058184	0.0059142	$0.959 \cdot 10^{-3}$
0.9	0.0022224	0.0022590	$0.366 \cdot 10^{-4}$

The technique by which we determine the mesh sizes h and k is related to the stability of the forward-difference method, that is, to its error propagation as t increases.

Let $\epsilon^{(0)} = (\epsilon_1^{(0)}, \epsilon_2^{(0)}, \ldots, \epsilon_{n-1}^{(0)})^T$ denote the initial error made in representing $u^{(0)}$ by $[f(x_1), f(x_2), \ldots, f(x_{n-1})]^T$, that is, the initial vector in the computational process is $u^{(0)} + \epsilon^{(0)}$. It may be seen that the error in $u^{(m)}$ is $A^m \epsilon^{(0)}$ and, to obtain stability, we request $\lVert A^m \epsilon^{(0)} \rVert \leq \alpha \lVert \epsilon^{(0)} \rVert$ for some natural matrix norm $\| \cdot \|$ and $\alpha > 0$ independent of m. By Theorem 7.4.7, a sufficient condition for stability is

$$\rho(A) < 1 \tag{9.4.13}$$

where $\rho(A)$ is the spectral radius of A. The eigenvalues of A in Eq. (9.4.11) can be shown to be

$$\lambda_i = 1 - 4a \left[\sin\left(\frac{i\pi}{2n}\right) \right]^2, \quad 1 \le i \le n - 1$$

and thus the condition for stability is

$$\rho(A) = \max_{1 \le i \le n-1} \left| 1 - 4a \left[\sin\left(\frac{i\pi}{2n}\right) \right]^2 \right| < 1$$

which implies $a \le 1/2$, that is, $ck/h^2 \le 1/2$ or

$$k \le h^2/(2c) \tag{9.4.14}$$

Thus, for given c and h, Eq. (9.4.14) determines an upper bound for the time step k. Beyond this bound the numerical solution is not expected to be stable. [If $\rho(A) > 1$ then $\|A^m\| \to \infty$ as $m \to \infty$ (why?) and instability occurs.]

Example 9.4.2.

In the previous example, if $h = 0.1$ the stability condition is $k \le 0.005$. For $k = 0.01$ the solution is instable, as shown in Table 9.4.2.   □

Table 9.4.2. The Forward-Difference Method, Using "Improper" Mesh Sizes h and k

x_i	$U_{i,50}$	$u_{i,50}$
0.1	0.0022224	703,387
0.3	0.0058184	1,837,738
0.5	0.0071919	2,264,060
0.7	0.0058184	1,825,586
0.9	0.0022224	695,887

The restriction imposed by Eq. (9.4.14) classifies the forward-difference technique as a *conditionally stable* method. The next finite difference scheme does not impose any restriction on h, k and is therefore *unconditionally stable*.

9.4.2   An Unconditionally Stable Method

Rather than using Eq. (9.4.5) to approximate $\partial u/\partial t$, let us express the time derivative by

$$\frac{\partial u}{\partial t}(x_i, t_j) = \frac{u_{i,j} - u_{i,j-1}}{k} + \frac{k}{2} \frac{\partial^2 u}{\partial t^2}(x_i, \eta_j) \tag{9.4.15}$$

where $\eta_j \in (t_{j-1}, t_j)$, thus obtaining

$$\frac{u_{i,j} - u_{i,j-1}}{k} = c \frac{u_{i-1,j} + u_{i+1,j} - 2u_{i,j}}{h^2} \tag{9.4.16}$$

leading to

$$u_{i,j} = \frac{u_{i,j-1} + (ck/h^2)(u_{i-1,j} + u_{i+1,j})}{1 + (2ck/h^2)} \tag{9.4.17}$$

This scheme is called a *backward-difference* method and its matrix representation is given by

$$Bu^{(j)} = u^{(j-1)}, \quad j \geq 1 \tag{9.4.18}$$

where

$$B = \begin{bmatrix} 1+2a & -a & 0 & \cdots & & 0 \\ -a & 1+2a & -a & & \ddots & \vdots \\ 0 & \ddots & \ddots & \ddots & & 0 \\ \vdots & & & & 1+2a & -a \\ 0 & \cdots & & 0 & -a & 1+2a \end{bmatrix} \tag{9.4.19}$$

The matrix B is tridiagonal and satisfies all the requirements of Theorem 7.3.4. Thus the system given by Eq. (9.4.18) is easily solved by factorizing B, using Eqs. (7.3.22) through (7.3.24). This is described in detail in Algorithm 9.4.1.

The eigenvalues of B can be shown to be

$$\lambda_i = 1 + 4a \left[\sin\left(\frac{i\pi}{2n}\right) \right]^2, \quad 1 \leq i \leq n-1$$

Thus $\lambda_i > 1$, $1 \leq i \leq n-1$. Because the eigenvalues of B^{-1} are simply $\{1/\lambda_i, \ 1 \leq i \leq n-1\}$ we have $\rho(B^{-1}) < 1$. By virtue of Eq. (9.4.18) we have

$$u^{(m)} = (B^{-1})^m u^{(0)}, \quad m \geq 1$$

Thus an initial error $\epsilon^{(0)}$ in $u^{(0)}$ propagates after m time steps to $(B^{-1})^m \epsilon^{(0)}$, which converges to zero because $\rho(B^{-1}) < 1$. The backward-difference method is therefore unconditionally stable, as demonstrated by the following example.

Example 9.4.3.

By applying the backward-difference method in the previous example, we obtain the results presented in Table 9.4.3.   □

Table 9.4.3. The Backward-Difference Method: Unconditional Stability

x_i	$U_{i,50}$	$u_{i,50}$	$\lvert U_{i,50} - u_{i,50}\rvert$
0.1	0.0022224	0.0028980	$0.676 \cdot 10^{-3}$
0.3	0.0058184	0.0075871	$0.177 \cdot 10^{-2}$
0.5	0.0071919	0.0093782	$0.219 \cdot 10^{-2}$
0.7	0.0058184	0.0075871	$0.177 \cdot 10^{-2}$
0.9	0.0022224	0.0028980	$0.676 \cdot 10^{-3}$

Algorithm 9.4.1.

(The backward-difference method for the heat equation). This algorithm approximates the solution to

$$\frac{\partial u}{\partial t} = c\frac{\partial^2 u}{\partial x^2}, \quad 0 < x < l, \quad t > 0, \quad c > 0$$

with boundary conditions

$$u(0,t) = 0, \quad u(l,t) = 0, \quad t > 0$$

and initial conditions

$$u(x,0) = f(x), \quad 0 \le x \le l$$

Input: End point l; maximum time T; equation constant c; integers n and N, which determine the discretization in x and t, respectively.

Output: Approximations $u_{i,j}$ to the exact solution at (x_i, t_j), $1 \le i \le n-1$, $1 \le j \le N$.

Step 1. Set $h = l/n$, $k = T/N$, $a = ck/h^2$, $x_i = ih$, $1 \le i \le n-1$.
Step 2. Set $u_{i,0} = f(x_i)$, $1 \le i \le n-1$ (initial conditions).

Steps 3 and 4 factorize the matrix B of Eq. (9.4.19), using Eq. (7.3.24).

Step 3. Set $\beta_1 = 1 + 2a$.
Step 4. For $2 \le i \le n-1$ set $\alpha_i = -a/\beta_{i-1}$, $\beta_i = (1+2a) + \alpha_i a$.

Steps 5 through 9 solve Eq. (9.4.18) for $1 \le j \le N$.

Step 5. For $1 \le j \le N$ do Steps 6 through 9.
Step 6. Set $t = jk$.
Step 7. Set $y_1 = u_{1,j-1}$ and for $2 \le i \le n-1$ set $y_i = u_{i,j-1} - \alpha_i y_{i-1}$.
Step 8. Set $u_{n-1,j} = y_{n-1}/\beta_{n-1}$ and for $i = n-2, n-3, \ldots, 1$ set

$$u_{i,j} = \frac{y_i + au_{i+1,j}}{\beta_i}$$

Step 9. Output (t) and for $1 \le i \le n-1$ output $(x_i, u_{i,j})$ and stop.

Although the backward-difference method is unconditionally stable, it has the disadvantage of having a local truncation error of order $O(k + h^2)$, which requires $k << h$, that is, time intervals much smaller than the spatial intervals.

The next finite difference scheme provides a method that is unconditionally stable but that also has a local truncation error of order $O(k^2 + h^2)$.

9.4.3   Crank–Nicolson Method

To obtain $u_{i,j+1}$ we may use the forward-difference scheme

$$\frac{u_{i,j+1} - u_{i,j}}{k} = c\frac{u_{i-1,j} + u_{i+1,j} - 2u_{i,j}}{h^2}$$

or the backward-difference scheme

$$\frac{u_{i,j+1} - u_{i,j}}{k} = c\frac{u_{i-1,j+1} + u_{i+1,j+1} - 2u_{i,j+1}}{h^2}$$

If we average both schemes we obtain

$$\frac{u_{i,j+1} - u_{i,j}}{k} = \frac{c}{2}\left(\frac{u_{i+1,j} + u_{i-1,j} - 2u_{i,j}}{h^2} + \frac{u_{i+1,j+1} + u_{i-1,j+1} - 2u_{i,j+1}}{h^2}\right) \tag{9.4.20}$$

The local truncation error associated with Eq. (9.4.20) can be shown to be of order $O(h^2 + k^2)$. This scheme presents the *Crank–Nicolson* method. Its matrix form is

$$Bu^{(j)} = Au^{(j-1)}, \quad j \geq 1 \tag{9.4.21}$$

where A, B are the $(n-1) \times (n-1)$ matrices

$$B = \begin{bmatrix} 1+a & -a/2 & 0 & \cdots & & 0 \\ -a/2 & 1+a & -a/2 & \ddots & & \vdots \\ 0 & -a/2 & \ddots & \ddots & & 0 \\ \vdots & & \ddots & & & -a/2 \\ 0 & & \cdots & 0 & -a/2 & 1+a \end{bmatrix} \tag{9.4.22}$$

$$A = \begin{bmatrix} 1-a & a/2 & 0 & \cdots & & 0 \\ a/2 & 1-a & a/2 & \ddots & & \vdots \\ 0 & a/2 & \ddots & \ddots & & 0 \\ \vdots & & \ddots & & & a/2 \\ 0 & & \cdots & 0 & a/2 & 1-a \end{bmatrix} \tag{9.4.23}$$

The matrix B is tridiagonal and satisfies all the requirements of Theorem 7.3.4. It is factorized using Eqs. (7.3.22) through (7.3.24) and the linear system of Eq. (9.4.21) is easily solved for $u^{(j)}$.

The Crank–Nicolson procedure can be shown to be unconditionally stable. This feature, together with having order of convergence $O(h^2 + k^2)$, offers an attractive tool for solving the heat equation. The complete algorithm is given next. Subroutines BDIF and CRNIC, which incorporate the backward-difference and Crank–Nicolson methods, are found on the attached floppy disk.

Algorithm 9.4.2.

(Crank–Nicolson method for the heat equation). This algorithm approximates the solution to

$$\frac{\partial u}{\partial t} = c\frac{\partial^2 u}{\partial x^2}, \quad 0 < x < l, \quad t > 0, \quad c > 0$$

with boundary conditions

$$u(0,t) = 0, \quad u(l,t) = 0, \quad t > 0$$

and initial conditions

$$u(x,0) = f(x), \quad 0 \le x \le l$$

Input: End point l; maximum time T; equation constant c; integers n and N, which determine the discretization in x and t, respectively.

Output: Approximations $u_{i,j}$ to the exact solution at (x_i, t_j), $1 \le i \le n - 1$, $1 \le j \le N$.

Step 1. Set $h = l/n$, $k = T/N$, $a = ck/h^2$; $x_i = ih$, $1 \le i \le n - 1$
Step 2. Set $u_{i,0} = f(x_i)$, $1 \le i \le n - 1$ (initial conditions).
Step 3. Set $c_1 = 1 + a$; $c_2 = 1 - a$; $c_3 = a/2$.

Steps 4 and 5 factorize the matrix B of Eq. (9.4.22), using Eq. (7.3.24).

Step 4. Set $\beta_1 = c_1$.
Step 5. For $2 \le i \le n - 1$ set $\alpha_i = -c_3/\beta_{i-1}$, $\beta_i = c_1 + \alpha_i c_3$.

Steps 6 through 13 solve Eq. (9.4.21) for $1 \le j \le N$.

Step 6. For $1 \le j \le N$ do Steps 7 through 13.
Step 7. Set $t = jk$.

Steps 8 through 10 calculate the vector $Au^{(j-1)}$.

Step 8. Set $f_1 = c_2 u_{1,j-1} + c_3 u_{2,j-1}$.
Step 9. For $2 \le i \le n - 2$ set $f_i = c_3(u_{i-1,j-1} + u_{i+1,j-1}) + c_2 u_{i,j-1}$.
Step 10. Set $f_{n-1} = c_3 u_{n-2,j-1} + c_2 u_{n-1,j-1}$.
Step 11. Set $y_1 = f_1$ and for $2 \le i \le n - 1$ set $y_i = f_i - \alpha_i y_{i-1}$.
Step 12. Set $u_{n-1,j} = y_{n-1}/\beta_{n-1}$ and for $i = n - 2, n - 3, \ldots, 1$ set $u_{i,j} = (y_i + c_3 u_{i+1,j})/\beta_i$.
Step 13. Output (t) and for $1 \le i \le n - 1$ output $(x_i, u_{i,j})$.
Step 14. Stop.

Example 9.4.4.

Consider the problem in Example 9.4.1. Implementation of the Crank–Nicolson method with $h = 0.1$, $k = 0.01$, $T = 0.5$ provides the results shown in Table 9.4.4.

A comparison between Tables 9.4.3 and 9.4.4 confirms that the Crank–Nicolson method converges faster than the backward-difference method.   □

Table 9.4.4. Crank–Nicolson Method: Unconditional Stability

x_i	$U_{i,50}$	$u_{i,50}$	$\lvert U_{i,50} - u_{i,50} \rvert$
0.1	0.0022224	0.0023051	$0.827 \cdot 10^{-4}$
0.3	0.0058184	0.0060349	$0.217 \cdot 10^{-3}$
0.5	0.0071919	0.0074595	$0.268 \cdot 10^{-3}$
0.7	0.0058184	0.0060349	$0.217 \cdot 10^{-3}$
0.9	0.0022224	0.0023051	$0.827 \cdot 10^{-4}$

PROBLEMS

1. Consider the heat equation

$$\frac{\partial u}{\partial t} = \frac{1}{2}\frac{\partial^2 u}{\partial x^2}, \quad 0 < x < 1, \quad 0 < t < 0.1$$

with boundary conditions

$$u(0,t) = 0, \quad u(1,t) = 0; \quad 0 \leq t \leq 0.1$$

and initial conditions

$$u(x,0) = x\sqrt{1-x}, \quad 0 \leq x \leq 1$$

(a) Approximate the exact solution $U(x,1)$ by using the forward-difference method with $h = 0.1$ and $k = 0.0005$.

(b) Given $h = 0.1, 0.05, 0.01$ determine the corresponding maximum values of k that guarantee stability.

2. Repeat part (a) of problem 1, using the backward-difference method with $h = 0.1$ and $k = 0.01$.

3. Repeat part (a) of problem 1, using the Crank–Nicolson method with $h = 0.1$ and $k = 0.1$.

4. Show that matrix A of Eq. (9.4.11) has the eigenvalues

$$\lambda_i = 1 - 4a\left[\sin\left(\frac{i\pi}{2n}\right)\right]^2, \quad 1 \leq i \leq n-1$$

and the corresponding eigenvectors

$$x^{(i)} = \left[\sin\frac{i\pi}{n}, \sin\frac{2i\pi}{n}, \ldots, \sin\frac{(n-1)i\pi}{n}\right]^T$$

5. Show that matrix B of Eq. (9.4.19) has the eigenvalues

$$\lambda_i = 1 + 4a \left[\sin \left(\frac{i\pi}{2n} \right) \right]^2, \quad 1 \le i \le n - 1$$

and the corresponding eigenvectors

$$x^{(i)} = \left[\sin \frac{i\pi}{n}, \sin \frac{2i\pi}{n}, \ldots, \sin \frac{(n-1)i\pi}{n} \right]^T$$

6. Change the backward-difference algorithm to treat the heat equation

$$\frac{\partial u}{\partial t} = c \frac{\partial^2 u}{\partial x^2}, \quad 0 < x < l, \quad 0 < t < T$$

subject to the boundary conditions

$$u(0, t) = g(t), \quad u(l, t) = h(t); \quad 0 \le t \le T$$

and the initial conditions

$$u(x, 0) = f(x), \quad 0 \le x \le l$$

7. Repeat problem 6, using the Crank–Nicolson method.

8. Apply the Crank–Nicolson method to

$$\frac{\partial u}{\partial t} = 2 \frac{\partial^2 u}{\partial x^2}, \quad 0 < x < 2, \quad 0 < t < 1$$
$$u(0, t) = 1, \quad u(2, t) = t; \quad 0 \le t \le 1$$
$$u(x, 0) = x(2 - x)^2, \quad 0 \le x \le 2$$

Calculate $u(x_i, 1)$, using $h = 0.2$ and $k = 0.2$.

9. Modify the Crank–Nicolson algorithm to accommodate

$$\frac{\partial u}{\partial t} = c \frac{\partial^2 u}{\partial x^2} + F(x), \quad 0 < x < l, \quad 0 < t < T$$
$$u(0, t) = 0, \quad u(l, t) = 0; \quad 0 \le t \le T$$
$$u(x, 0) = f(x), \quad 0 \le x \le l$$

10. Apply the modified Crank–Nicolson method to solve

$$\frac{\partial u}{\partial t} = 2 \frac{\partial^2 u}{\partial x^2} + 1, \quad 0 < x < 1, \quad 0 < t < 1$$

subject to homogeneous boundary conditions and initial conditions

$$u(x, 0) = \sin(\pi x) + x(1 - x), \quad 0 \le x \le 1$$

Calculate $u(x_i, 1)$, using $h = k = 0.1$.

9.5 Finite Differences: Hyperbolic Partial Differential Equations

We will now construct a finite difference approximate solution to a typical hyperbolic partial differential equation.

Consider the wave equation

$$\frac{\partial^2 u}{\partial t^2} - c^2 \frac{\partial^2 u}{\partial x^2} = 0, \quad 0 < x < l, \quad t > 0 \tag{9.5.1}$$

subject to the boundary conditions

$$u(0, t) = 0, \quad u(l, t) = 0; \quad 0 \le t \le T \tag{9.5.2}$$

and the initial conditions

$$u(x, 0) = f(x), \quad \frac{\partial u}{\partial t}(x, 0) = g(x); \quad 0 \le x \le l \tag{9.5.3}$$

where c is an arbitrary constant. The rectangular region is discretized and replaced by a grid of points (x_i, t_j), $0 \le i \le n$, $0 \le j \le N$ defined by

$$x_i = ih, \quad 0 \le i \le n$$
$$t_j = jk, \quad 0 \le j \le N \tag{9.5.4}$$

where

$$h = \frac{l}{n}, \quad k = \frac{T}{N} \tag{9.5.5}$$

are the mesh sizes in the x, t coordinates, respectively. At any internal point (x_i, t_j) (Fig. 9.5.1) the second order derivatives of Eq. (9.5.1) can be replaced by

$$\frac{\partial^2 u}{\partial x^2}(x_i, t_j) = \frac{u(x_{i+1}, t_j) + u(x_{i-1}, t_j) - 2u(x_i, t_j)}{h^2} - \frac{h^2}{12} \frac{\partial^4 u}{\partial x^4}(\xi_i, t_j) \tag{9.5.6}$$

$$\frac{\partial^2 u}{\partial t^2}(x_i, t_j) = \frac{u(x_i, t_{j+1}) + u(x_i, t_{j-1}) - 2u(x_i, t_j)}{k^2} - \frac{k^2}{12} \frac{\partial^4 u}{\partial t^4}(x_i, \eta_j) \tag{9.5.7}$$

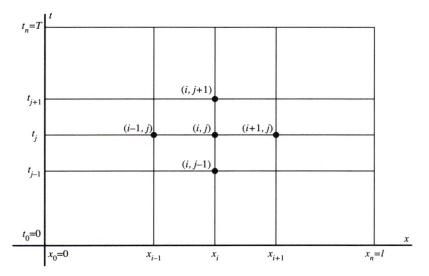

Figure 9.5.1.   Discretizing the region.

where $x_{i-1} < \xi_i < x_{i+1}$ and $t_{j-1} < \eta_j < t_{j+1}$. Substituting in Eq. (9.5.1) we obtain

$$\frac{u(x_i, t_{j+1}) + u(x_i, t_{j-1}) - 2u(x_i, t_j)}{k^2} - c^2 \frac{u(x_{i+1}, t_j) + u(x_{i-1}, t_j) - 2u(x_i, t_j)}{h^2}$$

$$= \frac{1}{12} \left[k^2 \frac{\partial^4 u}{\partial t^4}(x_i, \eta_j) - c^2 h^2 \frac{\partial^4 u}{\partial x^4}(\xi_i, t_j) \right] \qquad (9.5.8)$$

By neglecting the local truncation error

$$\tau_{ij} = \frac{1}{12} \left[k^2 \frac{\partial^4 u}{\partial t^4}(x_i, \eta_j) - c^2 h^2 \frac{\partial^4 u}{\partial x^4}(\xi_i, t_j) \right] = O(h^2 + k^2) \qquad (9.5.9)$$

we approximate the differential equation at (x_i, t_j) by

$$\frac{u_{i,j+1} + u_{i,j-1} - 2u_{i,j}}{k^2} - c^2 \frac{u_{i+1,j} + u_{i-1,j} - 2u_{i,j}}{h^2} = 0 \qquad (9.5.10)$$

where $u_{i,j}$, $0 \le i \le n$, $0 \le j \le N$ denote the approximate solution at (x_i, t_j). As before, the exact solution at (x_i, t_j) is denoted by $U_{i,j}$.

To apply Eq. (9.5.10) we rewrite it as

$$u_{i,j+1} = 2(1 - a^2)u_{i,j} + a^2(u_{i+1,j} + u_{i-1,j}) - u_{i,j-1} \qquad (9.5.11)$$

where $a = ck/h$, $1 \le i \le n - 1$, $1 \le j \le N - 1$. By virtue of the homogeneous boundary conditions and Eq. (9.5.11) we find

$$u^{(j+1)} = Au^{(j)} - u^{(j-1)} \qquad (9.5.12)$$

where

$$u^{(k)} = (u_{1,k}, u_{2,k}, \ldots, u_{n-1,k})^T, \quad 0 \le k \le N \tag{9.5.13}$$

and

$$A = \begin{bmatrix} 2(1-a^2) & a^2 & 0 & \cdots & & 0 \\ a^2 & 2(1-a^2) & a^2 & \ddots & & \vdots \\ 0 & a^2 & \ddots & \ddots & & 0 \\ \vdots & & \ddots & & & a^2 \\ 0 & & \cdots & 0 & a^2 & 2(1-a^2) \end{bmatrix} \tag{9.5.14}$$

To carry out the computation, one must know both $u^{(0)}$ and $u^{(1)}$. The first initial vector is

$$u^{(0)} = [f(x_1), \ldots, f(x_{n-1})]^T \tag{9.5.15}$$

The second vector is found by discretizing the second initial condition

$$\frac{\partial u}{\partial t}(x, 0) = g(x), \quad 0 \le x \le l \tag{9.5.16}$$

The most trivial way of doing it is to write

$$\frac{\partial u}{\partial t}(x_i, 0) = \frac{u(x_i, k) - u(x_i, 0)}{k} - \frac{k}{2} \frac{\partial^2 u}{\partial t^2}(x_i, \bar{\eta}_i) \tag{9.5.17}$$

where $\bar{\eta}_i$ is some interim point between 0 and k.
   Consequently we define

$$u^{(1)} = u^{(0)} + k[g(x_1), \ldots, g(x_{n-1})]^T \tag{9.5.18}$$

thus neglecting a local truncation error of order $O(k)$. To obtain a better approximate solution we write

$$u(x_i, k) = u(x_i, 0) + k\frac{\partial u}{\partial t}(x_i, 0) + \frac{k^2}{2}\frac{\partial^2 u}{\partial t^2}(x_i, 0) + \frac{k^3}{6}\frac{\partial^3 u}{\partial t^3}(x_i, \tilde{\eta}_i)$$

where $\tilde{\eta}_i$ is some interim point between 0 and k, and exchange Eq. (9.5.17) with

$$\frac{\partial u}{\partial t}(x_i, 0) = \frac{u(x_i, k) - u(x_i, 0)}{k} - \frac{k}{2}\frac{\partial^2 u}{\partial t^2}(x_i, 0) - \frac{k^2}{6}\frac{\partial^3 u}{\partial t^3}(x_i, \tilde{\eta}_i) \tag{9.5.19}$$

Assuming the differential equation to hold also for $t = 0$ we have

$$\frac{\partial^2 u}{\partial t^2}(x_i, 0) = c^2 \frac{\partial^2 u}{\partial x^2}(x_i, 0) = c^2 f''(x_i) \tag{9.5.20}$$

and by substituting in Eq. (9.5.19) we obtain

$$g(x_i) = \frac{u(x_i, k) - u(x_i, 0)}{k} - \frac{kc^2}{2}f''(x_i) + O(k^2) \tag{9.5.21}$$

or

$$u(x_i, k) = u(x_i, 0) + kg(x_i) + \frac{k^2c^2}{2}f''(x_i) + O(k^3) \tag{9.5.22}$$

This leads us to define

$$u^{(1)} = u^{(0)} + k[g(x_1), \ldots, g(x_{n-1})]^T + \frac{k^2c^2}{2}[f''(x_1), \ldots, f''(x_{n-1})]^T \tag{9.5.23}$$

The truncation error generated by this approximation is $O(k^2)$ [given by Eq. (9.5.21)]. Let us now impose an additional assumption: $f(x) \in C^4[0, l]$. We can then avoid a direct calculation of $f''(x)$ by writing

$$f''(x_i) = \frac{f(x_{i+1}) + f(x_{i-1}) - 2f(x_i)}{h^2} - \frac{h^2}{12}f^{(4)}(\bar{\xi}_i), \quad x_{i-1} < \bar{\xi}_i < x_{i+1}$$

and substituting in Eq. (9.5.23). The final expression for $u_{i,1}$ is

$$u_{i,1} = (1 - a^2)f(x_i) + \frac{a^2}{2}\left[f(x_{i-1}) + f(x_{i+1})\right] + kg(x_i) \tag{9.5.24}$$

and together with Eq. (9.5.15) can be used to initialize the computation of $u^{(j)}$, $2 \leq j \leq N$.

The numerical procedure given by Eq. (9.5.12) and initialized by Eqs. (9.5.15) and (9.5.24) is our next algorithm.

Algorithm 9.5.1.

(Wave equation—finite difference algorithm). This algorithm approximates the solution to

$$\frac{\partial^2 u}{\partial t^2} = c^2\frac{\partial^2 u}{\partial x^2}, \quad 0 < x < l, \quad 0 < t < T$$

given the boundary conditions

$$u(0, t) = u(l, t) = 0, \quad 0 \leq t \leq T$$

and the initial conditions

$$u(x, 0) = f(x), \quad \frac{\partial u}{\partial t}(x, 0) = g(x), \quad 0 \leq x \leq l$$

Input: End point l; maximum time T; equation constant c; integers n and N, which determine the discretization in x and t, respectively.

Output: Approximations $u_{i,j}$ to the exact solution at (x_i, t_j), $1 \leq i \leq n - 1$, $1 \leq j \leq N$.

Step 1. Set $h = l/n$, $k = T/N$, $a = ck/h$, $b = a^2$, $d = 2(1 - a^2)$, $e = b/2$, $f = d/2$

Step 2. Set $u_{i,0} = f(x_i)$, $1 \leq i \leq n - 1$; $u_{0,0} = u_{n,0} = 0$.

Step 3. Set $u_{i,1} = fu_{i,0} + e(u_{i-1,0} + u_{i+1,0}) + kg(x_i)$, $1 \leq i \leq n - 1$.

Step 4. For $j = 2, 3, \ldots, N$ do Steps 5 through 7 [implementing Eq. (9.5.12)].

Step 5. Set $u_{1,j} = du_{1,j-1} + bu_{2,j-1} - u_{1,j-2}$.

Step 6. For $i = 2, 3, \ldots, n - 2$ set $u_{i,j} = b(u_{i-1,j-1} + u_{i+1,j-1}) + du_{i,j-1} - u_{i,j-2}$.

Step 7. Set $u_{n-1,j} = bu_{n-2,j-1} + du_{n-1,j-1} - u_{n-1,j-2}$.

Step 8. For $0 \leq j \leq N$ set $t = jk$ and for $0 \leq i \leq n$ set $x = ih$. Output $(x, t, u_{i,j})$ and stop.

Example 9.5.1.

Consider the wave equation

$$\frac{\partial^2 u}{\partial t^2} = \frac{\partial^2 u}{\partial x^2}, \quad 0 < x < 1, \quad 0 < t < 1$$

subject to homogeneous boundary conditions and to the initial conditions

$$u(x, 0) = \sin(\pi x), \quad 0 \leq x \leq 1$$
$$\frac{\partial u}{\partial t}(x, 0) = 0, \quad 0 \leq x \leq 1$$

The exact solution is given by

$$U(x, t) = \sin(\pi x) \cos(\pi t)$$

Table 9.5.1 provides a comparison between $U_{i,N}$ and $u_{i,N}$ for the particular mesh $h = 0.1, k = 0.05$.

□

Table 9.5.1. A Wave Equation–Finite Difference Scheme
with a "Proper" Choice of h, k

x_i	$U_{i,20}$	$u_{i,20}$	$|U_{i,20} - u_{i,20}|$
0.1	−0.30902	−0.30900	$1 \cdot 10^{-5}$
0.3	−0.80902	−0.80898	$4 \cdot 10^{-5}$
0.5	−1.00000	−0.99995	$5 \cdot 10^{-5}$
0.7	−0.80902	−0.80898	$4 \cdot 10^{-5}$
0.9	−0.30902	−0.30900	$1 \cdot 10^{-5}$

It can be shown that Algorithm 9.5.1 is stable for

$$a = \frac{ck}{h} \leq 1 \tag{9.5.25}$$

and that its rate of convergence is $O(h^2 + k^2)$. An improper choice of h and k is given in the next example.

Example 9.5.2.

The choice $h = 0.025$, $k = 0.1$ in the previous example produces Table 9.5.2, which clearly indicates instability. $\qquad\square$

Table 9.5.2. Solving the Wave Equation
with Improper Mesh Sizes

x_i	$U_{i,20}$	$u_{i,20}$
0.1	−0.30902	−4.38
0.3	−0.80902	−6.07
0.5	−1.00000	−2.35
0.7	−0.80902	−9.76
0.9	−0.30902	22.75

The stability condition given by Eq. (9.5.25) has an interesting feature. It can be shown that for $a = 1$, the local truncation error of Eq. (9.5.9) is identically zero. Thus the *only* sources of error are the truncation error in calculating $u_{i,1}$, using Eq. (9.5.24), and round-off errors. For $a > 1$, Algorithm 9.5.1 is already unstable even if $a \approx 1$.

The subroutine WAVE, which incorporates Algorithm 9.5.1, is found on the attached floppy disk.

PROBLEMS

1. Use Algorithm 9.5.1 to solve

$$\frac{\partial^2 u}{\partial t^2} = 3\frac{\partial^2 u}{\partial x^2}, \quad 0 < x < 1, \quad 0 < t < 2$$
$$u(0,t) = u(1,t) = 0, \quad 0 \le t \le 2$$
$$u(x,0) = x(1-x), \quad 0 \le x \le 1$$
$$\frac{\partial u}{\partial t}(x,0) = \sin(\pi x), \quad 0 \le x \le 1$$

Aproximate the exact solution at $t = 2$, $0 < x < 1$ using $h = 0.1$, $k = 0.05$.

2. Use Algorithm 9.5.1 to solve

$$\frac{\partial^2 u}{\partial t^2} = \frac{1}{4}\frac{\partial^2 u}{\partial x^2}, \quad 0 < x < 2, \quad 0 < t < 1$$
$$u(0,t) = u(1,t) = 0, \quad 0 \le t \le 1$$
$$u(x,0) = 0, \quad 0 \le x \le 2$$
$$\frac{\partial u}{\partial t}(x,0) = x^2(2-x), \quad 0 \le x \le 2$$

Approximate the exact solution at $t = 1$, $0 < x < 2$ using $h = 0.1$, $k = 0.1$.

3. Repeat the computations of problem 1, using Eq. (9.5.18) instead of Eq. (9.5.24) for initialization.

4. Consider the wave equation

$$\frac{\partial^2 u}{\partial t^2} = \frac{\partial^2 u}{\partial x^2}, \quad 0 < x < 1, \quad 0 < t < 1$$

subject to the boundary conditions

$$u(0, t) = u(1, t) = 0, \quad 0 \le t \le 1$$

and the initial conditions

$$u(x, 0) = \sin(\pi x), \quad \frac{\partial u}{\partial t}(x, 0) = 0, \quad 0 \le x \le 1$$

(a) Approximate the exact solution at $t = 1$, $0 \le x \le 1$, using Algorithm 9.5.1 with $h = 0.25$ and $k = 0.05$.

(b) Repeat part (a), using $h = 0.1$ and $k = 0.02$.

(c) Repeat parts (a) and (b), using Eq. (9.5.18) instead of Eq. (9.5.24) for initialization.

5. Consider the problem of a vibrating elastic string with fixed end points, given $c = 1$, $l = 1$, and the initial conditions

$$u(x, 0) = f(x) = \begin{cases} x & , 0 \le x < 1/2 \\ 1 - x, & 1/2 \le x \le 1 \end{cases}$$

$$\frac{\partial u}{\partial t}(x, 0) = g(x) = \begin{cases} x & , 0 \le x < 1/4 \\ 1/4 & , 1/4 \le x < 3/4 \\ 1 - x, & 3/4 \le x \le 1 \end{cases}$$

(a) Approximate the exact solution at $t = 1$, using Algorithm 9.5.1 with $h = 0.1$ and $k = 0.05$.

(b) Compare your approximation with the Fourier series solution.

6. Show that the asymptotic local truncation error in Eq. (9.5.10) vanishes whenever

$$\frac{ck}{h} = 1$$

9.6   Finite Differences: Elliptic Partial Differential Equations

A popular elliptic partial differential equation that will be discussed in this section is the Poisson equation

$$\nabla^2 u = \frac{\partial^2 u}{\partial x^2} + \frac{\partial^2 u}{\partial y^2} = f(x, y), \quad (x, y) \in R \tag{9.6.1}$$

We will assume R to be a rectangle whose sides are parallel to the axes (Fig. 9.6.1), that is,

$$R = \{(x, y) \mid a < x < b, \ c < y < d\} \tag{9.6.2}$$

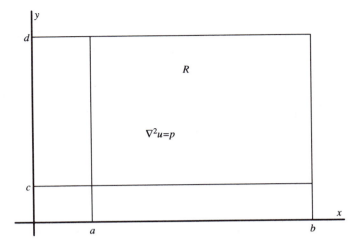

Figure 9.6.1. Solving the Poisson equation over a rectangular region.

Consider the boundary conditions

$$u(x, y) = g(x, y), \quad (x, y) \in S \tag{9.6.3}$$

where $g(x, y)$ is a continuous function defined over the boundary S of R. The boundary value problem defined by Eqs. (9.6.1) through (9.6.3) is known to possess a unique solution $U(x, y)$.

To approximate $U(x, y)$ we first discretize R and replace it by a grid of points (x_i, y_j) (Fig. 9.6.2), where

$$x_i = a + ih, \quad 0 \le i \le n$$
$$y_j = c + jk, \quad 0 \le j \le m$$

The mesh sizes are

$$h = \frac{b - a}{n}, \quad k = \frac{d - c}{m} \tag{9.6.4}$$

For each internal point (x_i, y_j), $1 \le i \le n - 1$, $1 \le j \le m - 1$ we obtain

$$\frac{\partial^2 u}{\partial x^2}(x_i, y_j) = \frac{u(x_{i+1}, y_j) + u(x_{i-1}, y_j) - 2u(x_i, y_j)}{h^2} - \frac{h^2}{12}\frac{\partial^4 u}{\partial x^4}(\xi_i, y_j) \tag{9.6.5}$$

$$\frac{\partial^2 u}{\partial y^2}(x_i, y_j) = \frac{u(x_i, y_{j+1}) + u(x_i, y_{j-1}) - 2u(x_i, y_j)}{k^2} - \frac{k^2}{12}\frac{\partial^4 u}{\partial y^4}(x_i, \eta_j) \tag{9.6.6}$$

where $x_{i-1} < \xi_i < x_{i+1}, y_{j-1} < \eta_j < y_{j+1}$. By adding Eqs. (9.6.5) and (9.6.6) and neglecting the local truncation error of $O(h^2 + k^2)$ one obtains [denoting $u_{i,j} = u(x_i, y_j), f_{i,j} = f(x_i, y_j)$]

$$\frac{u_{i+1,j} + u_{i-1,j} - 2u_{i,j}}{h^2} + \frac{u_{i,j+1} + u_{i,j-1} - 2u_{i,j}}{k^2} = f_{i,j} \tag{9.6.7}$$

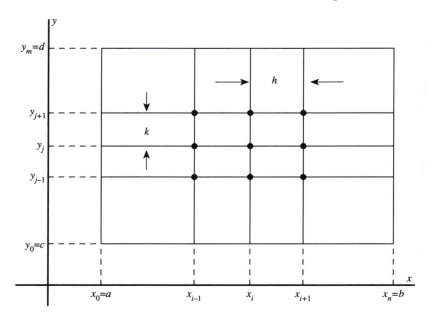

Figure 9.6.2. Discretization of the rectangle.

or

$$2\left(1 + \frac{h^2}{k^2}\right)u_{i,j} - (u_{i+1,j} + u_{i-1,j}) - \frac{h^2}{k^2}(u_{i,j+1} + u_{i,j-1}) = -h^2 f_{i,j} \tag{9.6.8}$$

for $1 \le i \le n-1$ and $1 \le j \le m-1$. This is a *central difference* approximating scheme to the Poisson equation at (x_i, y_j). We thus obtain a linear system of $(n-1)(m-1)$ equations with $(n+1)(m+1)-4$ unknowns. The missing $2n+2m-4$ equations are determined from the boundary conditions, that is,

$$
\begin{aligned}
u_{0,j} &= g_{0,j}, & 1 \le j \le m-1 \\
u_{n,j} &= g_{n,j}, & 1 \le j \le m-1 \\
u_{i,0} &= g_{i,0}, & 1 \le i \le n-1 \\
u_{i,m} &= g_{i,m}, & 1 \le i \le n-1
\end{aligned}
\tag{9.6.9}
$$

The four values at the four corners of R are irrelevant and do not participate in the computations. Each central difference equation [Eq. (9.6.8)] is a relation between the approximating values at five points—the internal point and its four immediate neighbors in the x, y directions (Fig. 9.6.3). Approximating the exact solution by solving Eqs. (9.6.8) and (9.6.9) is the *central finite difference* method. To solve its matrix form efficiently, it is suggested to label the mesh points (x_i, y_j) from left to right and from top to bottom, provided that $n \le m$. In other words, define

$$P_k = (x_i, y_j), \quad u_k = u_{i,j}, \quad f_k = f_{i,j}$$

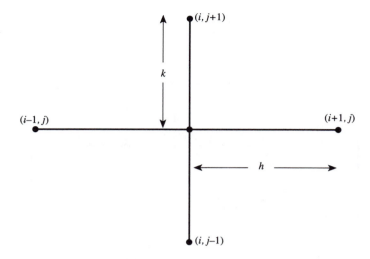

Figure 9.6.3.   The mesh points that participate in a central difference formula.

where

$$k = m - j + (i - 1)(m - 1), \quad 1 \le i \le n - 1, \quad 1 \le j \le m - 1$$

If $n > m$, label from top to bottom and from left to right.

Example 9.6.1.

For $n = 6$, $m = 4$ the labeling of the 15 mesh points is given in Fig. 9.6.4. ☐

By labeling the mesh points as in Fig. 9.6.4 one obtains a linear system whose matrix is a *banded* matrix with *bandwidth* $2m - 1$ at the most.

Example 9.6.2.

Consider the Poisson equation

$$\frac{\partial^2 u}{\partial x^2} + \frac{\partial^2 u}{\partial y^2} = 1, \quad 0 < x < 1, \quad 0 < y < 1$$

subject to the boundary conditions

$$u(x, y) = 1, \quad (x, y) \in S$$

and label the mesh points from left to right and from top to bottom. Let $n = m = 5$, that is, $h = k = 0.2$ (Fig. 9.6.5).

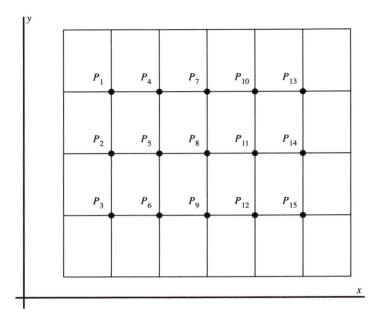

Figure 9.6.4. "Proper" labeling for efficient matrix calculations.

The linear system includes 16 equations with 16 unknowns given by

$$P_1 : \quad 4u_1 - u_2 - u_5 = u_{1,5} + u_{0,4} - h^2 f_1 = 1.96$$
$$P_2 : \quad 4u_2 - u_1 - u_3 - u_6 = u_{2,5} - h^2 f_2 = 0.96$$
$$\vdots$$
$$P_{16} : \quad 4u_{16} - u_{12} - u_{15} = u_{4,0} + u_{5,1} - h^2 f_{16} = 1.96$$

The right-hand side of each equation is determined by the boundary conditions and the function $f(x,y)$. The matrix form of the system is $Au = b$, where A is a 16×16 banded symmetric matrix with bandwidth 9. $\square$

The next algorithm applies the Gauss–Seidel iterative method for solving Eqs. (9.6.8) and (9.6.9).

Algorithm 9.6.1.

(Poisson equation—finite difference algorithm). This algorithm approximates the solution to

$$\frac{\partial^2 u}{\partial x^2} + \frac{\partial^2 u}{\partial y^2} = f(x,y), \quad a < x < b, \quad c < y < d$$

given the boundary conditions

$$u(x,y) = g(x,y), \quad (x,y) \in S \text{ (the boundary)}$$

Input: End points a, b, c, d; integers n and m, discretization parameters; tolerance ϵ; maximum number of iterations IT.

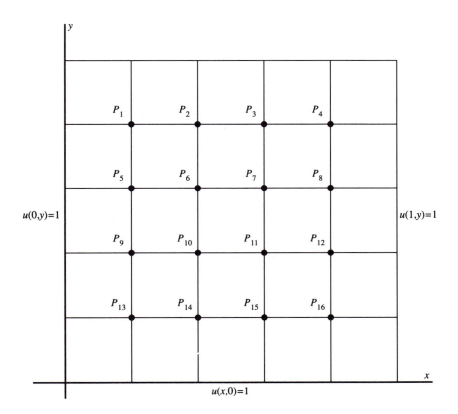

Figure 9.6.5. Labeling the mesh points for $\nabla^2 u = 1$, $u = 1$ on S, $n = m = 5$.

Output: Approximations $u_{i,j}$ to the exact solution at

$$(x_i, y_j), \quad 1 \le i \le n - 1, \quad 1 \le j \le m - 1$$

or a message that the maximum number of iterations IT was exceeded.

Step 1. Set $h = (b - a)/n, k = (d - c)/m, \alpha = h^2, \beta = h^2/k^2, \gamma = 2(1 + \beta), N = (n - 1)(m - 1)$.

Step 2. For $0 \le i \le n$ and $0 \le j \le m$ set $x_i = a + ih$ and $y_j = c + jk$, respectively.

Step 3. For $1 \le i \le n - 1, \ 1 \le j \le m - 1$ set $u_{i,j} = 0$ (initialization). Set $l = 1$; ERR $= 0$.

Step 4. For $l \le$ IT do Steps 5 through 17.

Steps 5–18 incorporate the Gauss–Seidel iterative method, assuming $n \le m$.

Step 5. Set

$$t = \frac{g_{0,m-1} + u_{2,m-1} + \beta(g_{1,m} + u_{1,m-2}) - \alpha f_{1,m-1}}{\gamma}$$
$$\text{ERR} = \text{ERR} + (t - u_{1,m-1})^2$$
$$u_{1,m-1} = t$$

Step 6. For $2 \le i \le n - 2$ set

$$t = \frac{u_{i-1,m-1} + u_{i+1,m-1} + \beta(g_{i,m} + u_{i,m-2}) - \alpha f_{i,m-1}}{\gamma}$$
$$\text{ERR} = \text{ERR} + (t - u_{i,m-1})^2$$
$$u_{i,m-1} = t$$

Step 7. Set

$$t = \frac{u_{n-2,m-1} + g_{n,m-1} + \beta(g_{n-1,m} + u_{n-1,m-2}) - \alpha f_{n-1,m-1}}{\gamma}$$
$$\text{ERR} = \text{ERR} + (t - u_{n-1,m-1})^2$$
$$u_{n-1,m-1} = t$$

Step 8. For $j = m - 2, m - 3, \ldots, 2$ do Steps 9 through 11.

Step 9. Set

$$t = \frac{g_{0,j} + u_{2,j} + \beta(u_{1,j+1} + u_{1,j-1}) - \alpha f_{1,j}}{\gamma}$$
$$\text{ERR} = \text{ERR} + (t - u_{1,j})^2$$
$$u_{1,j} = t$$

Step 10. For $2 \le i \le n - 2$ set

$$t = \frac{u_{i-1,j} + u_{i+1,j} + \beta(u_{i,j+1} + u_{i,j-1}) - \alpha f_{i,j}}{\gamma}$$

$$\text{ERR} = \text{ERR} + (t - u_{i,j})^2$$

$$u_{i,j} = t$$

Step 11. Set

$$t = \frac{u_{n-2,j} + g_{n,j} + \beta(u_{n-1,j+1} + u_{n-1,j-1}) - \alpha f_{n-1,j}}{\gamma}$$

$$\text{ERR} = \text{ERR} + (t - u_{n-1,j})^2$$

$$u_{n-1,j} = t$$

Step 12. Set

$$t = \frac{g_{0,1} + u_{2,1} + \beta(u_{1,2} + g_{1,0}) - \alpha f_{1,1}}{\gamma}$$

$$\text{ERR} = \text{ERR} + (t - u_{1,1})^2$$

$$u_{1,1} = t$$

Step 13. For $2 \leq i \leq n - 2$ set

$$t = \frac{u_{i-1,1} + u_{i+1,1} + \beta(u_{i,2} + g_{i,0}) - \alpha f_{i,1}}{\gamma}$$

$$\text{ERR} = \text{ERR} + (t - u_{i,1})^2$$

$$u_{i,1} = t$$

Step 14. Set **Step 14.** Set

$$t = \frac{u_{n-2,1} + g_{n,1} + \beta(u_{n-1,2} + g_{n-1,0}) - \alpha f_{n-1,1}}{\gamma}$$

$$\text{ERR} = \text{ERR} + (t - u_{n-1,1})^2$$

$$u_{n-1,1} = t$$

Step 15. If $\sqrt{\text{ERR}/N} < \epsilon$ do Step 16. Otherwise go to Step 17.

Step 16. For $1 \leq i \leq n - 1$, $1 \leq j \leq m - 1$, output $(x_i, y_j, u_{i,j})$ and stop.

Step 17. Set $l = l + 1$; ERR = 0. If $l < IT$ go to Step 4. Otherwise output "maximum number of iterations exceeded" and stop.

For large systems [i.e., large $N = (n-1)(m-1)$] the SOR method (Section 7.4) is recommended. Let the matrix A of the system be decomposed as

$$A = D + L + U$$

where D, L, and U are diagonal, lower triangular, and upper triangular matrices, respectively. It can be shown that the spectral radius of the Jacobi matrix

$$M_J = -D^{-1}(L + U)$$

is

$$\rho(M_J) = \frac{1}{2}\left[\cos\left(\frac{\pi}{n}\right) + \cos\left(\frac{\pi}{m}\right)\right] \tag{9.6.10}$$

The optimal choice for the relaxation factor w is therefore [Eq. (7.4.30)]

$$w = \frac{2}{1 + \sqrt{1 - \rho(M_J)^2}} = \frac{4}{2 + \sqrt{4 - [\cos(\pi/n) + \cos(\pi/m)]^2}} \tag{9.6.11}$$

Example 9.6.3.

For $n = m = 10$, the optimal choice of the relaxation factor is

$$w = \frac{4}{2 + \sqrt{4 - 4\cos^2(\pi/10)}} = \frac{2}{1 + \sin(\pi/10)} \approx 1.528 \qquad \square$$

Example 9.6.4.

Consider the Poisson equation

$$\frac{\partial^2 u}{\partial x^2} + \frac{\partial^2 u}{\partial y^2} = 2x, \quad 0 < x < 1, \quad 0 < y < 1$$

subject to the boundary conditions

$$u(x, 0) = e^x, \quad 0 \le x \le 1$$
$$u(x, 1) = e^x \cos(1) + x, \quad 0 \le x \le 1$$
$$u(0, y) = \cos y, \quad 0 \le y \le 1$$
$$u(1, y) = e \cos y + y^2, \quad 0 \le y \le 1$$

The exact solution is $U(x, y) = e^x \cos y + xy^2$. Let us discretize the region, using $n = m = 4$. Thus $h = k = 0.25$. The tolerance is taken as $\epsilon = 10^{-6}$ and the maximum number of iterations is IT $= 100$. Table 9.6.1 contains the approximations $u_{i,j}$, $1 \le i, j \le 3$ compared to $U_{i,j}$. The number of iterations needed for convergence is 22. $\qquad \square$

Table 9.6.1. Using Algorithm 9.6.1: Comparison of $U_{i,j}$ and $u_{i,j}$

i	j	$u_{i,j}$	$U_{i,j}$	$|U_{ij} - u_{ij}|$
1	1	1.2603	1.2597	0.0006
2	1	1.6296	1.6287	0.0009
3	1	2.0989	2.0981	0.0008
1	2	1.1901	1.1893	0.0008
2	2	1.5730	1.5719	0.0011
3	2	2.0463	2.0453	0.0010
1	3	1.0807	1.0801	0.0006
2	3	1.4884	1.4876	0.0008
3	3	1.9715	1.9709	0.0006

Example 9.6.5.

If we apply the SOR method in the previous example, the optimal choice for w is $w_{opt} = 2/[1 + \sin(\pi/4)] \approx 1.172$ and the number of iterations needed for convergence is 12. Any other w increases this number. For example, convergence for $w = 1.7$ occurs only after 40 iterations. □

The subroutine SOR, which solves the Poisson equation by using the successive overrelaxation method, is found on the attached floppy disk.

PROBLEMS

1. Use Algorithm 9.6.1 to solve

$$\frac{\partial^2 u}{\partial x^2} + \frac{\partial^2 u}{\partial y^2} = xy, \quad 0 < x < 2, \quad 0 < y < 1$$

subject to the boundary conditions

$$u(x,0) = x(2 - x), \quad 0 \le x \le 2$$
$$u(0,y) = u(2,y) = 0, \quad 0 \le y \le 1$$
$$u(x,1) = \sin(\pi x), \quad 0 \le x \le 2$$

Use $h = 0.2$, $k = 0.1$, $\epsilon = 10^{-6}$, IT = 100.

2. Solve problem 1 for $\epsilon = 10^{-5}$, 10^{-6}, and 10^{-7} and for each case obtain the number of iterations needed for convergence.

3. Use Algorithm 9.6.1 to solve

$$\frac{\partial^2 u}{\partial x^2} + \frac{\partial^2 u}{\partial y^2} = 0, \quad 0 < x < 1, \quad 0 < y < 1$$

subject to the boundary conditions

$$u(x,0) = x^2, \quad 0 \le x \le 1$$

$$u(x, 1) = x^2 - 1, \quad 0 \le x \le 1$$
$$u(0, y) = -y^2, \quad 0 \le y \le 1$$
$$u(1, y) = 1 - y^2, \quad 0 \le y \le 1$$

Use $h = 0.2$, $k = 0.2$, $\epsilon = 10^{-6}$, IT = 100 and compare the approximate solution with the exact solution $U(x, y) = x^2 - y^2$. Discuss the results.

4. Solve the partial differential equation

$$\frac{\partial^2 u}{\partial x^2} + \frac{\partial^2 u}{\partial y^2} = xy - y^2, \quad 0 < x < 2, \quad 0 < y < 1$$

subject to the boundary conditions

$$u(x, 0) = 2x, \quad 0 \le x \le 2$$
$$u(x, 1) = 0, \quad 0 \le x \le 2$$
$$u(0, y) = y(1 - y), \quad 0 \le y \le 1$$
$$u(2, y) = 4(1 - y), \quad 0 \le y \le 1$$

(a) Use Algorithm 9.6.1 with $h = 0.2$, $k = 0.1$, $\epsilon = 10^{-6}$, IT = 100.

(b) Use the SOR method, with optimal w, for the same mesh sizes.

(c) Use the SOR method with $w = 0.6 + 0.2i$, $0 \le i \le 6$ for the mesh sizes of part (a), and for each case obtain the number of iterations needed for convergence (use $\epsilon = 10^{-6}$).

5. Change Algorithm 9.6.1 to treat the Poisson equation with the following boundary conditions:

$$u(x, y) = g(x, y), \quad (x, y) \in S - \{(x, c) \mid a \le x \le b\}$$
$$\frac{\partial u}{\partial y}(x, c) = r(x), \quad a \le x \le b$$

For $(\partial u / \partial y)(x, c)$ use a finite difference formula with a local truncation error $O(k^2)$.

6. Use the algorithm of problem 5 to solve

$$\frac{\partial^2 u}{\partial x^2} + \frac{\partial^2 u}{\partial y^2} = 0, \quad 0 < x < 1, \quad 0 < y < 1$$

subject to the boundary conditions

$$u(x, 1) = e^x \cos(1), \quad 0 \le x \le 1$$
$$u(0, y) = \cos y, \quad 0 \le y \le 1$$
$$u(1, y) = e \cos y, \quad 0 \le y \le 1$$
$$\frac{\partial u}{\partial y}(x, 0) = 0, \quad 0 \le x \le 1$$

Use $h = k = 0.1$, $\epsilon = 10^{-6}$ and compare the computed and exact solutions $[U(x, y) = e^x \cos y]$.

7. Solve problem 6 using the SOR method, with optimal w, and compare the number of iterations needed for convergence ($\epsilon = 10^{-6}$) to that of problem 6.

9.7 Finite Elements

A method often used for solving partial differential equations and that has become popular in recent years is the *finite element method*. It provides the solution to a partial differential equation with boundary conditions as a function that minimizes an associated integral, the *functional*.

The main advantage of the finite element method over finite difference schemes is the way boundaries and boundary conditions are handled. Many problems in physics and engineering are defined over regions with irregularly shaped boundaries. In addition, the boundary conditions quite frequently involve derivatives. Finite difference schemes for treating such boundary conditions over complex boundaries are difficult to design. If a finite element procedure is used, the boundary conditions that involve derivatives are included as additional integrals to the minimized functional.

We start by introducing the Rayleigh–Ritz method, which is a one-dimensional finite element scheme.

9.7.1 Rayleigh–Ritz Method

Consider the linear differential equation

$$-\frac{d}{dx}\left[p(x)\frac{dy}{dx}\right] + q(x)y = r(x), \quad 0 \le x \le 1 \tag{9.7.1}$$

with homogeneous boundary conditions

$$y(0) = y(1) = 0 \tag{9.7.2}$$

This type of equation is called *self-adjoint* and, if some requirements are fulfilled, it will possess a unique solution that satisfies Eq. (9.7.2).

Theorem 9.7.1.
(Existence and uniqueness). *Let $p(x)$, $q(x)$, and $r(x)$ satisfy the following:*

1. $p(x) \in C^1[0, 1]$ *and* $q(x), r(x) \in C[0, 1]$.

2. *There exists a constant* $\rho > 0$ *such that*

$$0 < \rho \le p(x), \quad 0 \le x \le 1 \tag{9.7.3}$$

3. $q(x)$ *satisfies*

$$0 \le q(x), \quad 0 \le x \le 1 \tag{9.7.4}$$

Then Eqs. (9.7.1) and (9.7.2) have a unique solution.

Example 9.7.1.

The partial differential equation

$$-\frac{d}{dx}\left[(1+x^2)\frac{dy}{dx}\right] + |x|y = 1$$

satisfies the three conditions of Theorem 9.7.1 and thus has a unique solution with homogeneous boundary conditions.   □

The next result identifies the solution to Eqs. (9.7.1) and (9.7.2) as the function that minimizes an associated integral.

Theorem 9.7.2.

Let $p(x)$, $q(x)$, and $r(x)$ satisfy all the requirements of Theorem 9.7.1, and let $y(x)$ minimize the integral

$$F[f] = \int_0^1 \{p(x)[f'(x)]^2 + q(x)[f(x)]^2 - 2r(x)f(x)\}\, dx \qquad (9.7.5)$$

over

$$S = \{f(x) \mid f(x) \in C^2[0,1],\ f(0) = f(1) = 0\} \qquad (9.7.6)$$

Then $y(x)$ is the unique solution to Eqs. (9.7.1) and (9.7.2).

Note that each member of S satisfies the boundary conditions *a priori*. The integral of Eq. (9.7.5) is called the *functional* associated with Eqs. (9.7.1) and (9.7.2).

The opposite of Theorem 9.7.2 is also true, that is, a solution to the system given by Eqs. (9.7.1) and (9.7.2) minimizes the associated functional.

Example 9.7.2.

The unique solution to the problem

$$-\frac{d}{dx}\left[(2-x^2)\frac{dy}{dx}\right] + xy = x^2, \quad 0 \le x \le 1, \quad y(0) = y(1) = 0$$

minimizes the functional

$$F[f] = \int_0^1 [(2-x^2)f'^2 + xf^2 - 2x^2 f]\, dx \qquad \square$$

The *Rayleigh–Ritz* procedure, instead of minimizing F over S, performs a limited search over span $(f_1, \dots, f_n)$ where f_i, $1 \le i \le n$ are some given linearly independent functions in S called the *basis functions*, which satisfy

$$f_i(0) = f_i(1) = 0, \quad 1 \le i \le n$$

The search determines coefficients $c_1, \ldots, c_n$, which minimize $F[c_1 f_1 + c_2 f_2 + \cdots + c_n f_n]$ and the function

$$f(x) = \sum_{i=1}^{n} c_i f_i(x), \quad 0 \le x \le 1 \tag{9.7.7}$$

is called the *Rayleigh–Ritz approximation*, based on $f_1, \ldots, f_n$ to the exact solution $y(x)$. To compute it we first substitute $f(x)$ of Eq. (9.7.7) in Eq. (9.7.5) and obtain

$$F[f] = F\left[\sum_{i=1}^{n} c_i f_i\right]$$

$$= \int_0^1 \left\{ p(x) \left[\sum_{i=1}^{n} c_i f_i'(x)\right]^2 + q(x) \left[\sum_{i=1}^{n} c_i f_i(x)\right]^2 - 2r(x) \sum_{i=1}^{n} c_i f_i(x) \right\} dx \tag{9.7.8}$$

The right-hand side of Eq. (9.7.8) is a function of c_i, $1 \le i \le n$ and, for $f(x)$ to minimize F, it is necessary that

$$\frac{\partial F}{\partial c_j} = 0, \quad 1 \le j \le n \tag{9.7.9}$$

or

$$\sum_{i=1}^{n} \left\{ \int_0^1 \left[p(x) f_i'(x) f_j'(x) + q(x) f_i(x) f_j(x) \right] dx \right\} c_i = \int_0^1 r(x) f_j(x) \, dx \tag{9.7.10}$$

for $j = 1, \ldots, n$. We thus obtain a set of linear equations with the variables $c_1, \ldots, c_n$. The matrix form of Eq. (9.7.10) is $Ac = b$, and the elements a_{ij} and b_i, $1 \le i, j \le n$ of A and b are given by

$$a_{ij} = \int_0^1 \left[p(x) f_j'(x) f_i'(x) + q(x) f_j(x) f_i(x) \right] dx \tag{9.7.11}$$

$$b_i = \int_0^1 r(x) f_i(x) \, dx \tag{9.7.12}$$

Equation (9.7.11) implies that A is symmetric.

Example 9.7.3.

Consider the boundary value problem

$$-y'' + y = -x^2 + x + 2, \quad y(0) = y(1) = 0$$

and a single function $f_1(x) = \sin(\pi x)$ that clearly satisfies $f_1(0) = f_1(1) = 0$. Here, $p(x) = q(x) = 1$ and $r(x) = -x^2 + x + 2$. Therefore

$$a_{11} = \int_0^1 \left[\pi^2 \cos^2(\pi x) + \sin^2(\pi x) \right] dx = \frac{\pi^2 + 1}{2}$$

$$b_1 = \int_0^1 (-x^2 + x + 2) \sin(\pi x)\, dx = \frac{4(\pi^2 + 1)}{\pi^3}$$

$$c_1 = b_1/a_{11} = 8/\pi^3$$

The exact solution $y(x) = x(1-x)$ is thus approximated by

$$f(x) = c_1 f_1(x) = \frac{8}{\pi^3} \sin(\pi x)$$

A comparison between $y(x)$ and $f(x)$ is given in Table 9.7.1. Due to symmetry we consider only $0 \le x \le 0.5$.

Table 9.7.1. Applying the
Raleigh–Ritz Procedure with $n = 1$

x	$y(x)$	$f(x)$
0.1	0.09	0.080
0.2	0.16	0.152
0.3	0.21	0.209
0.4	0.24	0.245
0.5	0.25	0.258

In this particular example, a single-term approximation is already considerably close to $y(x)$.   □

It can be shown that Theorem 9.7.2 also holds if S is replaced by S_p, the set of all functions $f(x)$ that satisfy

$$f(x) \in C[0, 1], \quad f(0) = f(1) = 0$$

and possess two *piecewise continuous* derivatives. Let us construct a Rayleigh–Ritz scheme generated by basis functions that are *piecewise linear* polynomials.

Consider a partition of the interval $[0, 1]$ by x_i, $0 \le i \le n + 1$ that satisfy

$$0 = x_0 < x_1 < \cdots < x_n < x_{n+1} = 1$$

and denote $h_i = x_{i+1} - x_i$, $0 \le i \le n$. To this partition we associate the functions $f_i(x)$, $1 \le i \le n$ defined as

$$f_i(x) = \begin{cases} 0 & , \ 0 \le x \le x_{i-1} \\ (x - x_{i-1})/h_{i-1}, & x_{i-1} < x \le x_i \\ (x_{i+1} - x)/h_i & , \ x_i < x \le x_{i+1} \\ 0 & , \ x_{i+1} < x \le 1 \end{cases} \tag{9.7.13}$$

and shown in Fig. 9.7.1.

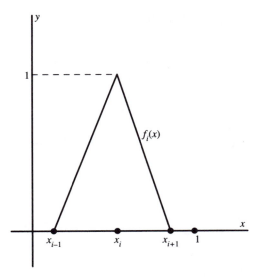

Figure 9.7.1. A piecewise linear basis function.

The derivative of each $f_i(x)$ is well defined everywhere over the interval $[0, 1]$ except for $x = x_{i-1}, x_i, x_{i+1}$, and is given by

$$f_i'(x) = \begin{cases} 0 & , \ 0 \le x < x_{i-1} \\ 1/h_{i-1}, \ x_{i-1} < x < x_i \\ -1/h_i \ , \ x_i < x < x_{i+1} \\ 0 & , \ x_{i+1} < x \le 1 \end{cases} \qquad (9.7.14)$$

It is easily seen that $f_i(x)$, $1 \le i \le n$ form a basis, that is, they are linearly independent. Also, because $f_i(x)$ and $f_i'(x)$ vanish outside the interval $[x_{i-1}, x_{i+1}]$, we have

$$f_i(x)f_j(x) = f_i'(x)f_j'(x) = 0, \quad 0 \le x \le 1$$

for all i, j that satisfy $|i - j| \ge 2$. The matrix A is thus tridiagonal and therefore easily manipulated. We now proceed and calculate the nonzero entries of A, which are a_{ii}, $1 \le i \le n$; $a_{i,i+1}$, $1 \le i \le n - 1$; $a_{i,i-1}$, $2 \le i \le n$; and b_i, $1 \le i \le n$. The diagonal entries are

$$\begin{aligned} a_{ii} &= \int_0^1 \{p(x)[f_i'(x)]^2 + q(x)[f_i(x)]^2\} \, dx \\ &= \int_{x_{i-1}}^{x_i} \left(\frac{1}{h_{i-1}}\right)^2 p(x) \, dx + \int_{x_i}^{x_{i+1}} \left(-\frac{1}{h_i}\right)^2 p(x) \, dx \\ &\quad + \int_{x_{i-1}}^{x_i} \left(\frac{x - x_{i-1}}{h_{i-1}}\right)^2 q(x) \, dx + \int_{x_i}^{x_{i+1}} \left(\frac{x_{i+1} - x}{h_i}\right)^2 q(x) \, dx, \quad 1 \le i \le n \quad (9.7.15) \end{aligned}$$

and the elements above this diagonal are given by

$$
\begin{aligned}
a_{i,i+1} &= \int_0^1 [p(x)f_i'(x)f_{i+1}'(x) + q(x)f_i(x)f_{i+1}(x)]\, dx \\
&= -\int_{x_i}^{x_{i+1}} \left(\frac{1}{h_i}\right)^2 p(x)\, dx + \int_{x_i}^{x_{i+1}} \left(\frac{1}{h_i}\right)^2 (x - x_i)(x_{i+1} - x)q(x)\, dx \\
&= \left(\frac{1}{h_i}\right)^2 \int_{x_i}^{x_{i+1}} [(x - x_i)(x_{i+1} - x)q(x) - p(x)]\, dx, \quad 1 \le i \le n-1
\end{aligned}
\tag{9.7.16}
$$

Due to symmetry $a_{i,i-1} = a_{i-1,i}$, $2 \le i \le n$ and finally

$$
\begin{aligned}
b_i &= \int_0^1 r(x)f_i(x)\, dx \\
&= \int_{x_{i-1}}^{x_i} \left(\frac{x - x_{i-1}}{h_{i-1}}\right) r(x)\, dx + \int_{x_i}^{x_{i+1}} \left(\frac{x_{i+1} - x}{h_i}\right) r(x)\, dx, \quad 1 \le i \le n
\end{aligned}
\tag{9.7.17}
$$

Example 9.7.4.

Let us consider the boundary value problem from the previous example and generate a three-element basis associated with the partition $x_i = 0.25i$, $0 \le i \le 4$. Here

$$
h_i = h = 0.25, \quad 1 \le i \le 4; \quad p(x) = q(x) = 1, \quad r(x) = -x^2 + x + 2
$$

By substituting in Eqs. (9.7.15) through (9.7.17) we obtain

$$
\begin{aligned}
a_{ii} &= 8.16667, \quad 1 \le i \le 3; \quad a_{i,i+1} = a_{i+1,i} = -3.95833, \quad 1 \le i \le 2 \\
b_1 &= b_3 = 0.544271, \quad b_2 = 0.559896
\end{aligned}
$$

and, consequently, the 3×3 linear system

$$
\begin{bmatrix}
8.16667 & -3.95833 & 0 \\
-3.95833 & 8.16667 & -3.95833 \\
0 & -3.95833 & 8.16667
\end{bmatrix}
\begin{bmatrix}
c_1 \\
c_2 \\
c_3
\end{bmatrix}
=
\begin{bmatrix}
0.544271 \\
0.559896 \\
0.544271
\end{bmatrix}
$$

whose unique solution is

$$
\begin{bmatrix}
c_1 \\
c_2 \\
c_3
\end{bmatrix}
=
\begin{bmatrix}
0.188393 \\
0.251185 \\
0.188393
\end{bmatrix}
$$

A comparison between the exact solution $y(x) = x(1 - x)$ and $f(x) = \sum_{i=1}^3 c_i f_i(x)$ is given in Table 9.7.2.

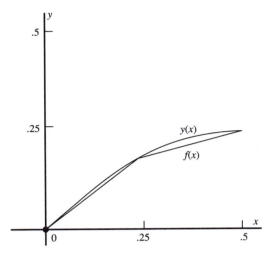

Figure 9.7.2. A piecewise linear Rayleigh–Ritz approximation vs. the exact solution.

Table 9.7.2. A Piecewise Linear
Rayleigh–Ritz Approximation to $x(1 - x)$

x	$y(x)$	$f(x)$
0.10	0.0900	0.0754
0.20	0.1600	0.1507
0.25	0.1875	0.1884
0.30	0.2100	0.2010
0.40	0.2400	0.2261
0.50	0.2500	0.2512

Note that $f(x)$ is most accurate at the nodes x_i. This is also shown in Fig. 9.7.2.   □

A subroutine RRPL (Rayleigh–Ritz piecewise linear), based on the Rayleigh–Ritz method, is found on the attached floppy disk.

9.7.2 Extension to Two-Dimensional Problems

The Rayleigh–Ritz procedure is designed for one-dimensional boundary value problems represented by self-adjoint equations. Its extension to two- and three-dimensional problems leads to the finite element method. In our discussion we will consider a general two-dimensional self-adjoint elliptic PDE with boundary conditions.

Definition 9.7.1. *The two-dimensional partial differential equation.*

$$L\phi \equiv \frac{\partial}{\partial x}\left(a\frac{\partial \phi}{\partial x} + b\frac{\partial \phi}{\partial y}\right) + \frac{\partial}{\partial y}\left(b\frac{\partial \phi}{\partial x} + c\frac{\partial \phi}{\partial y}\right) - d\phi = -p \qquad (9.7.18)$$

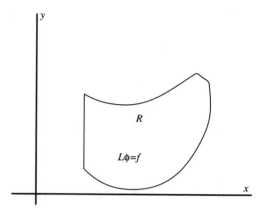

Figure 9.7.3. A general region where $L\phi = f$.

defined over a region R (Fig. 9.7.3) with a boundary B, where $a, b, c \in C^1(R)$ and $d, p \in C(R)$, is called self-adjoint.

If in addition

$$ac - b^2 > 0, \quad (x, y) \in R \tag{9.7.19}$$

the equation is called *elliptic* over R.

Example 9.7.5.

The Poisson equation

$$\nabla^2 \phi = -p(x, y)$$

is self-adjoint and elliptic over any region R. So is the equation

$$\frac{\partial}{\partial x}\left[(1 + x^2)\frac{\partial \phi}{\partial x} + xy\frac{\partial \phi}{\partial y}\right] + \frac{\partial}{\partial y}\left[xy\frac{\partial \phi}{\partial x} + (1 + y^2)\frac{\partial \phi}{\partial y}\right] + 2\phi = x^2 + y^2$$

because $a = 1 + x^2$, $b = xy$, $c = 1 + y^2$, and

$$ac - b^2 = (1 + x^2)(1 + y^2) - (xy)^2 = 1 + x^2 + y^2 \geq 1 > 0$$

On the other hand, the PDE

$$\frac{\partial}{\partial x}\left(x\frac{\partial \phi}{\partial x}\right) + \frac{\partial^2 \phi}{\partial y^2} = 1$$

has $a = x$, $b = 0$, $c = 1$. It is self-adjoint but not necessarily elliptic. It is elliptic if and only if $x > 0$. ☐

The requirements for a unique solution are specified next.

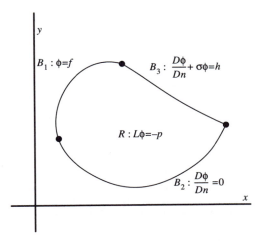

Figure 9.7.4. A general boundary value problem.

For Eq. (9.7.18) to possess a unique solution, boundary conditions (BCs) must be imposed on ϕ. Three types of BCs are usually considered (Fig. 9.7.4):

1. Dirichlet BC : The solution ϕ is specified over $B_1 \subset B$, that is,

$$\phi(x,y) = f(x,y), \quad (x,y) \in B_1 \tag{9.7.20}$$

2. Homogeneous Neumann BC: The "normal" derivative of ϕ vanishes over $B_2 \subset B$, that is,

$$\frac{D\phi}{Dn} \equiv (\phi_x, \phi_y) \begin{bmatrix} a & b \\ b & c \end{bmatrix} \begin{bmatrix} n_x \\ n_y \end{bmatrix} = 0, \quad (x,y) \in B_2 \tag{9.7.21}$$

where $\phi_x \equiv \partial\phi/\partial x$, $\phi_y \equiv \partial\phi/\partial y$, and n_x, n_y denote the x and y components of the outward normal vector to the boundary. Note that for $a = c = 1$ and $b = 0$ (Poisson equation), Eq. (9.7.21) has the familiar form $\partial\phi/\partial n = 0$.

3. Mixed BC: A linear combination of ϕ and $D\phi/Dn$ is specified over $B_3 \subset B$, that is,

$$\frac{D\phi}{Dn} + \sigma(x,y)\phi = h(x,y), \quad (x,y) \in B_3 \tag{9.7.22}$$

The requirements for the existence of a unique solution to a boundary value problem are given next.

Theorem 9.7.3.

(A uniqueness theorem). *Let* $a(x,y)$, $b(x,y)$, $c(x,y) \in C^1(R)$, $d(x,y)$, $p(x,y) \in C(R), f(x,y) \in C(B_1), \sigma(x,y), h(x,y) \in C(B_3)$ *and let* $ac - b^2 > 0$, $d \le 0$, $(x,y) \in R$. *Then, Eq. (9.7.18), subject to the BCs of Eqs. (9.7.21) through (9.7.23) with* $\sigma(x,y) > 0$, *yields a unique solution.*

Example 9.7.6.

Consider the Poisson equation

$$\frac{\partial^2 \phi}{\partial x^2} + \frac{\partial^2 \phi}{\partial y^2} = 1$$

defined at the interior of the unit circle, that is,

$$R = \{(x, y) \mid x^2 + y^2 < 1\}$$

and subject to the following boundary conditions :

$$\phi(x, y) = 1, \quad x^2 + y^2 = 1, \quad y > 0$$
$$\frac{\partial \phi}{\partial n}(x, y) = 0, \quad x^2 + y^2 = 1, \quad y \leq 0$$

This is a boundary value problem with two types of boundary conditions.   ☐

9.7.3   The Functional

Define the functional

$$F = \iint_R \left[a \left(\frac{\partial \phi}{\partial x} \right)^2 + 2b \frac{\partial \phi}{\partial x} \frac{\partial \phi}{\partial y} + c \left(\frac{\partial \phi}{\partial y} \right)^2 + d\phi^2 - 2p\phi \right] dxdy + \int_{B_3} (\sigma\phi^2 - 2h\phi) ds_3 \quad (9.7.23)$$

over

$$S = \{\phi \mid \phi \in C^2(R); \quad \phi(x, y) = f(x, y), \quad (x, y) \in B_1\} \tag{9.7.24}$$

that is, each ϕ satisfies *a priori* the Dirichlet BC. The following result is the extension of Theorem 9.7.2.

Theorem 9.7.4.

The functional F defined by Eq. (9.7.23) is uniquely minimized over S by the unique solution to the boundary value problem defined by Eqs. (9.7.18) through (9.7.22).

By virtue of Theorem 9.7.4, the homogeneous Neumann BCs are *natural boundary conditions*. They contribute no term to the functional or reduce the set S. Nevertheless they are automatically satisfied by ϕ_0, which minimizes F over S, that is,

$$\frac{D\phi_0}{Dn} = 0, \quad (x, y) \in B_2$$

To approximate ϕ_0 it is not necessary to evaluate $F(\phi)$ for every $\phi \in S$. Instead, we minimize F over a subset of S_p, the piecewise polynomials of degree n. Usually, n is chosen ≤ 4 because high-order piecewise polynomial approximations, like high-order interpolators, may produce ill-conditioned systems.

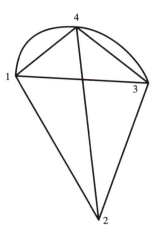

Figure 9.7.5. Curved elements.

The process of defining the set of piecewise polynomials, followed by the minimization of F, produces the *finite element method* whose basic features are given next.

9.7.4 The Finite Element Method

Implementation of a finite element scheme to minimize the functional F consists of two stages:

1. Triangulation: Dividing the region R to *finite elements* that are usually *regular* triangles.
2. Defining a subset S'_p of S_p such that each $\phi \in S'_p$ is a continuous function over R, a piecewise polynomial over each element, and satisfies the Dirichlet BC [Eq. (9.7.20)].
3. Minimizing of F by solving an associated system of linear equations.

9.7.5 Triangulation

Dividing the region R to triangles is called *triangulation* (mesh generation). A finite element mesh should satisfy, as much as possible, the following requirements:

1. Each element usually has three *straight* sides. If a side is curved (e.g., when the boundary B is curved), it should not produce large curvature (Fig. 9.7.5 illustrates an unacceptable curved element, which is therefore redivided).
2. Each element should be *regular*. By "regular" we mean a triangle, none of the angles of which is close to 0 or π.
3. Each pair of elements has either one common side, *adjacent elements*, one common vertex, or no common points.

Satisfying requirements 1 and 2 will generally increase the accuracy of the approximate solution. Requirement 3 is necessary for the generation of a simple and easily manipulated finite element algorithm.

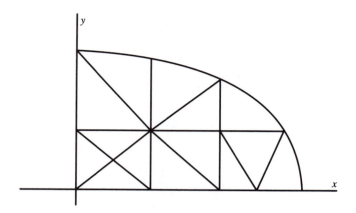

Figure 9.7.6.   A coarse finite element mesh.

If a low degree of accuracy is required a coarse finite element mesh is usually sufficient and the triangulation can be performed manually (Fig. 9.7.6). Otherwise, a fine mesh is necessary and an automatic mesh generator is needed. Such codes exist and use three input parameters: the region R, a standard side length d, and a specified density function $\rho(x, y)$, which determines the density of the elements at each point of the region (Fig. 9.7.7).

9.7.6   Shape Functions

The next step is defining a *trial function* and minimizing F over the set of all the trial functions. At this point we confine the discussion to approximations by piecewise *linear* polynomials (as in the Rayleigh–Ritz scheme). Let $P_1, P_2, \ldots, P_N$ denote all the vertices in the generated mesh. They are called the *node potentials* and the value of an arbitrary trial function at P_i is denoted by ϕ_i. Each trial function ϕ is a continuous function that satisfies the Dirichlet BC and over each element is a linear polynomial.

Let T denote an arbitrary triangle with vertices (for the sake of simplicity) P_1, P_2, P_3. Then the representation of ϕ over T *must* be

$$\phi = \sum_{i=1}^{3} \alpha_i(x, y)\phi_i \tag{9.7.25}$$

where the *shape functions* $\alpha_i(x, y)$ are linear polynomials that depend only on P_1, P_2, P_3 and satisfy

$$\alpha_i(x_j, y_j) = \delta_{ij}$$

Equation (9.7.25) determines ϕ as well as its derivatives

$$\frac{\partial \phi}{\partial x} = \sum_{i=1}^{3} \frac{\partial \alpha_i}{\partial x}\phi_i, \quad \frac{\partial \phi}{\partial y} = \sum_{i=1}^{3} \frac{\partial \alpha_i}{\partial y}\phi_i \tag{9.7.26}$$

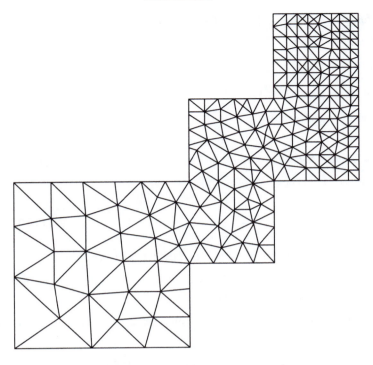

Figure 9.7.7. A fine inhomogeneous mesh.

The expressions given by Eqs. (9.7.25) and (9.7.26) are substituted in F, which becomes a bilinear form in $\{\phi_i\}_{i=1}^{N}$. The "Dirichlet nodes" are replaced by the specified boundary values and to minimize F we solve a system of linear equations

$$\frac{\partial F}{\partial \phi_{i_j}} = 0 \qquad (9.7.27)$$

where values of ϕ_{i_j} are all the remaining unknown node potentials.

Once the mesh generation is completed, all the shape functions are determined as well. Using the node potentials as the system variables guarantees the continuity of the solution along the common side of each two adjacent elements (why?). Thus the approximate solution is a *continuous* piecewise linear polynomial over R, given by Eq. (9.7.25) over an arbitrary element.

PROBLEMS

1. Write the following differential equations in the form of Eq. (9.7.1) and determine whether the conditions of Eqs. (9.7.3) and (9.7.4) hold.

   (a) $(1 + x^2)y'' + 2xy' - y\ln(2 + x) = 1,\ 0 \le x \le 1$

   (b) $e^x(y'' + y') + y - x = 0,\ 0 \le x \le 1$

   (c) $y'' + y' - y - \sin x = 0,\ 0 \le x \le 1$

2. Write the functional of Eq. (9.7.5) in the following cases:

   (a) $-y'' + y = 5x$

   (b) $y'' + y' - y - 1 = 0$

   (c) $(1 + \sin x)y'' + y' \cos x - |x| y = x$

3. Approximate the exact solution to
   $$-y'' + y = -x^2 + x + 2, \quad y(0) = y(1) = 0, \quad 0 \le x \le 1$$
   by using the Rayleigh–Ritz procedure and the basis $\{\sin(\pi x), \sin^2(\pi x)\}$

4. (a) Approximate the exact solution $[y(x) = x^3 - x]$ to

   $$-[(1 + x + x^2) y']' + 12y = -9x^2 - 16x + 1, \quad y(0) = y(1) = 0, \quad 0 \le x \le 1$$

   by using the Rayleigh–Ritz procedure and the basis of problem 3.

   (b) Repeat the computation using a basis of nine piecewise linear polynomials defined as in Eq. (9.7.13). Assume evenly spaced grid points.

5. Repeat part (b) of problem 4, using four piecewise linear polynomials and evenly spaced nodes

   $$x_i = 0.2i, \quad 0 \le i \le 5$$

6. Write the functional that is minimized by the solution to the following boundary value problem:

   $$\frac{\partial}{\partial x}\left[(1 + x^2)\frac{\partial \phi}{\partial x}\right] + \frac{\partial}{\partial y}\left(y^4 \frac{\partial \phi}{\partial y}\right) = 1, \quad (x, y) \in R$$

   where

   $$R = \{(x, y) \mid 0 < x < 2, \quad 1 < y < 2\}$$

   subject to the BCs

   $$\phi(x, 1) = \phi(x, 2) = 0, \quad 0 \le x \le 2$$
   $$\frac{D\phi}{Dn}(0, y) = 0, \quad 1 \le y \le 2$$
   $$\frac{D\phi}{Dn} + 2\phi = xy, \quad x = 2, \quad 1 \le y \le 2$$

7. The exact solution to

   $$\frac{\partial^2 \phi}{\partial x^2} + \frac{\partial^2 \phi}{\partial y^2} = 0, \quad 0 < x < 1, \quad 0 < y < \pi$$

   subject to the BCs

$$\frac{\partial \phi}{\partial n}(x,0) = \frac{\partial \phi}{\partial n}(x,\pi) = 0$$

$$\phi(0,y) = \cos y, \quad 0 < y < \pi$$

$$\phi(1,y) = e \cos y, \quad 0 < y < \pi$$

is $\phi_0 = e^x \cos y$. Let F denote the functional associated with this problem. Calculate $F(\phi_0)$ and compare with $F(\phi_1)$, where

$$\phi_1 = (1-x)\cos y + ex \cos y$$

(Note that ϕ_1 satisfies the Dirichlet BC)

8. For a given triangle express $\alpha_i(x,y), 1 \leq i \leq 3$ in terms of the vertices $(x_i, y_i), 1 \leq i \leq 3$.

Bibliography

1. Abramowitz, M. and Stegun, I. A. Eds., *Handbook of Mathematical Functions*, Dover Press, New York, 1964.

2. Ames, W. F., *Numerical Methods for Partial Differential Equations*, 2nd ed., Academic Press, New York, 1977.

3. Atkinson, K. E., *Elementary Numerical Analysis*, John Wiley & Sons, New York, 1985.

4. Atkinson, K. E., *An Introduction to Numerical Analysis*, John Wiley & Sons, New York, 1989.

5. Birkhoff, G. and Rota, G., *Ordinary Differential Equations*, John Wiley & Sons, New York, 1978.

6. Burden, R. L. and Faires, J. D., *Numerical Analysis*, 4th ed., PWS–KENT, New York, 1989.

7. Conte, S. and de Boor, C., *Elementary Numerical Analysis*, 3rd ed., McGraw-Hill, New York, 1980.

8. Davis, P. J. and Rabinowitz, P., *Methods of Numerical Integration*, 2nd ed., Academic Press, New York, 1975.

9. de Boor, C., *A Practical Guide to Splines*, Springer-Verlag, New York, 1978.

10. Forsythe, G. E. and Moler, C. B., *Computer Solution to Linear Algebraic Systems*, Prentice-Hall, Englewood Cliffs, New Jersey, 1967.

11. Forsythe, G. E., Malcolm, M. A., and Moler, C. B., *Computer Methods for Mathematical Computations*, Prentice-Hall, Englewood Cliffs, New Jersey, 1977.

12. Froberg, C. E., *Numerical Mathematics*, Benjamin/Cummings, Menlo Park, California, 1985.

13. Gear, C. W., *Numerical Initial-Value Problems in Ordinary Differential Equations*, Prentice-Hall, Englewood Cliffs, New Jersey, 1971.

14. Gerald, C. F. and Wheatley, P. O., *Applied Numerical Analysis*, 4th ed., Addison-Wesley, Reading, Massachusetts, 1989.

15. Gladwell, I. and Wait, R., *A Survey of Numerical Methods for Partial Differential Equations*, Oxford University Press, New York, 1979.

16. Golub, G. H. and Van Loan, C. F., *Matrix Computations*, Johns Hopkins University Press, Baltimore, Maryland, 1983.

17. Hageman, L. A. and Young, D. M., *Applied Iterative Methods*, Academic Press, New York, 1981.

18. Hamming, R. W., *Numerical Methods for Scientists and Engineers*, 2nd ed., McGraw-Hill, New York, 1973.

19. Henrici, P., *Discrete Variable Methods in Ordinary Differential Equations*, John Wiley & Sons, New York, 1962.

20. Henrici, P., *Error Propagation for Difference Methods*, John Wiley & Sons, New York, 1963.

21. Henrici, P., *Elements of Numerical Analysis*, John Wiley & Sons, New York, 1964.

22. Henrici, P., *Essentials of Numerical Analysis*, John Wiley & Sons, New York, 1982.

23. Hildebrand, F. B., *Introduction to Numerical Analysis*, 2nd ed., McGraw-Hill, New York, 1974.

24. Householder, A. S., *The Numerical Treatment of a Single Nonlinear Equation*, McGraw-Hill, New York, 1970.

25. Isaacson, E. and Keller, H. B., *Analysis of Numerical Methods*, John Wiley & Sons, New York, 1966.

26. Johnston, R. L., *Numerical Methods: a Software Approach*, John Wiley & Sons, New York, 1982.

27. Lawson, C. L. and Hanson, R. J., *Solving Least Squares Problems*, Prentice-Hall, Englewood Cliffs, New Jersey, 1974.

28. Maron, M. J., *Numerical Analysis—a Practical Approach*, Macmillan, New York, 1987.

29. Mitchell, A. R., *Computational Methods for Partial Differential Equations*, John Wiley & Sons, London, 1969.

30. Noble, B. and Daniel, J. W., *Applied Linear Algebra*, 2nd ed., Prentice-Hall, Englewood Cliffs, New Jersey, 1977.

31. Ortega, J. M., *Numerical Analysis—a Second Course*, Academic Press, New York, 1972.

32. Ortega, J. M. and Rheinboldt, W. C., *Iterative Solution of Nonlinear Equations in Several Variables*, Academic Press, New York, 1970.

33. Powell, M. J. D., *Approximation Theory and Methods*, Cambridge University Press, New York, 1981.

34. Prenter, P. M., *Splines and Variational Methods*, John Wiley & Sons, New York, 1975.

35. Ralston, A., *A First Course in Numerical Analysis*, McGraw-Hill, New York, 1965.

36. Ralston, A. and Rabinowitz, P., *A First Course in Numerical Analysis*, 2nd ed., McGraw-Hill, New York, 1978.

37. Rice, J. R., *Matrix Computations and Mathematical Software*, McGraw-Hill, New York, 1981.

38. Rice, J. R., *Numerical Methods, Software and Analysis*, IMSL Reference Edition, McGraw-Hill, New York, 1983.

39. Shampine, L. F. and Allen, R., *Numerical Computing*, W. B. Saunders, Philadelphia, Pennsylvania, 1973.

40. Smith, G. D., *Numerical Solution of Partial Differential Equations*, 2nd ed., Oxford University Press, New York, 1978.

41. Stewart, G. W., *Introduction to Matrix Computations*, Academic Press, New York, 1973.

42. Strang, W. G. and Fix, G. J., *An Analysis of the Finite Element Method*, Prentice-Hall, Englewood Cliffs, New Jersey, 1973.

43. Stroud, A. H. and Secrest, D., *Gaussian Quadrature Formulae*, Prentice-Hall, Englewood Cliffs, New Jersey, 1966.

44. Varga, R. S., *Matrix Iterative Analysis*, Prentice-Hall, Englewood Cliffs, New Jersey, 1962.

45. Vichnevetsky, R., Computer Methods for Partial Differential Equations, in *Elliptic Equations and the Finite Element Method*, Vol. 1, Prentice-Hall, Englewood Cliffs, New Jersey, 1981.

46. Wilkinson, J. H., *Rounding Errors in Algebraic Processes*, Prentice-Hall, Englewood Cliffs, New Jersey, 1963.

47. Wilkinson, J. H., *The Algebraic Eigenvalue Problem*, Clarendon Press, Oxford, 1965.

48. Young, D. M. and Gregory, R. T., *A Survey of Numerical Mathematics*, Vols. 1 and 2, Addison-Wesley, Reading, Massachusetts, 1972.

49. Zienkiewicz, O., *The Finite Element Method in Engineering Science*, McGraw-Hill, London, 1977.

List of Algorithms and Numerical Methods

SIM: Standard Iteration Method for $x = f(x)$ **(3.1.1)**

ATKN: Aitken's Δ^2 method for $x = f(x)$ **(3.3.1)**

STM: Steffensen's Method for $x = f(x)$ **(3.4.1)**

NRM: Newton Raphson Method for $F(x) = 0$ **(3.5.1)**

BISM: BISection Method for $F(x) = 0$ **(3.6.1)**

SECM: SECant Method for $F(x) = 0$ **(3.6.2)**

CNM: Cubic Newton's Method for $F(x) = 0$ **(3.6.3)**

MLR: Muller's method for $F(x) = 0$ **(3.6.4)**

GSIM: Generalized Standard Iteration Method for systems   **(3.7.1)**

GNMS: Generalized Newton's Method for Systems   **(3.7.2)**

The trapezoidal rule for integration   **(6.1)**

SIMP: SIMPson's rule for integration   **(6.1.1)**

Gauss integration method   **(6.2)**

ROM: ROMberg's method for integration   **(6.3.1)**

2-D Simpson's rule for integration   **(6.4)**

2-D Gaussian integration scheme   **(6.4)**

Gaussian elimination method for $Ax = b$ **(7.2)**

Doolittle's method for $Ax = b$ **(7.3)**

Solving $Ax = b$ for a tridiagonal matrix   **(7.3)**

Jacobi iterative method for $Ax = b$ **(7.4)**

GS: Gauss–Seidel method for $Ax = b$ **(7.4.1)**

Relaxation methods for $Ax = b$ **(7.4)**

The residual correction method for $Ax = b$ **(7.5)**

The power method for a single dominant eigenvalue   **(7.6)**

A modified power method for symmetric matrices   **(7.6)**

The deflation method and Wielandt's technique for calculating several eigenvalues   **(7.6)**

Euler's method for $y' = f(x, y)$ **(8.2)**

Taylor's method for $y' = f(x, y)$ **(8.3)**

Euler's modified method for $y' = f(x, y)$ **(8.4)**

The midpoint method for $y' = f(x, y)$ **(8.4)**

The classical Runge–Kutta method for $y' = f(x, y)$ **(8.4)**

RKF: Runge–Kutta–Fehlberg method for $y' = f(x, y)$ **(8.5.1)**

Adams–Bashforth Multistep methods for $y' = f(x, y)$ **(8.6)**

Adams–Moulton Multistep methods for $y' = f(x, y)$ **(8.6)**

532

PC4:   A fourth order predictor corrector for $y' = f(x, y)$ **(8.6.1)**

The Milne–Simpson predictor corrector method for $y' = f(x, y)$ **(8.6)**

Euler methods for systems   **(8.8)**

The separation of variables method for partial differential equations   **(9.3)**

The forward-difference method for the heat equation   **(9.4)**

BDIF: The Backward-DIFferences method for the heat equation   **(9.4.1)**

CRNIC: The CRank NIColson method for the heat equation   **(9.4.2)**

WAVE: A finite differences method for the WAVE equation   **(9.5.1)**

A Relaxation method for Poisson's equation   **(9.6.1)**

SOR: Successive OverRelaxation method for Poisson's equation   **(9.6)**

The Rayleigh–Ritz method for boundary value problems   **(9.7)**

A finite element method for 2-D elliptic self-adjoint problems   **(9.7)**

Answers and Solutions to Selected Problems

Chapter 1

Section 1.2

1. (b) $4 + i$

   (c) $z_1 = 1.0987 - 0.4551i, z_2 = -z_1$

2. Let $z = a + ib$. Then

$$\frac{1+i+z}{1-i+z} = \frac{1+i+a+ib}{1-i+a+ib} = \frac{(1+a)+i(b+1)}{(1+a)+i(b-1)}$$

$$= \frac{[(1+a)+i(b+1)][(1+a)-i(b-1)]}{(1+a)^2+(b-1)^2}$$

The imaginary part must vanish, i.e.,

$$(b+1)(1+a) - (1+a)(b-1) = 0$$

or $a = -1$. This also implies $b \neq 1$ (otherwise $1 - i + z = 0$).

Therefore

$$z = -1 + ib, b \neq 1$$

Section 1.3

2. (b) $2.4 - 1.8i$

6. Let $V_k = (1+i)^{1/3}$, $k = 1, 2, 3$; $W_j = (1-i)^{1/2}, j = 1, 2$. Then

$$V_1 = 2^{1/6}[\cos(15°) + i\sin(15°)] = 1.0842 + 0.2905i$$
$$V_2 = 2^{1/6}[\cos(135°) + i\sin(135°)] = -0.7937 + 0.7937i$$
$$V_3 = 2^{1/6}[\cos(255°) + i\sin(255°)] = -0.2905 - 1.0842i$$
$$W_1 = 2^{1/4}[\cos(-22.5°) + i\sin(-22.5°)] = 1.0987 - 0.4551i$$
$$W_2 = 2^{1/4}[\cos(157.5°) + i\sin(157.5°)] = -1.0987 + 0.4551i$$

The complex number $z = V_k + W_j$ with the smallest absolute value is

$$z = V_2 + W_1 = 0.3050 + 0.3386i, |z| = 0.4557$$

Section 1.4
1. (b) $z_1 = 1$, $z_2 = -i$ with multiplicities 2 and 3, respectively.

Section 1.5
4. The first three derivatives of $f(x) = x^{7/2}$ are

$$f'(x) = \frac{7}{2}x^{5/2}, \quad f''(x) = \frac{35}{4}x^{3/2}, \quad f'''(x) = \frac{105}{8}\sqrt{x}$$

At $x_0 = 0$ we have $f(0) = f'(0) = f''(0) = 0$. Therefore, $p_2(x) = 0$.

(a) Since $f(x)$ is "almost flat" near $x = 0$, $p_2(x)$ clearly approximates $f(x)$ at this neighborhood.
(b) The remainder is

$$R_2 = \frac{1}{3!}x^3 f^{(3)}(\zeta) = \frac{35}{16}x^3\sqrt{\zeta}, \quad 0 < \zeta < x$$

In this particular case we can easily obtain ζ, since

$$\frac{35}{16}x^3\sqrt{\zeta} = x^{7/2}$$

implies

$$\zeta = \frac{256}{1225}x$$

Section 1.6
2. (a) $-8.3585 + 18.2637i$

4. For $z = x + iy$ we have

$$e^{\bar{z}} = e^{x-iy} = e^x e^{-iy} = e^x[\cos(-y) + i\sin(-y)]$$
$$= e^x(\cos y - i\sin y) = \overline{e^x(\cos y + i\sin y)} = \overline{e^z}$$

Section 1.7

3. (b) $f_x = y\cos(xy) - 1/y$, $f_y = x\cos(xy) + x/y^2$

$$f_{xx} = -y^2\sin(xy), \quad f_{yy} = -x^2\sin(xy) - \frac{2x}{y^3}$$

$$f_{xy} = \cos(xy) - xy\sin(xy) + \frac{1}{y^2}$$

6. Let $f(z) = \cos z$, $z = x + iy$. Then

$$f(z) = \frac{e^{iz} + e^{-iz}}{2} = \cos x\left(\frac{e^y + e^{-y}}{2}\right) - i\sin x\left(\frac{e^y - e^{-y}}{2}\right)$$

Clearly

$$u_{xx} = -\cos x\left(\frac{e^y + e^{-y}}{2}\right), u_{yy} = \cos x\left(\frac{e^y + e^{-y}}{2}\right)$$

implying $\nabla^2 u = 0$. Similarly, $\nabla^2 v = 0$ as well.

Chapter 2

Section 2.1

3. (b) 506
   (c) 10650

7. (a) 101010111100110111101111

Section 2.2

5. (a)

$$x + y = (0.1002 + 0.0210) \cdot 10^3 = 0.1212 \cdot 10^3$$
$$xy = 0.02113218 \cdot 10^5 = \text{(after rounding)}\ 0.2113 \cdot 10^4$$

(c)

$$x = 0.3142 \cdot 10^1, y = 0.3142 \cdot 10^5$$
$$x + y = (0.0000 + 0.3142) \cdot 10^5 = 0.3142 \cdot 10^5$$
$$xy = 0.09872164 \cdot 10^6 = \text{(after rounding)}\ 0.9872 \cdot 10^5$$

Section 2.3

6. (a) Rewrite

$$\ln(x + a) - \ln x = \ln\left(\frac{x + a}{x}\right) = \ln\left(1 + \frac{a}{x}\right)$$

(If $a \ll x$ we can approximate the right-hand side by a/x).

(b) Rewrite

$$\sin(x + a) - \sin a = 2 \sin \frac{x}{2} \cos \left(a + \frac{x}{2} \right)$$

Example: Consider a five-digit mantissa with a rounding mode, and let $x = 0.001$, $a = 1$. Then

$$\sin(x + a) - \sin a = 0.84201 - 0.84147 = 0.00054 = 0.54 \cdot 10^{-3}$$

$$2 \sin \left(\frac{x}{2} \right) \cos \left(a + \frac{x}{2} \right) = 2 \cdot (0.5 \cdot 10^{-3}) \cdot 0.53988 = 0.53988 \cdot 10^{-3}$$

7. (a) Since $e^x \approx 1 + x + x^2/2! + \cdots + x^n/n!$ we have

$$\frac{e^x - 1}{x} \approx 1 + \frac{x}{2!} + \frac{x^2}{3!} + \cdots + \frac{x^{n-1}}{n!}$$

Example: Consider a five-digit mantissa with a rounding mode and let $x = 0.00001$. Then

$$\frac{e^x - 1}{x} = \frac{1.0000 - 1}{0.00001} = 0$$

For a quadratic Taylor approximation we get

$$\frac{e^x - 1}{x} \approx 1 + \frac{x}{2!} = 1 + 0.000005 = 1$$

Section 2.4

2.

$$\frac{x_t}{y_t} - \frac{x_c}{y_c} = \frac{x_t y_c - x_c y_t}{y_t y_c} = \frac{x_t y_c - x_t y_t + x_t y_t - x_c y_t}{y_t y_c}$$

$$= \frac{x_t(y_c - y_t) + y_t(x_t - x_c)}{y_t y_c} = \frac{x_t}{y_t y_c}(y_c - y_t) + \frac{1}{y_c}(x_t - x_c)$$

Thus

$$\left| \frac{x_t}{y_t} - \frac{x_c}{y_c} \right| \leq \left| \frac{x_t}{y_t y_c} \right| |y_t - y_c| + \left| \frac{y_t}{y_t y_c} \right| |x_t - x_c| \approx \frac{|x_c||y_t - y_c| + |y_c||x_t - x_c|}{|y_c|^2}$$

5. (d) For $f(x) = x^x$ we have

$$f(x_t) - f(x_c) \approx (x_t - x_c)f'(x_t) = (x_t - x_c)x_t^{x_t}[\ln(x_t) + 1]$$

$$\mathrm{rel}[f(x_c)] \approx \frac{(x_t - x_c)f'(x_t)}{f(x_t)} = (x_t - x_c)[\ln(x_t) + 1]$$

Chapter 3

Section 3.1

2. (a) $L = 13$

    (b) L-condition not satisfied.

    (c) $L = 3e$ (smallest)

4. Since $0 \leq f(x) \leq 1/2$ condition (1) of Theorem 3.1.3 is satisfied. Also $f(x) \in C(-\infty, \infty)$, and the derivative

$$f'(x) = -\frac{1}{2}[e^{-x}(\cos x + \sin x)]$$

yields

$$|f'(x)| \leq \frac{1}{2} \cdot 1 \cdot \left(\frac{\sqrt{2}}{2} + \frac{\sqrt{2}}{2} \right) = \frac{\sqrt{2}}{2} (= L)$$

Thus, all the requirements of Theorem 3.1.3 are fulfilled. If $x_0 = 1/4$ then $x_1 = 0.37729488$. Eq. (3.1.13) implies

$$|s - x_n| \leq \frac{L^n}{1 - L}|x_1 - x_0| = \frac{\left(\frac{\sqrt{2}}{2} \right)^n \cdot 0.12729488}{0.29289322}$$

and to guarantee $|x_n - s| \leq 10^{-5}$ we must use $N = 31$, i.e.,

$$\frac{\left(\frac{\sqrt{2}}{2} \right)^{31} \cdot 0.12729488}{0.29289322} \approx 10^{-5}$$

Section 3.2

1. (a)

$$\frac{\eta_{n+1}}{\eta_n} = \frac{x_{n+1} - x_n}{x_n - x_{n-1}} = \frac{f(x_n) - f(x_{n-1})}{x_n - x_{n-1}}$$

$$= \frac{(x_n - x_{n-1})f'(c_n)}{x_n - x_{n-1}} = f'(c_n)$$

where c_n is an interim point between x_{n-1} and x_n. Since $x_n \to s$ as $n \to \infty$ we get $c_n \to s$ and the continuity of $f'(x)$ guarantees

$$\lim_{n \to \infty} \frac{\eta_{n+1}}{\eta_n} = f'(s)$$

    (b) By Theorem 3.2.1 we obtain the rate of convergence of x_n to s by generating the sequence $\{\epsilon_{n+1}/\epsilon_n\}$ which approaches $f'(s)$ as $n \to \infty$. Since s and therefore ϵ_n are not known, one can instead consider $\{\eta_{n+1}/\eta_n\}$

(c) For any given n:

$$\epsilon_n = x_n - s = x_n - x_{n-1} + x_{n-1} - s = \eta_n + \epsilon_{n-1}$$

Thus, for large n

$$\epsilon_n \approx \eta_n + \frac{\epsilon_n}{f'(s)}$$

leading to

$$\epsilon_n \approx \frac{f'(s)\eta_n}{f'(s) - 1}$$

Section 3.3
1. For any given n

$$x_n' = x_n - \frac{(x_{n+1} - x_n)^2}{x_{n+2} - 2x_{n+1} + x_n}, x_n = \sum_{j=1}^{n} a_j$$

Thus

$$x_{n+1} - x_n = a_{n+1}, x_{n+2} - 2x_{n+1} + x_n = a_{n+2} - a_{n+1}$$

and

$$x_n' = \sum_{j=1}^{n} a_j - \frac{a_{n+1}^2}{a_{n+2} - a_{n+1}}$$

Section 3.5
1. The NRM scheme is

$$x_{n+1} = x_n - \frac{x_n^k - a}{kx_n^{k-1}} = \frac{(k-1)x_n^k + a}{kx_n^{k-1}}, n \geq 0$$

Let $s(k, a, x_0, \epsilon)$ denote the approximate solution for given a, k, an initial approximation x_0 and a tolerance ϵ. Then

$$s(2.5, 3, 1, 10^{-6}) = 1.551846$$
$$s(4, 3, 1, 10^{-6}) = 1.316074$$

6. The NRM scheme for solving $F(x) = x^3 - 2x^2 = 0$ is

$$x_{n+1} = x_n - \frac{x_n^3 - 2x_n^2}{3x_n^2 - 4x_n} = \frac{2x_n^3 - 2x_n^2}{3x_n^2 - 4x_n} = \frac{2x_n(x_n - 1)}{3x_n - 4}$$

For $x_0 = -1$ we get $x_1 = -0.5714$, $x_2 = -0.3143$, $x_3 = -0.1671$, $x_4 = -0.0867$, i.e., linear convergence to the exact solution $s = 0$. This is expected since $F'(x) = 3x^2 - 4x$ satisfies $F'(s) = 0$.

To speed the convergence we define

$$x_{n+1} = x_n - \frac{2F(x_n)}{F'(x_n)} = \frac{x_n^2}{3x_n - 4}$$

The generated sequence is $x_1 = -0.1429$, $x_2 = -0.004608$, $x_3 = -0.000005\ldots$ and it converges quadratically to $s = 0$.

Section 3.6

3. The maximal number of iterations n is the smallest integer that satisfies

$$\frac{9}{2^n} \leq 10^{-8}$$

i.e., $n = 30$.

7. The secant method for $F(x) = x^3 + x^2 - 1 = 0$ is

$$x_{n+1} = x_n - \frac{(x_n - x_{n-1})(x_n^3 + x_n^2 - 1)}{(x_n^3 + x_n^2 - 1) - (x_{n-1}^3 + x_{n-1}^2 - 1)}$$

Given $x_0 = 2$, $x_1 = 1$ we need 7 function evaluations to get $x_7 = 0.7548777$ and $|x_7 - x_6| < 10^{-6}$.

15. Here $F(x) = \sin x - \cos x$ and therefore

$$F'(x) = \cos x + \sin x, F''(x) = -\sin x + \cos x$$

The CNM method yields

n	x_n	F_n	F'_n	F''_n
0	0.80000000	0.02064938	1.41406280	−0.02064938
1	0.78539868	0.00000073	1.41421356	−0.00000073
2	0.78539816	0.00000000	1.41421356	0.00000000
3	0.78539816			

with a total of 9 function evaluations.

17. Let $a_n(x - x_n)^2 + b_n(x - x_n) + c_n$ denote the parabola through (x_{n-2}, F_{n-2}), (x_{n-1}, F_{n-1}), and (x_n, F_n) where $F_n = F(x_n)$, $n \geq 0$. The coefficients a_n, b_n, c_n are given by Eqs. (3.6.5)–(3.6.7). The intersection of the parabola with the x axis is denoted by x_{n+1} and is our next approximation to the solution s of $F(x) = 0$.

For $F(x) = x - e^{-x}$ and $x_0 = 0.9$, $x_1 = 0.8$, $x_2 = 0.7$ we have

n	x_n	a_n	b_n	$c_n (= F_n)$
0	0.900000			
1	0.800000			
2	0.700000	-0.224852	1.495049	0.203415
3	0.566617	-0.251359	1.564753	-0.000825
4	0.567144	-0.271468	1.567149	0.000001
5	0.567143			

with a total of 5 function evaluations.

Section 3.7

1. We have

$$L = \max_{0 \le x,y \le 1} \left[\left(\frac{\partial f_1}{\partial x} \right)^2 + \left(\frac{\partial f_1}{\partial y} \right)^2 + \left(\frac{\partial f_2}{\partial x} \right)^2 + \left(\frac{\partial f_2}{\partial y} \right)^2 \right]^{1/2}$$

where

$$f_1(x, y) = \left(\frac{x^2 + y^2}{2} \right)^{1/2}, f_2(x, y) = \frac{xy}{3}$$

Thus

$$\frac{\partial f_1}{\partial x} = \frac{1}{\left(\frac{x^2 + y^2}{2} \right)^{1/2}} \frac{x}{2}$$

$$\frac{\partial f_1}{\partial y} = \frac{1}{\left(\frac{x^2 + y^2}{2} \right)^{1/2}} \frac{y}{2}$$

$$\frac{\partial f_2}{\partial x} = \frac{y}{3} \le \frac{1}{3}, \frac{\partial f_2}{\partial y} = \frac{x}{3} \le \frac{1}{3}$$

and

$$L = \left(\frac{1}{2} + \frac{1}{9} + \frac{1}{9} \right)^{1/2} < 1$$

9. We have

$$F(x, y) = x - x^2 - y^2, \frac{\partial F}{\partial x} = 1 - 2x, \frac{\partial F}{\partial y} = -2y$$

$$G(x, y) = y - x^2 + y^2, \frac{\partial G}{\partial x} = -2x, \frac{\partial G}{\partial y} = 1 + 2y$$

Using Eq. (3.7.18) one gets $(x_4, y_4) = (0.771845, 0.419643)$ as the final approximation.

Chapter 4

Section 4.3

5. The characteristic polynomial $z^2 - z - 1 = 0$ has two distinct roots

$$z_1 = \frac{1 + \sqrt{5}}{2}, \quad z_2 = \frac{1 - \sqrt{5}}{2}$$

The general solution to the *LDE* is

$$x_n = c_1 \left(\frac{1 + \sqrt{5}}{2} \right)^n + c_2 \left(\frac{1 - \sqrt{5}}{2} \right)^n$$

If

$$\lim_{n \to \infty} x_n = \infty$$

we must have $c_1 \neq 0$ and since

$$|z_2| < |z_1|$$

we get

$$\lim_{n \to \infty} \frac{x_{n+1}}{x_n} = \frac{1 + \sqrt{5}}{2}$$

If $c_1 = 0$

$$\frac{x_{n+1}}{x_n} = \frac{1 - \sqrt{5}}{2}$$

for all n.

8. By substituting $x_n = \alpha$ in $x_n + a_1 x_{n-1} + a_2 x_{n-2} = A$ we get

$$\alpha = \frac{A}{1 + a_1 + a_2}$$

It provides a valid solution unless $1 + a_1 + a_2 = 0$. We then try $x_n = \alpha n$ to get

$$\alpha n + a_1 \alpha (n - 1) + a_2 \alpha (n - 2) = A$$

i.e.,

$$\alpha n (1 + a_1 + a_2) - \alpha (a_1 + 2a_2) = -\alpha (a_1 + 2a_2) = A$$

If $a_1 + 2a_2 = 0$ we must have $a_1 = -2, a_2 = 1$, but then $x_n = \alpha n^2$ is a particular solution. Indeed

$$\alpha n^2 + a_1\alpha(n-1)^2 + a_2\alpha(n-2)^2 = \alpha n^2(1 + a_1 + a_2)$$
$$- 2\alpha n(a_1 + 2a_2) + \alpha(a_1 + 4a_2)$$
$$= 2\alpha = A$$

i.e., $\alpha = A/2$.

Section 4.4

5. The characteristic polynomial $P_5(z) = (z-1)^3(z^2+1)$ has the root $z_1 = 1$ with multiplicity 3 and two simple roots $z_4 = i$, $z_5 = -i$. A basis to the solution set is, for example

$$\{1\}, \{n\}, \{n^2\}, \{i^n\}, \{(-i)^n\}$$

and the general solution

$$x_n = c_1 + c_2 n + c_3 n^2 + c_4 i^n + c_5(-i)^n$$

Section 4.5

5. The first five columns about $n = 0$ are

n	f_n	∇f_n	$\nabla^2 f_n$	$\nabla^3 f_n$	$\nabla^4 f_n$
-3	0		0		0
		0		0	
-2	0		0		ϵ
		0		ϵ	
-1	0		ϵ		-4ϵ
		ϵ		-3ϵ	
0	ϵ		-2ϵ		6ϵ
		$-\epsilon$		3ϵ	
1	0		ϵ		-4ϵ
		0		$-\epsilon$	
2	0		0		ϵ
		0		0	
3	0		0		0

Section 4.6

3. The polynomial $p(z) = z^3 - z^2 - 10z - 8$ corresponds to

$$x_n - x_{n-1} - 10x_{n-2} - 8x_{n-3} = 0$$

The coefficients are $a_0 = 1$, $a_1 = -1$, $a_2 = -10$, $a_3 = -8$ and Algorithm 4.6.2 provides

n	x_n	q_n	q'_n
-3	1		
-2	21		
-1		55 4.964	4.029
0		273 3.630	4.006
1		991 4.199	4.002
2	4,161	3.907	
3	16,255	4.048	
4	65,793	3.977	
5	261,631	4.012	
6	1,049,601	3.994	
7	4,192,255		

Eight evaluations of q_n and three of q'_n are needed to establish three accurate significant digits.

6. (a) The iteration stops at x_6 with $R \approx 2.236$ and $\cos\theta \approx -0.447$, i.e., $z_1 = -0.999 + 2.000i$ and $z_2 = -0.999 - 2.000i$. The exact zeroes are $z_1 = -1 + 2i$, $z_2 = -1 - 2i$.

    (b) The iteration stops at x_{22} with $R \approx 1.731$ and $\cos\theta = 0$, i.e., $z_1 = 1.731i$, $z_2 = -1.731i$. The exact values are $z_1 = \sqrt{3}i$, $z_2 = -\sqrt{3}i$.

Chapter 5

Section 5.1

3. The given data set is $(0, 0)$, $(\pi/4, \sqrt{2}/2)$, $(\pi/2, 1)$. Since $y_0 = 0$, $l_0(x)$ is not needed and one should only compute $l_1(x)$ and $l_2(x)$:

$$l_1(x) = \frac{(x-0)\left(x - \frac{\pi}{2}\right)}{\left(\frac{\pi}{4} - 0\right)\left(\frac{\pi}{4} - \frac{\pi}{2}\right)} = -1.6211x^2 + 2.5465x$$

$$l_2(x) = \frac{(x-0)\left(x - \frac{\pi}{4}\right)}{\left(\frac{\pi}{2} - 0\right)\left(\frac{\pi}{2} - \frac{\pi}{4}\right)} = 0.8106x^2 - 0.6366x$$

The interpolator is

$$p_2(x) = y_1 l_1(x) + y_2 l_2(x) = -0.3357x^2 + 1.1640x$$

and

$$\left\{ \frac{2}{\pi} \int_0^{\frac{\pi}{2}} \left[p_2(x) - \sin x\right]^2 dx \right\}^{1/2} = 0.0152$$

7. The function $f(x)$ must satisfy the system

$$a + b + c = 1$$
$$a + \frac{b}{2} + \frac{c}{4} = -1$$
$$a + \frac{b}{4} + \frac{c}{16} = 0$$

whose only solution is $a = 7/3$, $b = -12$, $c = 32/3$.

Section 5.2
1. Using Eq. (5.2.13) we get

$$|e^x - p_2(x)| \leq \frac{\sqrt{3}}{27} h^3 M_3$$

where

$$h = \frac{1}{2}, M_3 = \max_{0 \leq x \leq 1} |(e^x)'''| = e$$

Therefore

$$|e^x - p_2(x)| \leq 0.0218$$

5. Let $p_1(x)$ denote a linear interpolator between x_i and x_{i+1}. Then, by virtue of Eq. (5.2.12)

$$|\sqrt{x} - p_1(x)| \leq \frac{h^2}{8} \max |(\sqrt{x})''|, x_i \leq x \leq x_{i+1}$$

or

$$|\sqrt{x} - p_1(x)| \leq \frac{h^2}{32 x_i \sqrt{x_i}} \leq \frac{h^2}{32 a \sqrt{a}}$$

To guarantee an accuracy of $\epsilon = 10^{-4}$ one must satisfy

$$\frac{h^2}{32 a \sqrt{a}} \leq 10^{-4}$$

or

$$h \leq 0.05657(a^{3/4})$$

Clearly, this upper bound is independent of b.

Section 5.3
4. Since $|f^{(i)}(x)| \leq 1$, $i \geq 0$, $x \geq 0$ we have

$$|f(x) - p_n(x)| \leq \left| \frac{x(x-h)\dots(x-nh)}{(n+1)!} \right|, x \geq 0$$

Let m be an integer such that $(m-1)h \le x \le mh \le nh$. Then

$$|f(x) - p_n(x)| \le \frac{(nh)[(n-1)h][(n-2)h]\dots(mh)x^m}{(n+1)!}$$

$$\le \frac{(n!)h^{n-m+1}x^m}{(n+1)!} \le \frac{h^{n-m+1}x^m}{n+1}$$

and a sufficient condition for

$$\lim_{n\to\infty} p_n(x) = f(x), x \ge 0$$

is $h \le 1$.

Section 5.4

3. By definition

$$f[x_1, x_0, x_2] = \frac{f[x_0, x_2] - f[x_1, x_0]}{x_2 - x_1}$$

$$= \frac{f(x_2) - f(x_0)}{(x_2 - x_0)(x_2 - x_1)} - \frac{f(x_0) - f(x_1)}{(x_0 - x_1)(x_2 - x_1)}$$

$$= \frac{f(x_0)}{(x_0 - x_1)(x_0 - x_2)} + \frac{f(x_1)}{(x_1 - x_0)(x_1 - x_2)} + \frac{f(x_2)}{(x_2 - x_0)(x_2 - x_1)}$$

$$= f[x_0, x_1, x_2]$$

9. Let $p_{n+1}(t)$ be the polynomial that interpolates $f(x)$ at $x_0, x_1, \dots, x_n, x$. By Eq. (5.4.11)

$$p_{n+1}(t) - p_n(t) = (t - x_0)(t - x_1)\dots(t - x_n)f[x_0, x_1, \dots, x_n, x]$$

For $t = x$ and since $p_{n+1}(x) = f(x)$ we get

$$f(x) - p_n(x) = (x - x_0)(x - x_1)\dots(x - x_n)f[x_0, x_1, \dots, x_n, x]$$

Section 5.5

1. (a) $s'(1-) = 3$, $s'(1+) = 2$, i.e., $s(x)$ is not a cubic spline.
   (b) $s(1-) = s(1+) = 2$, $s'(1-) = s'(1+) = 2$, $s''(1-) = s''(1+) = 0$, i.e., $s(x)$ is a cubic spline.
   (c) $s''(2-) = 14$, $s''(2+) = 0$, i.e., $s(x)$ is not a cubic spline.

9. (a)

$$s_a(a+) = \lim_{x\to a+} (x - a)^3 = 0$$

$$s'_a(a+) = 3 \lim_{x\to a+} (x - a)^2 = 0$$

$$s''_a(a+) = 6 \lim_{x\to a+} (x - a) = 0$$

i.e., $s_a(x) \in C^2(-\infty, \infty)$. Since $s_a(x)$ is also a piecewise cubic polynomial, it is by definition a cubic spline.

(b) The sum

$$s(x) = \sum_{i=0}^{n} c_i s_{x_i}(x)$$

is clearly a piecewise cubic polynomial. For each x_j let us rewrite

$$s(x) = \sum_{i \neq j} c_i s_{x_i}(x) + c_j s_{x_j}(x)$$

The first term has an infinite number of continuous derivatives at x_j, while the second term [using the result of part (a)] possesses two continuous derivatives. Thus, $s(x)$ is a cubic spline.

Section 5.6

3. For $f(x) = 1/x$ we have

$$f^{(k)}(x) = (-1)^k \frac{k!}{x^{k+1}}$$

If we choose $a = 1$, $b = 3$ then

$$\frac{(b-a)^{n+1}}{2^{2n+1}(n+1)!} \max_{a \leq x \leq b} |f^{(n+1)}(x)| = \frac{2^{2n+1}}{2^{2n+1}(n+1)!} \frac{(n+1)!}{1^{n+2}} = 1$$

and the sequence of bounds does not converge to zero.

Section 5.7

5. (a) Clearly $x^n + p(x)$ is an nth order monic. Therefore, by Theorem 5.7.2

$$\rho_n(p) = \max_{-1 \leq x \leq 1} |x^n + p(x)| \geq \frac{1}{2^{n-1}}$$

(b) To get the smallest possible value of $\rho_n(p)$ we choose

$$x^n + p(x) = \frac{1}{2^{n-1}} T_n(x)$$

or

$$p(x) = \frac{1}{2^{n-1}} T_n(x) - x^n$$

Section 5.8

3. The four points at which each function $y = f(x)$ is interpolated are

$$x_k = \cos\left(\frac{2k+1}{8}\pi\right), \quad 0 \le k \le 3$$

i.e.,

$$x_0 = 0.9239, \; x_1 = 0.3827, \; x_2 = -0.3827, \; x_3 = -0.9239$$

The near minimax approximation is

$$p_3(x) = \sum_{i=0}^{3} l_i(x)y_i$$

where

$$l_0(x) = 0.7653x^3 + 0.7071x^2 - 0.1121x - 0.1036$$
$$l_1(x) = -1.8477x^3 - 0.7071x^2 + 1.5772x + 0.6037$$
$$l_2(x) = 1.8477x^3 - 0.7071x^2 - 1.5772x + 0.6037$$
$$l_3(x) = -0.7653x^3 + 0.7071x^2 + 0.1121x - 0.1036$$

Thus

$$\begin{aligned}
p_3(x) = \; & [0.7653(y_0 - y_3) + 1.8477(y_2 - y_1)]x^3 \\
& + 0.7071(y_0 + y_3 - y_1 - y_2)x^2 \\
& + [0.1121(y_3 - y_0) + 1.5772(y_1 - y_2)]x \\
& + 0.6037(y_1 + y_2) - 0.1036(y_0 + y_3)
\end{aligned}$$

(a)

$$y = \sin x, \; y_0 = 0.7980, \; y_1 = 0.3734, \; y_2 = -y_1, \; y_3 = -y_0$$
$$p_3(x) = -0.1584x^3 + 0.9989x$$

(b)

$$y = \tan x, \; y_0 = 1.3239, \; y_1 = 0.4025, \; y_2 = -0.4025, \; y_3 = -1.3239$$
$$p_3(x) = 0.5390x^3 + 0.9728x$$

(c)

$$y = x^4 + x^3 - 1, \; y_0 = 0.5173, \; y_1 = -0.9225, \; y_2 = -1.0346, \; y_3 = -1.0600$$
$$p_3(x) = x^3 + x^2 - 1.125$$

(d)

$$y = \ln(2 + x), \; y_0 = 1.0729, \; y_1 = 0.8682, \; y_2 = 0.4808, \; y_3 = 0.0733$$
$$p_3(x) = 0.0492x^3 - 0.1434x^2 + 0.4990x + 0.6956$$

Section 5.9

3. With $p(x) = 0.0001x^2 - 2x + 3$ the data set is best represented by a straight line.

7. Define

$$E = \sum_{i=1}^{n} \left(ae^{\alpha x_i} + be^{\beta x_i} - y_i \right)^2$$

Following Eq. (5.9.3) we demand

$$\frac{\partial E}{\partial a} = \sum_{i=1}^{n} 2e^{\alpha x_i} \left(ae^{\alpha x_i} + be^{\beta x_i} - y_i \right) = 0$$

$$\frac{\partial E}{\partial b} = \sum_{i=1}^{n} 2e^{\beta x_i} \left(ae^{\alpha x_i} + be^{\beta x_i} - y_i \right) = 0$$

$$\frac{\partial E}{\partial \alpha} = \sum_{i=1}^{n} 2ax_i e^{\alpha x_i} \left(ae^{\alpha x_i} + be^{\beta x_i} - y_i \right) = 0$$

$$\frac{\partial E}{\partial \beta} = \sum_{i=1}^{n} 2bx_i e^{\beta x_i} \left(ae^{\alpha x_i} + be^{\beta x_i} - y_i \right) = 0$$

The system of four nonlinear equations is then solved for a, b, α, β using Newton's method.

8. (a) Using Eq. (5.9.6) we get

$$y = 0.1896x + 29.79$$

where x is "homework" and y is "final".

(b) For $y = 75$ we get

$$x = \frac{75 - 29.79}{0.1896} \approx 238$$

(c) Students with homework grades "close" to 238 are expected to get final grades near 75.

Section 5.10

6. For any given $f(x)$ defined over $[-a, a]$ we can write

$$f(x) = \frac{f(x) + f(-x)}{2} + \frac{f(x) - f(-x)}{2}$$

The functions

$$g(x) = \frac{f(x) + f(-x)}{2}, \quad h(x) = \frac{f(x) - f(-x)}{2}$$

are even and odd, respectively.

7. **Step 1:** Set $\theta_0 = 1$.

   Step 2: Define $\theta_1 = x + \alpha_0 \theta_0 = x + \alpha_0$, where

$$\int_0^1 \theta_1 \theta_0 dx = \int_0^1 (x + \alpha_0) dx = 0$$

Therefore $\alpha_0 = -(1/2)$ and $\theta_1 = x - (1/2)$.

   Step 3: Define $\theta_2 = x^2 + \alpha_0 \theta_0 + \alpha_1 \theta_1$ such that

$$\int_0^1 \theta_2 \theta_0 dx = \int_0^1 x^2 dx + \alpha_0 \int_0^1 dx = 0$$

$$\int_0^1 \theta_2 \theta_1 dx = \int_0^1 x^2 \left(x - \frac{1}{2} \right) dx + \alpha_1 \int_0^1 \left(x - \frac{1}{2} \right)^2 dx = 0$$

Thus, $\alpha_0 = -(1/3)$, $\alpha_1 = -1$ and $\theta_2 = x^2 - x + (1/6)$.

Chapter 6

Section 6.1

2. (a) The integration nodes are $x_i = (3/8)i$, $0 \le i \le 8$. The trapezoidal rule yields

$$I(e^x; 0, 3) = \int_0^3 e^x dx \approx T_8(e^x; 0, 3) = 19.31$$

and

$$|I(e^x; 0, 3) - T_8(e^x; 0, 3)| \le \frac{h^2(b - a)}{12} \max_{a \le x \le b} |f''(x)| = \frac{0.375^2 \cdot 3}{12} e^3 = 0.71$$

(b) The integration nodes are $x_i = 1 + 0.125i$, $0 \le i \le 8$. The trapezoidal rule yields

$$I = \int_1^2 \frac{x^2}{2} \left(\ln x - \frac{1}{2} \right) dx \approx T_8 = -0.0462$$

The integrand $f(x) = (x^2/2)[\ln x - (1/2)]$ yields

$$f''(x) = 1 + \ln x$$

and thus, the error is bounded by

$$\frac{0.125^2}{12}(1 + \ln 2) = 0.0022$$

(The exact value is $I = -0.0480$.)

5. (a)

$$S_2 = 23.843, \qquad S_4 = 20.882$$
$$S_8 = 20.687, \qquad S_{16} = 20.675$$

(b) The exact value is $I = 20.674$. Thus

$$S_8 - I \approx 0.013$$
$$S_{16} - I \approx 0.001$$

in agreement with Eq. (6.1.25).

Section 6.2

2. By using Table 6.2.1 and Eq. (6.2.18) we get

$$n = 2: G_2 = 2.56125$$
$$n = 3: G_3 = 2.46763$$
$$n = 4: G_4 = 2.46950$$

The exact value is $I = 2.46948$ and the errors in absolute value are 0.092, 0.0019, and 0.000019, respectively.

6. (a) The integrand has an infinite discontinuity at $x = 0$ and

$$\frac{1}{\sqrt{x + x^3}} \approx \frac{1}{\sqrt{x}} \quad \text{near } x = 0$$

We should therefore rewrite

$$\int_0^1 \frac{dx}{\sqrt{x + x^3}} = \int_0^1 \left(\frac{1}{\sqrt{x + x^3}} - \frac{1}{\sqrt{x}} \right) dx + \int_0^1 \frac{dx}{\sqrt{x}}$$

$$= -\int_0^1 \frac{x^3 dx}{\sqrt{x}(\sqrt{x} + x^3)} + 2$$

The first integral is approximated using Gaussian quadrature with 2 and 3 nodes:

$$G_2 = -0.2222, \quad G_3 = -0.2235$$

Thus

$$\int_0^1 \frac{dx}{\sqrt{x} + x^3} = \begin{cases} 1.7778, \; n = 2 \\ \\ 1.7765, \; n = 3 \end{cases}$$

Section 6.3

4. To calculate

$$I(f; a, b) = \int_0^2 \sqrt{x}\, dx$$

using Romberg algorithm we start by computing

$$R_{1,1} = \frac{h}{2}[f(a) + f(b)], h = b - a$$

The computed $R_{k,i}$ arranged as in Figure 6.3.1 are

```
1.414214
1.707107   1.804738
1.819479   1.856937   1.860417
1.861473   1.875471   1.876707   1.876966
1.876891   1.882030   1.882468   1.882559   1.882581
1.882485   1.884350   1.884504   1.884537   1.884544   1.884546
```

The final approximation is 1.884546. The exact value is 1.885618.

Section 6.4

6. Let us rewrite

$$I = \int_0^1 \int_0^1 e^x \sin y \, dx \, dy = \int_0^1 \left[\int_0^1 e^x \sin y \, dy \right] dx$$
$$= \int_0^1 G(x)dx \approx 0.25 \left[\frac{G(0)}{2} + G(0.25) + G(0.5) + G(0.75) + \frac{G(1)}{2} \right]$$

where $G(x)$ is approximated by a four-interval trapezoidal rule. Thus

$$G(0) = 0.45730, \; G(0.25) = 0.58719, \; G(0.5) = 0.75396$$
$$G(0.75) = 0.96811, \; G(1) = 1.24307$$

and $I \approx 0.78986$. The exact value is 0.78989.

Section 6.5

2. (a)

$$f(x) = \sin(x^2),\ f'(x) = 2x\cos(x^2),\ x = 1$$

h	$D_c(f(x);\ 1, h)$
0.1	1.0566
0.05	1.0746
0.025	1.0791
0.0125	1.0802

The exact derivative is 1.0806.

(b)

$$f(x) = x^3,\ f'(x) = 3x^2,\ x = 1$$

h	$D_c(f(x);\ 1, h)$
0.1	3.01
0.05	3.0025
0.025	3.000625

The exact derivative is 3.

6. Let us approximate

$$f'(x) \approx \alpha f(x) + \beta f(x+h) + \gamma f(x+2h) = \alpha f(x)$$
$$+ \beta \left[f(x) + h f'(x) + \frac{h^2}{2!} f''(x) + \frac{h^3}{3!} f^{(3)}(\zeta_1) \right]$$
$$+ \gamma \left[f(x) + 2h f'(x) + \frac{4h^2}{2!} f''(x) + \frac{8h^3}{3!} f^{(3)}(\zeta_2) \right]$$

i.e.,

$$\alpha + \beta + \delta = 0$$
$$h(\beta + 2\gamma) = 1$$
$$\frac{h^2}{2!}(\beta + 4\gamma) = 0$$

and the solution is

$$\alpha = -\frac{3}{2h},\ \beta = \frac{4}{2h},\ \gamma = -\frac{1}{2h}$$

Thus

$$f'(x) \approx \frac{4f(x+h) - 3f(x) - f(x+2h)}{2h}$$

The error is

$$E = \frac{\beta h^3}{3!} f^{(3)}(\zeta_1) + \frac{8\gamma h^3}{3!} f^{(3)}(\zeta_2)$$

$$= \frac{h^2}{3} f^{(3)}(\zeta_1) - \frac{2h^2}{3} f^{(3)}(\zeta_2)$$

Since ζ_1, ζ_2 are close to x, we have

$$E \approx -\frac{h^2}{3} f^{(3)}(x)$$

Chapter 7

Section 7.1

5.

$$AB = \begin{pmatrix} a & b \\ c & d \end{pmatrix} \begin{pmatrix} 1 & -1 \\ 1 & 2 \end{pmatrix} = \begin{pmatrix} a+b & -a+2b \\ c+d & -c+2d \end{pmatrix}$$

$$BA = \begin{pmatrix} 1 & -1 \\ 1 & 2 \end{pmatrix} \begin{pmatrix} a & b \\ c & d \end{pmatrix} = \begin{pmatrix} a-c & b-d \\ a+2c & b+2d \end{pmatrix}$$

In order for AB and BA to be identical, we must have

$$a + b = a - c \tag{1}$$
$$-a + 2b = b - d \tag{2}$$
$$c + d = a + 2c \tag{3}$$
$$-c + 2d = b + 2d \tag{4}$$

From (4) or (1) we get $b = -c$ and from (2) or (3) $a - d = b$. Thus, a necessary condition is

$$A = \begin{pmatrix} a & b \\ -b & a-b \end{pmatrix}$$

and it is easily seen that the condition is also sufficient.

8.

$$\det(A) = \begin{vmatrix} 1 & -2 & 2 \\ 1 & 1 & 5 \\ 2 & -1 & 7 \end{vmatrix} = 0$$

Let us choose $x_3 = 1$. The first two equations are

$$x_1 - 2x_2 + 2 = 0$$
$$x_1 + x_2 + 5 = 0$$

i.e., $x_1 = -4$, $x_2 = -1$. The third equation is satisfied automatically by the solution

$$x = \begin{pmatrix} -4 \\ -1 \\ 1 \end{pmatrix}$$

Section 7.2

2. (a) The system is

$$\begin{pmatrix} 0.101 & -1 & 1 \\ 2 & 1 & -0.17 \\ 17.53 & -4 & 3 \end{pmatrix} \begin{pmatrix} x_1 \\ x_2 \\ x_3 \end{pmatrix} = \begin{pmatrix} 1.5 \\ 0 \\ 1 \end{pmatrix}$$

No pivoting:

$$\begin{pmatrix} 0.101 & -1 & 1 & 1.5 \\ 2 & 1 & -0.17 & 0 \\ 17.53 & -4 & 3 & 1 \end{pmatrix} \rightarrow \begin{pmatrix} 0.101 & -1 & 1 & 1.5 \\ 0 & 20.8 & -19.97 & -29.7 \\ 0 & 169.6 & -170.6 & -259.4 \end{pmatrix}$$

$$\rightarrow \begin{pmatrix} 0.101 & -1 & 1 & 1.5 \\ 0 & 20.8 & -19.97 & -29.7 \\ 0 & 0 & -7.8 & -17.2 \end{pmatrix}$$

and

$$x_3 = 2.205, \quad x_2 = 0.6889, \quad x_1 = -0.1594$$

Pivoting:

$$\begin{pmatrix} 0.101 & -1 & 1 & 1.5 \\ 2 & 1 & -0.17 & 0 \\ 17.53 & -4 & 3 & 1 \end{pmatrix} \rightarrow \begin{pmatrix} 17.53 & -4 & 3 & 1 \\ 2 & 1 & -0.17 & 0 \\ 0.101 & -1 & 1 & 1.5 \end{pmatrix}$$

$$\rightarrow \begin{pmatrix} 17.53 & -4 & 3 & 1 \\ 0 & 1.456 & -0.5123 & -0.1141 \\ 0 & -0.9770 & 0.9827 & 1.494 \end{pmatrix} \rightarrow \begin{pmatrix} 17.53 & -4 & 3 & 1 \\ 0 & 1.456 & -0.5123 & -0.1141 \\ 0 & 0 & 0.6389 & 1.417 \end{pmatrix}$$

and

$$x_3 = 2.218, \quad x_2 = 0.7019, \quad x_1 = -0.1624$$

The exact solution rounded to four significant digits is

$$x_1 = -0.1624, \quad x_2 = 0.7020, \quad x_3 = 2.218$$

4.
$$\left(B^{-1}A^{-1}\right)(AB) = B^{-1}\left(A^{-1}A\right)B = B^{-1}B = I$$

7. (a) Let

$$\begin{pmatrix} a & b \\ c & d \end{pmatrix}\begin{pmatrix} \alpha & \beta \\ \gamma & \delta \end{pmatrix} = \begin{pmatrix} 1 & 0 \\ 0 & 1 \end{pmatrix}$$

Then

$$a\alpha + b\gamma = 1, \; a\beta + b\delta = 0, \; c\alpha + d\gamma = 0, \; c\beta + d\delta = 1$$

i.e., a system of four linear equations. The unique solution is

$$\alpha = \frac{d}{\Delta}, \; \beta = -\frac{b}{\Delta}, \; \gamma = -\frac{c}{\Delta}, \; \delta = \frac{a}{\Delta}$$

where $\Delta = ad - bc$, provided that $\Delta \neq 0$. If $\Delta = 0$, the matrix is singular and has no inverse.

Section 7.3
5. (a) Let

$$A = \begin{pmatrix} a & b & c \\ b & d & e \\ c & e & f \end{pmatrix} = \begin{pmatrix} l_1 & 0 & 0 \\ l_2 & l_3 & 0 \\ l_4 & l_5 & l_6 \end{pmatrix}\begin{pmatrix} l_1 & l_2 & l_4 \\ 0 & l_3 & l_5 \\ 0 & 0 & l_6 \end{pmatrix}$$

Then

$$a = l_1^2, \; b = l_1 l_2, \; c = l_1 l_4$$
$$d = l_2^2 + l_3^2, \; e = l_2 l_4 + l_3 l_5, \; f = l_4^2 + l_5^2 + l_6^2$$

i.e.,

$$l_1 = \sqrt{a}, \; l_2 = \frac{b}{l_1}, \; l_4 = \frac{c}{l_1}$$
$$l_3 = \sqrt{d - l_2^2}, \; l_5 = (e - l_2 l_4)/l_3$$
$$l_6 = \sqrt{f - l_4^2 - l_5^2}$$

(b) For

$$A = \begin{pmatrix} 1 & -1 & 2 \\ -1 & 4 & 3 \\ 2 & 3 & 16 \end{pmatrix}$$

we have

$$l_1 = 1, \; l_2 = -1, \; l_4 = 2$$

$$l_3 = \sqrt{3}, \ l_5 = \frac{5}{\sqrt{3}}, \ l_6 = \sqrt{\frac{11}{3}}$$

10. Let

$$1 = x_1 < x_2 < \cdots < x_{n-1} < x_n = 2$$

where

$$x_{i+1} = x_1 + ih, \ 0 \le i \le n-1; \quad h = \frac{1}{n-1}$$

For $2 \le i \le n-1$ the differential equation is replaced by

$$\frac{\psi_{i-1} + \psi_{i+1} - 2\psi_i}{h^2} = x_i \psi_i$$

i.e.,

$$-\psi_{i-1} + \left(2 + h^2 x_i\right) \psi_i - \psi_{i+1} = 0, \ 2 \le i \le n-1$$

We thus get a set of $n-2$ linear equations with the boundary conditions $\psi_1 = 2, \ \psi_n = 3$. The system's matrix is tridiagonal, given by

$$A = \begin{pmatrix} 2 + h^2 x_2 & -1 & & & \\ -1 & 2 + h^2 x_3 & -1 & & 0 \\ & \ddots & \ddots & \ddots & \\ & & -1 & 2 + h^2 x_{n-2} & -1 \\ & 0 & & -1 & 2 + h^2 x_{n-1} \end{pmatrix}$$

and $A\psi = f$ where $f = (2, 0, \ldots, 0, 3)^T$, $\psi = (\psi_2, \ldots, \psi_{n-1})$. Since $x_i > 0$ the conditions of Theorem 7.3.4 are fulfilled and A can be factorized. For $n = 11$ we have, using the notation of Eq. (7.3.21),

$$b_i = 2.01 + 0.001i, \ 1 \le i \le 9$$
$$c_i = -1, \ 1 \le i \le 8$$
$$a_i = -1, \ 2 \le i \le 9$$

Let $A = LU$ as in Eq. (7.3.22) and let $y = (y_1, \ldots, y_9)^T$, $z = (z_1, \ldots z_9)^T$ solve

$$Ly = f, \ Uz = y$$

By applying Eqs. (7.3.24)–(7.3.26) we get

i	β_i	α_i	y_i	z_i
1	2.0110		2.0000	1.9708
2	1.5147	−0.4973	0.9945	1.9634
3	1.3528	−0.6602	0.6566	1.9795
4	1.2748	−0.7392	0.4853	2.0213
5	1.2306	−0.7844	0.3807	2.0914
6	1.2034	−0.8126	0.3094	2.1929
7	1.1860	−0.8310	0.2571	2.3295
8	1.1748	−0.8432	0.2168	2.5056
9	1.1678	−0.8512	3.1845	2.7269

and

$$\psi_i = z_{i-1}, 2 \leq i \leq 10$$

Section 7.4

2. To verify that

$$\|A\| = \sum_{i=1}^{n} \sum_{j=1}^{n} |a_{ij}|$$

is a matrix norm we must show that requirements 1 to 4 of Definition 7.4.3 are satisfied.

(1) By definition $\|A\| \geq 0$ for all A. Also

$$\|A\| = 0 \Leftrightarrow a_{ij} = 0, \quad 1 \leq i, j \leq n \Leftrightarrow A = 0$$

(2)

$$\|\alpha A\| = \sum_{i=1}^{n} \sum_{j=1}^{n} |\alpha a_{ij}| = |\alpha| \sum_{i=1}^{n} \sum_{j=1}^{n} |a_{ij}| = |\alpha| \|A\|$$

(3) Let

$$A = (a_{ij}), \ B = (b_{ij})$$

Then

$$\|A + B\| = \sum_{i=1}^{n} \sum_{j=1}^{n} |a_{ij} + b_{ij}| \leq \sum_{i=1}^{n} \sum_{j=1}^{n} (|a_{ij}| + |b_{ij}|) = \|A\| + \|B\|$$

(4) $AB = (c_{ij})$, where

$$c_{ij} = \sum_{k=1}^{n} a_{ik} b_{kj}$$

Thus

$$\|AB\| = \sum_{i=1}^{n}\sum_{j=1}^{n}\left|\sum_{k=1}^{n} a_{ik}b_{kj}\right| \le \sum_{i=1}^{n}\sum_{j=1}^{n}\sum_{k=1}^{n}|a_{ik}||b_{kj}|$$

$$= \sum_{i=1}^{n}\sum_{k=1}^{n}|a_{ik}|\sum_{j=1}^{n}|b_{kj}| \le \sum_{i=1}^{n}\sum_{k=1}^{n}\left(\sum_{l=1}^{n}|a_{il}|\right)\sum_{j=1}^{n}|b_{kj}|$$

$$= \sum_{i=1}^{n}\left(\sum_{l=1}^{n}|a_{il}|\right)\sum_{k=1}^{n}\sum_{j=1}^{n}|b_{kj}| = \|A\|\,\|B\|$$

9. (c) The optimal relaxation factor is

$$w = \frac{2}{1 + \sqrt{1 - [\rho(M_J)]^2}}$$

where

$$M_J = -D^{-1}(L + U) = -\begin{pmatrix} 1/4 & 0 & 0 & 0 \\ 0 & 1/4 & 0 & 0 \\ 0 & 0 & 1/4 & 0 \\ 0 & 0 & 0 & 1/4 \end{pmatrix}\begin{pmatrix} 0 & 1 & 0 & 0 \\ 1 & 0 & -1 & 0 \\ 0 & -1 & 0 & 1 \\ 0 & 0 & 1 & 0 \end{pmatrix}$$

$$= -\begin{pmatrix} 0 & 1/4 & 0 & 0 \\ 1/4 & 0 & -1/4 & 0 \\ 0 & -1/4 & 0 & 1/4 \\ 0 & 0 & 1/4 & 0 \end{pmatrix}$$

$$|\lambda I - M_J| = \begin{vmatrix} \lambda & 1/4 & 0 & 0 \\ 1/4 & \lambda & -1/4 & 0 \\ 0 & -1/4 & \lambda & 1/4 \\ 0 & 0 & 1/4 & \lambda \end{vmatrix} = \lambda^4 - \frac{3\lambda^2}{16} + \frac{1}{256}$$

Thus, $\rho(M_J) \approx 0.405$ and $w \approx 1.045$.

Section 7.5

1. (a)

$$\begin{pmatrix} 2.2 & 1.4 & 3.6 \\ 1.9 & 1.2 & 3.1 \end{pmatrix} \to \begin{pmatrix} 2.2 & 1.4 & 3.6 \\ 0 & 0 & 0 \end{pmatrix}$$

and we get a false indication that the matrix is singular.

(b)

$$\begin{pmatrix} 2.2 & 1.4 & 3.6 \\ 1.9 & 1.2 & 3.1 \end{pmatrix} \to \begin{pmatrix} 2.2 & 1.4 & 3.6 \\ 0 & -0.01 & -0.01 \end{pmatrix}$$

and $x_2 = 1$, $x_1 = 1$ in agreement with the exact solution.

4. Define a matrix norm as

$$||A|| = \sum_{i=1}^{n} \sum_{j=1}^{n} |a_{ij}|$$

and an ill-conditioned matrix as one for which cond $(A) \geq 10^5$. Then, for

$$A = \begin{pmatrix} a & 1 \\ 1 & a \end{pmatrix}$$

we have (provided that $a^2 \neq 1$)

$$A^{-1} = \frac{1}{a^2 - 1} \begin{pmatrix} a & -1 \\ -1 & a \end{pmatrix}$$

$$||A|| = 2(1 + |a|), \quad ||A^{-1}|| = \frac{2(1 + |a|)}{|a^2 - 1|}$$

and

$$\text{cond}(A) = \frac{4(1 + |a|)^2}{|a^2 - 1|}$$

The matrix is ill-conditioned whenever

$$\frac{4(1 + |a|)^2}{|a^2 - 1|} \geq 10^5$$

(a) $|a| > 1$: $4(1 + |a|) \geq 10^5(|a| - 1)$, $\quad 1 \leq |a| \leq 1.00008$

(b) $|a| \leq 1$: $4(1 + |a|) \geq 10^5(1 - |a|)$, $\quad 0.99992 \leq |a| < 1$

Section 7.6

3. If $B = TAT^{-1}$ then

$$|B - \lambda I| = |TAT^{-1} - \lambda I| = |TAT^{-1} - \lambda TT^{-1}|$$
$$= |T(A - \lambda I)T^{-1}| = |T||A - \lambda I||T^{-1}| = |A - \lambda I|$$

(since $|T||T^{-1}| = 1$).

5. Let $x^{(i)} = (x_1^{(i)}, \ldots, x_n^{(i)})$, $1 \leq i \leq m$ be mutually perpendicular, and let

$$y = \alpha_1 x^{(1)} + \alpha_2 x^{(2)} + \cdots + \alpha_k x^{(k)} + \cdots + \alpha_m x^{(m)} = 0$$

for some $\alpha_k \neq 0$. Then $x_1^{(k)} y_1 + \cdots + x_n^{(k)} y_n = 0$, but since $x^{(k)}$ is perpendicular to all $x^{(i)}$, $i \neq k$

we have

$$\alpha_k \left[\left(x_1^{(k)} \right)^2 + \left(x_2^{(k)} \right)^2 + \cdots + \left(x_n^{(k)} \right)^2 \right] = 0$$

Therefore $\alpha_k = 0$ in contradiction with the preliminary assumption.

6. (a)

$$R_1 = \{z \mid |z - 1| \leq 2\}$$
$$R_2 = \{z \mid |z| \leq 1\}$$
$$R_3 = \{z \mid |z - 3| \leq 2\}$$

All the eigenvalues are located within

$$\bigcup_{i=1}^{3} R_i$$

Therefore, they are also located within

$$R_4 = \{z \mid |z - 2| \leq 3\}$$

Chapter 8

Section 8.1

5. Substituting $y = x^\alpha$ in the differential equation yields

$$\alpha x^{\alpha-1} = x^{2\alpha} + x^{\alpha+1} - \frac{2}{x^2} - 1$$

i.e., $\alpha = -1$. Thus, $y_1(x) = 1/x$ is a particular solution. The general solution is given by

$$y(x) = y_1(x) + \frac{1}{v(x)}$$

and satisfies

$$-\frac{1}{x^2} - \frac{v'}{v^2} = \left(\frac{1}{x} + \frac{1}{v} \right)^2 + x \left(\frac{1}{x} + \frac{1}{v} \right) - \frac{2}{x^2} - 1$$

or

$$v' = - \left(\frac{2}{x} + x \right) v - 1$$

Thus

$$v(x) = \frac{- \int^x r(t)dt + C}{r(x)}$$

where

$$r(t) = \exp\left[\int^t \left(\frac{2}{u} + u\right) du\right] = t^2 e^{\frac{t^2}{2}}$$

and C an arbitrary constant.

9. (a)

$$A\sin(a) + B\cos(a) = \alpha$$
$$A\sin(b) + B\cos(b) = \beta$$

A unique solution exists if and only if

$$\begin{vmatrix} \sin(a) & \cos(a) \\ \sin(b) & \cos(b) \end{vmatrix} = \sin(a - b) \neq 0$$

i.e., $a - b \neq n\pi$.

(b) For $a = 0$, $b = \pi$, $\alpha = 0$, $\beta = 1$ there is no solution.

Section 8.2

2. Using Euler's method we get

$$y_{n+1} = y_n + he^{y_n}, \ y_0 = 0$$

and

h	$y_h(1)$
0.2	1.834
0.1	2.273
0.05	2.749

The exact solution $y = -\ln(1 - x)$ diverges as $x \to 1$. Therefore $y_h(1)$ does not converge as $h \to 0$.

6. (a) $Z' = x\sin(2Y)Z + (1/2)Y''$
   (b) $Z' = Z + (1/2)Y''$
   (c) $Z' = (2\sin x - x\sin Y)Z + (1/2)Y''$

where Y is the exact solution and Y'' can be expressed in terms of Y and x.

Section 8.3

1. (a)

$$f^{(1)}(x, y) = e^y + (xe^y)^2 = e^y + e^{2y}x^2$$
$$f^{(2)}(x, y) = 2xe^{2y} + (e^y + 2e^{2y}x^2)xe^y$$

$$= 2x^3 e^{3y} + 3xe^{2y}$$

(c)

$$f^{(1)}(x, y) = \cos(x + y) + \cos(x + y) \sin(x + y)$$
$$= \cos(x + y)[1 + \sin(x + y)]$$
$$f^{(2)}(x, y) = -\sin(x + y)[1 + \sin(x + y)]^2 + \cos^2(x + y)[1 + \sin(x + y)]$$

5.

p	$y(2)$
1	3.09005
2	3.08531
3	3.08261

The exact solution is $Y(2) = 3.08268$.

Section 8.4

3. Let y_{ME}, y_M, y_2 denote the modified Euler, midpoint and second order Taylor approximations to $Y(x)$.

x	y_{ME}	y_M	y_2	Y
0.1	0.1090	0.1095	0.11	0.11
0.2	0.2378	0.2389	0.24	0.24
0.3	0.3863	0.3881	0.39	0.39
0.4	0.5545	0.5572	0.56	0.56
0.5	0.7423	0.7461	0.75	0.75
0.6	0.9496	0.9548	0.96	0.96
0.7	1.1763	1.1831	1.19	1.19
0.8	1.4222	1.4311	1.44	1.44
0.9	1.6873	1.6987	1.71	1.71
1.0	1.9713	1.9857	2.00	2.00

In this particular case, the Taylor method yields no truncation error. Therefore $y_2 = Y$ for all x.

10. (a) The midpoint method is given by

$$y_{n+1} = y_n + hf\left[x_n + \frac{h}{2}, y_n + \frac{h}{2}f(x_n, y_n)\right]$$

If $y = Y_n$ then

$$Y_{n+1} = Y_n + hf\left[x_n + \frac{h}{2}, Y_n + \frac{h}{2}f(x_n, Y_n)\right] + R_n$$

$$\approx Y_n + h\left[f + \frac{h}{2}f_x + \frac{h}{2}ff_y + \frac{1}{2}\left(\frac{h^2}{4}f_{xx} + \frac{2h^2f}{4}f_{xy} + \frac{h^2f^2}{4}f_{yy}\right)\right] + R_n$$

$$= Y_n + hf + \frac{h^2}{2}(f_x + ff_y) + \frac{h^3}{8}\left(f_{xx} + 2ff_{xy} + f^2f_{yy}\right) + R_n$$

Alternatively

$$Y_{n+1} \approx Y_n + hY_n' + \frac{h^2}{2}Y_n'' + \frac{h^3}{6}Y_n'''$$

Since $Y_n' = f$, $Y_n'' = f_x + ff_y$ we get

$$R_n \approx \frac{h^3}{6}Y_n''' - \frac{h^3}{8}\left(f_{xx} + 2ff_{xy} + f^2f_{yy}\right)$$

Now

$$Y_n''' = f_{xx} + f^2f_{yy} + 2ff_{xy} + ff_y^2 + f_xf_y$$

Thus

$$R_n \approx \frac{h^3}{6}\left[f_{xx} + f^2f_{yy} + 2ff_{xy} + ff_y^2 + f_xf_y\right]$$
$$- \frac{h^3}{8}\left(f_{xx} + 2ff_{xy} + f^2f_{yy}\right)$$
$$= \frac{h^3}{24}\left(f_{xx} + f^2f_{yy} + 2ff_{xy}\right) + \frac{h^3}{6}\left(ff_y^2 + f_xf_y\right)$$

(b) For $f(x,y) = xy$: $f_{xx} = f_{yy} = 0$, $f_{xy} = 1$, $f_x = y$, $f_y = x$.

Thus,

$$R_n \approx \frac{h^3}{12}x_nY_n + \frac{h^3}{6}x_nY_n\left(1 + x_n^2\right)$$

Section 8.5

2. The Runge–Kutta–Fehlberg method provides in this particular case identical results for $\epsilon = 10^{-4}$, 10^{-5}, but not for $\epsilon = 10^{-6}$, as seen from the following table.

(a) $\epsilon = 10^{-4}$ or $\epsilon = 10^{-5}$

x_n	y_n	Y_n	$|Y_n - y_n|$
1.2	2.7667643	2.7667641	$2 \cdot 10^{-7}$
1.4	2.8965717	2.8965714	$3 \cdot 10^{-7}$
1.6	3.0956456	3.0956453	$3 \cdot 10^{-7}$
1.8	3.3609157	3.3609153	$4 \cdot 10^{-7}$
2.0	3.6945285	3.6945280	$5 \cdot 10^{-7}$
2.2	4.1022795	4.1022789	$6 \cdot 10^{-7}$
2.4	4.5929910	4.5929902	$8 \cdot 10^{-7}$
2.6	5.1783618	5.1783608	$1 \cdot 10^{-6}$
2.8	5.8730895	5.8730881	$1.4 \cdot 10^{-6}$
3.0	6.6951809	6.6951790	$1.9 \cdot 10^{-6}$

(b) $\epsilon = 10^{-6}$

x_n	y_n	Y_n	$|Y_n - y_n|$
1.13133337	2.73994124	2.73994120	$4 \cdot 10^{-8}$
1.25858357	2.79713773	2.79713767	$7 \cdot 10^{-8}$
1.41375605	2.90811776	2.90811767	$9 \cdot 10^{-8}$
1.60262447	3.09869771	3.09869759	$1.1 \cdot 10^{-7}$
1.80262447	3.36484155	3.36484140	$1.5 \cdot 10^{-7}$
2.00262447	3.69938272	3.69938251	$2.1 \cdot 10^{-7}$
2.20262447	4.10815883	4.10815853	$3.0 \cdot 10^{-7}$
2.40262447	4.60003035	4.60002990	$4.5 \cdot 10^{-7}$
2.60262447	5.18673421	5.18673355	$6.6 \cdot 10^{-7}$
2.78595048	5.82035659	5.82035572	$8.8 \cdot 10^{-7}$
2.95355635	6.49184342	6.49184233	$1.1 \cdot 10^{-6}$
3.0	6.69518010	6.69517897	$1.1 \cdot 10^{-6}$

Section 8.6

7. The modified Euler method yields

$$y_1 = y_0 + \frac{h}{2}\left\{ f(x_0, y_0) + f\left[x_1, y_0 + hf(x_0, y_0)\right]\right\}$$

and by substituting $f(x, y) = 1 + y$, $x_0 = y_0 = 0$, $h = 0.1$ we get $y_1 = 0.105$.
We can now apply the AB-2 method

$$y_{n+1} = y_n + h\left[\frac{3}{2}y'_n - \frac{1}{2}y'_{n-1}\right]$$

$$= y_n + h\left[\frac{3}{2}(1 + y_n) - \frac{1}{2}(1 + y_{n-1})\right]$$

$$= \left(1 + \frac{3h}{2}\right) y_n - \frac{h}{2}y_{n-1} + h = 1.15y_n - 0.05y_{n-1} + 0.1$$

and get

x_n	y_n	Y_n	$|Y_n - y_n|$
0	0	0	0
0.1	0.1050	0.1052	$2 \cdot 10^{-4}$
0.2	0.2208	0.2214	$7 \cdot 10^{-4}$
0.3	0.3486	0.3499	$1.2 \cdot 10^{-3}$
0.4	0.4899	0.4918	$2.0 \cdot 10^{-3}$
0.5	0.6459	0.6487	$2.8 \cdot 10^{-3}$
0.6	0.8183	0.8221	$3.8 \cdot 10^{-3}$
0.7	1.0088	1.0138	$5.0 \cdot 10^{-3}$
0.8	1.2192	1.2255	$6.4 \cdot 10^{-3}$
0.9	1.4516	1.4596	$8.0 \cdot 10^{-3}$
1.0	1.7084	1.7183	$9.9 \cdot 10^{-3}$

Section 8.7

3. Denote the approximate and the exact solutions to

$$y' = ky, \ 0 \le x \le 1, \ y(0) = 1$$

by $y^{(k)}$ and $Y^{(k)}$, respectively. Then $Y^{(k)} = e^{kx}$. The following tables provide a comparison between $y^{(k)}$ and $Y^{(k)}$.

x_n	$y_n^{(-1)}$	$Y_n^{(-1)}$	$\left|Y_n^{(-1)} - y_n^{(-1)}\right|$
0	1	1	0
0.2	0.800	0.819	$1.9 \cdot 10^{-2}$
0.4	0.640	0.670	$3.0 \cdot 10^{-2}$
0.6	0.512	0.549	$3.7 \cdot 10^{-2}$
0.8	0.410	0.449	$4.0 \cdot 10^{-2}$
1.0	0.328	0.368	$4.0 \cdot 10^{-2}$

x_n	$y_n^{(-10)}$	$Y_n^{(-10)}$	$\left|Y_n^{(-10)} - y_n^{(-10)}\right|$
0	1	1	0
0.2	-1	0.135	1.135
0.4	1	0.018	0.982
0.6	-1	0.002	1.002
0.8	1	0.000	1.000
1.0	-1	0.000	1.000

x_n	$y_n^{(-20)}$	$Y_n^{(-20)}$	$\left| Y_n^{(-20)} - y_n^{(-20)} \right|$
0	1	1	0
0.2	-3	0.018	3.018
0.4	9	0.000	9
0.6	-27	0.000	27
0.8	81	0.000	81
1.0	-243	0.000	243

Only the case $k = -1$ is within the region of absolute stability. It is expected from Eq. (8.7.23) and verified by the computed tables.

4. For the Adams–Moulton method of order 1 applied to Eq. (8.7.19) we have

$$y_{n+1} = y_n + \frac{h}{2} \left[ky_n + f(x_n) + ky_{n+1} + f(x_{n+1}) \right], \quad y_0 = Y_0$$

$$y_{\epsilon,n+1} = y_{\epsilon,n} + \frac{h}{2} \left[ky_{\epsilon,n} + f(x_n) + ky_{\epsilon,n+1} + f(x_{n+1}) \right], \quad y_0 = Y_0 + \epsilon$$

where Y_0, $Y_0 + \epsilon$ are the exact and perturbed initial conditions. Denote $z_{\epsilon,n} = y_{\epsilon,n} - y_n$. Then

$$z_{\epsilon,n+1} = z_{\epsilon,n} + \frac{h}{2} [kz_{\epsilon,n} + kz_{\epsilon,n+1}]$$

or

$$z_{\epsilon,n+1} = \frac{1 + (hk/2)}{1 - (hk/2)} z_{\epsilon,n}$$

We thus conclude

$$z_{\epsilon,n+1} = \epsilon \left[\frac{1 + (hk/2)}{1 - (hk/2)} \right]^{n+1}$$

The region of absolute stability is determined by

$$\lim_{n \to \infty} z_{\epsilon,n+1} = 0$$

i.e., by

$$\left| \frac{1 + (hk/2)}{1 - (hk/2)} \right| < 1$$

This inequality occurs if and only if

$$-\infty < hk < 0$$

Section 8.8

2. Denoting $y_1 = y$, $y_2 = y'$, $y_3 = y''$, $y_4 = y^{(3)}$, $y_5 = y^{(4)}$ we get

$$y_1' = y_2 \qquad\qquad , \quad y_1(1) = 0$$
$$y_2' = y_3 \qquad\qquad , \quad y_2(1) = 0$$
$$y_3' = y_4 \qquad\qquad , \quad y_3(1) = 0$$
$$y_4' = y_5 \qquad\qquad , \quad y_4(1) = 0$$
$$y_5' = 2y_3 - y_1 + x \ln x , \quad y_5(1) = 0$$

3. Implementing Euler's method for Problem 2 provides the following approximation:

x_n	y_1	y_2	y_3	y_4	y_5
1.0	0	0	0	0	0
1.2	0	0	0	0	0
1.4	0	0	0	0	0.044
1.6	0	0	0	0.009	0.138
1.8	0	0	0.002	0.036	0.288
2.0	0	0.000	0.009	0.094	0.501

5. (a) By using the modified Euler method we obtain

$$y_{1,n+1} = y_{1,n} + \frac{h}{2} \left\{ y_{1,n} y_{2,n} + (y_{1,n} + h y_{1,n} y_{2,n}) \left[y_{2,n} + h(y_{1,n} - y_{2,n}) \right] \right\}$$

$$y_{2,n+1} = y_{2,n} + \frac{h}{2} \left[y_{1,n} - y_{2,n} + y_{1,n} + h y_{1,n} y_{2,n} - y_{2,n} - h(y_{1,n} - y_{2,n}) \right]$$

based on the general scheme

$$y_{1,n+1} = y_{1,n} + \frac{h}{2} \left[f_1(x_n, y_{1,n}, y_{2,n}) + f_1(x_{n+1}, \bar{y}_{1,n+1}, \bar{y}_{2,n+1}) \right]$$

$$y_{2,n+1} = y_{2,n} + \frac{h}{2} \left[f_2(x_n, y_{1,n}, y_{2,n}) + f_2(x_{n+1}, \bar{y}_{1,n+1}, \bar{y}_{2,n+1}) \right]$$

where

$$\bar{y}_{1,n+1} = y_{1,n} + h f_1(x_n, y_{1,n}, y_{2,n})$$
$$\bar{y}_{2,n+1} = y_{2,n} + h f_2(x_n, y_{1,n}, y_{2,n})$$

(b)

$$y_{1,0.2}(1) = 3.2108, \ y_{2,0.2}(1) = 1.6108$$
$$y_{1,0.1}(1) = 3.2679, \ y_{2,0.1}(1) = 1.6289$$

Richardson's method yields

$$y_1(1) = 3.2869, \ y_2(1) = 1.6349$$

Chapter 9

Section 9.1

4. Let $u(x,t) = f(t) + g(x)$ solve the heat equation $u_t = u_{xx}$. Then

$$f'(t) = g''(x), \quad \text{for all } x, t$$

Therefore

$$f'(t) = g''(x) = A \text{ (const)}$$

leading to

$$f(t) = At + B$$
$$g(x) = \frac{Ax^2}{2} + Cx + D$$

where A, B, C, D are arbitrary constants.

Section 9.2

4. Since $\{1, \cos[(n\pi x)/l], \sin[(n\pi x)/l]\}$ is an orthogonal set, we get

$$\int_{-l}^{l} [f(x)]^2 dx = 2l \frac{a_0^2}{4} + \sum_{n=1}^{\infty} (la_n^2 + lb_n^2)$$

i.e.,

$$\frac{1}{l} \int_{-l}^{l} [f(x)]^2 dx = \frac{a_0^2}{2} + \sum_{n=1}^{\infty} (a_n^2 + b_n^2)$$

7. Let $f(x)$ be an odd function. Then $f(-x) = -f(x)$. In particular

$$f(0) = f(-0) = -f(0)$$

Thus $2f(0) = 0$ and clearly $f(0) = 0$. If $g(x)$ is even, then $g(-x) = g(x)$.

$$g'(0) = \lim_{h \to 0} \frac{g(h) - g(-h)}{2h} = \lim_{h \to 0} (0) = 0$$

10. (a) The Fourier cosine series is

$$\frac{a_0}{2} + \sum_{n=1}^{\infty} a_n \cos\left(\frac{n\pi x}{2}\right)$$

where

$$a_n = \frac{2}{2} \int_0^2 f(x) \cos\left(\frac{n\pi x}{2}\right) dx$$

$$= \int_1^2 \cos\left(\frac{n\pi x}{2}\right) dx = \frac{-2\sin\left(\frac{n\pi}{2}\right)}{n\pi}, \quad n \geq 1$$

and

$$a_0 = \int_0^2 f(x)dx = 1$$

Thus

$$f(x) = \frac{1}{2} - \frac{2}{\pi}\left[\cos\left(\frac{\pi x}{2}\right) - \frac{1}{3}\cos\left(\frac{3\pi x}{2}\right) + \frac{1}{5}\cos\left(\frac{5\pi x}{2}\right) - \cdots\right]$$

(b) For $n \geq 1$

$$a_n = \int_0^2 f(x) \cos\left(\frac{n\pi x}{2}\right) dx = \int_0^1 x\cos\left(\frac{n\pi x}{2}\right) dx + \int_1^2 \cos\left(\frac{n\pi x}{2}\right) dx$$

$$= \frac{4}{n^2\pi^2}\left[\cos\left(\frac{n\pi}{2}\right) - 1\right]$$

and

$$a_0 = \int_0^2 f(x)dx = \int_0^1 x\,dx + \int_1^2 dx = \frac{3}{2}$$

Therefore

$$f(x) = \frac{3}{4} + \frac{4}{\pi^2}\left[-\cos\left(\frac{\pi x}{2}\right) - \frac{2}{4}\cos\left(\frac{2\pi x}{2}\right) - \frac{1}{9}\cos\left(\frac{3\pi x}{2}\right) - \frac{1}{25}\cos\left(\frac{5\pi x}{2}\right) - \cdots\right]$$

Section 9.3

9. The general solution satisfying the homogeneous boundary conditions is

$$u(x,t) = \sum_{n=1}^{\infty} \sin\left(\frac{n\pi x}{l}\right)\left[a_n \cos\left(\frac{n\pi ct}{l}\right) + b_n \sin\left(\frac{n\pi ct}{l}\right)\right]$$

The initial conditions yield

$$f(x) = \sum_{n=1}^{\infty} a_n \sin\left(\frac{n\pi x}{l}\right)$$

$$g(x) = \frac{c\pi}{l} \sum_{n=1}^{\infty} nb_n \sin\left(\frac{n\pi x}{l}\right)$$

leading to

$$a_n = \frac{2}{l} \int_0^l f(x) \sin\left(\frac{n\pi x}{l}\right) dx$$

$$b_n = \frac{2}{nc\pi} \int_0^l g(x) \sin\left(\frac{n\pi x}{l}\right) dx$$

11. We substitute $a = b = 1$ and $f(x) = x$ in Eqs. (9.3.35)–(9.3.36) and obtain

$$u(x, y) = \sum_{n=1}^{\infty} b_n \sin(n\pi x) \sinh(n\pi y)$$

where

$$b_n = \frac{2 \int_0^1 x \sin(n\pi x) dx}{\sinh(n\pi)} = -\frac{2\cos(n\pi)}{(n\pi)\sinh(n\pi)}$$

15. We assume a solution $u(x, y) = X(x)Y(y)$ to the Laplace equation and obtain

$$\frac{X''}{X} = K, \ \frac{Y''}{Y} = -K$$

for some constant K. Only a negative K is possible, i.e.,

$$X = \alpha \cos\left(\sqrt{-K}x\right) + \beta \sin\left(\sqrt{-K}x\right)$$
$$Y = \gamma e^{\sqrt{-K}y} + \delta e^{-\sqrt{-K}y}$$

To satisfy $u(0, y) = u(a, y) = 0$ we must have

$$\alpha = 0, \ \sqrt{-K}a = n\pi$$

If we assume $u(x, y)$ to be bounded as $y \to \infty$ we get $\gamma = 0$. Thus, for each n, the associate solution is

$$u_n(x, y) = \sin\left(\frac{n\pi x}{a}\right) e^{-(n\pi y/a)}$$

Finally, coefficients b_n, $n \geq 1$ such that

$$\sum_{n=1}^{\infty} b_n \sin\left(\frac{n\pi x}{a}\right) = f(x)$$

must be found. They are given by

$$b_n = \frac{2}{a} \int_0^a f(x) \sin\left(\frac{n\pi x}{a}\right) dx$$

Section 9.4

5. For each i

$$\lambda_i - 1 - 2a = 4a\sin^2\left(\frac{i\pi}{2n}\right) - 2a = -2a\cos\left(\frac{i\pi}{n}\right)$$

Thus

$$(\lambda_i I - B)x^{(i)} = \begin{pmatrix} b & a & & & \\ a & b & a & & 0 \\ & \ddots & \ddots & \ddots & \\ & & a & b & a \\ 0 & & & a & b \end{pmatrix} \begin{pmatrix} \sin\left(\frac{i\pi}{n}\right) \\ \sin\left(\frac{2i\pi}{n}\right) \\ \vdots \\ \vdots \\ \sin\left[\frac{(n-1)i\pi}{n}\right] \end{pmatrix}$$

where $b = -2a\cos[(i\pi)/n]$. Since

(a)

$$b\sin\left(\frac{i\pi}{n}\right) + a\sin\left(\frac{2i\pi}{n}\right) = -2a\cos\left(\frac{i\pi}{n}\right)\sin\left(\frac{i\pi}{n}\right) + a\sin\left(\frac{2i\pi}{n}\right) = 0$$

(b)

$$a\sin\left[\frac{(k-1)i\pi}{n}\right] + b\sin\left(\frac{ki\pi}{n}\right) + a\sin\left[\frac{(k+1)i\pi}{n}\right] = a\sin\left[\frac{(k-1)i\pi}{n}\right]$$
$$-2a\cos\left(\frac{i\pi}{n}\right)\sin\left(\frac{ki\pi}{n}\right) + a\sin\left[\frac{(k+1)i\pi}{n}\right] = 0$$

(c)

$$a\sin\left[\frac{(n-2)i\pi}{n}\right] + b\sin\left[\frac{(n-1)i\pi}{n}\right] = a\sin\left[\frac{(n-2)i\pi}{n}\right]$$
$$-2a\cos\left(\frac{i\pi}{n}\right)\sin\left[\frac{(n-1)i\pi}{n}\right] = a\left\{\sin\left[\frac{(n-1)i\pi}{n}\right]\cos\left(\frac{i\pi}{n}\right)\right.$$
$$\left. - \sin\left(\frac{i\pi}{n}\right)\cos\left[\frac{(n-1)i\pi}{n}\right]\right\}$$
$$-2a\cos\left(\frac{i\pi}{n}\right)\sin\left[\frac{(n-1)i\pi}{n}\right]$$
$$= -a\sin\left[\frac{(n-1)i\pi}{n} + \frac{i\pi}{n}\right] = 0$$

we conclude that $(B - \lambda_i I)x^{(i)} = 0$, $1 \le i \le n-1$.

9. Instead of Eq. (9.4.20) we obtain

$$\frac{u_{i,j+1} - u_{i,j}}{k} = \frac{c}{2}\left[\frac{u_{i+1,j} + u_{i-1,j} - 2u_{i,j}}{h^2}\right.$$
$$\left. + \frac{u_{i+1,j+1} + u_{i-1,j+1} - 2u_{i,j+1}}{h^2}\right] + F_i$$

where $F_i = F(x_i)$. Thus, Eq. (9.4.21) is replaced by

$$Bu^{(j)} = Au^{(j-1)} + kF$$

where $F = (F_1, \ldots, F_{n-1})^T$. In Algorithm 9.4.2 we need to insert (between Step 10 and Step 11):
Step 10a. For $1 \le i \le n - 1$ set $f_i = f_i + kF_i$.

In addition we insert:
Step 2a. Set $F_i = F(x_i)$, $1 \le i \le n - 1$.

Also, the statement "Steps 8–10 calculate the vector $Au^{(j-1)}$," is replaced by "Steps 8–10a calculate the vector $Au^{(j-1)} + kF$."

Section 9.5
2. In this case we set

$$u_{i,0} = 0, \ \ u_{i,1} = kx_i^2(2 - x_i)$$

and apply Eq. (9.5.12). The results at $t = 1$ are

x	0.2	0.6	1.0	1.4	1.8
$u(x, 1)$	0.152	0.520	0.917	0.995	0.438

Section 9.6
1. Using Algorithm 9.6.1 for this problem with $IT = 100$ provides a message that the maximum number of iterations is exceeded. By increasing IT we find that 104 iterations are needed for convergence.

The approximate solution calculated at selected points is given below.

$x =$	0.4	0.8	1.2	1.6
$y = 0.2$	0.486	0.692	0.613	0.352
$y = 0.4$	0.409	0.505	0.320	0.101
$y = 0.6$	0.426	0.405	0.051	−0.158
$y = 0.8$	0.578	0.416	−0.231	−0.479

3. The use of Algorithm 9.6.1 in this case practically provides the exact solution $x^2 - y^2$. The reason is that for this particular problem the finite differences scheme is exact (i.e., it yields no truncation error). The number of iterations needed to approximate the exact solution within $\epsilon = 10^{-6}$ is 13.

Section 9.7

2. (a)

$$p(x) = 1, \; q(x) = 1, \; r(x) = 5x$$

$$I[f] = \int_0^1 \{[f'(x)]^2 + [f(x)]^2 - 10xf(x)\}dx$$

(c)

$$p(x) = 1 + \sin x, \; q(x) = |x|, \; r(x) = -x$$

$$I[f] = \int_0^1 \{(1 + \sin x) [f'(x)]^2 + |x| [f(x)]^2 + 2xf(x)\}dx$$

7. The functional is

$$F(\phi) = \iint\limits_R \left[\left(\frac{\partial \phi}{\partial x}\right)^2 + \left(\frac{\partial \phi}{\partial y}\right)^2 \right] dx\,dy$$

where

$$R = \{(x, y) \mid 0 < x < 1, \quad 0 < y < \pi\}$$

and $\phi \in C^2(R)$ and satisfies

$$\phi(0, y) = \cos y, \quad 0 < y < \pi$$
$$\phi(1, y) = e \cos y, \quad 0 < y < \pi$$

For the exact solution $\phi_0 = e^x \cos y$ we have

$$F(\phi_0) = \iint\limits_R (e^{2x} \cos^2 y + e^{2x} \sin^2 y)\, dx\,dy = \iint\limits_R e^{2x} dx\,dy$$

$$= \pi \int_0^1 e^{2x} dx = \frac{\pi}{2} (e^2 - 1) \approx 10.036$$

For $\phi_1 = (1 - x)\cos y + ex \cos y$ which satisfies the Dirichlet boundary conditions, we get

$$F(\phi_1) = \iint\limits_R \left[(e-1)^2 \cos^2 y + (1 - x + ex)^2 \sin^2 y \right] dx\, dy$$

$$= (e-1)^2 \int_0^\pi \cos^2 y + \int_0^1 (1 - x + ex)^2 dx \int_0^\pi \sin^2 y\, dy$$

$$= \frac{(e-1)^2 \pi}{2} + \frac{e^3 - 1}{3(e-1)} \frac{\pi}{2} \approx 10.454$$

Index

E

F